DATE DUE

DEMCO 38-296

Ultraviolet-Visible Spectrophotometry in Pharmaceutical Analysis

Ultraviolet-Visible Spectrophotometry in Pharmaceutical Analysis

Sándor Görög, Ph.D., D.Sc.
Member of the Hungarian Academy of Sciences
Director of Analytical Research
Chemical Works of Gedeon Richter Ltd.
Budapest, Hungary

CRC Press
Boca Raton   New York   London   Tokyo

Riverside Community College
Library
'01
MAR
4800 Magnolia Avenue
Riverside, CA 92506

Library of Congress Cataloging-in-Publication Data

Görög, S.
   Ultraviolet-visible spectrophotometry in pharmaceutical analysis /
Sándor Görög.
         p.   cm.
   Includes bibliographical references and index.
   ISBN 0-8493-8691-8
   1. Drugs—Analysis.   2. Spectrophotometry.   3. Pharmaceutical
chemistry.   I.   Title.
   [DNLM: 1. Drugs—analysis.   2. Spectrophotometry, Ultraviolet—
methods.   3. Chemistry, Pharmaceutical—methods.   QV 25 G672u 1995]
   RS189.G665   1995
   615'.1901—dc20
   DNLM/DLC
   for Library of Congress   95-14785
                                                                            CIP

PREFACE

This book is the partly updated English version of the volume, *Spektrofotometriás gyógyszeranalízis* (Akadémiai Kiadó, Budapest, Hungary, 1994).

Is it worthwhile to write a book on a relatively old analytical technique such as UV-VIS spectrophotometry in the age of modern spectroscopic and chromatographic methods? My answer to this question, naturally, is yes. The monographs of almost all bulk pharmaceuticals and formulations contain UV-VIS spectrophotometric measurements even in the latest editions of the leading pharmacopoeias. These are partly simple spectrophotometric tests and partly complex methods where the spectrophotometer is used as the monitor in HPLC instruments, automatic analyzers, etc. The several books available, most of them published many years ago, naturally do not reflect the new developments and the changing role of UV-VIS spectrophotometry in analytical chemistry. This especially applies to the pharmaceutical field, where to my best knowledge this book is the first to be fully devoted to the application of this classical yet, even up to the present time, continuously evolving method.

The book is primarily aimed at practicing analysts working in various fields of pharmaceutical analysis (industrial, clinical, and regulatory drug testing laboratories, etc.), but it is hoped that it will be useful for analysts working in other areas as well.

I wish to express my thanks to Dr. J. Géher-Glücklich, Dr. K. Kovács-Hadady, and Dr. Gy. Milch for contributing some of the chapters, to Mrs. M. Rényei for the spectra, to Dr. Gy. Jalsovszky for the translation of chapters 4, 5, 7, and 9, and to Mrs. E. Doktor for her help in the preparation of the manuscript. Thanks are also due to the referees of the Hungarian version, Professors K. Burger and L. Láng, for their careful work. The English text was revised by Dr. G. Carr. I am very grateful to him not only for carefully editing the manuscript but also for his useful advice.

Dr. Sándor Görög
Budapest
January 1995

LIST OF CONTRIBUTORS

Dr. Judit Géher-Glücklich
Chemical Works of Gedeon Richter Ltd.
Budapest, Hungary
(Chapter 4.D)

Dr. Katalin Kovács-Hadady
Institute of Inorganic Chemistry
Kossuth Lajos University
Debrecen, Hungary
(Chapters 10.C and 10.M)

Dr. György Milch
Chinoin Pharmaceutical Works
Budapest, Hungary
(Chapters 10.D, 10.G, and 10.H)

CONTENTS

Chapter 1

INTRODUCTION

A. HISTORICAL OVERVIEW*

The majority of compounds of pharmaceutical interest are colorless and, hence, studies of their light absorbance characteristics only developed when, during the second half of the nineteenth century, it was discovered that several alkaloids and other organic compounds of natural origin produced characteristic colors when treated with certain inorganic reagents. These discoveries launched rapid developments mainly in the field of forensic science. The methods developed were naturally only qualitative (or, at best, semi-quantitative) procedures for the detection of the compounds in question. At that time the instrumentation was not available for the further development of qualitative tests to provide quantitative applications.

The first two decades of the twentieth century brought about immense progress in the development of optical instrumentation. This created the basis for the development of spectrometric analysis of drugs, both qualitatively and quantitatively. Although the instrumentation for scanning ultraviolet and visible spectra were available in the 1910s and at that time spectra of several pharmaceutical compounds were becoming available, the operation of contemporary spectrophotometers was so clumsy that they were quite unsuitable for use in routine pharmaceutical analysis laboratories.

Filter photometers, however, found wide-ranging application. Of these, the best known was the Pulfrich photometer where the light beam emerging from the light source is split into two beams of equal intensity; these pass through the cells containing the sample solution to be measured and the pure solvent, respectively. Colored glass filters were used to select the wavelength range suitable for the determination. The intensity of the light beam was attenuated mechanically by using a diaphragm calibrated to absorbance units until its intensity was identical to that of the light beam passing through the sample solution which was established by the analyst using visual assessment. In order to make the measurement more comfortable, accurate and precise, photometers were introduced in which the function of the human eye was replaced by photocells for measuring the intensity of transmitted light. A common characteristic of these instruments was that they operated only in the visible spectral region. Thus it was a precondition of their application for samples which were transparent to visible irradiation that they were converted to colored derivatives prior to the measurement.

During the 1920s and 1930s this led to extensive research aimed at further development of classical, and discovery of newer, qualitative color tests to support quantitative photometric methods. As a result of this research, photometry became the first, highly efficient branch of instrumental analysis of drugs. This resulted in a revolutionary development in drug analysis which was, at that time, restricted to titrimetry and gravimetry. It was then that the first quantitative variants of several methods (subsequently updated several times) were described, e.g., the determination of salicylic acid in the form of its iron (III) complex,[10] the determination of aromatic amines, e.g., procaine after diazotization and coupling with β-naphthol,[14] determination of amino acids after reacting them with 1,2-naphthoquinone-4-sulfonic acid,[7] the measurement of alkaloids as picrates,[19] etc.

---

* See References 26 and 27.

A very important milestone in the development of spectrophotometric drug analysis was the introduction of the easy-to-handle Beckman DU ultraviolet-visible spectrophotometer at the beginning of the 1940s. This was designed to be used in more routine laboratories and, after the second World War, its use became more widespread, while at the same time similar instruments from other manufacturers started to appear.

With the advent of ultraviolet-visible spectrophotometers, a new opportunity appeared for the qualitative and quantitative estimation of a great variety of drugs in the UV region without preliminary transformation to colored derivatives. These years can be regarded as the heroic period of modern spectrophotometric drug analysis. Within a short time spectrophotometric methods became one of the most effective and most frequently used methods in this field. A vast number of publications appeared at that time, reporting increasingly selective and sensitive spectrophotometric methods and based on increasingly exacting chemical standards. This trend reached its apex in the 1950s. In the following decades, mainly due to the spread of chromatographic techniques, the researchers' interest in spectrophotometry diminished to some extent, but in practice—as will be explained in detail in the next section—spectrophotometry has remained one of the methods of predominant importance in the analysis of drugs.

As regards the development of methodology, it can be established that during the last 30–40 years no fundamental changes creating entirely new possibilities have occurred. Beckman, Hilger, Unicam, etc. spectrophotometers from the early period are still running in some laboratories, providing data suitable for the solution of a significant number of problems in contemporary drug analysis.

Instrument developments have led toward the direction of faster, more comfortable, selective and exact measuring possibilities and, additionally have opened new ways for coupling the method with chromatographic techniques and automatic analyzers (recording, fast-scanning, microprocessor-controlled instruments, the use of UV-VIS spectrophotometers as HPLC detectors, spreading of reflection spectrophotometric methods for the quantitative evaluation of chromatograms, etc.). Similar results were achieved by the development of spectrophotometric techniques and data processing (development of different methods for background correction, difference- and derivative spectrophotometry, algebraic and computer-based methods for the quantitative analysis of multicomponent systems). All these will be discussed in detail in the chapters to follow.

Finally, it should be mentioned that the nomenclature expressing the different stages of development of the method is not unambiguous in the literature. Originally, methods based on the comparison of color intensities, which were varied by dilution or changing of path lengths of the test and standard solutions, were termed "colorimetry," measurements with an instrument equipped with color filters and subjective or objective detection "photometry," and finally, the application of an instrument working with monochromatic light and supplied with an objective detector "spectrophotometry." Today, spectrophotometric measurements in drug analysis are almost exclusively carried out using the latter technique, filter photometers find their application field in clinical-toxicological laboratories but colorimetry, in the original sense of the word, is no longer in use in any form. The term "colorimetry" is, however, widely used as a generic term for methods where the measurement is based on converting the analyte into a colored derivative independent of the light source and the mode of detection.

B.   THE ROLE OF SPECTROPHOTOMETRY IN MODERN PHARMACEUTICAL ANALYSIS

As outlined in the preceding section, spectrophotometry can deservedly be regarded as one of the classical methods in drug analysis, although its importance has not decreased to any extent during modern times. Moreover—mainly as a result of its coupling with other methods—the field of application of spectrophotometry continues to increase.

Concerning qualitative applications of spectrophotometry, the presentation of spectra, or at least the wavelengths of their maxima and minima, is in many cases an important means for identification of pharmaceutical compounds, and this possibility continues to be widely applied even in recent revisions of pharmacopoeias. The determination of absorbance data of bulk drug materials affords information with a similar value as the measurement of other physical constants. UV-visible spectrophotometry has considerable importance as one of the spectroscopic methods supporting drug research. In certain cases, studying spectra provides information of primary importance in structure elucidation studies. Generally speaking, however, it can be stated that in the majority of cases these spectra indicate only the presence of certain functional groups and structural elements, rather than giving information about the molecule as a whole. This application of spectrophotometry can thus be regarded as a supplementary method to infrared, mass, and NMR spectroscopy, which are definitely more effective tools for the structure elucidation of potential drugs.

In contrast to the previously mentioned other spectroscopic techniques, it is the quantitative analytical application which is of greater importance in the use of UV-visible spectroscopy in drug analysis. Methods based on their natural absorption and on light absorption measured after chemical reactions are extensively applied for the determinations of active substances in bulk drugs to date. Since the selectivity of these methods is generally weaker than that of modern gas- or high-performance liquid chromatographic methods, they are often combined with preliminary column or thin-layer chromatographic separation to increase their selectivity. A further possibility for improving selectivity is the application of various derivative and difference spectrophotometric techniques, as well as of mathematical methods for the investigation of multicomponent systems and for the solution of spectral background problems.

A similar situation exists in relation to the application of the spectrophotometric method in the quantitative analysis of pharmaceutical preparations. It should be noted here that the importance of spectrophotometry in this field has greatly increased, due to the fact that this method can be very readily automated. Automatic analyzers equipped with spectrophotometric detectors are widely used for the serial analysis of formulated pharmaceutical preparations, especially for studying the content uniformity and dissolution characteristics of solid dosage forms.

The spectrophotometric method is applicable to stability testing of pharmaceutical products only if it can be made sufficiently selective for undegraded material or decomposition products. As this is the case only in fortunate situations, in general the spectrophotometric method is applicable to stability studies only with the inclusion of a chromatographic separation step. It must be stated, however, that due to their higher efficiency, gas chromatographic and HPLC methods are the preferred methods for stability testing in contemporary pharmaceutical analysis.

The possibilities of spectrophotometry in the determination of drugs and their metabolites in biological samples are very much limited as compared with its application to the analysis of bulk drugs and their formulations. Determinations on the basis of natural absorption can only be performed in exceptional cases. After the application of a chemical reaction transforming the drug to a (usually colored) product with intense and selective light absorption the spectrophotometric determination can be carried out, provided that the drug was administered at a high enough dose to reach adequate concentration levels in the biological samples, which is a prerequisite of the applicability of the method. In this field, however, chromatographic and immunological methods have much greater potential than spectrophotometric procedures.

The following points have to be taken into consideration when the possibilities of spectrophotometry are investigated in the preceding fields of drug analysis. The demands regarding the quality of drugs and the standards for their examination are continuously increasing. In those fields where the application of spectrophotometry may come into account, this means that the preceding demands are increasing: 1) regarding the selectivity, and 2) the sensitivity of the determination. As a consequence of these increasing demands, spectrophotometric methods are often replaced by more selective and sensitive techniques. It should be mentioned, however, that these methods are without exception more expensive than spectrophotometry. Consequently, especially in modestly equipped laboratories, the increasing demands are addressed by attempting to increase the sophistication of spectrophotometric methods chiefly by the application of one of the above-mentioned spectrophotometric techniques or chemical reactions which may improve the selectivity and sensitivity of the method. This is the reason and explanation for continuing research in this direction (reflected by high numbers of publications of this kind). On the basis of what has been described, however, it is not surprising that the majority of this research has been transferred in the last two decades to countries with moderate financial possibilities.

The somewhat decreasing importance of spectrophotometry (in the classical sense of the word) in pharmaceutical analysis is well compensated by the fact that its methodology and principles survive in highly effective modern methods where UV spectrophotometry is applied as an ancillary technique. For example in the majority of drug analytical problems solved by HPLC, UV spectrophotometers are used as the detectors. Another example is thin-layer densitometry where modern instruments are equipped with UV-visible spectrophotometric detectors, applicable in the reflection and transmission mode.

C. LITERATURE OF SPECTROPHOTOMETRIC PHARMACEUTICAL ANALYSIS

As is characteristic of classical methods which continue to be both widely used and subject to further development, the application of spectrophotometry in pharmaceutical analysis has a very broad literature.

All monographs and handbooks dealing with instrumental analysis and spectroscopy, taken in a broader sense, contain chapters on UV-visible spectrophotometry. Similarly, all clinical-analytical and pharmaceutical-analytical books include the applications of spectrophotometry or colorimetry in shorter or longer chapters. The listing of these chapters would be beyond the scope of this compilation. The reference section of this chapter, however, does contain the list of books published after 1960 on UV-visible spectrophotometry. Although these books deal with the general aspects of this subject, they contain several examples, plus chapters concerned with organic and pharmaceutical applications (see References 1–6, 8, 9, 13, 15, 16, 18, 21, 23, 25). References to many more books which specialize in certain fields, e.g., colorimetry or in certain families of drug compounds, can be found in reference sections of the subsequent chapters.

As regards publications in the literature, the number of articles dealing directly or indirectly with spectrophotometric determinations of drugs and related materials from the beginning to date exceeds 20,000. The main publication forum of articles in the early period were the journals of the various national pharmaceutical societies, the international analytical journals (*Analytical Chemistry, Zeitschrift für Analytische Chemie, Analytica Chimica Acta*, etc.) and periodicals which specialized in clinical-biochemical applications (*Journal of Biological Chemistry*, and *Biochemical Journal*). In recent years, *Analytical Chemistry* has specialized in the publication of new measuring principles. The main publication forums for the theory and applications of spectrophotometry in pharmaceutical analysis are the *Journal of Pharmaceutical and Biomedical Analysis, The Analyst, Journal of Pharmaceutical Sciences, Journal of the Association of Official Analytical Chemists, Analytical Letters* and several others. The national pharmaceutical journals continue to be important publication forums in this field, but in accordance with the above-described reasons Chinese, Indian, Russian, Ukrainian, Polish journals, for example, are increasingly involved in the publication of papers on the application of spectrophotometric methods in pharmaceutical analysis. Clinical applications are often published in *Clinical Chemistry*, and many other journals publish papers in this field from time-to-time.

D. SCOPE AND LIMITATIONS

As can be concluded from the literature survey in the preceding section, in spite of the high number of books on spectrophotometry and the enormous breadth of literature dealing with drug analysis, no book available is concerned with a pure drug analytical profile which covers aspects of contemporary spectrophotometric drug analysis. There are some relatively classical books concerned with a pure drug analytical profile which are invaluable sources of the early literature of spectrophotometric drug analysis; the treatment, however, is restricted to color reactions and which may be considered obsolete.[11,12,17,20,22]

It can be concluded that there is a rather wide gap in publications related to modern spectrophotometric pharmaceutical analysis, and it is the aim of the author to address this requirement within this book. The objective in this book is to begin with a brief summary of the basics of UV-visible spectrophotometry from the point of view of a pharmaceutical analyst taking examples for illustration of the principles described from this domain. The main chapters, however, are definitely focused on pharmaceutical analytical problems related to both qualitative analysis (structure–spectrum relationships) and techniques of quantitative analysis. Separate chapters deal with special problems of determinations of drugs in various matrices and with the possibilities of coupling chromatographic and spectrophotometric methods. Great emphasis is given to special problems associated with the use of chemical reactions prior to spectrophotometric measurements. At the end of the book the longest chapter gives a survey of the possibilities of spectrophotometric determinations for the main families of drugs emphasizing the achievements of the last decade.

The manner of the treatment of the material in this book is primarily to meet the requirements of practicing analysts. Theoretical and mathematical topics have been given a readily understandable treatment but at the same time do not compromise the approach required to discuss adequately the principles and techniques of the subject matter.

The limitations of the present book did not allow an exhaustive extraction of the data from the literature. It was not the aim of this book to provide a laboratory manual of recipes and detailed descriptions of methods for the reader to try out in practice. The aim of this book is no more or less than to help the reader become familiar with the fundamentals and modern possibilities of UV-visible spectrophotometry, to be of assistance in introducing the literature of drug analysis in order to find and apply the optimum solution for a given problem or to help develop a method of choice.

REFERENCES

1. Bauman, R. P. *Absorption Spectroscopy;* Wiley: New York, 1962.
2. Bousquet, P. *Spectroscopy and its Instrumentation;* Adam Hilger: London, 1971.
3. Burgess, C.; Knowles, A., Eds. *Standards in Absorption Spectrometry;* Chapman and Hall: London, New York, 1981.
4. de Galan, L. *Analytical Spectrometry;* Adam Hilger: London, 1971.
5. Derkosch, J. *Absorptionsspektralanalyse in ultravioletten und sichtbaren and infraroten Gebiet;* Akademische Verlagsgesellschaft: Frankfurt, 1967.
6. Edisbury, J. R. *Practical Hints on Absorption Spectroscopy;* Hilger and Watts: London, 1966.
7. Folin, O. *J. Biol. Chem.* 1922, *51*, 377, 386.
8. Hampel, A. *Absorptionsspektroskopie in ultravioletten und sichtbaren Spektrumbereich;* Vieweg: Braunschweig, 1964.
9. Jaffé, H. H.; Orchin, M. *Theory and Application of Ultraviolet Spectroscopy;* Wiley-Interscience: New York, 1962.
10. Jones, A. J. *Chemist and Druggist;* 1919, *91*, 402.
11. Kakac, B.; Vejdelek, Z. J. *Handbuch der Kolorimetrie, I,II (Kolorimetrie in der Pharmazie), III (Kolorimetrie in der Biologie, Biochemie and Medizin);* Gustav Fischer Verlag: Jena, 1962–1966.
12. Kakac, B.; Vejdelek, Z. J. *Handbuch der Photometrischen Analyse; organische Verbindungen;* Verlag Chemie: Weinheim, 1977.
13. Knowles, A.; Burgess, C., Eds. *Practical Absorption Spectrometry;* Chapman and Hall: London, New York, 1984.
14. Lauffs, A. *Apotheker-Ztg.* 1927, *42*, 621.
15. Nowicka-Jankowska, T.; Gorczynska, K.; Michalik, A.; Wieteska, E. *Analytical Visible and Ultraviolet Spectrometry;* Elsevier: Amsterdam, 1986.
16. Perkampus, H. H. *UV-VIS Spectroscopy and its Applications;* Springer Verlag: Berlin, 1992.
17. Pesez, M.; Poirier, P.; Bartos, J. *Pratique de l'Analyse Organique Colorimetrique,* Masson, Paris, 1966.
18. Rao, C. N. R. *Ultraviolet and Visible Spectroscopy, Chemical Application,* Butterworth: London, 1975.
19. Rojahn, C. A.; Seifert, R. *Arch. Pharm.* 1930, *268*, 499.
20. Sawicki, E. *Photometric Organic Analysis, I,II;* Wiley-Interscience: New York, 1970.
21. Schwedt, G., Ed. *Photometrische Analyseverfahren,* Wissenschaftliche Verlagsgesellschaft: Stuttgart, 1988.
22. Snell, F. D.; Snell, C. T.; Snell, C. A. *Colorimetric Methods of Analysis, III,IV (Organic Compounds) IIIA, IVA,* van Nostrand: Princeton, 1948–1970.
23. Sommer, L. *Analytical Absorption Spectrophotometry in the Visible and Ultraviolet;* Elsevier: Amsterdam, 1989.
24. Stearns, E. I. *The Practice of Absorption Spectrophotometry;* Wiley-Interscience: New York, 1969.
25. Stern, E. E.; Timmons, C. J. *Gillam and Stern's Introduction to Electronic Absorption Spectroscopy in the Visible and Ultraviolet;* Edward Arnold: London, 1970.

26. Szabadváry, F. *History of Analytical Chemistry;* Pergamon Press: Oxford, 1966.

27. Thorburn Burns, D. Aspects of the development of colorometric analysis and quantitative molecular spectroscopy in the ultraviolet-visible region, in *Advances in Standards and Methodology in Spectrophotometry;* C. Burgess; K. D. Mielenz, Eds. Elsevier: Amsterdam, 1987; 1–19. *Anal. Proc.* 1988, *25*, 253. *J. Anal. Atom. Spectrosc.* 1987, *2*, 343; 1988, *3*, 285.

28. Vejdelek, Z. J.; Kakac, B. *Farbreaktionen in der spektrophotometrischen Analyse organischer Verbindungen I, II, Ergänzungsband I,II;* Gustav Fischer Verlag: Jena, 1969–1973, 1980–1982.

Chapter 2

THE MEASUREMENT OF LIGHT ABSORPTION (AND REFLECTION)*

A. FUNDAMENTALS

The subject of this book is the application of UV-visible absorption spectrophotometry in qualitative and quantitative drug analysis. This technique is one of the spectroscopic methods based on the interaction of electromagnetic radiation with material. Methods of emission spectroscopy will not be dealt with at all and the treatise is restricted to cases where radiation from an external light source passes through a cell containing a dilute solution of the substance to be investigated and is partly absorbed. The method where the attenuation of the light beam takes place in the course of its reflection from the surface of the sample to be examined will also be briefly discussed.

In UV-visible spectrophotometry an approximately monochromatic light beam is generally employed. This is selected with the aid of a monochromator from the total emission spectrum of the light source as discussed in Section 2.D.2. In this region of the electromagnetic spectrum, the nature of the radiation is expressed as wavelength (nm); in theoretical papers, however, frequencies expressed as wavenumbers are used since the latter are directly related to radiation energy:

$$c = \nu \cdot \lambda \tag{2.1}$$

$$E = h \cdot \nu = h \cdot \frac{c}{\lambda} = 10^{-7} \cdot h \cdot c \cdot \nu^* \tag{2.2}$$

The relationship between wavelength, frequency, wavenumber and energy of the light radiation is expressed by Equations 2.1 and 2.2 where c = the velocity of light, ν = the frequency of light, λ = the wavelength of light (usually expressed in nanometers; 1 nm = 10^{-9} meters), E = energy of the light photon, h = Planck's constant, ν^* = wavenumber (usually expressed in cycles per centimeter). Thus the energy of the radiation is in inverse ratio to the wavelength.

The ultraviolet-visible region which is the subject of this book is a part of the electromagnetic spectrum. About 190 nm is generally regarded as the low wavelength–high energy end of the region. Below this is the so-called vacuum-ultraviolet region which can be studied only with special measuring techniques; this region is less important from the point of view of pharmaceutical analysis. The upper, high wavelength–low energy end of the region is at about 780 nm. This value corresponds to the upper wavelength limit of many of the spectrophotometers in general practice.

Moving further on the wavelength scale toward longer wavelengths, one passes through the moderately, but quite recently increasingly important near infrared (NIR) region and finally arrives at the very important infrared (IR) region.

The lower sensitivity limit of the human eye is at about 380 nm, which lies within the 190–780 region of the UV-visible spectrum although this value is of no importance either theoretically or for practical applications.

---

* As regards the basic principles and problems related to the instrumentation, the reader is referred to some of the books listed at the end of Chapter 1 as references.[1–6,8,9,13,15,16,18,23,24]

TABLE 2.1.   Colors and Their Complements

			Wavelength, nm			
	400	500		600		700
Color	violet	blue	green	yellow	orange	red
Complement	yellow	orange	red	violet	blue	green

In Table 2.1 the wavelength regions are presented which correspond to the colors of the visible spectrum together with their color complements, i.e., the colors which are observed when a solution absorbs light of a given wavelength from polychromatic (white) light. On the basis of this relationship the wavelength range in which the absorbance maximum of a solution may be expected can be estimated from the color of the solution.

For example, a solution of methyl orange indicator which has an orange color in acidic medium absorbs greenish-blue light (absorbance maximum around 500 nm).

B. THE MECHANISM OF LIGHT ABSORPTION IN THE UV-VISIBLE REGION

If an organic molecule absorbs electromagnetic radiation as mentioned in the preceding section, the absorbed energy can increase the energy content of the molecule in three ways. The rotational energy of the molecule is increased by absorption of the light with the lowest energy (highest wavelength). Higher energy is needed to increase the vibrational energy of the molecule. These energy transitions correspond to absorption of radiation in the IR region.

Conversely, absorption of radiation in the ultraviolet-visible region leads to excitation of electrons in the outer orbitals of the molecule i.e., they are promoted to orbitals of higher energy levels. The difference between the energy levels of the two orbitals (excited state and ground state) corresponds to the energy of the absorbed light. Accordingly, ultraviolet-visible spectroscopy is often termed electron excitation spectroscopy.

In order to explain the light absorption characteristics of different organic molecules it is necessary to take into consideration that the excitability of an electron in the molecule is determined not only by defining to which atom the electron belongs but also in which orbital it can be found in the ground state. This means that the excitable electrons stay in molecular orbitals and their excitability is predominantly determined by their environment in the molecule and by the bond in which they participate.

The quantumchemical basis of the origin of absorption bands is the fact that the energy levels of the electron orbitals are quantized, i.e., the electron can occupy orbitals of finite numbers having fixed energy contents both in the ground and excited states. Naturally, the number of possible transitions between these orbitals is also finite and a well defined energy transfer is associated with every transition.

The absorption bands correspond to such transitions. From the above-described principles it could be deduced that absorption spectra would consist of a series of sharp lines, each line corresponding to the difference between the energy levels of the two orbitals concerned. Such spectra, however, can only be obtained when atoms or simple molecules are investigated in the gas phase.

Solution spectra of organic pharmaceutical compounds generally consist of broad bands. The broadening of the lines into bands is the consequence of the vibrational and rotational energy transitions. As has already been mentioned in the introduction of this section, the energy absorption of the molecules not only results in the excitation of its electrons, but the excitation of the rotational and vibrational energy levels also takes place simultaneously. In the course of each electron excitation in the dissolved state, both the ground and excited states include a series of vibrational energy levels. Especially in the gas phase, each vibrational energy level has a series of rotational

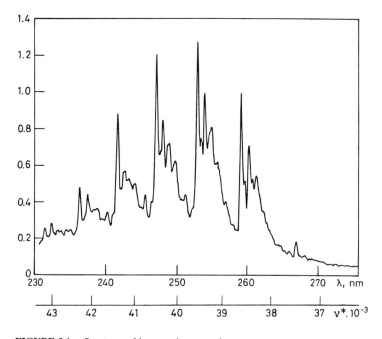

FIGURE 2.1.   Spectrum of benzene in vapor phase.

levels. Thus the energy needed to promote an electron to a higher orbital is influenced, albeit to a smaller extent, by the vibrational and rotational energy states of the molecule in the ground and excited states.

Although, as has already been mentioned, the energy absorption corresponding to vibrational energy transitions falls into the infrared spectral region (and rotational energy into the far infrared region) these are incorporated into spectral lines in the UV region corresponding to the excitation of electrons. In the majority of cases this causes lines to broaden to bands, or it appears in the form of inflections, especially in spectra taken in nonpolar solvents or in vapor phase vibrational fine structure appears where the energy differences between each secondary maxima of the band correspond to the difference between two vibrational energy levels. This is demonstrated by the vapor phase spectrum of benzene in Figure 2.1.

As shown in Figure 2.1, the distance between the principal lines of each secondary absorption group are about 5.5–5.5 nm on the wavelength scale. On the energy-related wavenumber scale this is equivalent to about 900 cm^{-1} and this is situated in the middle of the IR range. If the spectrum of benzene is scanned in the vapor phase using a high-resolution spectrophotometer, the rotational fine structure of the vibrational bands also appears. As seen in Figure 2.1, the spacing between these lines within each secondary absorption group is about 100–150 cm^{-1} and this corresponds to the far infrared region as described above.

It must be emphasized that the absence of vibrational fine structure or its appearance in the form of band broadening only in the majority of cases is not caused by instrumental limitations; rather it is the result of intermolecular or intramolecular interactions plus interactions with the solvent. In the vast majority of cases in UV-visible spectroscopy of pharmaceutical compounds, the spectra and the relevant information would not be influenced by the replacement of a conventional good resolution instrument by an ultra-high resolution spectrometer. This is a notable difference between UV-visible spectroscopy and some other branches of spectroscopy, mainly NMR and mass spectrometry.

Since in the majority of pharmaceutical compounds more than one electron pair can be excited in the UV-visible spectral region, their spectra usually consist of a band system where in some cases the bands are well resolved, but very often they fully or partly overlap each other.

Computations of spectra on the basis of quantum chemical considerations are possible in the case of simple molecules only. Nevertheless, partly on a theoretical basis, but even more by studying the UV-spectra of a great number of organic compounds, important conclusions can be

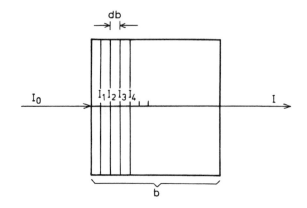

FIGURE 2.2. Scheme for the demonstration of Beer's Law. Reduction from I_0 to I of the intensity of a monochromatic light beam while passing through a cell with b pathlength.

drawn which may be of great value when the molecular structures of organic compounds are investigated by spectroscopic methods. This question is dealt with in a more detailed way in Chapter 3.

Finally, a brief answer is given to the question of the fate of the energy used for the excitation of electrons in the course of the absorption of UV-visible light. Molecules usually do not remain in the excited state for more than about 10^{-8} sec (compounds exhibiting phosphorescence are exceptions). The absorbed energy may cover the energy demands of photochemical reactions (e.g., especially in the short-wavelength UV range) the photodissociation of organic molecules. Usually the energy released when an electron returns from a higher orbital to a lower one is converted into heat: by collision of the excited molecule with the surrounding solvent molecules, the kinetic energy of the latter increases. Another mechanism of fundamental importance is when the energy is emitted as light.

This phenomenon, termed fluorescence, is the basis of an important analytical method. The theory and practical applications of this are beyond the scope of this book; it is however, important to note that the emitted light always has longer wavelengths than the irradiating light, the reason for this phenomenon being connected with the vibrational energy states of the excited molecule.

C. ABSORBANCE AND THE BEER-LAMBERT LAW

The fundamental law of spectrophotometry establishes a relationship between the reduction of light intensity caused by an absorbing solution and the concentration and pathlength of that solution.

The mathematical expression of the law may be derived by consideration of the cell represented in top-view in Figure 2.2 containing the absorbing solute at a concentration c through which passes a monochromatic (single-wavelength), parallel light beam with an intensity of I_0. The light beam passes through the cell of length b and emerges with an intensity of I. The cell length b is divided into a series of equal distances db. As the light passes through each section, the intensity decreases to $I_1, I_2, I_3, \ldots, I_{n-1}, I_n$. The rate of the reduction of the light intensity is constant along the total pathlength as expressed by Equation 2.3.

$$\frac{I_0}{I_1} = \frac{I_1}{I_2} = \frac{I_2}{I_3} = \frac{I_{n-1}}{I_n} = k \tag{2.3}$$

For example in the case of an attenuation of 5% per section:

$$I_1 = 0.95 \cdot I_0; \qquad I_2 = 0.903 \cdot I_0; \qquad I_3 = 0.857 \cdot I_0, \text{ etc.}$$

The absolute value of the attenuation of light per section is proportional to the intensity of light incident upon it and to the thickness of the layer, the above constant k being the proportionality coefficient:

$$-(I_{n-1} - I_n) = dI = k \cdot I_{n-1} \cdot db \tag{2.4}$$

A quantitative relationship may be established between the light intensities decreasing from I_0 to I and cell length b to produce the integral form of Equation 2.4:

$$\int_{I_0}^{I} \frac{dI}{I} = k \int_{0}^{b} db \tag{2.5}$$

After integration and changing from natural logarithm to base-10 logarithm, Equation 2.6 results.

$$\log \frac{I_0}{I} = k' \cdot b \tag{2.6}$$

Expressed in words this means that the logarithm of the ratio of the intensities of monochromatic light which enters and emerges from an absorbing body is independent of the absolute intensity of the incident light beam but is directly proportional to the thickness of that body. After its discoverer, this law is called Lambert's Law (Lambert 1760). It is noteworthy that this phenomenon had previously been described in 1729 by Bouguer.

The value of k' contains the conversion factor from natural to base-10 logarithms; it must be emphasized, however, that Equation 2.6 is valid only if the concentration c of the absorbing substance is kept constant. It was Beer (in 1852) who investigated the dependence of light absorption on concentration. Keeping the cell length at the constant (b) value, one can proceed similarly to the establishment of Equations 2.4–2.6 with the difference that b is substituted with c. This can be expressed in the following way. Consider the effect of increasing the concentration of the absorbing substance by a value of dc; the attenuation of the light intensity is proportional to the light intensity and the increase of the concentration. Integrating from 0 to c_1 the differential equation established to express the changes with the concentration yields Equation 2.7 in an analogous manner as Equation 2.6 was derived from Equation 2.5.

$$\log \frac{I_0}{I} = k'' \cdot c \tag{2.7}$$

This is the mathematical expression of Beer's Law. Expressing it in words, this means that in the case of a solution in a constant pathlength cell, the logarithm of the ratio of the intensities of incident and emerging light is directly proportional to the concentration of the absorbing solute in the cell.

It can be seen in Equations 2.6 and 2.7 that $\log I_0/I$ on the one hand is proportional to the pathlength and on the other hand to the concentration. It can thus be deduced that the two equations can reduce to Equation 2.8, which naturally requires a new coefficient a.

$$A = \log \frac{I_0}{I} = a \cdot b \cdot c \tag{2.8}$$

This is the fundamental equation of spectrophotometry. It is generally cited in the literature as the Beer-Lambert Law or Bouguer-Lambert-Beer Law (BLB). Since in spectrophotometric practice it is usually the dependency of the attenuation of light intensity on a concentration only which is of importance, it is often simply referred to as Beer's Law.

As is clear from the previous discussion, $\log I_0/I$ is a quantity of fundamental importance in spectrophotometry and therefore it has been given a separate name: absorbance (A) or, in the early literature, extinction (E). Absorbance is a linear function of the pathlength and the concentration of the absorbing solute but it is independent of the intensity of the incident light. Absorbance is the quantity that is usually directly read on the scale of spectrophotometers.

The constant a of Equation 2.8 has to be dealt with separately. This does not depend on the pathlength and the concentration, but it is a function of the wavelength of the incident monochromatic light and the character of the absorbing substance. If both the concentration of the dissolved absorbing substance and the cell thickness are selected as unity (1 g/100 ml or 1 mol/liter and 1 cm, respectively), then according to Equation 2.8:

$$A = a \qquad (2.9)$$

The value of a depends on the mode of presentation of the concentration. If it is expressed as w/v percent, the proportionality factor a is termed specific absorbance ($A_{1\,cm}^{1\%}$) or, in the early literature, specific extinction ($E_{1\,cm}^{1\%}$). If the concentration is expressed as molarity, the name of a is molar absorptivity (ε) or, in the early literature, molar extinction coefficient. The relation between the two coefficients can be given by Equation 2:10:

$$\varepsilon = \frac{A_{1\,cm}^{1\%} \cdot M}{10} \qquad (2.10)$$

where M is the molecular weight.

A third possibility is to use the term: absorptivity (a). This relates to concentration expressed as g/liter; $10a = A_{1\,cm}^{1\%}$.

Equation 2.8 expressing the Beer-Lambert Law furnishes a key for the qualitative and quantitative applications of spectrophotometry. From the data in this equation, A is a measurable quantity, b is the length of the cell. Provided that $A_{1\,cm}^{1\%}$ is a known value, the unknown c can be calculated, i.e., the method can be used for the measurement of concentrations or e.g., the determination of the active ingredient content of bulk drug materials and finished products. The details of this question as well as deviations from the Beer-Lambert Law will be discussed in Chapter 4.

When the concentration is known in Equation 2.8, i.e., by weighing and diluting pure substance, this equation enables the specific absorbance or the molar absorptivity to be calculated. These values at the wavelengths of the maxima of the absorption bands are characteristic of the analyte and can be considered as physical constants eminently suitable for its characterization and identification. Of the different absorption coefficients, ε is used in the case of applying the spectral data to qualitative analysis and structural investigations while in quantitative analytical applications $A_{1\,cm}^{1\%}$ or a are predominantly used.

It is worth mentioning that, especially in the early literature, the term transmittance (T) was often used. The definition of transmittance and its relation with absorbance are given in Equations 2.11 and 2.12:

$$T = \frac{I}{I_0} \qquad (2.11)$$

$$A = \log \frac{1}{T} \qquad (2.12)$$

Today, transmittance is usually used to characterize the colorlessness of solutions in the visible range. In these cases the transmittance expressed as percentage (100T) is presented as the minimum requirement.

D. MAIN PARTS OF THE SPECTROPHOTOMETER[3]

1. Light Sources

The tungsten filament bulb is generally used as the light source in the visible region. Its maximum emission is in the NIR region and it decreases rapidly at shorter wavelengths; it is not applicable below 350 nm. Recently, a variant of this has come into use where some iodine vapor is mixed with argon in the lamp bulb (tungsten-halogen lamp). In this way the formation of tungsten deposits on the cool parts of the bulb can be hindered, thus prolonging the lifetime of the lamp.

In the ultraviolet region the deuterium arc lamp is about almost exclusively used (replacing the formerly used hydrogen lamp). The light emitted by ionization of the gas due to the effect of an electrical discharge across a quartz envelope, filled with deuterium at a pressure of about 0.01 atm, produces a continuous emission spectrum below 375 nm. Therefore this lamp can be used in the entire UV region. The spectrum in the visible region is a line-spectrum: of the lines available, those appearing with high intensity at 486.02 and 656.10 nm are eminently suitable for the calibration of the wavelength scale of the spectrophotometer.

Instruments operating in the full UV-visible region are equipped with both lamps: the change over from one lamp to the other takes place by means of a simple switch (usually automatically).

The xenon discharge lamp, which has a very high emission intensity in both the UV and visible regions, would be an ideal light source. However, as a consequence of its relatively short lifetime, the hazards of the high-pressure lamp, and the intensive dissipation of heat and ozone emission, it is used only in those cases when high intensities are really required: measurement of the difference of light absorptions of strongly absorbing solutions but chiefly in the fluorimetric analysis.* The above listed problems can be reduced by using low-pressure xenon lamps with pulsing light emission.

2. Monochromators

For the spectrophotometric measurements (with the exception of diode-array instruments to be discussed later) monochromatic or at least approximately monochromatic light with a bandwidth restricted to the narrowest possible wavelength region is needed. The simplest way to produce this would be to use a light source with line-emission but the possibilities of its practical application would naturally be very much restricted. It should be mentioned that mercury vapor lamps are widely used as light sources in UV detectors of high-performance liquid chromatographs. The highly intensive line of this lamp at 253.7 nm is excellently suitable for monitoring the majority of organic compounds with good sensitivity. In this case it is naturally not necessary to use a monochromator: only radiation with wavelengths related to the other emission lines of mercury has to be eliminated from the light beam by the application of suitable filters.

By applying appropriate filters, a more or less narrow wavelength region can be selected from the light beam of sources emitting continuous visible radiation, too, while the rest of the light with different wavelengths is absorbed by the filter. If simple colored glass filters are used for this purpose, characterizable by the color complement of the light to be obtained, then the bandpass is too large and the transparency of the filter is poor even at its maximum wavelength. Better efficiency and narrower wavelength range (15–20 nm bandpass) is obtainable by using interference filters. Photometers designed for the routine serial analysis of colored compounds or colored derivatives of colorless compounds are usually equipped with such filters. The most suitable filter for the measurement of a given product is selected on the basis of the nominal wavelength of the filter and the absorption maximum of the analyte or empirically.

Recording spectra as well as quantitative measurements with more demanding requirements can only be performed by using a spectrophotometer; the light applied here for the measurement is produced by a monochromator. Here the beam of rays from the light source emitting mixed light with different wavelengths enter through the entrance slit. Before or after the slit, mirrors or lenses are placed, making the beam parallel which then reaches the dispersing element of the monochromator where the radiation with different wavelengths is deflected over different angles, the latter being the function of the wavelength. The deflected light leaves the monochromator

---

* As is obvious from Equation 2.8 absorbance in spectrophotometry does not depend on the intensity of the incident light (I_0). In contrast to this, in fluorimetry the intensity of the emitted fluorescence is a linear function of the intensity of the light used for excitation, i.e., the increase of the latter is a very important method for increasing the sensitivity of the measurement.

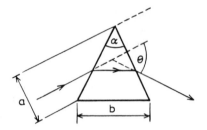

FIGURE 2.3. Prism as a dispersion element. α: angle of refraction of the prism; θ: angle of deviation; a: diameter of the incident light beam; b: length of the base of the prism.

through the exit slit. The approximately monochromatic light thus obtained is used in the spectrophotometric measurements.

As the dispersion element, a prism or a diffraction grating is used. The functioning of the prism monochromator is based on the fact that the refractive index (n) of the prism material (glass or quartz) is a function of the wavelength. The dispersion of the prism is defined as the derivative with respect to wavelength of the function of the angle of deviation (see Figure 2.3). As seen in Equation 2.13 this depends on the geometry of the monochromator and on the dispersion of the refractive index ($dn/d\lambda$):

$$\frac{d\theta}{d\lambda} = \frac{b}{a} \cdot \frac{dn}{d\lambda} \tag{2.13}$$

where b is the base length of the prism and a is the maximum diameter of the light beam.

The dispersion of the refractive index has no constant value: it diminishes as the wavelength increases. For this reason the resolution of prism monochromators in the short-wavelength UV region is very good, but it is poorer in the long-wavelength visible region. The consequences of this are seen in Figure 2.6.

Accordingly, there are significant disadvantages in applying the prism as a dispersion element. It is inconvenient in itself that the resolution changes with the wavelength (i.e., passing along the wavelength scale from short wavelengths toward long ones, the scale becomes more and more crowded). Additionally, the resolution in the long-wave ultraviolet and visible regions is definitely poorer than with grating monochromators. This, as well as the dependence of the dispersion on temperature, are the reasons why prism monochromators have been gradually replaced in practice. Although more expensive, modern spectrophotometers are almost exclusively equipped with grating monochromators (see Figure 2.4).

The gratings applied in the monochromators of spectrophotometers all work in the reflection mode. The grating is a highly polished glass plate with a large number of grooves on the surface (100–2500 grooves/millimeter). In the course of mass production of diffraction gratings replica gratings are made by copying from ruled masters in a similar way that gramophone records are pressed. The surface obtained is covered by a thin layer of aluminum. The light reflected at this surface undergoes diffraction and contains, therefore, the full spectrum several times as the first, second, etc., order spectra.

As in the case of the prism (see Equation 2.13) an equation can be established for dispersion by the grating (Equation 2.14).

$$\frac{d\theta}{d\lambda} = \frac{m}{d \cdot \cos\theta} \sim \frac{m}{d} \tag{2.14}$$

where m is the order of the spectrum (1,2,3, etc.) and d is the distance between two adjacent grooves. The inverse dependence of the dispersion on the latter means that the resolution is increased in relation with the number of grooves on a given area. Cos θ could be neglected because the angle of deviation differs from 10 only to a small extent (e.g., cos 10° = 0.985). In its simplified form Equation 2.14 expresses the very important fact that, contrary to the case of prisms, the

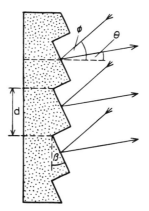

FIGURE 2.4. Grating as a dispersion element. ϕ: angle of incidence; θ: angle of reflection; d: distance between two adjacent grooves; β: angle between the grating plane and the wall of a groove.

dispersion of the diffraction grating does not depend on the wavelength and, as seen in Figure 2.6, it has a constant value along the full spectrum.

Recently, the technology of manufacturing diffraction gratings has greatly developed (holographic gratings, which make use of ion beams and lasers in their production) and hence gratings working at high and standard resolution in a broad spectral range have become readily available. Accordingly, grating monochromators are almost exclusively used in modern spectrophotometers in spite of the fact that their application also raises special problems. These include difficulties originating from the partial overlapping of the spectra of different orders and which can be overcome by using a special filter, a second grating or a prism-grating combination. Another problem is that the spread of the energy among the spectra of different orders diminishes the light energy available to the desired order. In order to solve this problem the (blaze) angle between the walls of single grooves and the grating plane is optimized: in this way reflection of 70% of the total energy in the first order can be achieved (see Figure 2.4).

Monochromators may be evaluated first on the basis of the spectral purity of the light which leaves them. This is characterized from the extent that this light approaches ideal monochromatic radiation and that is determined by the widths of both the entrance and exit slits and the dispersion characteristics of the monochromator. Investigating the spectral distribution of the light beam leaving through the exit slit it can be seen that in the case of a very small slit width an idealized triangle depicted in Figure 2.5 can be obtained where the intensity of the radiation is plotted as a function of the wavelength. Such a symmetrical distribution can only be achieved if the widths

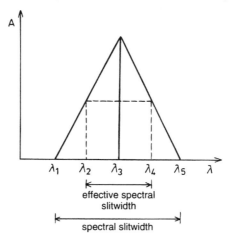

FIGURE 2.5. Schematic representation of spectral bandwidth and spectral slitwidth.

of the entrance and exit slits are set to the same value and, of course, the dispersion of the grating is constant along the whole spectrum range.

It is the wavelength pertaining to the maximum intensity (λ_3) that can be read on the wavelength scale of the spectrophotometer. λ_1 and λ_5, respectively, characterize the shortest and longest wavelengths leaving the monochromator, while the difference of the wavelengths corresponding to half of the maximum intensity ($\lambda_4 - \lambda_2$) is termed the effective spectral slitwidth of the monochromator. ($\lambda_5 - \lambda_1$) is termed as the spectral slitwidth; (it is given in nm in contrast to the mechanical slitwidth given in mm). As seen in Figure 2.5, on the basis of elementary geometrical considerations, it is evident that in an ideal case the energy defined by the effective spectral slitwidth equals 75% of the total energy leaving the exit slit. In practice the energy distribution in the case of real slitwidths is trapezoid rather than triangle-shaped; the above considerations, however, are essentially valid in these cases, too.

The effective spectral slit width is given by the product of the mechanical slitwidth and the reciprocal of the linear dispersion* of the dispersive element. The transformed version of Equation 2.15 expressing this relation contains the angular dispersion introduced already in Equations 2.13 and 2.14:

$$\lambda_4 - \lambda_2 = \Delta\lambda = w \cdot \frac{1}{ds/d\lambda} = \frac{w}{f} \cdot \frac{1}{d\theta/d\lambda} \tag{2.15}$$

where w is the mechanical slitwidth, $ds/d\lambda$ is the linear dispersion,* and f is the focal length of the lens or concave mirror mapping the light leaving the dispersion element onto the exit slit.

The spectral purity of the radiation which can be characterized by the spectral bandwidth is of great importance for recording the spectra and their quantitative evaluation. As is pointed out in Sections I and J, when the effective spectral slitwidth is too wide, spectra of distorted shape may be obtained and deviations from Beer's law may also occur. Since there is a direct relationship between the resolution of the monochromator and the effects on spectral slitwidth ($\lambda_4 - \lambda_2$), it follows that it is advantageous to perform the measurement using a small effective slitwidth.

As indicated in Equation 2.15 this may be achieved in two ways: by decreasing the slitwidth or by increasing the dispersion of the dispersive element. Decreasing the slitwidth is limited by the fact that the quantity of energy available also decreases simultaneously, and this results in deterioration of the signal-to-noise ratio (see Section I). According to Equation 2.13 it is by increasing the base of the prism that provides the possibility of increasing dispersion; these possibilities are, of course, limited. There is no limitation of this kind in the case of grating monochromators: as seen in Equation 2.14, dispersion can be proportionally increased by decreasing the grating constant (d), i.e., by increasing the number of grooves/mm on the surface of the grating.

The monochromators of commercially available spectrophotometers are generally characterized by their effective spectral slitwidths (spectral bandwidths). In the case of grating instruments this is generally constant across the full spectral range if the slitwidth is also kept constant. In some instruments the spectral bandwidth is adjustable; in other cases, however, it is fixed. In the case of inexpensive instruments used for quantitative analytical purposes only it is typically 4 or 5 nm. Numerous instruments possess adjustable spectral bandwidths at e.g., 0.25, 1, 2, 4 nm. An instrument working with a spectral bandwidth of 1 nm is suitable for the majority of the tasks in drug analysis and recording high quality spectra. Higher resolution than this is necessary only in exceptional circumstances in the practice of drug analysis and in structure elucidation studies (e.g., recording spectra with ultrafine structures in the gas phase).

To illustrate the described features of monochromators, the reciprocal of the linear dispersion of some types is depicted as a function of the wavelength (Figure 2.6). According to Equation 2.15 the spectral bandwidth can be calculated from the values taken from the graphs in Figure 2.6 by multiplying them by the mechanical slitwidth.

---

* In the term $ds/d\lambda$ of the linear dispersion, ds is the distance of two light beams characterized by the two ends of the wavelength difference $d\lambda$ after being mapped onto the plane of the exit slit.

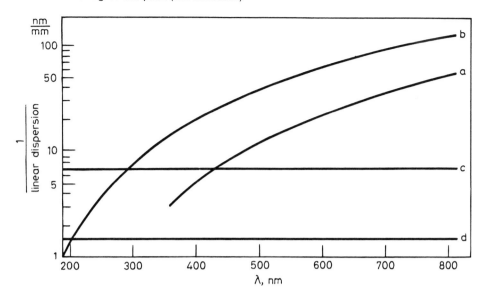

FIGURE 2.6. Reciprocal of the linear dispersion of different monochromators as a function of the wavelength; a: Glass prism monochromator (Opton M4 G II); b: Quartz prism monochromator (Opton M4 Q III); c: Grating monochromator (600 grooves/mm); d: Grating monochromator (1200 grooves/mm).

3. Cell Holder With the Cells

The cell holder should accommodate cells of different lengths fulfilling the requirement that the light beam should be incident perpendicularly upon the optical surface of the cell. The standard cell most generally used is the rectangular cell with 10 mm pathlength, but others with pathlengths of 1, 2, 5, 20, 40 up to 100 mm are also available. Microvariants of the cells with medium pathlengths and which only require about 0.1 ml are also available.

Since absorbance is related to path length according to Lambert's law (Equation 2.6), in order to obtain accurate and reproducible results, it is very important to use cells with accurate pathlengths. Thus, one can purchase precision cells of 10 mm which are guaranteed by the manufacturer to 0.1% accuracy. In the case of cells with smaller path length this accuracy decreases: the application of these cells should be avoided in the quantitative analysis as far as possible.

Cells used in UV-visible spectrophotometry are made of glass or quartz. Cells made of fused quartz can be used down to 210 nm. Their applicability can be enhanced down to 180 nm if a synthetic variant (fused silica) is used which is prepared by the high-temperature dissociation of purified silicon compounds. Substantially less expensive glass cells can also be applied in the visible region down to about 330 nm (and in the case of special optical glass cells, down to 310 nm). Disposable plastic cells can be used for less sophisticated applications in the same spectral region.

To avoid solvent vapors getting into the inner part of the instrument and the evaporation of the solution being measured, cells are furnished with glass stoppers or at least with lids.

The cell holder is designed to make it possible to measure several solutions in succession. This can be achieved in two ways:

1. The cell housing is shaped to enable a series of cells to be drawn into the path of light in succession, in the case of modern instruments automatically.
2. In modern instruments specially shaped cells enable the user to empty and wash the cells after use and fill them up with a new solution without removing them from the instrument.

Finally, for many practical applications of spectrophotometry, e.g., in the dissolution testing of tablets (Section 9.C), in automatic analyzers (9.J) and in UV-visible spectrophotometers used as HPLC detectors (5.C) flow-cells are used.

An important accessory of spectrophotometers used for kinetic analysis (7.J), enzymatic analysis (7.K) and precision measurements is a temperature regulator suitable for keeping constant temperature by heating or cooling of the cell housing or the cell itself.

4. Detectors

As seen in Equation 2.8, the determination of absorbance is based on the measurement of light intensity. This can be achieved most accurately and simply by transforming it to an electrical signal. The photocell (phototube) used for this purpose is a vacuum tube with a half cylinder-like cathode and an anode rod at the axis of the latter. The voltage between the electrodes is about 100 V. The surface of the cathode is covered by a thin layer of material which emits electrons upon illumination. The electrons are drawn to the anode, thus producing a photocurrent the intensity of which is proportional to the intensity of the light. After appropriate amplification the current is measured. The so-called blue-sensitive cell where the cathode is covered with cesium antimonide is used in the UV and visible region below 620 nm and a cesium oxide red-sensitive layer for higher wavelengths.

Electronic amplification of the photocurrent also results in the amplification of the noise level, and hence there are limitations in the use of photocells in spectrophotometry. It is not suitable for measuring very low light intensities and low differences of intensities, since these applications would require broad spectral bandwidths which in turn would compromise spectral resolution. As a consequence of the problems outlined above, photocells are used today only in simple single-beam spectrophotometers.

In the majority of spectrophotometers an electron multiplier (photomultiplier) is used as the detector. This is in fact a special form of vacuum photocell but has a much greater sensitivity, and consequently, a better spectral purity is attainable by its use than with the conventional photocell. This is achieved in such a way that electrons emitted by the photocathode do not impinge directly on the anode but on a further electrode called a dynode, which is also coated with cesium antimonide; thus, secondary electrons are emitted from its surface (4–5 secondary electrons/one electron). These are attracted to the surface of a second dynode, etc. By using 8–10 dynodes in sequence with increasingly positive potential as high as 10^6-fold, amplification of the photocurrent is attainable, thus creating an excellent basis for measuring low light intensities.

An alternative to the electron multiplier is the silicon cell which can be used in the full UV-visible region but with a lower sensitivity than the former. Its function is based on a special property of the semiconductor silicon, namely that its conductance is directly proportional to the level of illumination.

Semiconductor silicon photodiodes have found a new application field in diode-array detectors introduced into the practice in the early 1980s. Making use of their easy miniaturization, several hundreds of these diodes are mounted as an array in the detector. The dispersed light coming from the grating is monitored across the full spectrum simultaneously in this way.

In the ultraviolet range the density of the diodes is about one diode/nm while in the visible, two. All diodes are connected to small condensers which are in the charged state at the beginning of the measurement. When light strikes the diode, the condenser will be discharged to an extent which is proportional to the intensity of the light. The electrical current necessary to recharge the condenser periodically at a high frequency is measured continuously, and this is the basis for the absorbance measurement and hence the simultaneous scanning of the spectrum at all wavelengths. The spectrophotometer constructed on this basis is described on p. 22.

5. Monitors

The electrical signal generated by the detector is subjected to electronic amplification; the extent depends on the type of the detector. In older manual spectrophotometers the photocurrent is compensated by a current generated from a potentiometer, the latter being scaled into absorbance units. This type of instrument is no longer manufactured, but they are still in use in some laboratories.

Modern instruments are equipped with digital readout enabling direct absorbance reading; moreover, in the case of instruments coupled with a microcomputer the absorbance values can be

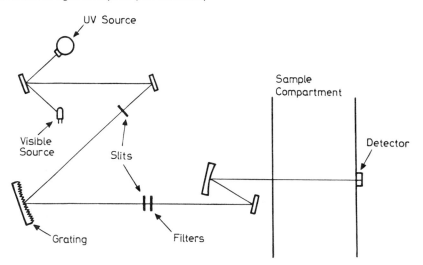

FIGURE 2.7. Optical scheme of the single-beam Beckman DU-60 spectrophotometer.

obtained in printed form. Monitors designed for direct recording of spectral are discussed in Section 2.F.

E. SINGLE-BEAM SPECTROPHOTOMETERS

As has already been mentioned in Section 1.A, the Beckman DU was the first, widely used spectrophotometer in the course of the historical development of spectrophotometry. Its principal optical scheme has served as the basis for most of the subsequent instruments, some of them being still in use.

The optical scheme of the current Beckman DU-60 series is shown in Figure 2.7. As seen, the light beam of either the UV or the visible source enters the monochromator through the entrance slit. Dispersion takes place with the aid of a concave holographic grating blazed at 220 nm. The beam leaves the monochromator through the exit slit and passes through stray light cutting filters before it reaches the cell in the sample compartment and finally the detector. It is worth mentioning that modern, computer-assisted instruments do not use movable cell holders to pull the reference and sample cells consecutively into the light beam in the sample compartment to obtain the absorbance. For example in the Beckman DU-60 series, the sample compartment accommodates only one cell. The reference solution is first introduced and the detector signal is stored in the computer of the instrument. This is followed by the sample solution in the same cell and the absorbance value is calculated by the computer from the two stored detector signals. This kind of absorbance measurement requires stabilized energy output of the light source ("stable beam" technology). Although the main application field of the single-beam instruments is the absorbance measurement mainly for the purpose of quantitative analysis, the stable-beam technology enables single-beam instruments to be used for scanning spectra: in this case the full "spectrum" of the reference is stored in the computer, and after scanning the sample solution, the former is subtracted from the latter to obtain the spectrum.

Another approach to eliminate the effect of variations of the energy output of the light source can be exemplified by the Shimadzu UV-1200 single-beam spectrophotometer. In this instrument a fixed beam-splitter is inserted between the grating of the monochromator and the exit slit splitting the beam into sample (about 95%) and monitor beams (about 5%). The latter reaches the surface of the "monitor detector" (independent of the detector of the instrument where the sample beam is focused). The intensities of the two beams will vary with the changes of the energy output of the light sources but their ratio will not. Consequently, with the aid of the monitor detector the effects of the above-mentioned changes can be eliminated.

Many of the existing single-beam instruments are filter photometers. Since the majority of single-beam instruments with grating monochromators are used for quantitative analysis and not for scanning spectra, the resolution of their monochromators is usually not very high (spectral bandwidth 2–5 nm), and the aim of the extra options of more sophisticated instruments is to ensure rapid and accurate quantitative measurements (automatic cell changer, programmable devices to fill, empty and wash the cells, computer to store and calculate the data and to calibrate the wavelength and absorbance scale).

F. DOUBLE-BEAM SPECTROPHOTOMETERS

Although recently several simple, manual, non-recording variants of double-beam instruments have become available for quantitative analytical purposes, the main application of double-beam instruments is the recording of spectra.

In modern "ratio-recording" double-beam instruments, the absorbance (log I_0/I) is measured in the following way. The light beam leaving the monochromator through the exit slit is split into two beams of equal intensities with the aid of a beam splitter. The two beams pass through the reference and sample cells, respectively, which are located in the sample compartment at fixed positions. The beams are then focused onto the same detector with the aid of independent mirror systems. The detector thus receives the reference beam (I_0) and the attenuated sample beam (I) alternately, the frequency of the change from one to the other being determined by the frequency of the rotation of the sector mirror which is usually used as the beam splitter. Accordingly, the electrical signal thus obtained in the detector also changes periodically and is directly proportional to transmittance (I/I_0) or, after logarithmic transformation, the absorbance.

In practice the magnitude of the signal, related to the intensity of the reference beam (I_0) is kept constant. I_0 changes as a function of the wavelength during the recording of the spectrum. These changes are due to the dependence of the energy output of the light source on the wavelength and the light absorption of the optical system. These changes must be compensated either electronically (by adjusting of the sensitivity of the electron multiplier and the potential of the dynodes) or mechanically (by changing automatically the slitwidth of the monochromator).

As a typical and up-to-date double-beam instrument, the Perkin-Elmer Lambda 9 spectrophotometer is presented in Figure 2.8. The optical system of Lambda 19 is the same, differing from the former only in the higher degree of computerization.[6]

As light sources, a deuterium (DL) and a halogen-tungsten (HL) lamp, respectively, are used. These are accommodated in a fixed position. When measuring in the visible or near-infrared (NIR) region, the light of the lamp, HL is reflected by the mirror, M1 on mirror M2 and at the same time it blocks the light path of the lamp, DL. When measuring in the UV range the mirror ML is lifted and the light of the lamp DL is directly reflected on the mirror M2. From here the light beam is reflected with the aid of mirrors M3, M4 and the collimating mirror, M5 to the entrance slit, SA of the monochromator (monochr. I). A disc filter system (FW) is located between M3 and M4. By automatically rotating FW, the light is made to pass through the filter corresponding to the spectral region to be investigated and thus it is ensured that prefiltered light passes through the entrance slit.

In the monochromator there is a separate holographic grating for the UV-visible region (1440 grooves per millimeter) and another one for the NIR region (360 grooves per millimeter). The monochromator is of the Littrow type in which case the light beam returns from the grating to the collimating mirror M5 and then it passes through the exit slit. The latter serves as the entrance slit of an identical second monochromator (monochrom. II). It should be noted that a second monochromator aimed at improving the resolution and decreasing the stray light is only built into the most sophisticated instruments. Several routine instruments of high quality are in use with one monochromator only.

The synchronized rotation of the gratings of the monochromators assures the high level of the monochromaticity of the light leaving through the exit slit of the monochromator. It is the automatic adjustment of slits in slit unit SA that ensures constant energy to be incident on the detector during the recording of the spectrum.

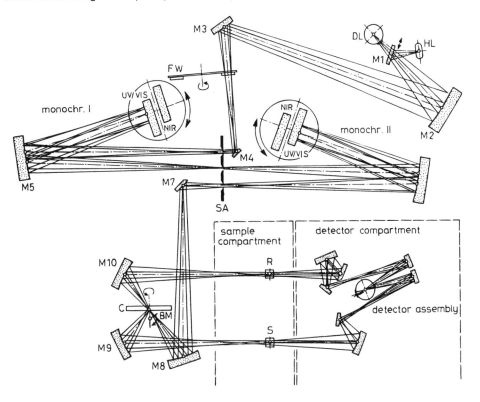

FIGURE 2.8. Optical scheme of the double-beam Perkin-Elmer Lambda 19 spectrophotometer.

From the exit slit of the second monochromator, the monochromatic light beam is reflected with the aid of the toroidal mirrors M7 and M8 on the beam splitter C. This is a motor driven rotating sector mirror with a window-like sector, a sector covered with mirror, and a dark sector to block the light path. As the open, window-like sector gets into the light path, the beam passes through it and falls on mirror M10, which reflects the beam to reach the reference cell, R. When the light beam strikes the mirror sector of the beam splitter, it is then reflected to the mirror, M9, from where it is reflected to reach the sample cell, S. The dark sector of the beam splitter absorbs all light while it is in the light path and the dark current is thus obtained in the detector. The light strikes the reference and sample cells periodically at a frequency determined by the rotation speed of the beam splitter. From the cells, the beams get to the detector with the aid of a mirror system. The detector is an electron multiplier in the UV-visible spectrum range and a lead sulphide cell in the NIR region.

The above-described instrument—just like several others on the market in this category—takes advantage of the benefits provided by the microcomputer-microprocessor technology. Automatic control of the structure elements includes the change over switch of the lamps, the slits, the gratings determining the wavelength scale and the beam splitter. By suitable programming, the instrument performs, e.g., the calibration of the wavelength scale (using the line of the deuterium lamp at 656.1 nm), the correction of the base line shift by means of the dark sectors in the beam splitter, etc. Spectra can be recorded either directly by means of a recorder or by using a line printer after having displayed the spectrum on the screen of the computer (together with all recording parameters). In addition to solving many other computer-related problems the instrument performs calibration and—if necessary—correction of the absorbance or concentration scale by applying suitable factors. The instrument is capable of performing derivative (even higher order derivative) spectra, too.

The majority of spectrophotometers currently on the market are essentially similar to the above described instrument but may vary in the geometry and resolution of the monochromator as well as in the level of computerization. On the other hand the reversed optics of the Hewlett-Packard

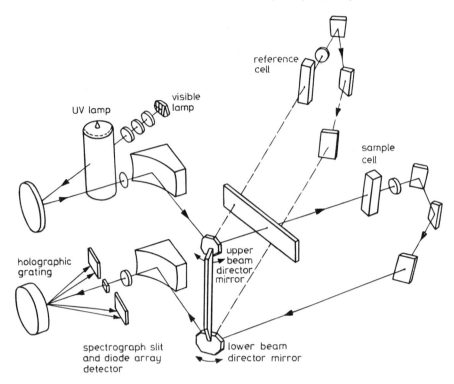

FIGURE 2.9.   Optical scheme of the diode-array Hewlett-Packard 8452 spectrophotometer.

8452 spectrophotometer, as introduced in the following paragraph, represent an entirely new principle possessing numerous advantages.[5]

As seen in Figure 2.9 the light beam from the source strikes the reference and sample cell, respectively, as polychromatic radiation prior to dispersion by a monochromator. In this case the beam splitter (beam director) is a pair of vibrating mirrors from which the upper one reflects the light alternately to one of the two cells. (Apart from the cooling fan the vibrating mirror is the only moving part of the instrument.) The light beams passing through the cells are reflected by the lower beam director mirror to the holographic grating. The light dispersed here then reaches the diode-array detector consisting of 316 microdiodes (see p. 18). Detecting the full spectrum, at the same time the registration speed depends only on the data collection and data handling by the computer. In this way the spectrum is displayed on the monitor within a time interval in the order of 10 milliseconds. This high speed is especially beneficial when the spectrophotometer is used as a detector in HPLC (high performance liquid chromatography; see Section 5.C) or FIA (flow injection analysis; see Section 9.J) by applying flow cells in the sample compartment. The importance of the ability of the diode-array spectrophotometers to scan spectra in the order of ten milliseconds is self-explanatory in the above-mentioned flowing systems. In addition, diode-array spectrophotometers can be successfully used in all cases where the high recording speed associated with sophisticated computerization may be advantageous.[16]

G. THE MEASUREMENT OF LIGHT ABSORPTION

If the absorbance of a solution in a cell is defined according to Figure 2.2 and Equation 2.8 the absorbance consists of two components even if the experimental conditions are selected in the most careful way (carefully cleaned cell, manufactured from the most suitable material and containing the most suitable solvent): 1) the absorbance of the dissolved substance (A_{real}) and, 2) the light attenuation caused by the cell and the solvent therein which can also be expressed as absorbance (A_{void}).

$$A_{total} = A_{real} + A_{void} \tag{2.16}$$

A_{void} includes components originating from the light absorption, reflection and scattering of the cell, light absorption of the solvent and light scattering of any particles present in the solution. Although its value is low related to A_{real} if the experimental conditions are properly selected, it is by no means negligible.

Since in the case of the analytical applications of absorptiometry from both the qualitative and quantitative points of view, it is A_{real} that should be regarded as the absorbance of the dissolved material, experimental conditions have to be selected which allow its simple and direct determination. This is why in the course of the measurements, in addition to the cell containing the sample solution, a reference cell is also used which is similar to the former but contains only pure solvent. (As described in Sections 2.E and 2.F when a single-beam instrument is used, the cells are placed into the beam path successively and in the case of double-beam spectrophotometers they are positioned into two independent beam paths.)

As monochromatic light beam of intensity I_0 strikes both cells, its intensity will be reduced to I in the case of the sample cell and I_0' in the case of the reference cell. The following equation (Equation 2.17) can therefore be established by substituting these values into Equation 2.16:

$$A_{real} = A_{total} - A_{void} = \log \frac{I_0}{I} - \log \frac{I_0}{I_0'} = \log \frac{I_0'}{I} \tag{2.17}$$

As can be seen in Equation 2.17 it is the intensity of the light beam attenuated by the reference cell that is taken into account when measuring the absorbance rather than the intensity of the original light beam.

Simple spectrophotometers allow absorbance measurements in the A = 0–2 range but in more expensive, modern instruments the range is 0–3, or even in some cases 0–4. As A = 4 means that 99.99% of the incident light energy is absorbed, for measuring the very low intensity of the attenuated light beam, a highly sensitive detector should be used. It is important to notice that the capability of some instruments to measure high absorbances can only be exploited if the value of stray light in the instrument is sufficiently low (see Section 2.J).

The values of absorbance in the range of which the absorbance measurements can be performed with maximum precision, i.e., the relative error ($\Delta A/A$) of the absorbance measurement is minimum, can be calculated by differentiation of the Beer equation (Equation 2.8). When semiconducting detectors are used where the detector noise (ΔA) is independent of the light intensity, this value is 0.43 while with instruments using a photocell or electron multiplier as the detector, where the detector noise is proportional to the square root of the light intensity the optimum absorbance is 0.86. Since both ΔA vs. A curves, especially the latter, show rather broad minima, the relative error of the absorbance measurement can be decreased in a broad absorbance range below 0.5%.

Further possibilities for reducing the error of the absorbance measurement will be shown in Section 4.G. As for the detailed treatment of this question and the derivation of the above values the monographs listed in the introduction of this chapter are referred to.

H. EFFECT OF CELLS AND SOLVENTS ON ABSORBANCE

Some of the sources of error of absorbance measurement originate from the incorrect selection or handling of the cells. The discussion of this problem is the subject of this section. The effect of slitwidth and stray light on measurement errors is discussed in Sections 2.I and 2.J.

Since the material, design and exact path length of cells depend on the manufacturer even in the case of virtually identical cells, exclusively matched cells from the same source must be used for reference and sample solutions. Departing from this could be a source of major errors especially in the short-wavelength ultraviolet range as a consequence of differences in light absorption and reflection. It may also be a source of error if the light does not fall perpendicularly on the optical surface of the cell, possibly because of the imperfect finishing and wearing of the cell holder. Due to the repeated removal and insertion of the cell, exact repositioning into their original location cannot be assured. It is therefore advisable that emptying, rinsing and refilling of cells during a series of measurements be performed without removing them from the spectrophotometer with the aid of an automatic device or with a simple pipette.

Even if appropriate cells and cell holders are used, the incorrect treatment of cells is a very serious source of error. The optical surfaces of the cells should be protected from effects which may eventually lead to permanent alterations (scratching, etching caused by strong alkaline solutions) or contaminations. After use the cells must be rinsed without delay with the solvent used for the measurement followed by water or alcohol depending on the miscibility of the solvents. In the case of contaminations which are difficult to remove, rinsing with neutral detergents or as a final measure treatment with concentrated nitric acid or chrome sulphuric acid should be carried out. It must also be taken into consideration that possible sediments often have absorption in the ultraviolet region only: consequently, a cell looking clean at first glance is not necessarily suitable for the production of spectra. For this reason it is advisable to repeat from time to time the above described treatment even if no visible contamination is observed.

To control the cells to be used as reference and sample cells they are filled up with pure solvent and the apparent absorbance of the sample cell is measured at the wavelength for the subsequent analytical measurement. When spectra are produced the baseline is recorded in a similar way. If the values obtained exceed ±0.003–0.005 absorbance units, then either unmatched cells are being used or they need to be cleaned. Cell correction is often applied: the above values are subtracted from or added to the measured absorbance values. In principle this is a fairly good solution to the problem but since such corrections could be poorly reproducible, ill-defined values this method should be avoided as far as possible.

When not in use it is advisable to store cells in water or alcohol. While using the cells, their optical surfaces must be highly glossy. For wiping, a clear, soft cloth, free from fiber (preferably lens tissue) should be used in a very careful manner avoiding extensive rubbing. Useful advice concerning the treatment and cleaning of cells can be found in the book of Burgess and Knowles.[2]

A further source of error originates from the improper selection of solvents. The question of solvents for spectrophotometry is treated in Sections 2.I and 2.K. Although in ideal cases the solvent would be completely transparent over the spectral range in question, as can be concluded from Equation 2.17 the measurements can be performed even in cases where solvents or their contaminants have some (but not very intensive) light absorption within the wavelength range of interest. Consequently, in the case of simple measurements it is not absolutely necessary to use "spectroscopic" grade solvents. It is, however, a source of very serious errors if the reference and sample cells are not filled up with a solvent from the same lot (moreover from the same bottle) as even minor differences of content of solvent contaminants may cause significant differences in the light absorption. This source of error is illustrated with a problem often occurring when absolute alcohol is used as the solvent. This usually contains benzene below 0.01%. Its specific absorbance is 26 at the maximum of the α-band at 254 nm. If ethanol containing 0.005% benzene is filled into the reference cell (10 mm) and ethanol containing 0.01% benzene is used for the preparation of the sample solution, the measured absorbance will be higher than the true value by 0.13 absorbance units.

This problem is significant even if distilled water is used as the solvent especially if it is stored in plastic containers: it may contain a considerable quantity of contamination which absorbs or scatters light. (For this reason it is often recommended that distilled water for UV work should not be stored in plastic containers but only in glass.)

I. EFFECT OF SLITWIDTH ON ABSORBANCE

The absorbance can be significantly influenced under certain conditions by the width of the exit slit of the monochromator and so the improper selection of the slitwidth can be a source of major errors.

As has already been mentioned in Section 2.D.2, dealing with monochromators, the spectral bandwidth is directly proportional to the slitwidth (see Equation 2.15). Since on the one hand the spectral slitwidth determines the spectral purity of the measurement according to an inverse relation and on the other hand the signal-to-noise ratio of the absorbance measurement is in direct relation with the spectral bandwidth, it is evident that in selecting the spectral bandwidth (or the slitwidth determining it) a compromise should be reached. From the viewpoint of spectral purity as narrow a slit as possible would be naturally optimal, but by reducing the slitwidth, the signal-to-noise ratio declines and thus the precision of the measurement deteriorates. The analyst can be constrained to increase the slitwidth by the following factors:

1. Poor condition of the spectrophotometer. Every factor which reduces the light energy reaching the detector for a given slitwidth take effect in this direction. A factor of this type may be the deviation of the light path which causes only a part of the light beam to reach the detector through the cell. Another possibility can be contamination of the optical elements (mirrors, etc.) which is a particularly important problem around or below 200 nm where the energy output of the deuterium lamp is relatively low. A contamination of this type can occur especially in those laboratories where chemical work is performed in the room where the spectrophotometer is operating, thus exposing the optical surfaces to the effect of vapors of solvents and chemicals. The cautious cleaning of contaminated surfaces from time to time and their polishing after prolonged use can again guarantee the measuring capabilities according to the original specification of the instrument.*

2. Difficulties in the selection of solvents for the measurement. While colorless solvents of all kinds can be used in the course of spectrophotometric investigations in the visible spectral region, in the ultraviolet and particularly in the short-wavelength ultraviolet range, the range of suitable solvents is strongly limited. Of the solvents generally applicable in spectrophotometric drug analysis only water, methanol, ethanol, hexane, cyclohexane and acetonitrile can be used at the lower wavelength limit of spectrophotometers (down to 190 nm). Below 210 nm, particularly pure solvents, labeled as "spectroscopic" should be used. In this spectral range absorbed carbon dioxide, moreover oxygen content also contribute to the light absorption of the solvents. Because of these difficulties—insofar as the solubility of the substance examined permits—it is reasonable to perform spectroscopic examinations in this spectral range in a more concentrated solution using 1 mm cells. (This is advantageous from the point of view of stray light, too; see next section.) Using old instruments it is often difficult to make measurements in this spectral range even when all precautions are exercised. The cut off wavelengths (lower practical wavelength limits of the applicability) of other solvents are as follows: diethyl ether, 205 nm; dioxane, 215 nm; tetrahydrofuran, 240 nm; chloroform, 245 nm; carbon tetrachloride, 260 nm; dimethylformamide, acetic acid, dimethylsulphoxide, 270 nm; benzene, 280 nm; pyridine, 305 nm; and acetone, 330 nm. Close to these limits, similar energy-related problems may occur as described above.

3. Problems associated with the use of chemical reactions prior to the spectrophotometric measurement (see Chapters 6 and 7). When the determination is based on a preliminary chemical reaction excess reagents may lead to serious problems if they have significant absorption at the absorption maximum of the reaction product. If the molar absorptivities are comparable it may be necessary to remove the excess of the reagent from the reaction mixture prior to the measurement. Especially serious slitwidth-related problems arise in difference spectrophotometric variants of methods based on chemical reactions (see Chapter 8).

The nature and extent of errors caused by excessively increasing the slitwidth depend on the quality of the spectrophotometer and the spectrum of the substance examined. The dependence on the specification of the spectrophotometer presents itself in relation to the monochromator and the detector. According to Equation 2.15 the spectral slitwidth is directly proportional to the mechanical slitwidth and inversely proportional to the linear dispersion of the monochromator. This means that in the case of a monochromator with twofold dispersion the slitwidth could be doubled without

---

* The use of sealed optics in modern instruments greatly overcomes this problem. It is then only necessary to clean entrance and exit windows to the sealed system.

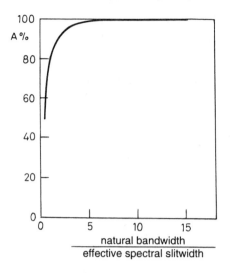

FIGURE 2.10. Percentage of absorbance related to the theoretical value as a function of the ratio of natural bandwidth and spectral bandwidth.

increasing the spectral bandwidth as compared with the value attained with an instrument equipped with a monochromator of lower resolution. On the other hand, of course, the sensitivity of the detector is also of great importance: if the detector is more sensitive a narrower slit can be used.

The question of whether the increase of the slitwidth (and consequently the spectral bandwidth) causes error and, if so, to what extent depends on the spectrum of the compound in question or more exactly on the sharpness of its spectral bands. The latter can be characterized by the natural bandwidth of the band. (The natural bandwidth is the difference of two wavelengths on the left and right sides of the band maximum where the absorbances have the half value of that in the maximum.) In the case of fused peaks the absorbance is calculated from the baseline as can be seen in Figure 2.11. Theoretically the true absorbance of the band can only be measured at infinitesimal narrow bandwidth; practically the error caused by the finite bandwidth is within the experimental error if the natural bandwidth is at least 5 to 10 times higher than the spectral bandwidth. This is illustrated in Figure 2.10 where the percentage value of the measured absorbance at the absorption maximum of the band related to the theoretical absorbance is plotted as a function of the ratio of natural bandwidth and effective spectral slitwidth (spectral bandwidth). When drawing the idealized curve in Figure 2.10 it was supposed that the spectral band is an ideal Gaussian curve and the composition of the light energy leaving the exit slit is determined by the idealized triangle in Figure 2.5. Practical cases are more or less close to the relationship represented in Figure 2.10. In the course of the solution of a new problem, especially in those cases when the resolution of the monochromator of the spectrophotometer is not sufficient and the spectral band is sharp (particularly if a prism monochromator is used in the visible range) it is advisable to study these slitwidth–absorbance relations. Curves of this kind can be seen in Figure 2.12. The absorbance values corresponding to very broad slits can be obtained in such a way that the light beam is mechanically attenuated (e.g., by placing a bronze mesh filter into the light path). Reliable absorbance values can only be obtained if the measurement takes place at such a slitwidth range where the absorbance is independent of the slitwidth. If the low energy does not permit this the measurement can be performed on the descending section of the slitwidth–absorbance curve; in this case, however, it is very important to keep the slit at a constant value during serial measurements. The effect described above is illustrated by the spectrum of vitamin B_{12} as shown in Figure 2.11 and the slitwidth–absorbance curves plotted at the absorbance maxima in Figure 2.12.

For the evaluation of the curves in Figure 2.12 the natural bandwidths of the spectrum of vitamin B_{12} in Figure 2.10 are herewith presented. These are 15 nm at the maximum at 278 nm, 20 nm at the main maximum at 360 nm, and finally 80 nm at the long-wavelength maximum at 549 nm. These values are 5, 6.7, and 27 times greater than the spectral bandwidth of 3 nm obtainable by the fully open slit of the instrument Pye-Unicam SP-1800 equipped with a grating monochromator

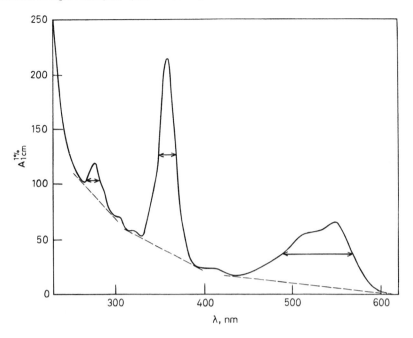

FIGURE 2.11. Spectrum of vitamin B_{12}. Solvent: water.

having constant spectral bandwidth over the full spectral range. (The spectral bandwidth of 3 nm corresponds to a mechanical slitwidth of 1 mm.) On the basis of these data it is not surprising that the curves in Figure 2.12 do not show a greater decrease than 1% related to the ideal absorbance value even at a slitwidth of 1 mm.

Entirely different is the situation with the prism spectrophotometer Opton PMQ-2 (curves "b"). At 278 nm where the dispersion of the prism is good, the absorbance does not alter until 0.5 mm and with the fully open slit (in this case 2 mm) it decreases only by 6%. The band at 360 nm is rather sharp; here in addition the dispersion of the prism is inferior to that at the previous wavelength. Accordingly the decrease of the absorbance begins even at a slitwidth of 0.1 mm and it reaches 50% at the fully open slit position. The dispersion of the prism is very low at 548 nm; the linear dispersion obtainable from curve "b" in Figure 2.6 is 28 nm/mm. This is balanced to some extent by the great natural bandwidth of the band. Therefore the absorbance decreases by 30% at the fully open slit position.

J. EFFECT OF STRAY LIGHT ON ABSORBANCE

Apart from that of the slitwidth discussed in the previous section, it is the stray light that causes the most problems for the analyst during the spectrophotometric measurement. This is a consequence of the light passing through the exit slit including energy outside the wavelength range determined by the spectral slitwidth of the monochromator: this light is termed stray light. The stray light originates from different sources. The most important sources are imperfections and defects of the dispersing elements (e.g., the uneven density and altering depth of the grooves in the grating), irregularities and contaminations on the optical surfaces and the higher order spectra produced by the grating.

Since in the course of the absorbance measurement ($A = \log I_0/I$) the stray light (S) is superimposed on both the reference beam (I_0) and the sample beam attenuated by the investigated solution (I) (see Equation 2.8) its effect on the absorbance can be very significant in some cases. This means that the observed absorbance (A_0) which is subject to the effect of stray light is always lower than the theoretical value.

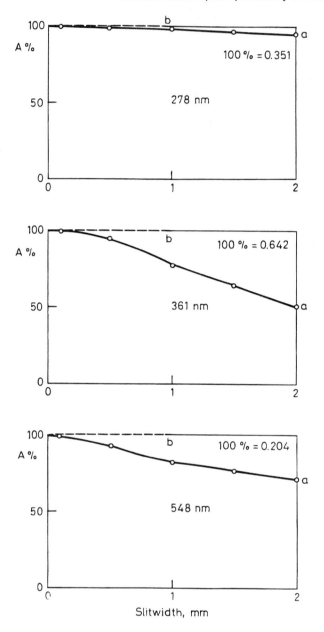

FIGURE 2.12. The change of absorbance with mechanical slitwidth at the maximum wavelengths in the spectrum of vitamin B_{12}. Curves "a": quartz prism monochromator (Opton PMQ-2); Curves "b": grating monochromator (Pye-Unicam SP-1800).

$$A_0 = \log \frac{I_0 + S}{I + S} \qquad (2.18)$$

The stray light detectable in spectrophotometers is usually expressed as a percentage of the intensity of the incident light beam (I_0). The absorbance values calculated by means of Equation 2.18 taking into account 0–5% of stray light in the absorbance range of 0–4 are summarized in Table 2.2.

As can be seen from the values in Table 2.2 the differences between the theoretical and observed absorbances are higher at high absorbance values. In the case of stray light of 1 and 5% the deviation is significant even at the very beginning of the absorbance scale. If the stray light is

TABLE 2.2. Absorbance Values Calculated by Equation 2.18 (The Effect of Stray Light on Absorbance)

I_0	I	Absorbance at different stray light values (stray light calculated as % of I_0)					
		0%	0.001%	0.01%	0.1%	1%	5%
1.0	0.8	0.097	0.097	0.097	0.097	0.096	0.092
1.0	0.6	0.222	0.222	0.222	0.222	0.219	0.208
1.0	0.4	0.398	0.398	0.398	0.397	0.392	0.368
1.0	0.2	0.699	0.699	0.699	0.697	0.682	0.623
1.0	0.1	1.000	1.000	1.000	0.996	0.963	0.845
1.0	0.01	2.000	2.000	1.996	1.959	1.703	1.243
1.0	0.001	3.000	2.996	2.959	2.699	1.963	1.314
1.0	0.0001	4.000	3.959	3.699	2.959	2.000	1.321

0.1%, the error is negligible up to about absorbance value 1; if it is 0.01%, then up to 2–3. The absorbance scale ranging between 0 and 4 available in some modern instruments can only be fully utilized if the stray light is reduced to 0.001%.

From the data in Table 2.2 it follows that in the course of measuring the absorbance of solutions with higher concentrations, after a linear section with a length depending on the value of the stray light, the calibration line changes to a curve (negative deviation from Beer's Law; see Section 4.B). It should be mentioned, however, that Equation 2.18 and subsequent calculation contain the simplifying assumption that the sample does not absorb the stray light. If it is partly absorbed, the deviation will be less than the calculated value.

The relative quantity of the stray light depends on the wavelength of the light emerging from the monochromator. It causes particularly serious problems at the wavelength limits of the energy output of the light sources where the light intensity (I_0) changes rapidly with the wavelength. With tungsten lamps the wavelength ranges of this kind are between 320 and 400 nm as well as above 800 nm. In the case of deuterium lamps the critical range is below 220 nm.

The problems in the visible range can be solved relatively easily by using suitable stray light filters. However in the ultraviolet range no filters are available which absorb above 220 nm but are transparent below this wavelength. This is why in a wide range of spectrophotometers used in practice, stray light of 0.1–1% is an inherent parameter of the instrument in the short wavelength UV range and rapidly increasing in the direction of 190 nm. The absorbance can be reduced by this rapid increase to such an extent that it can overcompensate the effect of the increase of the absorbance towards the short wavelength end of the UV region which is characteristic of the majority of spectra of organic compounds. As a consequence of this the short-wavelength side of absorption bands appearing around 220 nm are "cut off" by the instrument and the shape of the band is distorted. As this effect naturally occurs to a greater extent at higher concentrations, the apparent position (wavelength) of the maximum may become concentration dependent on the sample. This effect is particularly sharply observable when solvents are used (e.g., ethanol) the absorbances of which rapidly increase with decreasing wavelengths around 200 nm. The above mentioned overcompensation may lead to the displacement of the absorption maxima and even to the appearance of false maxima around 200 nm.

To reduce the error it is advisable to apply aqueous media and cells with shorter pathlength if the solubility of the analyte permits. As below 200 nm, the light absorption of oxygen also contributes to the error, the conditions of the measurement can be improved by flushing the instrument with nitrogen. The development of the technology of optical elements, especially the introduction of holographic gratings has resulted in the dramatic reduction of stray light in modern instruments which are often equipped with double monochromators. The declared stray light of the most sophisticated spectrophotometers is in the range of 0.001–0.01% and it does not exceed 0.1% even in the most critical wavelength range.

This is illustrated by the spectra of benzyl alcohol in aqueous and ethanolic solutions as presented in Figures 2.13 and 2.14, with absorption maxima in the critical short wavelength range. Curves "a" are spectra of solutions of 20 mg/100 ml concentration with pathlength of 1 mm while curves

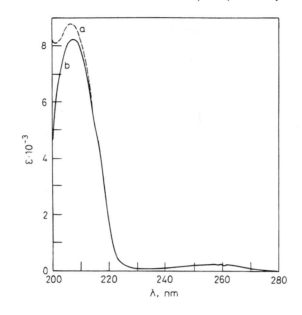

FIGURE 2.13. Spectra of benzyl alcohol. Solvent: ethanol. a: c = 20 mg/ml, pathlength: 1 mm; b: c = 2 mg/ml, pathlength: 10 mm.

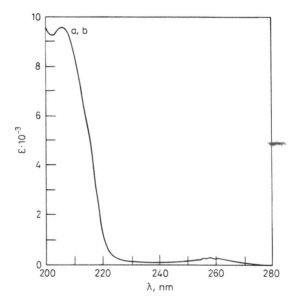

FIGURE 2.14. Spectra of benzyl alcohol. Solvent: water. a: c = 20 mg/ml, pathlength: 1 mm; b: c = 2 mg/ml, pathlength: 10 mm.

"b" are spectra of solutions with tenfold dilution in 10 mm cells. The coincidence of the two curves in Figure 2.14 indicates that under the given experimental conditions, in aqueous medium the effect of stray light is not apparent above 200 nm. In Figure 2.13 only a minor distortion of the band is observable in ethanolic medium when the path length is 1 mm but a major distortion is visible when a 10 mm cell is used because of the cut off toward shorter wavelengths and consequently the apparent position of the p band is shifted from 206 to 208 nm also.

It should be noted that the spectra in Figures 2.13 and 2.14 were produced with a recording spectrophotometer, Varian DMS-200 where the declared value of stray light is less than 0.02% at 220 nm.

Had a less sophisticated instrument been used, or one of a suitable standard but in a poor technical condition, or solvents used having spectra with steep ascendent section in the direction of the short wavelengths because of their own absorption or that of their contaminants—much greater distortions, or shifts of the maxima (moreover appearance of false maxima) could occur.[7]

As demonstrated, different problems arise in the wavelength range below 220 nm. In addition to the reduction of the intensity of light (described in Section 2.I) contamination of the optical surfaces leads to serious increases of stray light. This requires great caution by the analyst in the evaluation of spectrophotometric data measured in this region. This means that reliable quantitative analytical methods cannot be based on absorbances measured below 215–220 nm and even in qualitative analysis the maxima detected in this region are to be treated cautiously. On the one hand this is because in this region several, basically different functional group and structure elements possess absorption bands (see Chapter 3) and consequently the maxima detected here are of very limited value in structure elucidation research. On the other hand, as has been described previously, very misleading maximum shifts or even appearance of false maxima may occur in this region. If it is absolutely necessary to obtain spectral data from this region, only high-level instruments equipped with double monochromator should be used and the stray light should be controlled by the methods to be described at the end of this section. Spectra taken from the literature should also be handled very cautiously. In a large number of such spectra published in the literature, the region below 220 nm is very unreliable and in many cases evidently false.

As for measurement of the stray light, the various possibilities are usually based on the quantitative estimation of those phenomena which present the above described problems during spectrophotometric measurement. This is, for example, the determination of the absorption of glass filters or solutions with a sharp cut off and very intensive absorption in the spectral range where, without stray light, total absorption of light ought to be registered while in the case of strong stray light false maxima are seen. A solution recommended for the critical region of 195–223 nm is the aqueous solution of sodium bromide at a concentration of 10 g/liter in a 10-mm cell. Another possibility for the quantitative characterization of the stray light is to measure the extent of the deviation from Beer's Law in such systems which obey it if the effect of the stray light is eliminated.

Finally the reader's attention is drawn to the excellent summarization of the problems related to stray light from both theoretical and practical points of view in the book of Burgess and Knowles.[2]

K. SCANNING THE SPECTRUM

1. The Importance of Spectral Scans

As has already been stated in the course of the treatment of Equation 2.8, the constant of the Lambert-Beer Law, the molar absorptivity depends on the wavelength. This relation is expressed by the absorption spectrum. In the practice of spectrophotometric drug analysis recording of spectra is needed in many cases:

1. Although UV-visible spectroscopy is only a moderately effective tool in the structural elucidation of native or synthetic organic compounds, the spectra of potential drugs, their intermediates, degradation products and metabolites, in many cases afford useful information, and they are indispensable parts of the analytical documentations of drugs.
2. Simpler qualitative analytical problems, i.e., identification of pharmaceutical compounds also necessitates the production of spectra in order to be able to compare them with those of standard substances, or at least to compare the maximum and minimum wavelengths in the spectra with the data taken from the literature.
3. The production of spectra is often needed even in such cases when the only aim of the spectrophotometric measurement is quantitative analysis. Thus the spectra allow the analyst to select the most suitable wavelengths for the quantitative measurement.
4. The production of spectra is also often necessary in cases when quantitative analysis is performed and the wavelengths suitable for measuring the components in question are known. Namely, when pharmaceutical compounds are measured in various matrices (pharmaceutical preparations, biological samples, etc.) the comparison of the spectra of the

pure components with those of the solutions obtained after dissolution, dilution or extraction of the sample concerned affords information about the possible interferences caused by other components, excipients, contaminants, background absorption, etc. On this basis, it is possible to judge if a direct measurement is possible or not. Some examples are in Section 4.C.

2. Techniques for Scanning Spectra

For the production of an absorption spectrum the first step is to select a suitable solvent. Two factors are to be taken into account: the ability of the solvent to dissolve the analyte and its transmittance. The former generally does not cause major problems since the measurement is usually carried out in dilute solutions: at concentrations in the order of 0.001–0.01 or in exceptional cases 0.1%. The majority of substances of pharmaceutical interest are soluble in water, methanol, ethanol, 2-propanol, etc., at such concentrations. Other solvents such as hexane, cyclohexane, diethyl ether, dioxane, etc., are used only in special cases.

The transmittance of solvents can be characterized by the lowest wavelength at which measurements can be reliably performed. Some data can be found in Section 2.I. The relation between the selection of the solvent and problems which may occur during the measurement of the spectrum (spectral purity, stray light) are discussed in Sections 2.I and 2.J.

A solution of 0.001–0.01% is usually prepared for recording the spectrum using the solvent selected for the measurement. In the case of spectrophotometrically poorly active compounds, or when recording less intense sections of the spectra of active compounds, 10–100 times more concentrated solutions may be used. Another possibility is the variation of cell thickness; its reduction can be especially useful in the region around 200 nm to diminish the effect of stray light.

Spectra are generally measured on a double-beam recording instrument (see Section 2.F). Before starting, the instrument should be zeroed either in such a way that pure solvent is poured into both cells or the cells containing the blank and sample solutions are placed into the cell holder and the instrument is zeroed at a (long) wavelength where the sample solution has no absorption.

Apart from the instruments equipped with diode-array detector where the full spectrum can be recorded within less than one second, the time necessary to measure the spectrum by traditional instruments is usually adjustable between about 1 min and 1 hour. The determination of the minimum time needed for the recording is a complex problem. Here the time constant of the recorder should also be taken into account. This is the time necessary to obtain 98% of the final amplitude; on modern instruments stepwise adjustment is available.

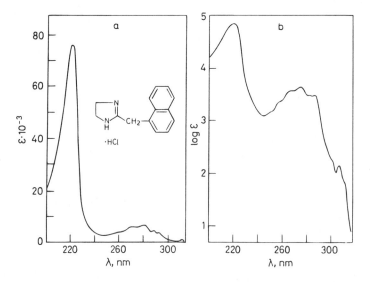

FIGURE 2.15. Spectra of naphazoline hydrochloride. Solvent: water. a: Ordinate in molar absorptivity; b: Ordinate in the logarithm of molar absorptivity.

By decreasing the time constant the recording speed is increased but the signal-to-noise ratio is reduced. In addition to this the optimum recording speed depends on the shape of the spectrum, on demands regarding the resolution and, hence naturally, on the slitwidth, too. Too rapid scanning of complex spectra consisting of narrow bands can lead to the distortion of the band system and the displacement of the maxima. The recording speed can be considered ideal if it does not exceed 10% of the natural bandwidth during the time corresponding to the time constant according to the above definition.

If a recording spectrophotometer is not available, a fully identical spectrum can be obtained using a non-recording single-beam instrument with the only major difficulty that the recording time and the labor consumption increases by orders of magnitude. As in the 1940s and 1950s, the classical period of UV-visible spectrophotometry, the reliability of recording instruments did not reach that of single-beam instruments, the majority of spectra published in the literature at that time which are regarded as fundamental reference spectra even now were produced by this technique. In the case of manual measurements of spectra with a single-beam instrument, the absorbance of the solution is determined in the usual way at many wavelengths and the results are plotted as a function of the wavelength. Points have to be measured at an adequate density: in the flat region of the spectrum at a frequency of 2–5 nm and around the maxima or inflections at a frequency of 0.5–1 nm depending on their sharpness.

3. Methods for the Presentation of Spectra

In the spectra in the literature prepared for quantitative analytical purposes, the abscissa is exclusively presented in a linear-wavelength form and this mode of presentation is also used in the majority of studies related to structure elucidation and identification. In theoretical publications spectra are often plotted as a function of the frequency expressed as wavenumbers of the light since these data are directly related to the energy of the light (see Section 2.A). Using the latter mode of presentation the bands are more symmetrical than in the case of the linear-wavelength presentation.

The light absorption is recorded on the ordinate almost exclusively in the form of absorbance or some other related function. (In the older literature, spectra with transmittance values on the ordinate can often be found.) If the measured absorbance is plotted directly as a function of wavelength, the exact concentration of the solution should be presented. In more advanced modes of presentation of spectra, the specific absorbance or the molar absorptivity is recorded on the ordinate, the $A_{1\,cm}^{1\%}$ values for quantitative analysis, while ε for studying the structure–spectral relations. In the latter case it is often necessary to show bands of very low intensity which are of no importance from the point of view of quantitative analysis. As the intensity of these may be lower by 2–4 orders-of-magnitude as compared to the main maximum, these sections of the spectra (measured at relatively high concentrations) cannot be plotted together with the bands of higher intensity in the same spectrum. Accordingly, the different spectral sections have to be presented either separately with different scalings on the ordinate or more preferably by applying a logarithmic scale. The latter allows for a single plot of the complete spectrum, even though the bands differ from each other by 2–4 orders-of-magnitude. This is demonstrated in the example of naphazoline in Figure 2.15.

When a spectrum is being prepared it is important to select the wavelength range before recording the spectrum. If it is being prepared for private use only or as a preliminary experiment before a quantitative measurement, it is sufficient to scan and represent it only over wavelength range of interest. If, however, the aim of preparing the spectrum is structural elucidation or the spectrum is intended for publication in the literature, the full spectrum has to be taken and represented. One has to be very cautious, however, with the spectral region below 220 nm: this can only be published if the parameters of taking the spectrum (especially the low value of stray light) enable a spectrum free of distortions and false maxima to be taken (see Section 2.J). Similar caution has to be exercised in the production of bands with low intensities ($\varepsilon < 1000$) since this can be done reliably only if a highly pure preparation is available as sample. (This question will be dealt with in detail in the subsequent section.) For the same reasons one has to be critical when evaluating spectra taken from the literature or using them for one's own purposes. Namely, several more or less distorted, erroneous spectra can be found in the literature due to the use of insufficiently pure materials and improper recording conditions. It is therefore advisable to rely upon spectrum atlases referred to in Chapter 3 or other reliable sources or to prepare the spectrum yourself.

4. Factors Influencing the Spectra

The absorption spectrum of a substance, the position and intensity of the bands are influenced by several factors. Some of these factors can be considered as real influences since they may lead to several, more or less different spectra for the same substance even if the recording parameters are properly selected. Other factors, however, can be regarded as improperly selected parameters for producing the spectra: their effect on the spectra can be regarded as distortion rather than real influence. The majority of factors in both groups have been or will be discussed in previous and subsequent sections, respectively: in these cases we only refer to the corresponding sections.

Factors influencing spectra:

1. Solvent effect. In certain cases there are significant differences between spectra produced in polar and apolar solvents. Since important conclusions can be drawn from these differences regarding the structure of the molecule of the examined material, this question is treated in detail in Section 3.B.
2. Effect of pH. The spectra of substances, the acidic or basic groups of which are in conjugation with the electron system establishing the spectrum of the substance, often show dependence on the pH of the solution. This question is discussed in Section 3.P in detail. It is noted that for quantitative determinations of substances of this kind or for documentation purposes, such conditions should be applied which guarantee either the total ionization of the molecule or the total exclusion of ionization. This can be achieved by taking the spectrum in apolar solvents or if using polar solvents, especially water, 0.01–1 M hydrochloric acid or sodium hydroxide has to be used in the solvent, depending on the acid or basic strength of the analyte.
3. Effect of concentration. On the basis of the considerations described in Section C, the molar absorptivity, and hence the spectrum itself, is independent of the concentration of the solution. In the practice of drug analysis this applies in the majority of cases. There are, however, cases where the shift of the possible association equilibrium caused by dilution results in spectral shift, too. Cases of this type leading to deviations from Beer's Law are dealt with in Section 4.B.
4. Effect of temperature. Measurement of the spectrum in a suitable solvent at low temperature (e.g., at the temperature of liquid air) leads to the fine structure of spectra consisting of complex band systems to be more pronounced than at room temperature. The structural conclusions which can be drawn from the dependence of spectra on temperature are, however, of minor importance as compared with other spectroscopic techniques, mainly NMR spectroscopy.

In some cases (e.g., in the case of dyes in associative equilibrium[13]) there exist changes within narrower temperature ranges, too. In the case of pharmaceutical compounds these effects can be generally neglected and spectra are usually presented without specifying the temperature. However, when precision measurements are carried out, or extreme alteration of temperature can be expected, it is advisable to use a thermostatted cell housing or cell holder. In this way not only the slight shift of the spectrum can be prevented but the absorbance differences originating from the heat of dilution of the solvents can be compensated as well.

A very intensive temperature-dependence was found by Sakai[8–12] on investigating the spectra of the ion-pair complexes of different alkaloids and related bases with tetrabromophenolphthalein ethyl ester. As the temperature-dependence of the different bases is significantly different and the correlation between absorbance and temperature is linear, spectrophotometric measurements at different temperatures enable the simultaneous determination of these bases and this was applied in the analysis of such preparations.

Factors distorting the spectra:

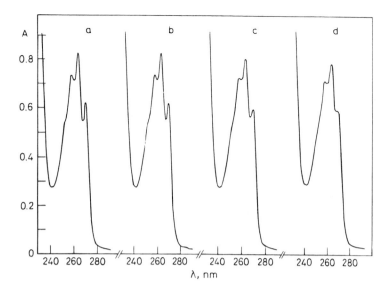

FIGURE 2.16. Spectra of flumecinol measured at different spectral bandwidths. a: 0.6 nm; b: 1.0 nm; c: 2.0 nm; d: 3.0 nm. Solvent: hexane; c = 0.27 mg/ml.

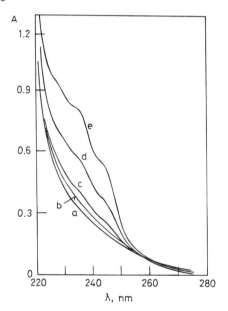

FIGURE 2.17. Spectra of ethynodiol diacetate spiked with the impurity 17α-ethinyl-3,5-oestradiene-17β-ol acetate. Solvent: ethanol. a: 0.085%; b: 0.17%; c: 0.43%; d: 0.85%; e: 1.7%.

1. Incorrectly selected slitwidth and spectral bandwidth. This problem is treated in detail in the sections dealing with monochromators (2.D.2) and the slitwidth (2.I). Instruments with a spectral bandwidth of 2 nm are suitable for the examination of the majority of drug analytical problems but in case of spectra with narrow bands and especially with fine structure higher resolution may be necessary as shown on the example of flumecinol in Figure 2.16.

   Many of the spectrophotometers used in practice mainly for quantitative measurement have spectral bandwidths of 5 nm or even more. The spectra taken with these instruments are often distorted as compared to the true ones. They may be suitable for selecting the wavelengths for the quantitative measurements but by no means for documentation purposes.

2. Stray light. The distorting effect of this factor on the absorption spectrum is dealt with in Section J.
3. Improperly selected time constant and too rapid scanning speed. See Section 2.K.2.
4. Signal-to-noise ratio too low. This question and its relation to other parameters has already been discussed in the previous sections.
5. When the spectrum is recorded point-to-point using a manual single beam instrument, a distorting factor can be insufficient density of the measuring points.
6. Application of non-calibrated or incorrectly calibrated instruments. This question will be treated in the following section.
7. Recording the spectrum of impure sample. The presence of small but strongly absorbing impurities in the sample may cause a major error especially when the spectrum, or at least a portion of it, is of low intensity. An illustration of this is presented in Figure 2.17 where it is clear that the distorting effect of as little as 0.1% of 17α-ethinyl-3,5-oestradiene-17β-ol acetate impurity in ethynodiol diacetate generates a distinct distortion and the impurity at the level of 0.5–1% produces a well defined additional band in the spectrum of ethynodiol diacetate. The typical spectrum of the impurity is presented in Figure 3.6a.

L. CALIBRATION OF SPECTROPHOTOMETERS

1. Calibration of the Wavelength-Scale

The accuracy of the wavelength scale to around 0.1 nm is guaranteed by the manufacturers. The correctness of the wavelength reading, however, has to be checked periodically especially after prolonged use. Two general ways are used for this calibration: application of light sources emitting light at well defined wavelengths or standards absorbing at exactly known wavelengths.

The simplest use of the former possibility is to exploit the sharp lines of the deuterium lamp of the instrument at 486.02 and 656.10 nm. The lines detectable at 243.7, 364.9, 404.5, 435.8, 546.1, 576.9 and 579.0 nm of the widely used medium-pressure mercury lamp present excellent opportunities for the calibration as well.

Glass filters containing rare earth oxides such as holmium glass and didymium glass with several bands in the ultraviolet-visible region are widely used for checking the wavelength scale. By using solutions more reproducible and sharper bands are obtainable. Ten percent w/v solution of holmium (III) perchlorate in 17.5% perchloric acid, proposed by Burgess[1] has absorption maxima at 241.1, 249.7, 278.7, 287.1, 333.4, 345.5, 361.5, 385.5, 416.3, 450.8 and 485.8 nm in a 10 mm cell and spectral slitwidth of 1 nm. Metal film nichrome filters are also available with calibrations in the UV region even down to about 200 nm.

2. Calibration of the Absorbance Scale

As for the wavelength scale discussed in the preceding section, several standards in the solid form and as solutions are available for the calibration of the absorbance scale. Many kinds of glass filters are obtainable as accessories for spectrophotometers and also independently. These have well defined absorbance values at selected wavelengths. Some analysts, however, prefer to prepare their own absorbance standards using solutions of suitable substances because they can be measured under identical conditions as used for samples to be analyzed. In principle, many kinds of substances could be used as absorbance standards provided their absorbance bands are not so sharp, to avoid dependence on the slitwidth nor is their molar absorptivity very high, to avoid errors originating from repeated dilutions. Further prerequisites are stability against heat and light as well as availability in pure form. Another advantage of calibration using solution is that the linearity of the detector of the instrument can also be checked by the determination of the absorbances of a series of dilutions providing, of course, that the standard material obeys Beer's Law.

Potassium dichromate is the most frequently used calibration substance. The commercial material of very high purity is available which after drying for 1 hour at 110°C is ready for use. Solutions of 0.02–0.1 g/liter in 0.005 M sulphuric acid are generally prepared to obtain the calibration standard. Under these conditions the molar absorptivities at the maximum at 257 nm and 350 nm are 4210 and 3170, respectively. The correct adjustment of the acid concentration is a very important factor as the spectrum is pH-dependent due to dissociation and dimerization equilibria. Such a problem is not encountered if the measurement is performed in 0.05 M potassium hydroxide, since under such conditions only CrO_4^{2-} ions are present in the solution ($\varepsilon_{273 \text{ nm}} = 3,700$; $\varepsilon_{372 \text{ nm}} = 4,810$). The above molar absorptivities were calculated from the large quantity of data collected in the book of Burgess and Knowles.[3] This book contains many further methods for the calibration of the wavelength and absorbance scales.

M. THE MEASUREMENT OF LIGHT REFLECTION

The discussion in the previous sections of this chapter has been restricted to those cases where the test substance is present in solution during the investigation. In the practice of drug analysis, from time to time solid samples also have to be investigated spectrophotometrically. The absorption spectra of organic substances in the solid phase can be taken by applying the measuring technique generally used in infrared spectroscopy.[14] The test substance is triturated with potassium chloride then a transparent disc is prepared using a sufficiently high pressure. After placing the disc into the light beam of the spectrophotometer its absorbance can be measured in a similar way to the solutions in the cell. This technique is not widely used in the pharmaceutical analysis.

The study of the light reflection of solid samples is especially important because it opens a possibility for the qualitative and quantitative evaluation of thin-layer chromatograms. To facilitate the treatment of this subject in Section 5.E, the main rules of light reflection are briefly discussed here.

The layer of 100–250 μm thickness on a thin-layer chromatographic (TLC) plate consists of particles of silica gel, alumina, etc. The particle size is in the order of 10 μm. The separated compounds to be investigated are located on the surface of these particles within the spots of the chromatogram. Light is not reflected from such irregular shaped particles following the law of reflection, but instead undergoes diffuse reflection. This means that in the course of 1) multiple reflection from the surface of the particles, 2) penetration into the particles and 3) deflection among the particles the light penetrates into the layer to a certain thickness. As a consequence of these phenomena (generally referred to as light scattering) the direction of the light is changed, moreover reversed to some extent at some points between the surface and the greatest depth of penetration. The reflected light then leaves the surface of the layer at various directions. The intensity of the reflected light and the spatial distribution of the intensity depend on the angle of the irradiation. In the case of perpendicular irradiation, maximum intensity and approximately hemispherically symmetrical distribution are obtained; this is the usual arrangement for analytical applications of this phenomenon.

It has to be noted that the thickness of the layer of the TLC plate is usually less than the above mentioned maximum depth of light penetration. This means that a considerable proportion of the incident light reaches the opposite side of the layer. This is the basis of the possibility of the evaluation of TLC plates in the transmission mode providing that visible light is used and the layer is applied on glass or a colorless plastic material. In this case Beer's Law is more or less obeyed and on this basis, quantitative analytical measurements can be carried out (see Section 5.E).

As for the intensity of reflected light from a TLC plate containing an analyte on its surface Beer's Law is not applicable directly in this case. The correlation between the intensity of the reflected light (R_∞)*, the absorption (K) and scattering (S) is expressed by the Kubelka-Munk equation[4] (Equation 2.19):

$$\frac{K}{S} = \frac{(1 - R_\infty)^2}{2R_\infty} \tag{2.19}$$

While S depends only on the number, volume, shape and refractive index of the particles and

---

* The index ∞ in Equation 2.19 expresses that the equation relates to the case when the layer is thick enough to preclude the possibility of light reflection from the backside of the layer, i.e., the reflected light is entirely of diffuse light character.

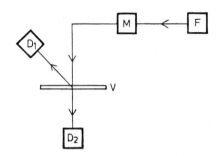

FIGURE 2.18. Optical scheme of a chromatogram spectrophotometer working in reflection and transmission modes. F: light source; M: monochromator; V: thin-layer chromatogram; D_1: detector (reflection mode); D_2: detector (transmission mode).

is independent of the relatively small amount of chromatographically separated substances on the surface of the particles, the value of K does depend on their quantity, molar absorptivity and hence, of course, on the wavelength, too.

On the analogy of *transmittance* (see Equation 2.11) which is the ratio of the intensities of the attenuated light beam emerging from the absorbing body (solution, thin-layer, etc.) and the incident beam the term *reflectance* can be introduced:

$$R = \frac{I_R}{I_0} \qquad (2.20)$$

where R is the reflectance, I_0 is the intensity of the incident light beam and I_R is the intensity of the diffusely reflected light.

The measurement of reflectance is to some extent analogous to that of the transmittance but there are also notable differences. During the measurement of transmittance, the total attenuated light beam reaches the detector. In the course of reflectance measurements, however, when the incident light beam strikes the layer perpendicularly, the diffusely reflected fraction of the total light intensity is radiated in all directions of a hemisphere. This means that the detector positioned at 45° to the plate detects only a fraction of the total reflected light (see Figure 2.18). R can be determined in such a way that at first the intensity of the reflected light, I_R, is determined with the TLC spot at the place where the light beam strikes the plate, followed by repeating the measurement in the same way but with an empty area of the TLC plate at the same place. I_0 is considered to be the intensity of the reflected light in the latter measurement. Using this technique the effects of all factors are taken into account which are not associated with the chromatographic spot but which would contribute to the attenuation of the light. Of these effects, the most significant is the change of the reflectance of the adsorbent layer with the wavelength: it strongly decreases with decreasing wavelength below 400 nm. This approach can be considered as analogous to the procedure resulting in cancelling out the light absorption of the solvent in the course of the measurement of transmittance or absorbance when the instrument is zeroed to the cell containing the pure solvent (see Equation 2.17).

Many commercially available spectrophotometers are equipped with adapters enabling the study of diffuse light reflectance of solid samples. A special device is necessary for the evaluation of chromatograms which includes a highly sensitive detector and an optical and electronic system are required enabling the detection and quantitative evaluation of the separated substances which may be present only in the order of microgram or submicrogram quantities. To obtain an acceptable signal-to-noise ratio is especially problematic in the UV region where the light reflection of adsorbents greatly decreases.

Densitometers, of which the most developed ones are often called chromatogram-spectrophotometers, are constructed principally from the elements discussed in Section 2.D. Deuterium, tungsten and mercury lamps are used as the light sources. Beside cheaper instruments using color filters, excellent chromatogram-spectrophotometers with prism or grating monochromators are also available. An electron multiplier is usually used as the detector. For measuring reflectance the arrange-

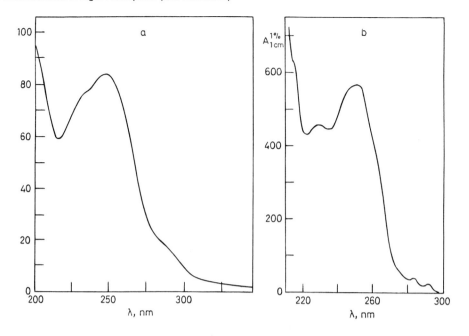

FIGURE 2.19. Spectra of cinnarizine. a: Reflection spectrum after TLC separation; b: Absorption spectrum. Solvent: ethanol.

ment shown in Figure 2.18 is usually applied, but the most developed instruments are also equipped with adapters for the measurement of transmittance and fluorescence, and even for the simultaneous monitoring of two of the above three parameters. Both single- and double-beam instruments are in use.

Even single-beam instruments equipped with a suitable monochromator enable the reflectance spectra of separated compounds to be taken with high sensitivity (down to the order of nanogram/spot). In this case point-to-point manual recording is necessary; using double-beam instruments this can be done more conveniently by automatic recording.

Reflection spectra correlate closely with absorption spectra taken in solution as shown in the example of cinnarizine in Figure 2.19. This is naturally of great importance in the identification and structure elucidation of separated compounds, impurities, degradation products, etc.

The quantitative evaluation of chromatograms is of even greater importance. Several attempts have been made in the past years aimed at the transformation of the Kubelka-Munk equation for the purposes of the quantitative evaluations of chromatograms and the computerized linearization of the concentration—response curves. Modern TLC techniques and the most developed, highly sensitive densitometers enable the quantitative determination of the separated materials down to the nanogram range. In this low concentration range linear calibration graphs are often obtained when peak height or peak area in the chromatogram is plotted vs. the quantity of the material in the spot. As for the measuring techniques, see Section 5.E.

Regarding the literature of reflection spectrophotometry, the monographs of Wenlant and Hecht[17] as well as Stearns[13] are recommended, while an excellent summary of the application of the theory of reflection spectroscopy to thin-layer chromatography can be found in the book of Touchstone and Sherma.[15]

REFERENCES

1. Burgess, C. *UV Spec. Grp. Bull.* 1977, *5*, 77.
2. Burgess, C.; Knowles, A. *Standards in Absorption Spectrometry;* Chapman and Hall: London, New York, 1981.
3. Knowles, A.; Burgess, C., Eds. *Practical Absorption Spectrometry;* Chapman and Hall: London, New York, 1984.
4. Kubelka, P.; Munk, F. *Z. Tech. Phys.* 1931, *12*, 593.

5. Owen, A. J. *The Diode-Array UV Advantage in UV-Visible Spectroscopy;* Hewlett-Packard: Waldbronn, 1988, 11.

6. *Perkin Elmer Lambda 9 UV/VIS/NIR Spectrometer. Operator's Manual,* 4–5.

7. Ruecker, G. *Dt. Apoth. Ztg.* 1975, *115,* 935.

8. Sakai, T. *J. Pharm. Sci.* 1979, *68,* 875.

9. Sakai, T.; Ohno, N. *Chem. Pharm. Bull.* 1979, *27,* 2846.

10. Sakai, T.; Ohno, N. *Analyst* 1981, *106,* 584.

11. Sakai, T. *Analyst* 1982, *107,* 640.

12. Sakai, T.; Ohno, N. *Analyst* 1982, *107,* 634.

13. Stearns, E. I. *The Practice of Absorption Spectrophotometry;* Wiley-Interscience: New York, 1969, 73.

14. Suzuki, H. *Electronic Absorption Spectra and Geometry of Organic Molecules;* Academic Press: New York, 1967, 115–118.

15. Touchstone, J.; Sherma J., Eds. *Densitometry in Thin-Layer Chromatography;* Wiley: New York, 1979.

16. Ueda, H.; Pereira-Rosario, R.; Riley, C.M.; Perrin, J.H. *Drug. Dev. Ind. Pharm.* 1985, *11,* 833.

17. Wenlandt, W.W.; Hecht, H.G. *Reflectance Spectroscopy;* Wiley-Interscience: New York, 1966.

Chapter 3

QUALITATIVE ANALYSIS: RELATIONSHIP BETWEEN THE STRUCTURE AND SPECTRA OF PHARMACEUTICAL COMPOUNDS

A. IMPORTANCE OF STRUCTURE–SPECTRA CORRELATION STUDIES

There are two points necessitating that drug analysts become acquainted with the bases of the correlation between the structures and ultraviolet spectra of drugs and related organic substances.

1. As has already been pointed out in Section 1.B, besides the more effective spectroscopic methods such as infrared, nuclear magnetic resonance and mass spectroscopy as well as X-ray diffractometry, UV-VIS spectroscopy also affords important supplementary data for the structure elucidation of the products of drug research and their intermediates, degradation products, isolated impurities, etc.

2. Even if UV-VIS spectroscopy is used exclusively as a quantitative analytical method, a sufficient level of knowledge about structure-spectra correlation is important for the practical analyst. With this knowledge and being aware of the formula of the compound to be determined it can be predicted with fairly good certainty if the given compound can be determined spectrophotometrically in the given sample matrix: whether it has sufficiently strong absorption band(s) and, if so, in which spectral region and what kind of interference can be expected from the side other components of the sample.

   The objective of this chapter is to summarize the fundamental knowledge which is necessary for the drug analyst working in these fields.

In accordance with the importance of this field an enormous literature is available for studying theoretical and practical aspects of structure-spectra correlations. Some books dealing with individual families of drugs are discussed in the introductions to the respective sections of Chapter 10. The general books and other publications can be divided into three groups:

1. Collections containing spectral data including wavelengths of absorption maxima together with the absorption coefficients but not the spectral curves. Two fundamental books can be recommended. Hirayama[13] presents data in tabulated form on selected organic compounds the basis of the organization of the material being the functional groups and structural elements of the compounds. The 27 volumes of *Organic Electronic Spectral Data* published so far include practically all literature data described about the UV characteristics of organic compounds the contents being organized on the basis of the formulae of the compounds.[20]

2. Of the several spectrum atlases and series of atlases the 24-volume series of Láng[17] excels from the points of view of both the systematic selection of the compounds included and the reliability and good organization of the data. Although not specializing to drugs, this series contains spectra of a high number of drugs and other materials of pharmaceutical interest. The *UV Atlas of Organic Compounds*[21] and the spectrum collections published by Sadtler[25] and Hershenson[12] also contain important data from the point of view of pharmaceutical analysis.

Of the collections specializing in drugs the 3-volume atlas compiled by Dibbern[9] is the most important: it contains the spectra (in various media) of almost all important drugs together with their numerical spectral data. The books of Clarke[8] and Sunshine[30] were compiled with the requirements of toxicological analysis in mind but they also contain numerous data of general analytical interest.

3. The classical books of Scott,[26] Rao,[24] and Stern and Timmons[28] are the most important of those dealing with the correlation between the UV-visible spectra and the structures of organic compounds because the style of discussion in them correlates with the way of thinking of drug analysts working in practice. Several monographs deal with theoretical questions in this field.[4,5,14,31]

The limitations of this book do not allow exhaustive discussion of the enormous number of papers published in this field. The series of Krácmar and Krácmarová, does however, merit attention: it contains evaluated spectra of almost all important drug substances. Of the 51 papers published in *Die Pharmazie* and *Ceskoslovenska Farmacie*—for the sake of shortness—only the first and last members of the series are referred to here.[15,16]

Of the spectra presented in this chapter, those which do not contain a reference in the figure caption were prepared in the author's laboratory using a Varian DMS-200 spectrophotometer and in the case of some earlier spectra, a Pye-Unicam SP-1800 instrument. The same also applies to a few spectra in the other chapters in this book. Unless otherwise stated the spectral data (λ_{max}, ε and $A^{1\%}_{1cm}$ values) in the text are from the books referred to earlier.[8,13,26,28,30]

The mode of treatment of the question of the correlation between structure and spectra will be basically qualitative and empirical. A treatment based on quantum chemical approach would exceed the limitations of this book and the objectives of the book could be attained even with this simplified treatment. On the other hand it can be stated that the quantum chemical treatment affords convincing results in the case of simple molecules only. The mathematical apparatus necessary for the quantitative treatment of more complex structures characteristic of the majority of pharmaceutical molecules rapidly increases with increasing complexity of the molecules and hence the results of this kind of treatment are hardly accessible for practical analysts.

B. CHARACTERIZATION OF ABSORPTION BANDS

In order to be familiar with the bases of structure–spectrum relationships some factors have to be briefly discussed which are related to the structural elements in the molecule. These are the origin of the bands (their relation to the electrons excited in the course of the light absorption), the position, intensity and shape of bands and factors influencing the above listed characteristics of bands.

1. The Origin and Position of Bands

The origin of ultraviolet-visible spectra of organic molecules based on the excitation of electrons to orbitals of higher energy was already briefly introduced in Section 2.B. These electrons are of three types:

1. σ-electrons. These are the electrons forming single bonds. The σ orbitals can be characterized by their low energy level such electrons are strongly held and are therefore difficult to be excited.
2. π-electrons. These form carbon-carbon and carbon-hetero atom double and triple bonds. The energy levels of π-orbitals are higher; the π-electrons are more loosely held and are therefore easier to excite.
3. n-electrons. These are non-bonding unshared electron pairs of hetero atoms in organic compounds (e.g., oxygen, sulphur, nitrogen, halogens).

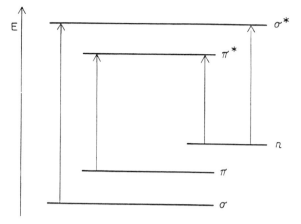

FIGURE 3.1. Energy scheme of various types of electron excitations.

By the effect of light absorption in the ultraviolet-visible region, these three types of electrons are excited to σ^* and π^* orbitals which can be characterized by their higher energy levels. The energy relations of the transitions are depicted on the scheme in Figure 3.1. The highest energy is absorbed during the σ–σ^* transition. Transitions of this type are observed in the vacuum ultraviolet region. The π–π^* transitions require lower energy difference while the n–π^* transitions can be characterized by an even lower energy consumption. Accordingly, the bands belonging to the π–π^* and n–π^* transitions are observed in the short-wavelength and long-wavelength ultraviolet region, respectively. The n–σ^* bands of the carbon–hetero atom single bonds appear at various places in the ultraviolet region. The position (wavelength) of the band is naturally influenced by the environment of the electron pair in the molecule. Of these influences conjugation effects are especially important (see Section 3.E).

2. The Intensity of Bands

While the wavelength of the absorption band is determined by the difference between the energy levels of the ground and excited states of the electron, the intensity of the band is governed by the number of transitions per time unit. On the one hand this depends on the size (cross-section) of the molecule during its collision with the photon and on the other hand on the probability of the transition. This is expressed by Equation 3.1:

$$\varepsilon = k \cdot P \cdot a \tag{3.1}$$

where ε is the molar absorptivity (see p. 12), k is a constant of the order of 10^{20}, P is the probability of the transition with a value between 0 and 1, and a is the above mentioned cross-section which is on the order of about 10^{-15} cm^2 in the case of average absorbing molecules.

If the value of P is 1, every collision results in absorption and the molar absorptivity is on the order of 10^5. The highest molar absorptivities observed in spectroscopic practice are indeed on the order of a few hundreds of thousands. The great differences in the ε values of different compounds and different groups are principally due to the differences of the P values. Bands with molar absorptivities higher than 10,000 are usually considered to be high-intensity bands, those between 1000 and 10,000 bands of medium intensity while bands with $\varepsilon < 1000$ are termed low-intensity bands. The weak bands ($P < 0.01$) are usually forbidden bands, i.e., they correspond to forbidden electron transitions according the selection rules of quantum chemistry.

3. The Shape of the Bands: Solvent Effects

As is described in Section 2.B, in ideal cases absorption bands ought to be Gaussian curves with very narrow natural bandwidths possessing vibrational and rotational fine structure. However,

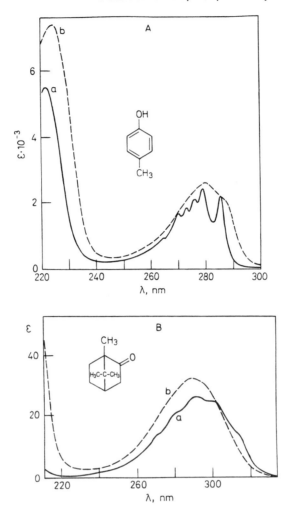

FIGURE 3.2. The effect of solvent on the fine structure of absorption bands. A. p-cresol. Solvent: a. isooctane; b. ethanol. (From N.D. Coggleshall; E.M. Lang *J. Am. Chem. Soc.* 1948, *70,* 3283.) B. Camphor. Solvent: a. hexane; b. methanol.

rotational fine structure can usually only be investigated in the vapor phase since rotation is hindered in the solution phase and hence the rotational side-bands usually merge with the vibrational side-bands or, if the latter are not observable either, with the main band itself. As far as the vibrational fine structure is concerned its occurrence or lack is the function of several factors both internal and external to the molecule. Of the internal factors the most important is the presence or absence of polar substituents (hydroxy, amino, etc. groups); these diminish or even completely eliminate the fine structure. Of the external factors the decrease of temperature favors the appearance and the sharpness of the fine structure. However, the effect of the solvent is much more pronounced. Apolar solvents which do not get into close interaction with the molecules of the dissolved material favor the fine structure while in polar solvents, capable of hydrogen bonding, etc., it is usually degraded to inflections or even fully merged with the main band. As illustrations the spectra of p-cresol and camphor in apolar and polar solvents are shown in Figure 3.2.

In addition to the effect of solvents on the fine structure of the spectra, the position (wavelength) of the band may also depend on the solvent and the direction of the shift on the effect of changing the polarity of the solvent can furnish important data on the correlation of the absorption band with various types of electronic transitions.

The shifts of the absorption bands caused by various effects (changes of the solvent including pH changes in aqueous media, temperature, introduction of a functional group in the vicinity of

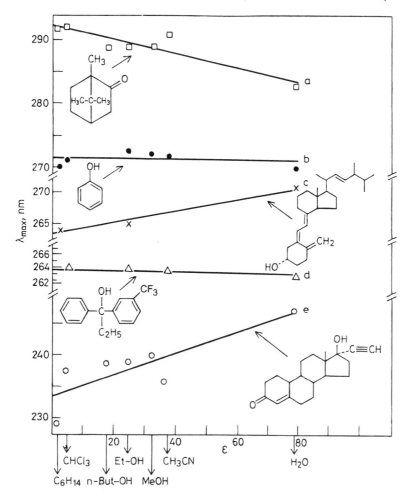

FIGURE 3.3. The effect of solvent on the position of absorption bands. The wavelength of the absorption maximum of some drug substances as a function of the dielectric constant of the solvent. a. camphor; b. phenol; c. vitamin D_2; d. flumecinol; e. norethisterone.

the chromophoric group, chemical reactions resulting in the change of the electronic system of the molecule, etc.) are termed as follows.

- bathochromic (red) shift = shift of λ_{max} to longer wavelengths
- hypsochromic (blue) shift = shift of λ_{max} to shorter wavelengths
- hyperchromic shift = intensity increase of the band
- hypochromic shift = intensity decrease of the band

For example the absorption bands corresponding to n–π transitions undergo hypsochromic shifts when the polarity of the solvent is increased while in the case of π–π* bands bathochromic shifts are observable under the same conditions. There are also exceptions to this rule: no significant solvent effect is observable in the case of the principle band of aromatic compounds, around 260 nm.

Figure 3.3 shows the effect of the change of the dielectric constant of the solvent on the absorbance maximum of various drug substances. The hypsochromic shift of the n–π* band of camphor with increasing dielectric constant, the weak bathochromic shift of the π–π* band of Vitamin D_2 containing a conjugated triene system as well as the strong bathochromic shift of the same band in the spectrum of norethisterone possessing an α, β-unsaturated keto group under the same conditions are good illustrations of the rules outlined above. The independence of the benzenoid band on the solvent is illustrated by the examples of phenol containing and flumecinol not containing an auxochromic hydroxy group in the aromatic nucleus.

From the above described considerations it is evident that the spectra of organic compounds generally consist of more or less broad bands (natural bandwidth 20–40 nm). In the majority of cases even a medium-sized drug molecule with average complexity contains more electrons excitable in the UV-visible region, or there are more than one possibility for an electron pair to be excited. Consequently, the spectra consist of more than one band which are usually not completely resolved from each other. It often occurs that an intense band overlaps a less intense one to such extent that the latter is hardly observable or totally unobservable. Even more typical is the appearance of a fused, asymmetrical peak complex consisting of partially overlapping peaks. The evaluation and correlation of these peaks is often a difficult task. The computer-assisted spectral convolution-deconvolution technique which is capable of building up a complex spectrum from several Gaussian curves (each representing an individual band) or eliminating the contribution of well defined bands from the complex is a useful tool for the evaluation of peaks of this type.[6]

C. SATURATED COMPOUNDS: AUXOCHROMES

The σ electrons of C–H and C–C bonds of saturated hydrocarbons are excitable by radiation within the vacuum ultraviolet region, σ–σ* bands appearing below 160 nm. Accordingly, saturated hydrocarbons (hexane, cyclohexane, etc.) are excellent solvents for ultraviolet spectrophotometric investigations. Their application is especially advisable if the aim of the study is structure elucidation. As a consequence of their limited interaction with the solute molecules, bands become sharper and spectra are more structured than is the case in polar solvents. Their wide application is, however, hindered by their limited capability to dissolve pharmaceutical compounds.

If substituents containing hetero atoms (oxygen, sulphur, nitrogen, halogens, etc.) but not containing double bonds are attached to the molecules of saturated hydrocarbons then n–σ* bands appear in their spectra which correspond to the excitation of the non-bonding n electrons of the hetero atoms.

In the case of alcohols and dialkyl ethers the n–σ* band is not higher than around 180 nm either and since the transition is forbidden, its intensity is low. For this reason simple alcohols and ethers are good solvents for the spectrophotometric investigation of drug substances. (The unfavorable consequences of the band around 180 nm regarding the stray light in the region below 220 nm is discussed in detail in Section 2.J.) Pharmaceutical compounds containing only hydroxy or ether groups on their hydrocarbon skeleton for practical purposes are spectrophotometrically quite inactive (e.g., glycerol, menthol, sugars and sugar alcohols). The absorption maxima of organic peroxides are also below 220 nm.

Alkyl mercaptans and diaklyl sulphides have their n–σ* band below 215 nm, however, their intensity is higher than that of their oxygen analogs with molar absoptivities on the order of 1,000. Disulphides, especially those which are cyclic absorb at longer wavelengths (up to 350 nm) but the intensities of the bands are only on the order of 100.

Saturated primary, secondary and tertiary amines possess n–σ* bands of medium intensity between 190 and 200 nm (ε ~ 1000–4000). In the case of cyclic tertiary amines this band is shifted to about 215 nm. Of the simple amino compounds used in pharmacy, the absorption maximum of hexamethylenetetramine is at 195 nm (ε = 1,370) and of l-aminoadamantane at 194 nm (ε = 870). It is important to note that these data relate to the free bases. If the non-bonding electron-pair of the nitrogen atoms is bonded to a hydrogen ion, as in the course of salt formation, strong hypsochromic and hypochromic shifts occur due to the reduced delocalization of the lone electron pair.

The positions of the n–σ* transitions of halogens depend on the nature of the halogen. The atomic radius increases while the electronegativity decreases with increasing atomic number, and consequently the n electrons of bromine and especially iodine are more readily excited than in the case of lower halogens. Accordingly, absorption maxima of monofluoro and monochloro alkanes are in the vacuum ultraviolet region, but that of bromomethane is at 204 nm (ε = 200) and of iodomethane is at 258 nm (ε = 365).

Increasing the number of halogen atoms attached to one carbon atom causes pronounced bathochromic shifts. In the case of chloroform and carbon tetrachloride, this results in their non-applicability as solvents below 240 and 265 nm, respectively, iodoform (CHI_3), however, is yellow ($\lambda_{max} = 349$ nm; $\varepsilon = 230$).

The above discussed saturated groups containing hetero atoms are termed auxochromic groups. The origin of this name is that, although they posses no or only very slight spectrophotometric activity, their strong influence on the color of dyestuffs was observed even in the classical period of colorimetry. This is because of the interaction of the unshared electron-pair of the auxochromic group with aromatic rings or other conjugated systems. Influences of this kind frequently occur in pharmaceutical compounds, too, as will be demonstrated with several examples in subsequent sections.

D. CHROMOPHORES

While the minor spectrophotometric activities discussed so far based on the transitions of σ and n electrons are of very small importance in both qualitative and quantitative drug analysis, the chromophoric groups to be discussed in this section are of greater importance. They contain carbon–carbon, carbon–hetero atom or hetero atom–hetero atom double bonds and, hence, in their spectra n–π^* and π–π^* bands appear. If they occur as isolated structural features within the molecule only weak bands are observed in the analytically important UV region above 220 nm. The term "chromophore" relates to the fact that the conjugation of these groups results in intense bands in the ultraviolet and extending into the visible regions.

The most important chromophoric group is naturally the carbon–carbon double bond. The π–π^* band of ethylene in the gas phase is at 170 nm ($\varepsilon \sim 10,000$) accompanied by another one at 200 nm, the intensity of which is an order-of-magnitude lower. The latter is not observed in substituted derivatives, thus substitution of the hydrogens of ethylene results in the bathochromic shift of the intense π–π^* band. Alkyl substitution or the appearance of the double bond in cycloalkenes shift the maximum up to 180–200 nm. The absorption maxima of sterols containing isolated (Δ^5) double bond (cholesterol and related compounds) have their maxima at 193–205 nm ($\varepsilon = 6,000$–$12,000$).[10] The effects of halogen or alkoxy substitution are not considerable either. However, the bathochromic effect of nitrogen-substitution is more pronounced. This is illustrated by the example of the enamine, 1-(1-piperidyl)-buten-1 ($\varepsilon_{228\ nm} = 10,000$).

It is remarkable that the absorption band of acetylene containing two π-electron-pairs is at even shorter wavelength than that of ethylene. Consequently, ethinyl derivatives show practically no spectrophotometric activity at 200 nm. The same also applies to the carbon–nitrogen triple bond. Thus, acetonitrile is an excellent solvent for short-wavelength UV spectrophotometric measurements.

Beside the carbon–carbon double bond the other important chromophoric group of pharmaceutical compounds is the carbonyl group which contains the carbon–oxygen double bond. The most characteristic band in the spectra of carbonyl derivatives is the weak n–π^* band. (The n–σ^* band appears around 190 nm while the π–π^* band is in the vacuum UV region.)

The n–π^* bands of some characteristic carbonyl derivatives are shown in Figure 3.4. It is to be noted that as a consequence of ring strain, the maximum of the alicyclic ketone camphor is at higher wavelength than that of acetone and, in the apolar solvent hexane, even the fine structure appears (as in the case of acetaldehyde which is not included in Figure 3.4—$\varepsilon_{293\ nm} = 12$). The fine structure disappears in polar solvents. (For further details regarding the solvent effect see Section 3.B and Figure 3.3.)

It can also be seen in Figure 3.4 that attaching auxochromic groups to the carbonyl group (carboxy and ester groups) results in strong hypsochromic shifts. Similar conclusions can be drawn from the spectra of some related derivatives not included in Figure 3.4: the absorption maximum of acetyl chloride in hexane is at 220 nm, that of acetic anhydride in the same solvent is at 223 nm, while acetamide shows maximum absorption in methanol at 207 nm. The reason for this is because the energy level of the n electrons in the ground state is not influenced by the interaction

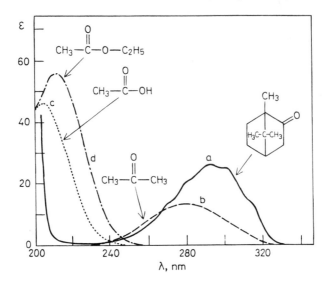

FIGURE 3.4. Spectra of carbonyl compounds. Solvent: hexane. a. camphor; b. acetone; c. acetic acid; d. ethyl acetate.

with the electrons of the auxochromic group while energy level of the π^* orbital occupied in the excited state is increased by this interaction.

The replacement of the oxygen atom by sulphur in the carbonyl group causes a very strong bathochromic shift of all bands. For example in the spectrum of thiocamphor in cyclohexane the $n-\pi^*$ band is at 493 nm ($\varepsilon = 12$) while the $n-\sigma^*$ and $\pi-\pi^*$ bands appear at 244 and 214 nm, respectively ($\varepsilon = 11{,}500$ and $4{,}200$). Aliphatic sulphoxides exhibit a band of medium intensity around 215 nm while the analogous sulphones do not possess significant absorption in the ultraviolet range.

Of the nitrogen containing chromophores the $n-\pi^*$ band of the nitro group is around 270 nm; the intensity of the band is very low. Azomethane and diazomethane exhibit very weak $n-\pi^*$ bands at 347 and 400 nm, respectively ($\varepsilon = 3$ and 5). It is interesting that even this weak color enables the endpoint of the esterification reaction between fatty acids, bile acids and diazomethane to be detected in the course of the derivatization of these acids for gas chromatographic determinations.

When the diagnostic value of the spectra of auxochromic and chromophoric groups in evaluated in structural elucidation studies, the following observation can be made. Although the principles outlined and the data summarized in the preceding paragraphs are of some value in this field, especially in the interpretation of the spectra of complex molecules where these groups are the building blocks of the structures carrying spectrophotometric activity, simple molecules containing auxochromic and isolated chromophoric groups are usually considered to be spectrophotometrically inactive for the practicing drug analyst, because these compounds exhibit significant light absorption only below 220 nm and this is a problematic region (see Section 2.J).

There are, however, two circumstances necessitating the study of such weak or "problematic" light absorption of compounds possessing isolated chromophores.

1. It is a common situation in pharmaceutical analysis that a small quantity of spectrophotometrically active compound has to be determined in a matrix which contains spectrophotometrically inactive or only poorly active materials at a concentration of 2–4 orders higher than the analyte. In such cases the estimation of even these low absorptions which cause spectral background is an important task. An example of this was shown in Figure 2.17. In this example a spectrophotometrically highly active conjugated diene impurity is determined in the "inactive" ethynodiol diacetate. In the course of this measurement the background caused by the weak absorption of the isolated double bond and the acetoxy goups of the latter has to be taken into account. This problem is discussed in detail in Section 4.C.

2. The importance of low intensity bands and especially of those appearing below 220 nm

has greatly increased since the advent of high-performance liquid chromatography. Many compounds exhibiting only low intensity bands can only be monitored during the HPLC assay with the aid of these bands. For details see Section 5.C.

E. CONJUGATED CHROMOPHORES

If the chromophoric groups described in Section 3.D are present in a molecule isolated from each other (i.e., they are separated by at least two single bonds), no interaction usually takes place between them: the spectrum of the compound will be approximately the sum of the spectra of the individual chromophoric groups. No significant increase of the intensity can be observed in the case of cumulated double bonds either. For example diethylketene containing the C=C=O bonding system exhibits absorption bands at 227 and 375 nm (ε = 360 and 20, respectively). Diethylcarbodiimide containing the N=C=N bonding system has maxima at 230 and 270 nm (ε = 200 and 25).

A fundamentally new situation exists, however, if the double bonds are conjugated, i.e., they occur alternately with single bond(s). In this case delocalization of the π-electrons takes place and a more or less uniformly distributed π-electron system forms along the bonding chain. In addition to the well known increased reactivity, this also results in the easier excitability of the electrons. As a consequence of the conjugation, a strong bathochromic and in the majority of cases hyperchromic shift is exhibited.

In the spectrum of the simplest conjugated diene, 1,3-butadiene the highly intense π–π^* band is found at 217 nm (ε = 21,000). It has to be noted that in this case the position of the two ethylenic linkages is *trans* related to the C_2–C_3 single bond. *Cis* dienes exhibit maximum absorptions at considerably higher wavelengths but the intensities of these bands are lower. For example the λ_{max} of 1,3-cyclohexadiene is at 256 nm (ε = 8,000). If further conjugated double bonds, auxochromic groups or even simple alkyl groups are attached to one of these two basic types of conjugated dienes, their maxima are shifted toward longer wavelengths. There are several possibilities for the calculation of the extent of the wavelength displacement. The Woodward rule[33] modified by Fieser[11] and Scott[26a] is suitable especially for the prediction of the maxima of steroid di- and trienes.

The usefulness of this mode of calculation, summarized in Table 3.1 is demonstrated with the calculation of the absorption maxima of a drug, a potential drug and a natural steroid which is the starting material of the synthesis of vitamin D_2.

TABLE 3.1.   Calculation of the Absorption Maxima of Conjugated Dienes and Trienes[26a]

Parent heteroannular diene	214 nm
Parent homoannular diene	253 nm

Increments to be added to one of the above values	
Double bond extending conjugation	30 nm
Alkyl substitution or ring residue	5 nm
Exocyclic double bond	5 nm
O-acyl (enol ester)	0 nm
O-alkyl (enol ether)	6 nm
S-alkyl (thioenol ether)	30 nm
F, Cl, Br	5 nm
N(alkyl)$_2$ (enamine)	0 nm

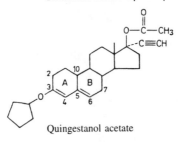

Quingestanol acetate

		λ_{max}
3,5-heteroannular diene		214 nm
2–3, 5–10, 6–7 single bonds	$3 \times 5 =$	15 nm
5–6 double bond (exocyclic related to ring A)		5 nm
3-cyclopentyloxy group		6 nm
	Calculated:	240 nm
	Observed:	242 nm

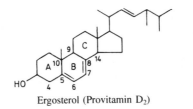

Ergosterol (Provitamin D_2)

		λ_{max}
5,7-homoannular diene		253 nm
4–5, 5–10, 8–9, 8–14 single bonds	$4 \times 5 =$	20 nm
5–6 double bond (exocyclic related to ring A)		5 nm
7–8 double bond (exocyclic related to ring C)		5 nm
	Calculated:	283 nm
	Observed:	282 nm

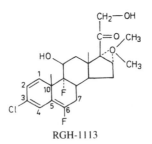

RGH-1113

		λ_{max}
1,3-homoannular diene		253 nm
5–6 double bond		30 nm
1–10, 5–10, 6–7 single bonds	$3 \times 5 =$	15 nm
5–6 double bond (exocyclic related to ring A)		5 nm
3-Cl, 6-F	$2 \times 5 =$	10 nm
	Calculated:	313 nm
	Observed:	312 nm

In the spectra of compounds containing more than two double bonds, additional double bonds cause further bathochromic shifts. The increment of 30 nm given in Table 3.1 for the calculation of the effect of one more conjugated double bond is approximately valid up to about 7 double bonds. As it is seen from the data of Table 3.2, above seven double bonds the increment decreases. It is also clear from the data in Table 3.2 that the intensity of the bands greatly increases with increasing numbers of double bonds. From the data in Table 3.2 and the spectral data of other substituted polyenes it is evident that above $n = 5-6$ polyenes are colored compounds. For example the color of carotinoids and related compounds where n may be up to 11 can be from yellow to red.

It is important to note that in the overwhelming majority of natural polyenes the configuration of the double bonds in the chain is in the more stable *trans* configuration.

The spectra of all-trans dienes and polyenes usually exhibit fine structure. In various data collections (also in Table 3.2) the wavelength of the last band is presented.

The data in Table 3.2 naturally are for orientation purposes only. In order to interpret the maximum wavelengths, the shapes, and intensities of the spectra of polyene-type drugs, several factors have to be taken into account such as alkyl substitution causing bathochromic shifts, steric effects causing hypsochromic and hypochromic shifts and the loss of fine structure (e.g., the interruption of the *all-trans* chain by one *cis* bond). The spectra of the two polyene-type vitamins shown in Figure 3.5 can also be interpreted taking into consideration steric effects. This is why

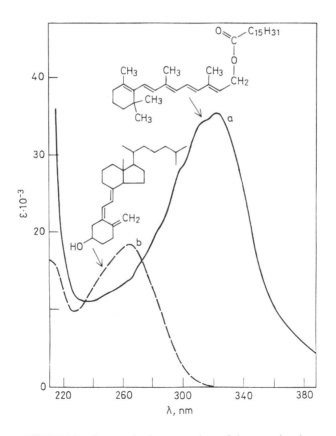

FIGURE 3.5.   Spectra of polyene-type drugs. Solvent: methanol. a. vitamin D_3; b. vitamin A palmitate.

TABLE 3.2. Longest Wavelength Absorption Maxima of the All-trans Polyenes: $H(CH=CH)_nH$ (Solvent: 2,2,4-trimethylpentane[27])

n	λ_{max}, nm	ε*
2	217	21,000
3	268	34,600
4	304	
5	334	121,000
6	364	138,000
7	390	136,000
8	410	108,000
10	447	

*The spectra of polyenes usually consists of four bands, the last one being the most intense.

both vitamins A and D_3 exhibit maximum absorptions at shorter wavelengths than would be expected, the intensities are also lower than expected and fine structure is lost completely in the case of vitamin D_3 and degraded to inflections in the spectrum of vitamin A.

Attaching polar substituents to the carbon atom of the double bond also gives rise to the loss of the fine structure as is shown in Figure 3.6 by the example of steroid 3,5-dienes.

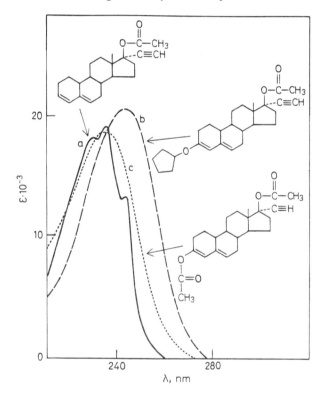

FIGURE 3.6. The effect of polar substituents on the spectra of steroidal conjugated dienes. Solvent: methanol. a. 17α-ethinyl-3,5-oestradiene-17-ol acetate (unsubstituted diene type). b. 17α-ethinyl-3-cyclopentyloxy-3,5-oestradiene-17-ol acetate (quingestanol acetate) (enol ether type). c. 17α-ethinyl-3,5-oestradiene-3,17-diol diacetate (enol ester type). (From S. Görög; Gy. Szász *Analysis of Steroid Hormone Drugs;* Akadémiai Kiadó: Budapest, Elsevier: Amsterdam, 1978, 176.)

In a great number of important drug substances the carbon–carbon double bond is in conjugation with a carbonyl double bond. In the spectra of the simplest α, β-unsaturated aldehyde and ketone, acrylaldehyde and methylvinylketone, both the π–π* and the n–π* bands are displaced toward long wavelengths as a result of the conjugation; (acrylaldehyde: $\varepsilon_{208.5\ nm} = 12{,}600$; $\varepsilon_{328\ nm} = 13$, methyvinylketone: $\varepsilon_{219\ nm} = 3{,}980$; $\varepsilon_{324\ nm} = 25$). Especially important α, β-unsaturated ketones are among various groups of steroid hormones. The wavelength of their intensive π–π* band (often termed K band referring to the conjugation causing its appearance at the relatively long wavelength region) can also be calculated with the aid of Woodward's rule.[33] A version of this, modified by Scott,[26b] is summarized in Table 3.3.

TABLE 3.3. Calculation of the Absorption Maxima of Conjugated Unsaturated Ketones[26b] (Solvent: ethanol)

$$\begin{array}{cc} \overset{\beta\quad\alpha}{\underset{|\ \ |\ \ |}{\beta\text{-C}=\text{C}-\text{C}=\text{O}}} \quad \text{or} & \overset{\delta\quad\gamma\quad\beta\quad\alpha}{\underset{|\ \ |\ \ |\ \ |\ \ |}{\delta\text{-C}=\text{C}-\text{C}=\text{C}-\text{C}=\text{O}}} \end{array}$$

6-membered cyclic or acyclic parent enone		215 nm
5-membered parent enone		202 nm

Increments to be added to one of the above values

Double bond extending conjugation		30 nm
Alkyl substituent or ring residue	α	10 nm
	β	12 nm
	γ and higher	18 nm
Exocyclic double bond		5 nm
Homoannular diene component		39 nm
OH	α	35 nm
	β	30 nm
	δ	50 nm
O-acyl		6 nm
O-methyl	α	35 nm
	β	30 nm
	γ	17 nm
	δ	31 nm
S-alkyl	β	85 nm
Cl	α	15 nm
	β	12 nm
Br	α	25 nm
	β	30 nm
N(alkyl)₂	β	95 nm

The applicability of this rule is illustrated by examples taken from the circle of steroid hormone drugs.

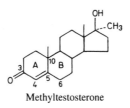

Methyltestosterone

		λ_{max}
Six-membered cyclic enone		215 nm
β-single bonds (5–6, 5–10)	2 × 12 =	24 nm
4–5 double bond (exocyclic related to ring B)		5 nm
	Calculated:	244 nm
	Observed:	241 nm

In connection with the above example it is to be noted that the introduction of a second double bond into the molecule of methyltestosterone at the position 1–2 does not lengthen the conjugation chain. Accordingly the maximum wavelength of this "cross-conjugated" derivative differs only slightly from that of methyltestosterone (λ_{max} = 244 nm). Much greater displacement is caused, however, by further double bonds if they are linearly conjugated; see the next two examples.

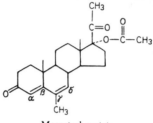

Megestrol acetate

	λ_{max}
Six-membered cyclic enone	215 nm
6–7 double bond	30 nm
5–10 single bond (β)	12 nm
6-methyl (γ)	18 nm
7–8 single bond (δ)	18 nm
4–5 double bond (exocyclic related to ring B)	5 nm
Calculated:	298 nm
Observed:	291 nm

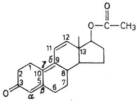

Trenbolone acetate

		λ_{max}
Six-membered cyclic enone		215 nm
9–10 and 11–12 double bonds	2 × 30 =	60 nm
5–6 single bond (β)		12 nm
1–10 single bond (δ)		18 nm
8–9 and 12–13 single bonds	2 × 18 =	36 nm
4–5 double bond (exocyclic related to ring B)		5 nm
9–10 double bond (exocyclic related to rings A and C)		5 nm
Calculated:		351 nm
Observed:		345 nm

Due to the conjugation with the double bond, the n–π* band of α, β-unsaturated carbonyl compounds are also shifted towards longer wavelengths (up to about 320 nm) but their intensity remains low (ε < 100).

α-Dioxo compounds (diacetyl, glyoxal) exhibit very low absorption in the entire ultraviolet region. It is interesting, however, that their n–π* band is shifted up to 450 nm. This unusually high wavelength maximum is the reason for the yellow color of these materials. Cyclic α-diketo compounds are capable of enolization. The enol form (α-hydroxy substituted α, β-unsaturated ketone) exhibits naturally strong absorption. If the enol form is stabilized by the acetylation of the enolic hydroxy group and hence the interaction between the conjugated bonds and the unshared electron pair of oxygen is eliminated, a strong hypsochromic shift occurs. This effect can be clearly illustrated by the spectra of 4-hydroxy-4-ene-3-keto and 4-acetoxy-4-ene-3-keto steroids derivable from 3,4-diketo-steroids. The absorption maxima of these derivatives can be calculated with good approximation with the aid of Woodward's rule (see Table 3.3):

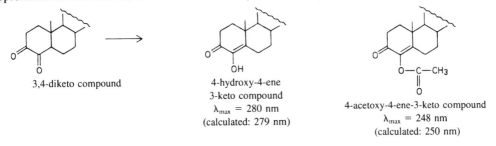

3,4-diketo compound

4-hydroxy-4-ene
3-keto compound
λ_{max} = 280 nm
(calculated: 279 nm)

4-acetoxy-4-ene-3-keto compound
λ_{max} = 248 nm
(calculated: 250 nm)

α, β-Unsaturated carboxylic acids absorb at shorter wavelengths than the analogous aldehydes or ketones. The absorption maxima of such derivatives mono-, di-, and tri-substituted with alkyl groups are around 208, 217 and 225 nm, respectively. Double bonds involved in further conjugation (in the same way as in the case of ketones in Table 3.3) show further bathochromic shift of 30 nm (or in the case of exocyclic ketones 35 nm).[19] The spectra of alkyl esters show only slight differences as compared to those of the free acids in undissociated form.

The α, β-unsaturated lactones, e.g., digoxin exhibit an intense absorption band around 215–220 nm while in the spectrum of ascorbic acid the presence of two enolic hydroxy groups gives rise to a strong bathochromic shift (λ_{max} = 245 nm). The change of this spectrum with the pH will be discussed in Section 3.Q (see Figure 3.27).

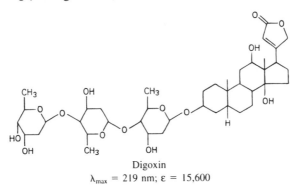

Digoxin
λ_{max} = 219 nm; ε = 15,600

F. AROMATIC COMPOUNDS

The widespread occurrence of aromatic rings among medicinal compounds makes the study of their characteristic light absorption very important for structure-spectra correlations.

The 6 delocalized electrons, uniformly distributed in the symmetrical, coplanar, six-membered benzene ring, give rise to great stability as compared with the apparently similar conjugated trienes. Consequences of this stability are the slight reactivity of benzene and in parallel with this the reduced excitability of its π-electrons.

1,3,5-Hexatriene exhibits an intense absorption band (with fine structure) at 258 nm (ε = 79,000) while the maximum of 1,3,5-cyclooctatriene containing *cis* double bonds in accordance with the discussion in Section 3.E is at longer wavelengths with decreased intensity ($\varepsilon_{265 \, nm}$ = 4,000). In contrast to these, as can be seen in Figure 3.7, the only band of high intensity in the spectrum of benzene is in the vacuum UV region ($\varepsilon_{185 \, nm}$ = 68,000). According to the nomenclature introduced by Clar for condensed aromatic hydrocarbons this band is termed β band. Of the two bands accessible to practical spectroscopy the one appearing at 204 nm (ε = 8,800) is the p band and the low intensity band with a very characteristic fine structure at 254 nm (ε = 250) is termed α band. The position of these bands is hardly influenced by the polarity of the solvent (see Figure 3.3). It is very characteristic, however, that the sharpness of the fine structure decreases in polar solvents and it is distorted to inflections by polar substituents. (The vapor phase spectrum of benzene with its vibrational and even rotational fine structure was shown in Figure 2.1 and discussed in Section 2.B.)

It is worth noting that in addition to the above-mentioned Clar nomenclature, several others are also used to denote the bands of aromatic compounds. Some of them are summarized (after Rao[24a]) in Table 3.4.

TABLE 3.4.   Nomenclature of the Absorption Bands of Benzene (After Rao[24a])

	Band	
around 180 nm	around 200 nm	around 260 nm
—	short wavelength benzenoid	long wavelength benzenoid
$A_{1g} \rightarrow E_{1u}$	$A_{1g} \rightarrow B_{1u}$	$A_{1g} \rightarrow B_{2u}$
E_1	E_2	B
—	K	B
A	B	C
Primary (second)	Primary (first)	Secondary
β	p	α
^1B	^1L_a	^1L_b

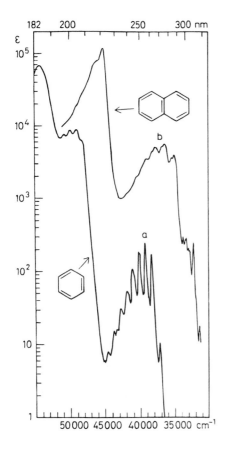

FIGURE 3.7.   Spectra of benzene (a) and naphthalene (b). Solvent: hexane. (From H.M. Hershenson *UV and Visible Absorption Spectra*, Academic Press, New York, 1961.)

TABLE 3.5. Spectral Data of Monosubstituted Benzene Derivatives $(C_6H_5–R)$[26,28]

R	Compound	p band*		K band*		α band*		n–π* band*		Solvent
		λ_{max},nm	ε	λ_{max},nm	ε	λ_{max},nm	ε	λ_{max},nm	ε	
–H	benzene	203	7,400			254	204			water
–CH$_3$	toluene	206	7,000			261	225			water
–F	fluorobenzene	204	6,200			254	900			ethanol
–Cl	chlorobenzene	210	7,500			257	170			ethanol
–Br	bromobenzene	210	7,500			257	170			ethanol
–I	iodobenzene	226	13,000			256	800			ethanol
–OH	phenol	211	6,200			270	1,450			water
–O$^-$	phenolate ion	235	9,400			287	2,600			water
–OCH$_3$	anisole	217	6,400			269	1,500			water
–SH	thiophenol	236	10,000			269	700			hexane
–NH$_2$	aniline	230	8,600			280	1,430			water
–N$^+$H$_3$	anilinium ion	203	7,500			254	160			water
–N(CH$_3$)$_2$	dimethylaniline	251	14,000			299	2,100			ethanol
–NH–COCH$_3$	acetanilide			242	14,400	280	500			ethanol
–C≡N	benzonitrile			224	13,000	271	1,000			water
–SO$_2$NH$_2$	benzenesulphonamide			218	9,700	265	740			water
–NO$_2$	nitrobenzene			252	9,500	280	1,000			heptane
–CH=CH$_2$	styrene	214	14,000	248	14,000	282	760			ethanol
–COCH$_3$	acetophenone	200	20,000	243	13,000	279	1,200	315	55	ethanol
–CHO	benzaldehyde			242	14,000	280	1,400	328	55	ethanol
–COOH	benzoic acid			230	10,000	270	800			water
–COC$_6$H$_5$	benzophenone			254	18,000	270	1,700	330	160	ethanol
–C$_6$H$_5$	biphenyl	201	46,500	247	17,000					hexane
–N=N–C$_6$H$_5$	azobenzene (trans)	230	16,000	313	25,000			450	400	2,2,4-trimethylpentane
–N=N–C$_6$H$_5$	azobenzene (cis)	245	10,000	280	4,000			432	1,200	2,2,4-trimethylpentane

*In the case of conjugated systems containing hetero-atom, the unequivocal assignment of the bands is not always possible.

In spite of its low intensity, studying the position and shape of the α band as a function of the substituents of the benzene ring furnishes important data for the interpretation of the spectra of more complicated aromatic derivatives. The α band is based on forbidden transitions because of the high symmetry of the benzene molecule in the ground state. The appearance of the band, even if a low intensity, can be ascribed to the distortion of the symmetry of the benzene molecule due to bending vibrations. Since the velocity of the electronic transitions is much higher than that of the vibrations such transition may take place between the temporarily distorted states giving rise to the fine structure of the α band.

Beside the spectrum of benzene that of naphthalene is also shown in Figure 3.7. It can be seen that due to the extended π-electron system of naphthalene all the three bands of benzene (β, p and α) undergo strong bathochromic (about 40, 70 and 50 nm) and hyperchromic shifts. This trend continues in the three-ringed systems (anthracene, phenanthrene). Analogs with higher ring numbers (of no importance from the point of view of pharmaceutical chemistry) are colored.

G. DERIVATIVES CONTAINING UNSUBSTITUTED PHENYL GROUPS: SUBSTITUTION WITH AUXOCHROMIC GROUPS

In the case of monosubstituted benzene derivatives the high symmetry of the benzene molecule decreases and this results in the bathochromic displacement of its bands. As seen from the data of Table 3.5, only small shifts are observable in the case of alkyl and halogen substitution and the intensities are also only slightly altered. Much greater effect can be found when the substituents are electron donating auxochromic (hydroxy, amino) and chromophoric groups.

The interaction of the aromatic ring with chromophoric groups gives rise to a conjugation (K) band, which is in many cases not easily distinguishable from a p band shifted toward longer wavelengths. A further difficulty in the assignment of the bands originates from the facts that the weak α-bands are often overlapped by the intense K bands and, in the case of chromophoric groups containing hetero atoms, n–π^* band also occurs. Since these families of aromatic compounds contain several important drug substances, this field will be discussed in detail in the subsequent part of this section.

Many of the aromatic drug substances are monosubstituted benzene derivatives where the benzene ring is linked to the other parts of the molecule through a methylene bridge. These compounds containing isolated, bare benzene rings can be regarded as toluene derivatives. Whether the presence of this phenyl group can be recognized in the UV spectrum depends on the other chromophoric groups in the same molecule.

There are only a few compounds in the circle of medicinal compounds (mainly auxiliary materials) possessing no more chromophoric groups (e.g., benzyl alcohol, β-phenylethanol, etc.). The spectra of these compounds are closely related to that of toluene.

Much higher is the number of those derivatives where in addition to the isolated benzene ring other chromophores or absorbing groups are also present. If this group is ester (methylphenidate, scopolamine, atropine, hyoscyamine, etc.), carboxamide (primidone, etc.), primary amine (amphet-amine, etc.), secondary amine (ephedrine, phenmetrazine, methylphenidate, prenylamine, phendili-nium, etc.), tertiary amine (scopolamine, atropine, hyoscyamine, etc.) or quaternary ammonium group (methylhomatropinium bromide, etc.) the α-band with its characteristic fine structure is readily detected around 257 nm (ε = 200–400). The effects of primary amino and ester groups on the p band are negligible: it appears around 205 nm ($\varepsilon \sim$ 7,000). In the case of secondary and tertiary amines (especially the cyclic derivatives) the p band and the n–π^* band of the amino group partially overlap and this may give rise to the displacement of the maximum up to about 210 nm accompanied with an increase of intensity. In some cases the p band appears as an inflection only. As an illustration the spectrum of hyoscyamine is shown in Figure 3.8, curve "a."

If chromophoric systems exhibiting more intense absorption than the above discussed groups are present isolated from the benzene ring the α band of the latter appears only in the form of side maxima or inflections on the descending slope of the spectrum of the chromophoric system. Systems of this kind are e.g., the thiazolidine-lactam group in benzylpenicillin (see Figure 3.8, curve "b") or the –N=C–NH– group in tolazoline, etc.

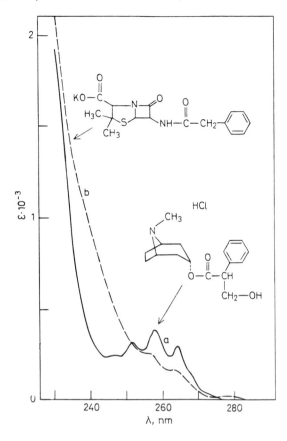

FIGURE 3.8.   Spectra of drugs containing an isolated benzene ring. Solvent: water. a. hyoscyamine sulphate; b. potassium penicillin-G.

It often occurs that the α bands of isolated phenyl groups cannot be recognized at all since they are entirely overlapped and therefore masked by much more strongly absorbing conjugated systems. For example, the conjugation band of the α, β-unsaturated 3-keto group of nandrolone phenylpropionate ($\varepsilon_{240 \text{ nm}}$ = 17,000) possesses strong enough absorption even at 260 nm to completely mask the α band of the isolated phenyl group in the phenylpropionyl unit which cannot therefore be detected in the spectrum. Another example is the spectrum of tripelennamine where the α band of the isolated phenyl group is not detectable because of the strong absorption of the α-aminopyridine moiety ($\varepsilon_{245 \text{ nm}}$ = 16,000 and $\varepsilon_{306 \text{ nm}}$ = 4,700).

As can be seen in Table 3.5, halogen substitution does not markedly influence either the position or the intensity of the benzenoid bands. As an example miconazole is mentioned, which contains two dichlorophenyl units (λ_{max} = 272 nm; ε = 720). As regards the weak effect of the often occurring trifluoromethyl group the spectrum of flumecinol (Figure 2.16) is referred to where the α bands of the phenyl and trifluoromethylphenyl groups give rise to an overlapping complex band system.

In contrast to the slight effects discussed so far, the effect of auxochromic groups on the spectrum of the benzene ring is dramatic. The reason for the strong bathochromic and hyperchromic shifts is the interaction of the unshared electrons of oxygen, sulphur and nitrogen in the auxochromic groups with the π-electron sextet of the aromatic ring. Consequently, the fine structure of the α band partly or completely disappears.

In the spectra of phenols the *p* band is around 220 nm while the α band is shifted to about 275 nm. The intensity of the latter is markedly increased with a molar absorptivity of about 2000. The formation of phenol ether derivatives does not significantly influence the spectra of phenols. Much more intense is the bathochromic and hyperchromic shift resulting from alkalization of the solvent when both maxima of phenols are shifted by about 20 nm toward longer wavelengths accompanied with an increase of about 50% of the intensities due to the formation of the phenolate anion in strongly alkaline solution. In contrast, strong hypsochromic and hypochromic shifts are obtained when phenolic hydroxy groups are transformed to their carboxylic esters. The reason for this is that the electron attracting ester-carbonyl group decreases the above mentioned interaction between the unshared electron pair of the phenolic hydroxy group with the aromatic ring. These effects are illustrated by Figure 3.9 where the spectra of oestrogens are shown. Free phenols are represented by ethinyloestradiol in neutral and alkaline media, as the ether and ester types the spectra of mestranol and oestradiol dipropionate are shown.

Di- and tri-phenols and their esters exhibit similar spectra with occasional displacements to longer wavelength of the maxima by a few nanometers. For example, the α band of emetine is at 282 nm ($\varepsilon = 5,280$; two diphenol ether units!). Methyldopa representing the adrenaline-like free diphenolic structure exhibits maximum absorption at 283 nm ($\varepsilon = 2,790$).

The spectral data of aniline and especially those of N,N-dimethylaniline in Table 3.5 indicate that the interaction of the π-electron system of the aromatic ring with the unshared electron pairs of the amino nitrogen gives rise to bathochromic dispalcement of both the p and α bands. If, however, this unshared electron pair is bound by the addition of strong acid to the solution the interaction ceases and a benzene-like spectrum is obtained; see Figure 3.31. Of their drug derivatives the spectrum of fluspirilene based on dimethylaniline is shown in Figure 3.10. As can be seen the positions and intensities of both the p and α bands are close to those of N,N-dimethylaniline in Table 3.5. It is noteworthy that the inflection and maximum between the two bands (at 273 nm) is due to the p-fluorophenyl groups at the other end of the molecule.

From the pharmaceutical point of view the acylated derivatives of aniline are of greater importance. As can be seen from the data in Table 3.5 the interaction of the C=O group with the aromatic amino group causes a further bathochromic shift of the p band; the partially overlapped α band

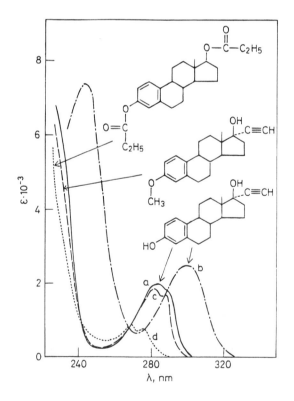

FIGURE 3.9. Spectra of phenolic-steroids. Solvent: methanol–water 4:1. a. ethinyl oestradiol (free phenol type); b. ethinyl oestradiol in 0.2 M sodium hydroxide (phenolate type); c. mestranol (phenol ether type); d. oestradiol dipropionate (phenol ester type). (From S. Görög; Gy. Szász *Analysis of Steroid Hormone Drugs;* Akadémiai Kiadó: Budapest, Elsevier: Amsterdam, 1978, 178.)

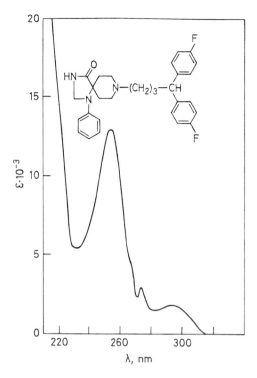

FIGURE 3.10. Spectrum of fluspirilene. Solvent: methanol.

is difficult to recognize. Of the drugs containing this type of chromophoric system the spectra of strychnine and phenacetin are shown in Figure 3.11.

The further displacement of the *p* band of phenacetin toward a longer wavelength is the consequence of the ethoxy group in the 4 position. The same reason accounts for the apparently high intensity of the α band appearing as an inflection around 280–290 nm as compared to that of unsubstituted acetanilide. The analogous nature of the spectrum of strychnine shows that the chromophoric system of this dihydroindole alkaloid possessing rather complicated structure is also the acetanilide unit. The spectra of acetanilide derivatives are greatly influenced by steric effects. This is dealt with in detail in Section 3.P.

H. BENZENE DERIVATIVES CONJUGATED WITH CHROMOPHORIC GROUPS

Of the derivatives of nitrobenzene used in the therapy, the spectrum of chloramphenicol is shown in Figure 3.12. In the polar solvent used, the rather intense and broad conjugation bands of both the α band and the n–π* band of the nitro group overlap.

In Figure 3.13 the spectra of three drug sustances are shown in which the chromophoric system are carbon-carbon double bonds in conjugation with the benzene rings. The most intense band in their spectra is the *K* band originating from this conjugation. It can be seen from the comparison of the spectral data of the three compounds with those in Table 3.5 that the position and intensity of the *K* band depend on the substituents and steric effects and may greatly differ from the data

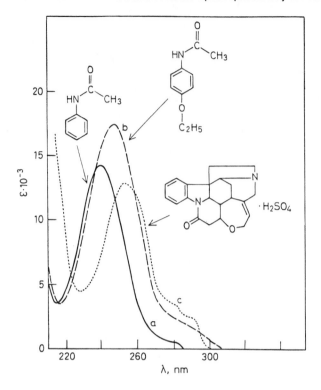

FIGURE 3.11.   Spectra of acylated aniline derivatives. Solvent: methanol. a. acetanilide; b. phenacetin; c. strychnine sulphate.

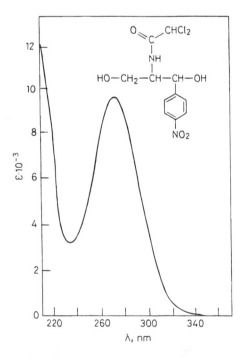

FIGURE 3.12.   Spectrum of chloramphenicol. Solvent: methanol.

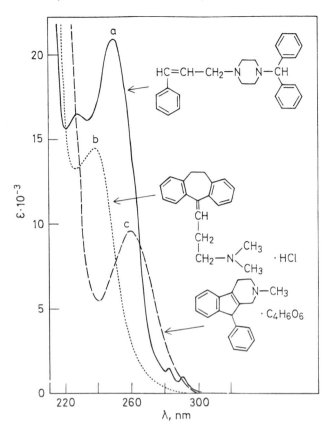

FIGURE 3.13.   Spectra of styrene derivatives. Solvent: methanol. a. cinnarizine; b. amitriptyline hydrochloride; c. phenindamine tartrate.

of styrene which is the parent material of this group of aromatic compounds. The difference is relatively the smallest in the case of cinnarizine which is a β-substituted styrene derivative. The hypsochromic shift observable in the spectrum of amitriptyline is due to the steric hindrance of the conjugation between the benzene rings in the condensed ring system with the exocyclic ethylene group. In contrast to this, a remarkable bathochromic shift can be seen in the spectrum of phenindamine. The reason for this is that the conjugation is stabilized by the rigid ring system; (the absorption maximum of indene which represents the parent compound in this series is at 249 nm). The further bathochromic shift is due to the alkyl substituents. The α band can be recognized in the first case only.

Of the derivatives of 1,2-diphenylethylene (stilbene) diethylstilbestrol is mentioned here. It exhibits an absorption maximum at 236 nm in methanolic solution ($\varepsilon = 11{,}000$). This is an amazingly short wavelength for the maximum and the intensity is also rather low if comparison is made with the spectrum of trans-stilbene ($\varepsilon_{296\ nm} = 29{,}000$). The reason for this enormous hypsochromic and hypochromic shift is that the two ethyl group hinder the coplanarity of the phenyl groups and the ethylene bridge (see Section 3.P).

Triphenylethylene exhibits a conjugation band at 298 nm. In accordance with this the long wavelength band of clomiphene citrate is at 295 nm ($\varepsilon = 11{,}400$). It is interesting that in the spectrum of dienestrol which possesses one more double bond in conjugation than diethylstilboestrol the absorption occurs at a lower wavelength than the latter ($\varepsilon_{226\ nm} = 32{,}900$). This compound can be considered to be the 2,3-diphenyl derivative of 1,4-dimethylbutadiene. The calculated value of the absorbance maximum based on the Woodward rule (see Section 3.E) is 224 nm. This clearly indicates that the phenolic rings of dienoestrol do not participate in the conjugation.

As can be seen from the data of Table 3.5 the conjugation of an aromatic ring with a carbonyl group gives rise to an intense K band around 240–260 nm. In spite of its remarkable bathochromic and hyperchromic shift the α band is often difficult to recognize. The same applies to a greater extent to the weak n–π* band. The intensity of the K band can be strongly decreased by steric hindrance (see Section 3.P).

As typical examples of aromatic ketones, the spectra of tolperisone and haloperidol are shown in Figure 3.14.

From the spectral data of benzoic acid in Table 3.5 it is evident that the K band of the carbonyl group of the carboxyl moiety in conjugation with the benzene ring is at a lower wavelength (around 230 nm) than that of the analogous ketone derivatives and hence in these spectra, the α band possessing fine structure can be recognized around 270 nm. The hypsochromic shift taking place in alkaline media where the carboxylic acids are present as the anions is discussed in Section 3.Q. The spectra of aromatic carboxylic esters are very similar to the spectra of the corresponding undissociated acids.

In Figure 3.15 the spectra of benzoic acid and its ester derivative (cocaine) is reproduced. It is interesting to compare the spectrum of aspirin (acetylsalicylic acid) with that of benzoic acid. Their close similarity can be attributed to the fact that the acetylation of the phenolic hydroxy group greatly decreases the interaction between its unshared electron pair and the π-electron system of the aromatic ring (cf., with the spectral data of salicylic acid in Table 3.7). A similar phenomenon was discussed on p. 60 in connection with the spectrum of oestradiol dipropionate.

I. DISUBSTITUTED BENZENE DERIVATIVES

Several disubstituted benzene derivatives have already been discussed in Sections 3.G and 3.H. In those instances, however, the substituents (alkyl groups, chlorine, etc.) only slightly influenced the spectra originating from the interaction of the electron system of the aromatic ring with those of chromophoric or auxochromic groups. In this section those cases will be discussed where the interaction of the two substituents with the aromatic ring or with each other gives rise to entirely new spectra, as compared to the spectrum of the monosubstituted derivatives. This effect is especially pronounced if the substituents are in para positions.

The spectra of nitroanilines are good examples to demonstrate the aforesaid effect. The conjugation band of nitrobenzene is at 268 nm and it overlaps the α band ($\varepsilon_{268\,nm} = 7,800$). In the spectrum of aniline the p and α bands are at 230 and 280 nm, respectively ($\varepsilon = 8,600$ and $1,430$). In contrast to these the spectrum of 2-nitroaniline contains bands at 245, 282.5 and 412 nm ($\varepsilon = 7,000$, $5,400$ and $4,500$), 3-nitroaniline absorbs at 280 and 358 nm ($\varepsilon = 4,800$ and $1,450$) while the intense-conjugation band of the 4-nitro derivative of aniline is shifted up to 381 nm ($\varepsilon = 13,500$). Similar phenomena can also be observed among nitrophenols.

An empirical method is available (similar to the Woodward rule) for the calculation of the maximum wavelengths of aromatic carbonyl derivatives where the position of the band depends on the type of the carbonyl group (ketone, aldehyde, carboxylic acid or its ester), the nature and position of the second substituent. Table 3.6 contains the data (taken from the monograph of Scott[26c]) for the calculation of the shifts. These data which include those of pharmaceutically important derivatives clearly demonstrate the exceptional situation of the para substituents.

In connection with the data in Table 3.6 it has to be noted that the position and especially the intensities of the bands may greatly be influenced by steric hindrance. This is treated in detail in Section 3.P.

Some characteristic spectra of disubstituted benzene derivatives are reproduced in Figure 3.16.Nipagin representing p-hydroxybenzoic acid esters (curve "a") exhibits a maximum at 256 nm in good agreement with the value of 255 nm calculated on the basis of Table 3.6. Procaine (p-aminobenzoic acid type) absorbs at 296 nm (calculated value: 288 nm). Sulphaguanidine in curve "c" represents sulphonamides which are also para-disubstituted derivatives with the general formula $H_2N–C_6H_4–SO_2–NHR$. It has to to be noted that in those cases when R is a spectrophotomet-

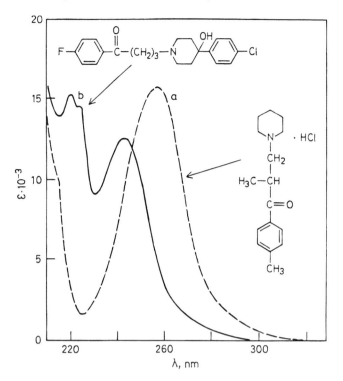

FIGURE 3.14.   Spectra of aromatic ketones. Solvent: ethanol. a. tolperisone hydrochloride; b. haloperidol.

rically active group, its spectrum is superimposed on the above spectrum leading to displacements of maxima.

In the spectra of the para-disubstituted derivatives discussed so far the conjugation bands appearing at rather long wavelengths in all cases obscure the α band. A different situation exists in the case of an ortho-substituted benzoic acid (salicylic acid) where the conjugation band is shifted up to only 236 nm (calculated value 237 nm) and hence the α band appearing at 303 nm is well separated. The spectrum of its acetyl derivative (aspirin) is shown in Figure 3.15.

Since all of the above discussed compounds contain ionizable groups their spectra exhibit characteristic pH dependence (see Section 3.Q).

J.   TRI- AND TETRA-SUBSTITUTED BENZENE DERIVATIVES

Multiple substitution also occurs among pharmaceutically important benzene derivatives. Their spectra can usually be successfully interpreted on the basis of the principles laid down in connection with the mono- and disubstituted derivatives. However, it has to be taken into consideration that as a consequence of the multiple substitution the environment of the aromatic ring is rather crowded and, especially in the case of bulky substituents, steric effects can hinder the influence of some substituents on the spectrum (see Section 3.P).

Of the trisubstituted derivatives it is interesting to make comparison among the spectral data of p-aminosalicylic acid and two disubstituted derivatives (p-aminobenzoic acid and salicylic acid) which both contain an auxochromic group together with the carboxyl group (see Table 3.7).

Interesting conclusions can be drawn from the comparison of the spectra of the closely related tri- and tetra-substituted benzene derivatives clopamide and furosemide, too. As can be seen in Table 3.5 the sulphonamide group is a weak chromophore and hence in the spectrum of clopamide the benzoic acid character is predominant (see curves "b" in Figure 3.17 and "a" in Figure 3.15). The spectrum of furosemide (Figure 3.17; curve "a") is entirely different. The great differences

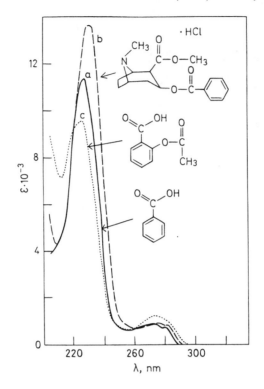

FIGURE 3.15. Spectra of benzoic acid and derivatives. Solvent: methanol. a. benzoic acid; b. cocaine hydrochloride; c. acetylsalicylic acid.

TABLE 3.6. Calculation of the Absorption Maxima of the Derivatives of Benzaldehyde, Acetophenone and Benzoic Acid (Solvent: ethanol[26c])

Parent compound C_6H_5-COR		
R=alkyl		246 nm
H		250 nm
OH or O-alkyl		230 nm

Increments to be added to one of the above values		
Alkyl	o,m	3 nm
	p	10 nm
OH, O-alkyl	o,m	7 nm
	p	25 nm
O^-	o	11 nm
	m	20 nm
	p	78 nm
Cl	o,m	0 nm
	p	10 nm
Br	o,m	2 nm
	p	15 nm
NH_2	o,m	13 nm
	p	58 nm
NHAcyl	o,m	20 nm
	p	45 nm
$NHCH_3$	p	73 nm
$N(CH_3)_2$	o,m	20 nm
	p	85 nm

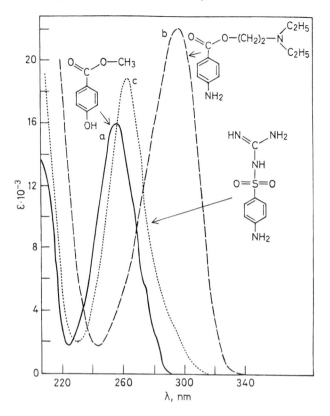

FIGURE 3.16. Spectra of disubstituted benzene derivatives. Solvent: methanol. a. nipagin; b. procaine hydrochloride; c. sulphaguanidine.

in the number of bands and their intensities are attributable to the presence of the substituted amino group in spite of the fact that the contribution of the isolated furyl group in this group to the spectrum is almost negligible. For the interpretation of the spectrum that of sulphaguanidine in curve "c" of Figure 3.16 is referred to and in addition it should be noted that in the spectrum of 2-aminobenzoic acid the following bands can be found (with their molar absorptivities in parentheses): 217 nm (18,500), 248 nm (3,900) and 327 nm (1,940).

The interpretation of the spectrum of trimetozine, which is also a tetra-substituted benzene derivative (2,3,4-trimethoxybenzoic acid morpholide), is an easier task. The conjugation band in the spectrum of benzamide is at 225 nm; as a result of the three methoxy groups, this is shifted to 250 nm in the spectrum of trimetozine.

TABLE 3.7. Spectral Data of Some Benzoic Acid Derivatives

	λ_{max}, nm	ε	Solvent
Benzoic acid	228	11,400	ethanol
	273	910	
	280	750	
4-Aminobenzoic acid	219	8,560	methanol
	289	18,200	
Salicylic acid	236	7,000	ethanol
	303	3,700	
4-Aminosalicylic acid (PAS)	237	7,660	ethanol
	279	12,300	
	305	13,200	

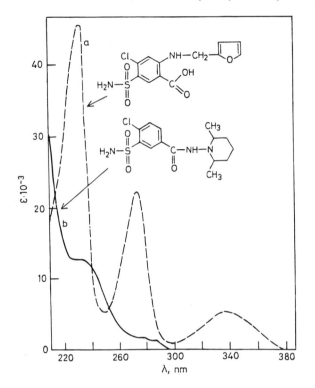

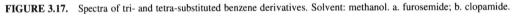

FIGURE 3.17.   Spectra of tri- and tetra-substituted benzene derivatives. Solvent: methanol. a. furosemide; b. clopamide.

K. FIVE-MEMBERED HETEROCYCLIC DERIVATIVES

Saturated and unsaturated heterocycles have already occurred several times in the molecules of the compounds discussed in the preceding sections. Their spectral characteristics only slightly differ from those of their aliphatic analogs. All the more different are the spectra of aromatic heterocycles. Because of their great pharmaceutical importance the following three sections are devoted to these derivatives.

The spectral data of some pharmaceutically important five-membered heterocycles (taken from the book of Stern and Timmons[28]) are collected in Table 3.8. It can be seen that the three heterocycles containing only one single heteroatom (furan, pyrrole and thiophene) absorb at lower wavelengths than the analogous cyclopentadiene (λ_{max}/hexane = 238 nm; ε = 3,400). The reason for this is that the four π-electrons of the conjugated diene together with the unshared electron pair of the heteroatom form a delocalized, aromatic type electron sextet. The contribution of the n electrons to this is the reason that no n–π* band appears in the spectrum of the above compounds.

It is also evident from the data of Table 3.8 that (with the exception of sulphur derivatives possessing the lowest ionization energy) the maxima appear at wavelengths too short to be really useful for analytical purposes, nor are the intensities sufficiently high. For this reason those drug substances which contain these rings without chromophoric or auxochromic groups exhibit only slight spectrophotometric activity. For example the absorption maximum of pilocarpine hydrochloride in methanolic medium is at 216 nm (ε = 5,940) and this corresponds to the spectral data of dialkylsubstituted imidazole derivatives. Another example of this kind is histamine hydrochloride ($\varepsilon_{208\ nm}$ = 5,760). The related amino acid histidine is therefore not considered to be a spectrophotometrically active amino acid.

More typical are the spectra of those derivatives where strong chromophoric systems isolated from the heteroaromatic ring overlap and hence more or less mask the spectra of the latter. This is the case, e.g., with sandostene where the band of the thiophene ring is hidden under that of the N,N-disubstituted aniline moiety (λ_{max} = 244 nm). Of course the band of the furyl group cannot be recognized in the spectrum of furosemide (Figure 3.17) containing a tetrasubstituted benzene

TABLE 3.8. Spectral Data of Some Pharmaceutically
Important 5-Membered Heterocycles[28]

	λ_{max}, nm	ε	Solvent
Furan	207	9,100	cyclohexane
Thiophene	231	7,100	cyclohexane
Pyrrole	208	7,700	hexane
Imidazole	206	4,800	water
Pyrazole	210	3,650	ethanol
Oxazole	205	4,100	water
Thiazole	209	2,750	heptane
	232	3,550	

ring with much stronger absorption along the full wavelength scale. Even the isolated benzene rings play a predominant role in the spectrum of prenoxdiazine containing an isolated oxadiazine ring ($\varepsilon_{258\ nm}$ = 460).

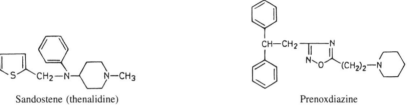

Sandostene (thenalidine)   Prenoxdiazine

In contrast to the poor absorption characteristics of unsubstituted five-membered heterocyclic derivatives, their derivatives containing auxochromic or chromophoric groups show a high diversity of rather intense and characteristic spectra. The introduction of just one formyl group into position 2 of the furan ring (furfural) causes a very strong bathochromic and hyperchromic shift ($\varepsilon_{278\ nm}$ = 15,800). The 5-nitro-substituted hydrazones of this are yellow exhibiting maxima in the range of 350–380 nm. Of these pharmaceutically very important derivatives the spectrum of nitrofurantoin is shown in Figure 3.18.

Of the substituted derivatives of imidazole methimazole is mentioned: as a consequence of the presence of the auxochromic sulphydryl group its maximum is shifted up to 251 nm (ε = 15,800). Of the several 5-nitroimidazole derivatives used in therapy the spectrum of metronidazole is shown in Figure 3.34. In the spectrum of sulphamethoxazole ($\varepsilon_{270\ nm}$ = 20,400) the oxazole ring has only little influence on the spectrum of the 4-aminobenzene-sulphonamide moiety; (for comparison: the absorption maximum of unsubstituted sulphanilamide is at 262 nm (ε = 19,200)).

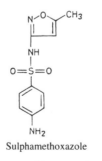

Sulphamethoxazole

Tautomerism (the simultaneous presence of keto and enolic forms) is a general phenomenon among the hydroxy derivatives of the heterocycles discussed so far; it is of decisive importance in the field of pharmaceutically important pyrazolones. The spectrum of antipyrine (Figure 3.18, curves "b" and "c") is also influenced by tautomerism: the long wavelength band almost completely disappears in strongly acidic solution. It merits attention that the 4-dimethlyamino derivative of antipyrine (aminopyrine) has a very similar spectrum: λ_{max} (methanol) = 236 and 268 nm; ε =

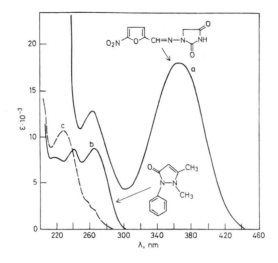

FIGURE 3.18.   Spectra of furan and pyrazole derivatives. a. nitrofurantoin. Solvent: water containing 0.6% v/v dimethyl-formamide; b. antipyrine. Solvent: methanol; c. antipyrine. Solvent: 0.1 M hydrochloric acid.

10,200 and 8,800. This indicates that the unshared electron pair of the nitrogen atom of the dimethylamino group does not take part in the chromophoric system.

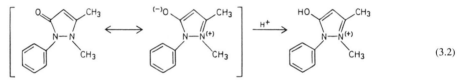

$$(3.2)$$

Tautomerism and Protonation of Antipyrine

The closely related phenylbutazone exhibits a maximum in ethanol at 239 nm ($\varepsilon = 15,900$). The very strong bathochromic shift of this band which is due to the dissociation of the enolic form in alkaline medium is discussed in Section 3.Q.

L.   SIX-MEMBERED HETEROCYCLIC DERIVATIVES

Of the six-membered heterocyclic compounds, the nitrogen containing derivatives are of prominent importance in therapy. The spectra of these show close analogy to those aromatic hydrocarbons from which they can be derived by replacing one or more –CH= groupings by nitrogen.

It can be seen from the data in Table 3.9 the p and α bands of pyridine appear at about the same wavelengths as those of benzene. An important difference is, however, that the intensity of the latter is higher by more than one order than that of the α band of benzene. The reason for this is the lower symmetry of the molecule of pyridine as compared with benzene as a consequence of which the electronic transition corresponding to the α band is less forbidden. Another important feature of the spectrum is derived from the fact that in the case of pyridine and its derivatives (in contrast to the five-membered analogs) the unshared electron pair of the nitrogen atom does not take part in the π-electron sextet. This is why, in this case, the n–π^* bands do appear, although in the spectrum of pyridine as a low-intensity inflection on the descending, long wavelength arm of the α band. In the case of analogs containing two or more nitrogens the n–π^* bands are more easily recognizable. These bands exhibit fine structure and they undergo hypsochromic shift upon increasing the polarity of the solvent (see Section 3.B).

Other data in Table 3.9 indicate the influence of the polarity of the solvent on the intensity of the α band of pyridine: a twofold increase can be observed when the solvent is changed from

TABLE 3.9.   Spectral Data of Nitrogen Containing 6-Membered Aromatic Heterocycles[28]

Compound	p band λ_{max}, nm		α band λ_{max}, nm		$n-\pi^*$ band λ_{max}, nm		Solvent
Benzene	204	8,800	254	250		—	hexane
Pyridine	198	6,000	251	2,000	270	infl.	hexane
	195	5,400	257	3,050		overlapped by α band	water
			256	5,300		overlapped by α band	5 M HCl
Pyrazine	194	6,100	260	6,000	328	1,050	hexane
			261	6,270	302	800	water
Pyridazine	192	5,400	251	1,400	340	315	hexane
			242	1,400	299	320	water
Pyrimidine	189	10,000	244	2,050	298	325	hexane
			244	3,160	267	320	water

hexane to alcohols or water which are capable of forming hydrogen bonding with the unshared electron pair of the nitrogen. Further increase takes place when the spectrum is prepared in strongly acidic medium where pyridine exists in the protonated form.

Of the simple derivatives of pyridine which do not contain auxochromic or chromophoric groups the spectrum of pyridinol carbamate is represented in Figure 3.19a (curve "b"). As it can be seen this spectrum is very similar to that of unsubstituted pyridine (curve "a"). It is noteworthy that the spectra of N-alkyl quaternary derivatives are also similar (e.g., cetylpyridinium bromide: ε_{260} $_{nm}$ = 4,840). The reason for the somewhat increased intensity is that the effect of the formation of a quaternary ammonium ion is the same as that of the salt formation in acidic media. Bisacodyl also exhibits a similar spectrum ($\varepsilon_{261\ nm}$ = 5,670); the reason for the increased intensity is the contribution of the two phenol acetate-type rings to the spectrum (cf., Section 3.G).

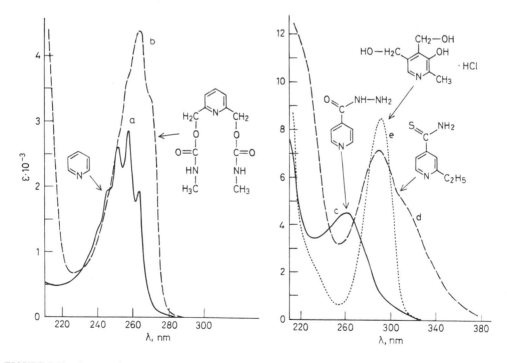

FIGURE 3.19.   Spectra of pyridine derivatives. Solvent; methanol. a. pyridine; b. pyridinol carbamate; c. isonicotinoyl hydrazide; d. ethionamide; e. vitamin B_6 (pyridoxine hydrochloride).

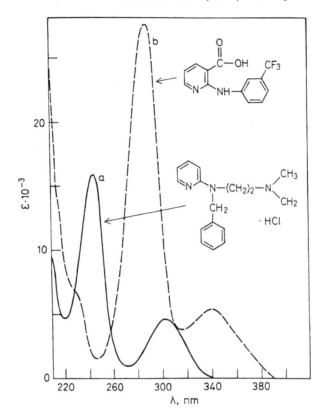

FIGURE 3.20. Spectra of α-aminopyridine derivatives. Solvent: methanol a. tripelennamine hydrochloride; b. niflumic acid.

The position and intensity of the α band which serves as the basis for the analytical determination of substituted pyridine derivatives are not significantly influenced by the introduction of carboxyl and carboxamide group to positions 3 or 4. Some characteristic spectral data are as follows: nicotinic acid $\varepsilon_{263\ nm} = 2{,}590$; nicotinamide $\varepsilon_{262\ nm} = 2{,}950$ and nikethamide $\varepsilon_{263\ nm} = 3{,}270$.

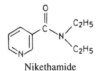

Nikethamide

Hydrazide derivatives exhibit absorption bands at the same place but with higher intensity; for example the maximum in the spectrum of isonicotinoyl hydrazide (Figure 3.19b, curve "c") is at 262 nm ($\varepsilon = 4{,}580$). Curve "d" in the same figure is the spectrum of ethionamide where the oxygen atom in the carboxamide moiety is replaced by sulphur. The spectrum can be characterized by strong bathochromic and hyperchromic shifts, the asymmetrical shape of the band, and in connection with this an inflection (attributable to the n–π* band) between 310 and 340 nm.

It is principally the phenolic hydroxyl group which is responsible for the shift of the α band up to 292 nm in the spectrum of vitamin B_6. Its intensity is also the highest among the derivatives discussed so far (see curve "e" in Figure 3.19b).

Another important family of derivatives is related to α-amino-pyridine. Because of the strong bathochromic shift of the p and α bands both appear in the easily accessible UV region at 235 and 296 nm. In the spectrum of tripelennamine which belongs to this family a further bathochromic shift is observable, due to the replacement of the hydrogens of the amino group by alkyl groups (see Figure 3.20). Niflumic acid where the α-amino group is substituted with a phenyl group exhibits absorption maximum at 342 nm ($\varepsilon = 5{,}410$).

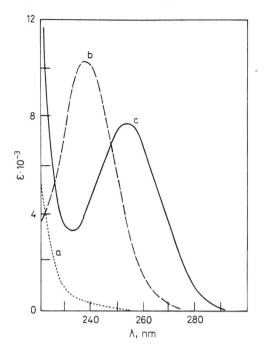

FIGURE 3.21. Spectra of barbital. Solvent: water. a. 0.1 M sulphuric acid; b. pH 10; c. 0.1 M sodium hydroxide.

Of the six-membered heterocycles containing two nitrogens, pyrimidines are most widely used in therapy, especially in the form of barbiturates. Since the latter contain two substituents at position 5, they cannot adopt an aromatic structure. As a result of lactam–lactim tautomerism and the presence of two exchangeable hydrogens, 5-disubstituted barbiturates can be present in three forms:

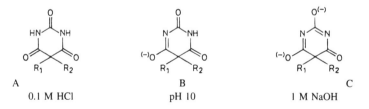

A	B	C
0.1 M HCl	pH 10	1 M NaOH

In strongly acidic medium where barbiturates are present in the tricarbonyl form, they exhibit an n–π* band only at about 212 nm. The band appearing at 240 nm in slightly alkaline medium and that around 255 nm in strongly alkaline solution belong to the monolactim and dilactim forms, respectively. The R_1 and R_2 substituents have no remarkable effect on the spectrum, except in the spectrum of phenobarbital which contains one isolated phenyl ring and exhibit some fine structure on the ascending section of the spectrum around 260 nm.

The spectra of barbital ($R_1 = R_2 = C_2H_5$) are characteristic of the above listed three forms (see Figure 3.21).

If in addition to the two substituents at position 5 the nitrogen at position 1 also carries a substituent (e.g., hexobarbital), there is no possibility for the second tautomeric equilibrium and thus the third form and the third spectrum above pH 10 does not exist. If, however, the C=O group at position 2 is exchanged to C=S (thiobarbiturates, e.g., thiopental sodium) a very strong bathochromic and hyperchromic shift is obtained ($\varepsilon_{305\,nm} = 20,400$).

Sulphonamides containing pyrimidine or pyridazine units exhibit only slightly different spectra as compared to those without these units. For example sulphadimidine: $\varepsilon_{269\,nm} = 22,600$; sulphamethoxydiazine: $\varepsilon_{271\,nm} = 20,700$; sulphamethoxypyridazine: $\varepsilon_{268\,nm} = 21,700$ (cf., with the spectrum of unsubstituted sulphanilamide: $\varepsilon_{262\,nm} = 19,200$).

TABLE 3.10. Spectral Data of Nitrogen Containing
Condensed Aromatic Heterocycles[28]

Compound	λ_{max}' nm	ε	Solvent
Naphthalene	221	117,000	hexane
	275	5,600	
	311	250	
Quinoline	226	34,000	methanol
	281	3,600	
	308	3,850	
Isoquinoline	216	82,000	heptane
	266	4,150	
	318	3,650	
Indole	215	35,500	heptane
	267	6,600	
	287	4,250	
Purine	263	7,600	methanol
Pteridine	210	11,000	cyclohexane
	302	7,400	
	390	75	

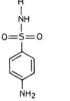

Sulphadimidine   Sulphamethoxydiazine   Sulphamethoxypyridazine

M. CONDENSED HETEROCYCLIC DERIVATIVES

Many drug substances contain two or more condensed (aromatic or non-aromatic) heterocyclic rings. The spectral data of the most important types of these ring systems are collected in Table 3.10.

1. Quinolines and Isoquinolines

Comparing the data in Table 3.10 with the spectra in Figure 3.7 reveals that similar to the close relationships of the spectra of benzene and pyridine, discussed in Section 3.L the spectra of quinoline and isoquinoline also show similarities to that of naphthalene.

Of the derivatives of quinoline, curve "a" in Figure 3.22 represents the spectrum of 8-hydroxyquinoline. The main band of high intensity is derivable from the band of naphthalene at 220 nm and that of quinoline at 226 nm with a bathochromic shift caused by the hydroxy substituents. The situation is similar with the spectrum of quinine (curve "b") but with a smaller bathochromic effect. Cinchonine and cinchonidine not containing methoxy group exhibit entirely quinoline-like spectra. The spectrum of iodochlorhydroxyquinoline can be seen in Figure 8.1.

Isoquinoline alkaloids are represented by the spectrum of papaverine hydrochloride in curve "c." The statements made in connection with quinoline derivatives essentially apply to this family, too. The extremely high intensity of the main band is remarkable. Even the less intense bands around 280–330 nm are intense enough to serve as the basis for quantitative analytical measurements. In

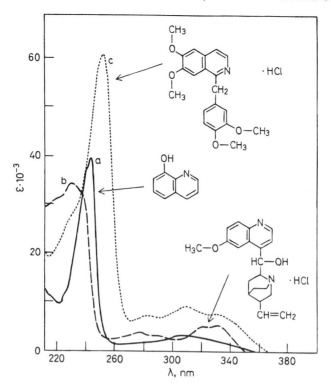

FIGURE 3.22. Spectra of quinoline and isoquinoline derivatives. a. 8-hydroxyquinoline. Solvent: ethanol; b. quinine hydrochloride. Solvent: methanol; c. papaverine hydrochloride. Solvent: 0.01 M hydrochloric acid.

this spectrophotometrically highly active molecule even the dimethoxybenzyl moiety can be considered as a "spectrophotometrically inactive" group: it hardly contributes to the spectrum.

2. Indole Derivatives

Of the two-ring systems containing one nitrogen the derivatives of benzopyrrole (indole) are of outstanding importance in pharmaceutical chemistry. With the contribution of the unshared electron pair of nitrogen a π-electron system consisting of 10 electrons, similar to that of naphthalene applies in the case of indole, too; therefore, its spectrum is also similar to that of naphthalene. This spectrum is characteristic of tryptophan (see Figure 10.8), which, after tyrosine, is the second spectrophotometrically active amino acid, as well as several alkaloids such as vincamine (see Figure 3.23) and yohimbine. In these derivatives no auxochromic or chromophoric groups are bound to the indole ring system. A typical auxochromic group in the indole family is the methoxy group. This causes a relatively small spectral shift as demonstrated by pindolol ($\varepsilon_{265\ nm} = 7,950$ and $\varepsilon_{288\ nm} = 4,200$). In the case of reserpine which also contains a methoxy group the position and intensities of the bands are influenced by the presence of the isolated trimethoxybenzoyl group ($\varepsilon_{266\ nm} = 16,700$ and $\varepsilon_{294\ nm} = 10,200$).

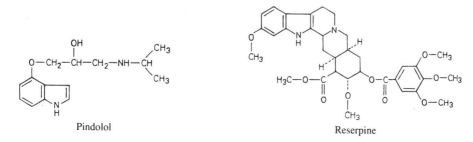

Pindolol   Reserpine

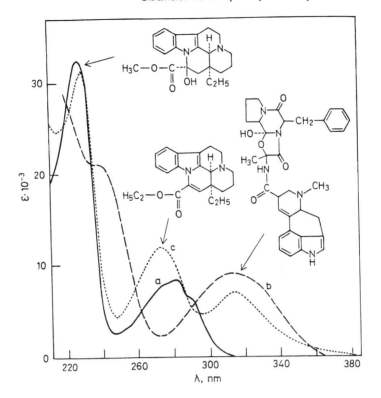

FIGURE 3.23.   Spectra of indole alkaloid derivatives. a. vincamine. Solvent: ethanol; b. ergotamine. Solvent: methanol; c. vinpocetine. Solvent: ethanol.

In several indole derivatives there is a double bond in conjugation with the indole nucleus causing pronounced bathochromic and hyperchromic shifts. A typical example for this is ergotamine in curve "b", Figure 3.23. In the semisynthetic derivative vinpocetine (curve "c") the conjugated double bond is in a different position.

If the additional double bond of these derivatives is saturated (hydrogenated ergot alkaloids) the original indole-like spectrum reappears. If, however, the double bond in the five-membered ring in the indole nucleus is also saturated the aromatic character is restricted to the six-membered ring. In these cases aniline-type spectra are obtained instead of the characteristic indole spectra. The spectra of such acylated dihydroindole derivatives were discussed in Section 3.G where the spectrum of strychnine was also shown as a characteristic example (see Figure 3.11).

Another type of the conjugation of chromophoric groups with the indole ring system is represented by e.g., indomethacin, where the indole nitrogen is 4-chlorobenzoylated. This exhibits a highly intensive band at 230 nm ($\varepsilon = 20,800$), an inflection at 260 nm ($\varepsilon = 16,200$) and a band of medium intensity at 319 nm ($\varepsilon = 6,290$). On the basis of the principles described earlier, this compound could also be considered a diarylketone: its spectrum really shows some similarity with that of 1-naphthyl-phenylketone ($\lambda_{max} = 221, 251, 280, 289$ and 307 nm; $\varepsilon = 50,000, 16,600, 6,600, 5,900$ and $5,100$).

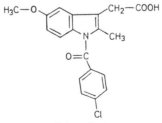

Indomethacin

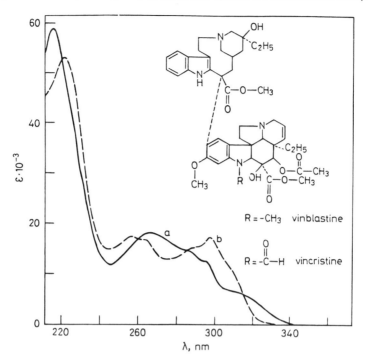

FIGURE 3.24.   Spectra of bis-indole alkaloids. Solvent: methanol. a. vinblastine sulphate; b. vincristine sulphate.

Bis-indole alkaloids constitute an interesting group of indole derivatives characterized by the simultaneous presence of an indole and a dihydroindole (indoline) unit in the molecule. As seen in Figure 3.24 vinblastine exhibits a very intense band at 214 nm: this is produced jointly by the indole and indoline units. The band at 267 nm is the *p* band of the indoline unit as is the inflection at 310 nm (α band). The indole spectrum with its characteristic maxima at 286 and 295 nm is between these two bands. It is interesting that the spectrum is dominated by the less conjugated indoline unit. In connection with this it is worth recalling the spectral data of N,N-dimethylaniline in Table 3.4: *p* band $\varepsilon_{351\ nm} = 14,000$; α band $\varepsilon_{299\ nm} = 2,100$).

The spectrum of vincristine can be assigned in a similar manner. The band at 255 nm is the *p* band of the acetanilide-type formylindoline unit; the long wavelength fused band complex is produced jointly by the α-band of the formylindoline unit and by the indole-type band of the other part of the molecule.

3.   Benzazepines and Benzodiazepines

The spectrum of carbamazepine betrays strong conjugation since this molecule can be considered simultaneously a *cis*-stilbene and diphenylamine derivative (see Figure 3.25)

The spectra of two characteristic members of the benzodiazepine family are also represented in Figure 3.25. The spectrum of diazepam with its two maxima and one inflection indicates that the diazepine ring is on the one hand of carboxanilide type and on the other hand it is linked to the aromatic ring by a conjugated azomethine-like bond. If the latter is eliminated by the hydrogenation of the C=N bond, an unambiguously carboxanilide-type spectrum is obtained: $\varepsilon_{252\ nm} = 13,400$. The third spectrum in Figure 3.25 shows that the nitro group in the molecule of nitrazepam results in the increase of the long wavelength band.

4.   Purine and Pterine Derivatives

Of the two closely related families purines are represented by the pharmaceutically very important xanthine group. As a consequence of the non-enolizable carbonyl groups in their molecules, the

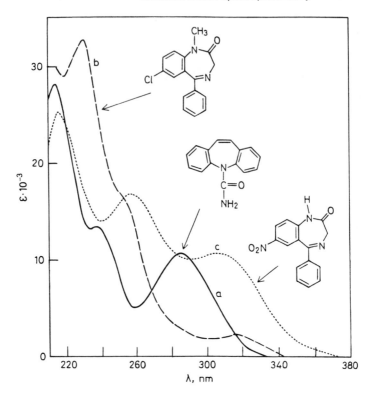

FIGURE 3.25. Spectra of benzazepine and benzodiazepine derivatives. Solvent: methanol. a. carbamazepine; b. diazepam; c. nitrazepam.

spectra of theobromine, theophylline and caffeine are very similar (see Table 3.11). Their band around 270 nm does not show fine structure. The spectrum of caffeine is shown in curve "a" of Figure 3.26. The spectral data of allopurinol containing the isomeric allopurine skeleton are also included in Table 3.11.

Of the aromatic pterine derivatives Table 3.11 contains the data of folic acid and its antagonist, methotrexate. The yellow and orange color, respectively, of these derivatives is due to the diamino-pterine unit substituted with auxochromic groups: the contribution of the isolated 4-dimethylamino-benzamide unit does not contribute to this.

TABLE 3.11. Spectral Data of Purine and Pterine Drugs (Solvent: methanol[9])

Compound	λ_{max}, nm	ε
Theophylline	269	9,020
Theobromine	271	9,580
Caffeine	273	9,220
6-Mercaptopurine	328	16,890
Allopurinol	249	7,700
Folic acid	260	23,500
	284	28,970
Methotrexate	263	16,820
	299	24,370

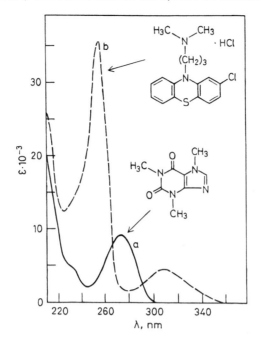

FIGURE 3.26. Spectra of purine and phenothiazine derivatives. Solvent: methanol. a. caffeine; b. chlorpromazine hydrochloride.

5. Phenothiazines

1,4-Thiazine condensed with two benzene rings (phenothiazine) exhibits a characteristic spectrum consisting of two bands. The majority of its numerous derivatives contain non-chromophoric substitutents; the spectra of these derivatives are naturally very similar to that of unsubstituted phenothiazine. As a characteristic example the spectrum of chlorpromazine is shown in curve "b" of Figure 3.26.

It is worth mentioning that methylene blue, which is the oxidation product of phanothiazine-3,7-bis-dimethylamine, exhibits absorption maximum at 655 nm ($\varepsilon = 89,000$).

6. Miscellaneous

The xanthene derivative propantheline bromide possesses diphenylether-like spectrum ($\lambda_{max} = 243$ and 281 nm; $\varepsilon = 4,560$ and 2,230). In the long wavelength section of the thioxanthene derivative methixene hydrochloride, one intense band appears at 269 nm ($\varepsilon = 11,300$); this band obscures the benzenoid α band. If this ring system is conjugated with a double bond such as in the case of chlorprothixene, a strong bathochromic and hyperchromic shift takes place ($\lambda_{max} = 230, 270, 326$ nm; $\varepsilon = 30,000, 14,000, 3,150$).

Of the derivatives of acridine acriflavine (tripaflavine) with its intense yellow color is an excellent example ($\varepsilon_{262\,nm}$ 45,400; $\varepsilon_{462\,nm} = 45,600$).

Propantheline bromide   Methixene

Chlorprothixene   Tripaflavin

Several pharmaceutically active benzopyran derivatives exhibit intense spectra. Acenocoumarol representing the coumarins has intense bands at 282 and 305 nm (ε = 22,500 and 18,000). The nitrobenzene unit in the molecule contributes only slightly to this. Benzo-γ-pyrone-type flavonoids also belong to this family. For example rutoside exhibits intensive absorption at 264 and 382 (ε = 23,000 and 17,700). Generally speaking the bathochromic and hyperchromic shift causing the intensive yellow color of flavonoids is due to the presence of the phenyl group linked to the benzo-γ-pyrone unit together with the effect of phenolic and enolic hydroxy groups.

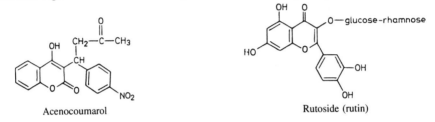

Acenocoumarol   Rutoside (rutin)

N. DYES

In the preceding section some pharmaceutically important dyes such as methylene blue and tripaflavine have already been mentioned. Several other dyes are also used in therapy and as diagnostics. A common feature of these is the presence of an aromatic ring or ring system with auxochromic or chromophoric groups as substituents. Since carboxyl and amino groups are among the substituents in almost all cases, the spectra are usually pH-dependent. Consequently, pharmaceutically important dyes usually possess acid-base indicator properties.

The most important dyes used in therapy are derivatives of triphenylmethane. In these there is a pH-dependent equilibrium between the triphenylmethyl cation and its quinoidal form. Of the aromatic rings, two or three may contain amino, methylamino or dimethylamino groups at position 4. The protonation of the latter leads to further equilibria.

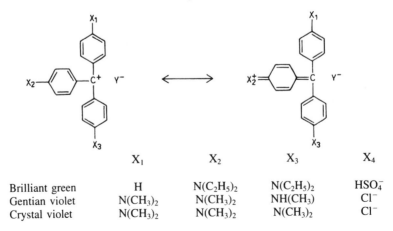

	X_1	X_2	X_3	X_4
Brilliant green	H	$N(C_2H_5)_2$	$N(C_2H_5)_2$	HSO_4^-
Gentian violet	$N(CH_3)_2$	$N(CH_3)_2$	$NH(CH_3)$	Cl^-
Crystal violet	$N(CH_3)_2$	$N(CH_3)_2$	$N(CH_3)_2$	Cl^-

Of the above derivatives the spectra of crystal violet in the free base, mono- and dicationic forms are represented in Figure 3.33 of Section 3.Q.

O. METAL COMPLEXES

The interaction between metal ions and ligand anions or molecules capable of the formation of complexes often results in the development or change of colors suitable for the characterization or quantitative determination of the metal or the ligand. Metals containing easily excitable d or f electrons are especially suitable for the formation of colored complexes. Section 7.G contains several examples of applications of complex formation in the spectrophotometric analysis of drugs.

Of the metal complex type drugs the ultraviolet-visible spectrum of vitamin B_{12} is shown in Section 2.I (Figure 2.11).

P. STERIC HINDRANCE

When the effect of the conjugation of chromophores on the absorption spectrum was introduced in Section 3.E, one important question was not touched: the effect of steric factors on conjugation and, through this, on the spectrum itself. As a matter of fact even the effect of cis-trans isomerism on the spectra of dienes and polyenes discussed in Section 3.E belongs to this category. The differences of the steric position of geometrical isomers can be expressed by drawing their formulae in the plane of the paper sheet. In many other cases, however, conjugated systems and on the basis of this intense spectra could be expected from the formula of the compound projected on the plane of the paper sheet, however, in practice, the band appears at shorter (sometimes much shorter)

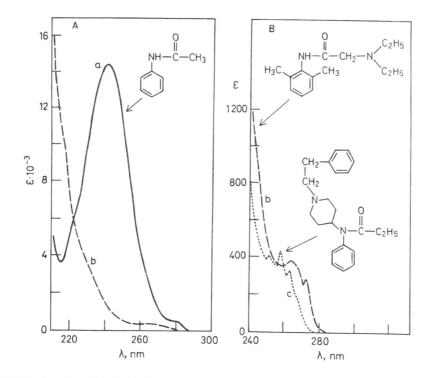

FIGURE 3.27.   The effect of steric hindrance on the spectra of aniline derivatives. a. acetanilide; b. lidocaine; c. fentanyl citrate.

wavelength and with (much) lower intensity than expected. These phenomena can usually be traced back to steric hidrance. This means that in addition to prerequisites of the formation of a conjugated system, described in Section 3.E, one further important point is the coplanarity of the system. In many cases the conditions to fulfill this additional requirement do not exist: because the coplanarity can be hindered by certain groups (especially by bulky ones) in the molecule.

A classical example of the effects of steric hindrance on the spectrum is the case of biphenyl and its derivatives. In the case of unsubstituted biphenyl the two benzene rings are positioned almost entirely coplanarly. Its intense conjugation band without fine structure ($\varepsilon_{252\ nm} = 18,300$) submerges the α band which would possess fine structure. Step-by-step introduction of methyl groups into the four ortho positions in the benzene rings gradually increases the torsion angle along the C–C bond between the two benzene rings, thus decreasing the conjugation between the electron systems of the two rings. This results in the displacement of the absorption maximum and especially in the decrease of the intensity of the band. In the case of four methyl groups at $2,2',6,6'$ positions the planes of the two benzene rings are almost perpendicular to each other and as a result of this the conjugation ceases and the conjugation band disappears. The absorption band of the above tetramethylbiphenyl ($\varepsilon_{262\ nm} = 500$; fine structure) is characteristic of the sum of the spectra of two isolated toluene-like units.

Several examples of this kind can be found among pharmaceutical compounds, especially aromatic derivatives. Three spectra can be seen in Figure 3.27a. Curve "a" is the spectrum of acetanilide. As described in Section 3.G, this intensive spectrum is the result of the conjugation of the aromatic ring, the unshared electron pair of the nitrogen and of the C=O group linked to the latter. In this case the coplanarity of these elements is assured. On the basis of the representation of the structural formula of lidocaine in the plane of the paper a similar spectrum could be expected. Oppositely, no conjugation band is observable in the spectrum of lidocaine: its spectrum is similar to that of m-xylene (curve "b"), moreover—as is seen in Figure 3.27b which is an enlarged version of the spectrum in the wavelength range between 240 and 280 nm—even the fine structure of the α band appears. This is because the two methyl groups at the 2,6-positions hinder the coplanar positioning of the benezene ring and the bulky acylamino group. Similar phenomena can be observed in the case of the analogous mepivacaine: $\varepsilon_{263\ nm} = 420$.

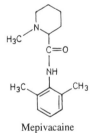

Mepivacaine

One more spectrum can also be found in Figure 3.27b: the long wavelength section of the spectrum of fentanyl. The weak band with fine structure is the fused peak of the α bands of the isolated phenyl and propionylanilide groups. In this case the anilide moiety does not contain substituents to cause the steric hindrance; the coplanar positioning of the above mentioned three structural elements is hindered by the bulky substituent linked to the nitrogen atom.

The interpretation of steric effects hindering the coplanar positioning of aromatic rings and carbonyl groups or aromatic rings and carbon–carbon double bonds can be quite similar. The consequences of the lack of conjugation caused by this steric hindrance were demonstrated in Section 3.H by the examples of amitriptyline, diethylstilbestrol and dienestrol.

Q. THE EFFECT OF pH ON SPECTRA

1. General Considerations

Those compounds which can reversibly dissociate or associate proton, i.e., acids and bases change their spectra upon changing the hydrogen ion concentration of the solution. Notable spectral changes are only observed if the acidic or basic groups in molecules which take part in the above mentioned protolytic equilibria contribute with their n and/or π-electrons to the conjugated electron systems of the molecules. The variations of the weak and not very characteristic bands of the isolated carboxyl and amino groups below 220 nm upon changing the pH are not very characteristic either. In contrast to this the salt formation in the case of the spectra of e.g., benzoic acid (upon basification) or aniline (upon acidification) causes profound differences in their spectra. This is caused by the conjugation of their acidic and basic groups with the benzene rings. If, however, spectral changes of phenylacetic acid and benzylamine are investigated on basification or acidification only minor changes are observable. This is because in these cases the acidic and basic groups, respectively, are separated from the benzene rings by methylene units. The fact that the spectra of codeine and morphine practically do not change upon acidification can be similarly explained since their tertiary amino groups are isolated from both the aromatic ring and the double bond. At the same time the dissociation of the phenolic hydroxy group of morphine results in profound change of the spectrum. This change naturally does not take place in the case of codeine where the phenolic hydroxy group is in the form of a phenolic ether.

The ratio of the protonated and deprotonated forms of an acid or base at a given pH or the pH necessary to obtain a certain degree of protonation or deprotonation can be calculated from the K_a value using Equation 3.3 or its logarithmic form, Equation 3.4.

$$K_a = \frac{[H^+][D]}{[P]} \qquad (3.3)$$

where K_a is the dissociation constant of the acid or the protonated base, [D] and [P] are the concentrations of the deprotonated and protonated forms possessing separate spectra.

$$pK_a = pH - \log \frac{[D]}{[P]} \qquad (3.4)$$

For example, in the case of a typical carboxylic acid $pK_a = 5$, which means that at pH 5, [D]/[P] $= 1$. In other words at pH 5, 50% of the material shows the spectrum of the protonated and the other 50% that of the deprotonated form. The contribution of the deprotonated form in the overall spectrum will be about 10% at pH 4, 1% at pH 3 and 0.1% at pH 2. Thus, at pH 2 the spectrum is of the practically pure protonated form and no further change takes place upon further acidification. If the pH is raised, the relative quantity of the deprotonated form is about 90% at pH 6, 99% at pH 7 and 99.9% at pH 8. The spectrum is not changed upon further increase of the pH.

With the exception of those cases where the aim of the absorbance measurement is to study protolytic equilibria (especially in the case of quantitative analysis), measurements are carried out at pH values which assure the presence of either the fully protonated or fully deprotonated form. In the majority of cases this means that it is advisable to work in 0.01 M hydrochloric acid or sodium hydroxide solutions, that is at pH 2, where weak and moderately strong acids ($pK_a > 5$) are present in the undissociated form and strong and moderately strong bases (pK_a of the protonated base is above 5) in the protonated form, while at pH 12 weak and moderately strong bases (pK_a of the protonated base is less than 9) are present in the deprotonated form and strong and moderately strong acids ($pK_a < 9$) exist in the dissociated state. Higher acid concentration is necessary to achieve total dissociation of very weak acids, e.g., phenols and higher acid concentration is required in the case of some very weak bases to reach their fully protonated states.

If for any reason a spectrum which is sensitive to pH is taken between pH 2 and 12 a buffered solution is used to assure the reproducibility of the measurement (naturally using spectrophotometrically inactive buffers). In the case of compounds of this type it is not advisable to use pure aqueous medium as the solvent. Because in this case the pH of the solution is not well defined and in addition to this the spectrum may depend on the dilution since the degree of dissociation increases in unbuffered solutions with increasing dilution. This may cause deviations from the Beer's Law (see Section 4.B).

As an illustration two spectra of papaverine hydrochloride are shown in Figure 3.28 using water from two different sources as the solvent. The pK_a value of the protonated base is 5.93. According to Equation 3.4 at about pH 6 the protonated (λ_{max} = 251 nm) and the deprotonated (λ_{max} = 237 nm) forms of papaverine base are present in approximately identical concentrations. As can be seen in the figure no remarkable hydrolysis of the hydrochloride of papaverine takes place if highly purified water stored in a quartz bottle is used as the solvent. However, using ion-exchanged water stored in a glass bottle, the base form is predominantly present in the solution.

If for any reason difficulties are encountered when measuring the spectra in strongly acidic, alkaline or buffered solutions, there is one more possibility to consider: When a series of spectra based on acid-base equilibria is investigated, there exist one or more wavelength(s) where all

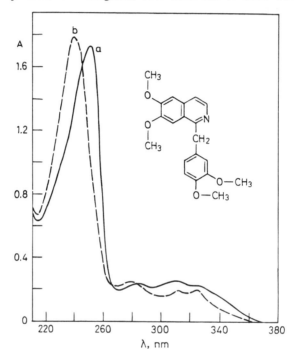

FIGURE 3.28. Spectra of papaverine hydrochloride. a. Solvent: water distilled and stored in quartz apparatus. c = 1.208 mg/100 ml; b. Solvent: ion exchanged water stored in glass vessel. c = 1.2528 mg/100 ml.

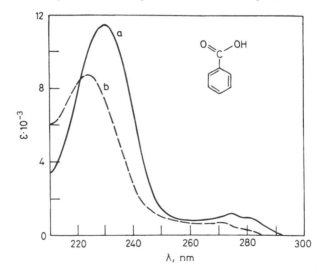

FIGURE 3.29. Dependence of the spectrum of benzoic acid on the pH. Solvent: water. a. 0.01 M hydrochloric acid; b. 0.01 M sodium hydroxide.

spectra cross one another, i.e., their absorbance values are equal (isosbestic point). In other words, the absorbance at this wavelength does not depend on the pH of the solution. As a result, reliable absorbance measurements can be made at this wavelength even in unbuffered media. Since, however, the isosbestic point is usually far from the maximum, using this wavelength leads to reduced sensitivity of the measurement.

Investigations of the dependence of the spectra on pH affords useful data for the analyst. Important conclusions on compound structures can be drawn from the character and extent of the change.

The selectivity and sensitivity of quantitative measurements can also be increased if the measurement is carried out at a pH where the light absorption is the highest and the measurement is not exposed to interferences.

2.   Dependence of the Spectra of Acids on pH

Figure 3.29 represents the spectra of benzoic acid in undissociated (pH 2) and dissociated (pH 12) forms. The reason for the hypsochromic and hypochromic shifts in alkaline media (characteristic of the aromatic acids) is that the carboxylate anion where the distribution of the electrons is symmetrical is less capable of conjugation with the aromatic ring than the carbonyl group of the undissociated carboxyl group.

More pronounced differences can be observed in the spectra of enols and phenols as a consequence of the ionization of these groups. The stepwise dissociation of barbital and its three spectra at different pH values are discussed in Section 3.L and the spectra shown in Figure 3.21. A further example of the effect of the ionization of enols on spectra is ascorbic acid which contains two ionizable enolic hydroxyls ($pK_1 = 4.17$; $pK_2 = 11.57$). Two of the three spectra are shown in Figure 3.30. The spectrum of the dianionic form exhibiting maximum at 300 nm has been omitted because of the instability of the strongly alkaline solution.

Similar to the changes found in the case of enols, the spectra of phenols also undergo bathochromic shifts in alkaline media. As shown in Figure 3.9 on the example of ethinyloestradiol the intensity of the α band also increases upon basification but the fine structure disappears. Because of

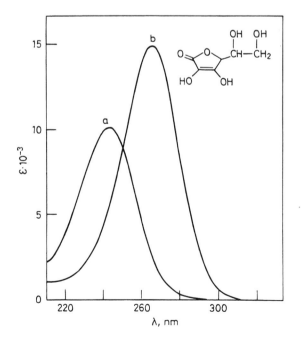

FIGURE 3.30.   Dependence of the spectrum of ascorbic acid on the pH. Solvent: water. a. 0.01 M hydrochloric acid; b. pH 8 (phosphate buffer).

the weak acidity of phenols strongly alkaline media (pH > 13) are necessary to obtain full ionization.

Phenylbutazone and its derivatives are good models for the investigation of keto-enol tauto-merism and the effect of the ionisation of the enolic form on their spectra. The transformation of the dioxo form to the mesomeric anion above pH 8 results in bathochromic and hyperchromic shift.

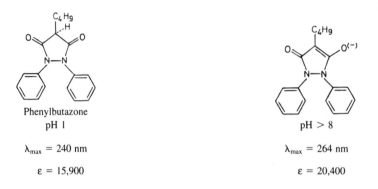

Phenylbutazone
pH 1

λ_{max} = 240 nm

ε = 15,900

pH > 8

λ_{max} = 264 nm

ε = 20,400

In contrast to the previous example, oxymetholone exists in the enolic form in neutral organic solvents. It exhibits an absorption maximum at 285 nm (ε = 8,130) in methanolic solution. The longer wavelength of this conjugation band related to the α, β-unsaturated steroidal ketones is due to the hydroxyl substituent at the double bond. In 0.1 M sodium hydroxide solution, the enol dissociates and the band is shifted to 315 nm (ε = 18,000).

The deprotonation of imino groups results in various kinds of spectral shifts. Xanthine derivatives (theobromine and theophylline) show only slight change in alkaline medium. In contrast to this the maximum displacement and the increase of the intensity is very strong in the case of 2-methyl-5-nitroimidazole (intermediate in the synthesis of metronidazole). In neutral solution $\varepsilon_{311\ nm}$ is 6,780 while in the ionized state (above pH 12) the maximum is displaced to 366 nm (ε = 12,300).

The $-SO_2-NHR$ group of sulphonamides is also of acidic character. Salt formation in alkaline media, however, causes the spectrum to shift toward the shorter wavelengths (as it was shown in the case of aromatic carboxylic acids).

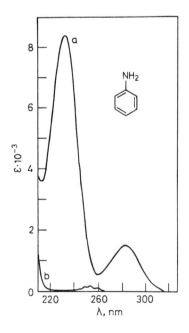

FIGURE 3.31. Dependence of the spectrum of aniline on the pH. Solvent: ethanol–water 9:1 v/v. a. neutral; b. 0.1 M hydrochloric acid.

3. Dependence of the Spectra of Bases on pH

On the basis of the dependence of their spectra on pH compounds containing basic nitrogen can be classified into two groups:

1. the unshared electron pair of the nitrogen is in conjugation with the π-electron system of the molecule (usually an aromatic ring)
2. nitrogen containing aromatic heterocycles where the unshared electron pair does not take part in the aromatic π-electron sextet

The most characteristic representative of the first group is aniline. As noted in Section 3.G and seen in Figure 3.31, a strong bathochromic shift of both band p and α is observable in the spectrum of aniline relative to that of benzene, and the intensity of the latter also greatly increases. The changes result from the effect of the unshared electron pair of the nitrogen on the electron system of the aromatic ring. Also, as seen in Figure 3.31, in strongly acidic medium which is necessary for the protonation of the weak base aniline this effect disappears: as a consequence of a very strong hypsochromic and hypochromic shift the intensity, and even the fine structure of the α band in acidic medium, reappears and is entirely benzene-like.

No general rules of this kind can be established for the effect of protonation on the spectra of nitrogen containing aromatic heterocycles. Both bathochromic and hypsochromic shifts can be observed depending on the ring and the character of the substituent and in many cases the position of the α band remains unchanged but its intensity increases as a result of salt formation. This is the case, for example, with pyridine which exhibits absorption maximum at 257 nm with a molar absorptivity of 3,050; this intensity increases at practically the same wavelength (256 nm) to 5,300 in acidic medium. Of the drug derivatives of pyridine the behavior of nicotinamide is exactly the same (see Figure 3.32).

An example of the above mentioned hypsochromic shift is the series of spectra in Figure 3.34: the change of the spectrum of the imidazole derivative metronidazole as a result of protonation while the bathochromic shift can be exemplified by iodochloroxyquinoline (see Figure 8.1) and papaverine (see Figure 3.28).

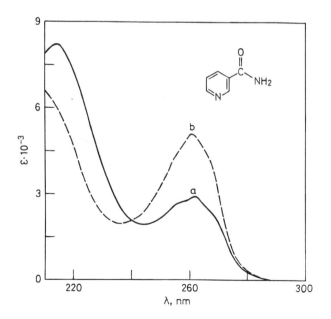

FIGURE 3.32. Dependence of the spectrum of nicotinamide on the pH. a. neutral; b. 0.1 M hydrochloric acid.

Of the indicator-type basic dyes discussed in Section 3.N, the dependence of the spectrum of crystal violet on the pH is demonstrated in Figure 3.33. Curve "a" is the spectrum of the free base, curve "b" represents predominantly the monocation while curve "c" is the spectrum of the dicationic species. Accordingly, the colors of the above solutions are violet, blue and yellow.

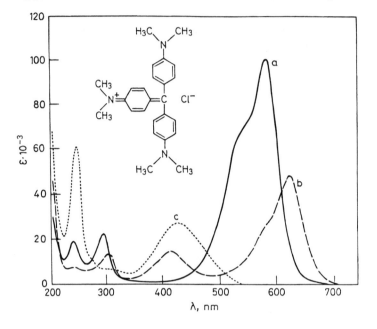

FIGURE 3.33. Dependence of the spectrum of crystal violet on the concentration of hydrochloric acid. Solvent: water. a. neutral; b. 0.1 M hydrochloric acid; c. 3 M hydrochloric acid.

4. Determination of Acid–Base Dissociation Constants

Equation 3.3 is not only suitable for the calculation of the expected pH range of the protonation–deprotonation, but it can be used for the determination of the dissociation constants of spectrophotometrically active acids and bases where the change of the pH causes considerable changes in their spectra.

In practice, a series of dilutions of the same concentration is prepared, the pH of which is set to different values by means of suitable buffers. Water is used as the solvent which contains—if necessary—the minimum quantity of organic solvent for the solubilization of the material. The pH range is selected in such a way that the values necessary for the full protonation and deprotonation are represented and in addition to this 6–8 further pH values determined with high accuracy where the protonated and deprotonated forms are present at commensurable concentrations. After preparing the spectra, one or two suitable wavelengths are selected for the calculations (usually the maximum of the spectrum of the protonated or deprotonated form or both). The absorbances at the selected wavelength(s) are plotted as a function of the pH. The curves thus obtained begin with a constant section (protonated form), followed by ascending or descending section and, at the end, another constant section representing the deprotonated form can be found. The simplest method for the calculation of the pK value is to identify the inflection point of the curve, i.e., the pH value where the absorbance is just the mathematical mean of the absorbances of the fully protonated and deprotonated forms.

This means that at this pH, the concentrations of the two forms are equal. According to Equation 3.4 this pH value directly represents the pK_a of the acid or the protonated base. As an illustration the spectra of metronidazole at various pH values and the absorbance–pH plots at the maxima of the protonated and deprotonated forms are presented in Figures 3.34 and 3.35.

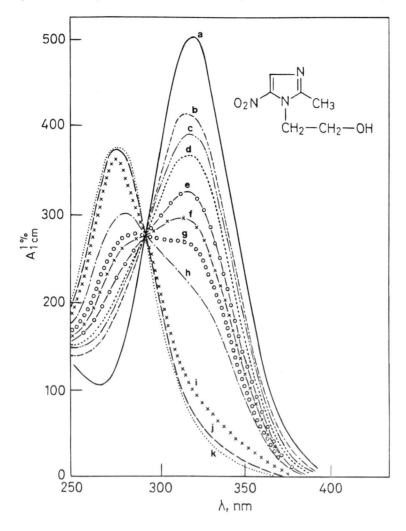

FIGURE 3.34. Dependence of the spectrum of metronidazole on the pH. Solvent: water. a. neutral; b. pH 3.171; c. pH 3.037; d. pH 2.878; e. pH 2.745; f. pH 2.596; g. pH 2.457; h. pH 2.271; i. 0.03 M hydrochloric acid; j. 0.1 M hydrochloric acid; k. 1 M hydrochloric acid.

A more accurate calculation method is to compute the pK value at all pH values separately where the two forms are present in commensurable concentrations. For the calculation Equation 3.5, derivable from Equation 3.3 can be used:

$$pK_a = pH - \log \frac{[D]}{[P]} = pH - \log \frac{A - A_P}{A_D - A} \tag{3.5}$$

where [D] and [P] are the concentrations of the deprotonated and protonated forms, A is the absorbance at a given pH, A_P and A_D are the absorbances of the fully protonated and deprotonated forms. The pK values calculated at different pH values should be constant within the limits of the experimental error: the pK value may then be taken as the mean of the values thus obtained. If, however, the pK values systematically increase or decrease with increasing pH the mean must not be computed: in such cases the acid-base equilibrium is probably being disturbed by another equilibrium.

The ionic strength should be kept constant during the measurement by adding a fixed concentration of a spectrophotometrically inactive salt (usually 0.1 M) to all solutions.

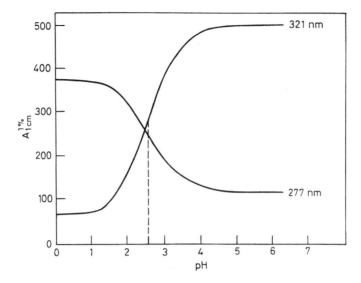

FIGURE 3.35. Determination of the pK_a value of metronidazole by the spectrophotometric method on the basis of the data in Figure 3.34.

Since under pH 2 and above 12 the reliability of the pH values measured by glass electrode decreases considerably, in the case of the determination of the pK values of very weak acids or bases the pH is calculated from the analytical concentrations of the acid or base used for the acidification or basification taking into account their activity coefficients.

Several further methods are available in addition to this simple method of measurement and calculation for the spectrophotometric determination of dissociation constants with special respect to more complicated cases such as for those molecules containing more than one acidic and/or basic group, treatment of overlapping dissociation equilibria, etc. The discussion of these would be beyond the scope of this book. For this reason two monographs of fundamental importance[1,29] and some important papers[2,18,23,37] as well as recent papers on the determination of the dissociation constants of isonicotinic acid[3] and flurazepam[22] are mentioned.

REFERENCES

1. Albert, A.; Serjeant, E. P. *The Determination of Ionization Constants;* Chapman and Hall: London, 1971.
2. Asuero, A. G.; Jiménez-Trillo, J. L.; Navas, M. J. *Talanta.* 1986, *33,* 531.
3. Asuero, A. G.; Navas, M. J.; Herrador, M. A.; Recamales, A. F. *Int. J. Pharm.* 1986, *34,* 81.
4. Bauman, R. P. *Absorption Spectroscopy;* Wiley: New York, 1962.
5. Beaven, G. H.; Johnson, E. A.; Willis, H. A.; Miller, R. G. J. *Molecular Spectroscopy;* Heywood: London, 1961.
6. Blass, W. E.; Halsey, G. W. *Deconvolution of Absorption Spectra;* Academic Press: New York, 1981.
7. Clar, E. Aromatische Kohlenwasserstoffe; Springer: Berlin, 1952.
8. Clarke, E. G. C. *Isolation and Identification of Drugs;* The Pharmaceutical Press: London, 1975.
9. Dibbern, H.-W. *UV- und IR-Spektren wichtiger pharmazeutischer Wirkstoffe,* I–III; Editio Cantor: Aulendorf, 1980.
10. Ellington, P. S.; Meakins, G. D. *J. Chem. Soc.* 1960, 697.
11. Fieser, L. F.; Fieser, M. *Steroids;* Reinhold: New York, 1959; 15–24.
12. Hershenson, H. M. *UV and Visible Absorption Spectra;* Academic Press: New York, 1961.
13. Hirayama, K. *Handbook of Ultraviolet Absorption Spectra of Organic Compounds;* Plenum Press: New York, 1967.
14. Jaffé, H. H.; Orchin, M. *Theory and Applications of Ultraviolet Spectroscopy;* Wiley: New York, 1962.
15. Krácmar, J.; Krácmarová, J. *Cesk. Farm.* 1966, *15,* 16.
16. Krácmar, J.; Krácmarová, J.; Bokoviková, T. N.; Ciciro, V. E.; Nesterová, G. A.; Suranová, A. V.; Trius, N. V. *Pharmazie,* 1991, *46,* 253.
17. Láng, L. *Absorption Spectra in the Ultraviolet and Visible Region,* I–XXIV; Akadémiai Kiadó: Budapest; Academic Press: New York, 1959–1982.

18. Navas, M. J.; Jiménez-Trillo, J. L.; Asuero, A. G. *Boll. Chim. Fam.* 1985, *124,* 439.

19. Nielsen, A. T. *J. Org. Chem.* 1957, *22,* 1539.

20. *Organic Electronic Spectral Data,* I–XXVII; Wiley-Interscience: New York, 1960–1991.

21. Perkampus, H.; Timmons, C. J.; Sandeman, I. *UV Atlas of Organic Compounds;* Butterworths: London; Verlag Chemie: Weinheim, 1977.

22. Pfendt, L. J.; Janjic, T. L.; Popovic, G. V. *Analyst,* 1990, *115,* 1457.

23. Pfendt, L. J.; Stadic, D. M.; Janjic, T. J.; Popovic, G. V. *Analyst* 1990, *115,* 383.

24. Rao, C. N. R. *Ultraviolet and Visible Spectroscopy;* Butterworths: London, 1975, a. p.60.

25. *Sadtler Handbook of UV Spectra;* Heyden and Sons: London, 1979.

26. Scott, A. I. *Interpretation of the Ultraviolet Spectra of Natural Products;* Pergamon Press: Oxford, 1964. a. p. 50; b. p. 58; c. p. 109.

27. Sondheimer, F.; Ben Efraim, D. A.; Wolowsky, R. *J. Am. Chem. Soc.* 1961, *83,* 1675.

28. Stern, E. S.; Timmons, C. J. *Electronic Absorption Spectroscopy in Organic Chemistry;* Edward Arnold: London, 1970.

29. Sucha, L.; Kotrly, S. *Solution Equilibria in Analytical Chemistry;* Van Nostrand Reinhold: London, 1972.

30. Sunshine, I. *Handbook of Spectrophotometric Data of Drugs;* CRC Press: Boca Raton, 1981.

31. Suzuki, H. *Electronic Absorption Spectra and Geometry of Organic Molecules;* Academic Press: New York, 1967.

32. Thamer, B. J.; Voigt, A. F. *J. Phys. Chem.* 1952, *56,* 225.

33. Woodward, R. B. *J. Am. Chem. Soc.,* 1942, *64,* 72.

Chapter 4

QUANTITATIVE ANALYSIS ON THE BASIS OF NATURAL LIGHT ABSORPTION

A. ANALYSIS OF SINGLE-COMPONENT SYSTEMS

1. Determination on the Basis of the Beer-Lambert Law

This chapter deals with the quantitative analysis of single- and multicomponent systems, with this section dealing with single-component systems. From the spectrophotometric point of view, those systems can be regarded as single-component ones for which, at the wavelength selected for the measurement, the determination of the analyte is not influenced by either another substance with well defined spectrum or undefined, so called "background" absorption. In such cases the determination can be based on the Beer-Lambert Law (Equation 2.8). Expressing concentration from Equation 2.8 and replacing a by the specific absorbance, we obtain

$$\text{concentration (g/100 ml)} = \frac{\text{absorbance}}{A_{1\,cm}^{1\%} \cdot \text{cell thickness (cm)}} \tag{4.1}$$

In drug analysis, the determination of the drug content of solid samples (drug raw materials, intermediates, drug preparations) is an important field of application as well as the determination of the concentrations of solutions. In such cases a stock solution is prepared from the solid sample by dissolution or extraction; this solution is usually diluted, and the absorbance of the resulting solution is measured. Drug content can then be determined by Equation 4.2:

$$\text{drug content (\%)} = \frac{\text{dilution} \cdot \text{stock volume (ml)} \cdot \text{absorbance}}{A_{1\,cm}^{1\%} \cdot \text{sample weight (g)} \cdot \text{cell thickness (cm)}} \tag{4.2}$$

As to the conditions of the applicability of Equations 4.1 and 4.2 and the optimization of analysis, several questions should be answered.

For the selection of determination wavelength, it is usually preferable to perform the measurement at the absorption maximum, both because this is the point where the method is the most sensitive and since the reproducibility of wavelength setting is better at the maximum than on the edges of the absorption band (particularly if they are steep). It is a further advantage that for a given slitwidth, the variation of absorbance within the spectral bandwidth is the smallest at this point. If the spectrum has several maxima, in the selection of measuring wavelength it should also be considered in addition to the aspects of sensitivity that at very low wavelengths (below 220 nm) a number of perturbing effects may be encountered (see Sections 2.I and 2.J).

The problems connected with absorbance measurement and the optimization of the measurement were discussed in Sections 2.G to 2.J. The significance of the appropriate choice of solvent (Sections 2.K and 3.B) and the correct pH setting (Section 3.Q) was also pointed out.

In connection with the specific absorbance, the following facts should be considered. Data taken from the literature may be used only if the cited experimental conditions (solvent, pH, etc.) can be reproduced. However, as the literature contains a large amount of unreliable data, extreme care should be exercised in their use; this possibility may be utilized for measurements where high accuracy is not required, if there is absolutely no chance for the use of a standard sample of sufficient purity, although in most cases this provides the best solution. From this carefully dried standard with chromatographically checked purity, a standard solution should be prepared to determine the specific absorbance with the instrument to be used for the analysis. Care should be taken to ensure that the parameters (cell thickness, slitwidth, etc.) used for the measurement of absorbance in the analysis are the same as those for the measurement of specific absorbance. The latter should be checked regularly, but the safest solution is to determine it simultaneously with every series of measurements, i.e., to take the series of measurement together with the standard.

A very important task is to check the applicability of Beer's Law, i.e., the linearity of the absorbance vs. concentration relationship, on the system to be measured. This can be done by preparing a dilution series from the stock solution of the standard, and determining the absorbances of all solutions against the same reference solution at the selected wavelength. The measured absorbances are then plotted as a function of concentration. With instruments of appropriate quality and maintenance, a straight line passing through the origin should be obtained with the vast majority of drug substances. If the measurement is based on natural light absorption, any intercept (either positive or negative) on the absorbance axis must be due to experimental error (application of unsuitable reference solution). The situation is slightly more problematic with analyses based on chemical reactions; this problem is discussed separately in Section 6.C. A separate section (4.B) deals with the various reasons for deviations from linearity. Since the majority of the deviations discussed there occur only at high absorbances, and since quantitative measurements with demands of higher precision should be performed in the 0.2 to 1 absorbance range, calibration lines are usually determined and given for this range only. More recently, in the literature, instead of the calibration line and the measurement points, the equation of the regression line ($A = a + bc$) calculated from the latter by means of the method of least squares is given. If concentration is given in g/100 ml, regression coefficients a and b of this equation are the intercept and specific absorbance, respectively, and the correlation coefficient (r), which is usually also given, characterizes the deviation of the individual measurement points from the regression line (for good spectrophotometric methods $r > 0.999$).

As an example, Figure 4.1 shows the calibration lines taken at the two absorption maxima of the spectrum of papaverine hydrochloride, together with the corresponding regression equations (the spectrum is shown in Figure 3.22).

2. Problems of Selectivity

Even after the optimization of all factors discussed so far, there remains a question of fundamental importance that basically determines the correctness of the result obtained by the spectrophotometric method: namely, the question of the selectivity of the method. The simplest way of checking whether the system investigated is really a single-component system from spectrophotometric aspects (i.e., the determination of the analyte does not include contributions by other substances) is to take the spectra of the sample to be analyzed and of the pure component (standard) to be determined. If the two spectra are similar, but the spectrum of the sample is to a certain extent distorted relatively to that of the standard, a background problem is probably responsible, which makes the direct analysis impossible, but with the use of mathematical methods (see Sections 4.C and 4.F) there is a chance for a solution of appropriate precision. If a well defined, extra band occurs in the spectrum, there is another, spectrophotometrically active component in the system. These problems can be solved by methods discussed in Sections 4.D and 4.E. However, even identical spectra of the sample and the reference do not guarantee the satisfactory selectivity of measurement. The analyte may very often be accompanied by compounds with similar chromophores (starting materials or side products of synthesis, decomposition products, etc.), the spectra of which may be the same as or at least closely resembling that of the analyte. For instance, with the example shown in Figure 4.1, at any of the maxima of papaverine, its oxidation product, papaverinol containing a hydroxy group on the methylene group linking the isoquinoline skeleton and the dimethoxyphenyl ring, is measured together with papaverine.

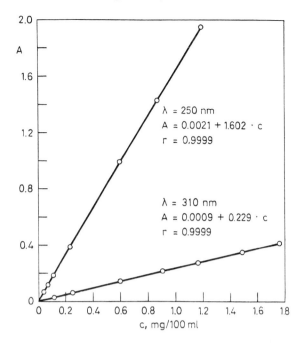

The graph shows:

$\lambda = 250$ nm
$A = 0.0021 + 1.602 \cdot c$
$r = 0.9999$

$\lambda = 310$ nm
$A = 0.0009 + 0.229 \cdot c$
$r = 0.9999$

c, mg/100 ml

FIGURE 4.1.   Calibration lines for the determination of papaverine hydrochloride. Solvent: 0.01 M hydrochloric acid (see spectrum on p. 75)

The spectrophotometric measurement of prednisolone at 242 nm, as prescribed by the pharmacopoeia, can indicate better quality than is the reality, since the most important contaminants of the product prepared by fermentation dehydrogenation, i.e., the hydrocortisone precursor and the by-products of fermentation: the 6β- and 20β-hydroxy-derivatives all give the same characteristic α,β-unsaturated ketone spectrum with a maximum at ca. 242 nm and very similar molar absorption coefficients.

Prednisolone   Hydrocortisone

The situation is similar with mestranol, which has a phenol-ether type spectrum (λ_{max} 278 nm; $A^{1\%}_{1\,cm} = 60$). The spectral properties of the impurities arising from the synthesis of this product, namely ethinyl oestradiol containing a free phenolic hydroxy group and oestrone methyl ether containing an isolated keto group instead of the ethinyl group, are very close to those of mestranol.

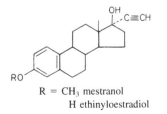

R = CH₃ mestranol
H ethinyloestradiol

With a third steroid-type compound, norethisterone, the case is more favorable. The spectral feature used for determinations is the band due to the 4-ene-3-keto group (λ_{max} 240 nm,

$A_{1\ cm}^{1\%}$ = 570), and, of the impurities, aromatic mestranol causes only a slight interference and the non-conjugated 4,5-dihydro derivative and the isomeric norethynodrel do not interfere with the analysis at all.

Norethisterone   Norethynodrel

Consequently, the selectivity of direct spectrophotometric methods should always be evaluated, and in the majority of cases it is insufficient. To improve the situation, either selective absorption is produced by means of chemical reactions (see Chapters 6 and 7), or the spectrophotometric measurement is made more selective by a preceding chromatographic separation (see Chapter 5).

B. DEVIATIONS FROM THE BEER-LAMBERT LAW

As mentioned in the previous section, in some cases deviations from the Beer-Lambert Law may occur, which can be traced back in part to optical and in part to chemical reasons.

Of the optical reasons the use of inadequately dispersed light is one possibility. As discussed in detail in Section 2.I, a wide spectral bandwidth decreases measured absorbance relatively to its true value. As the relative absorbance reduction is larger at high absorbances than at low ones, it may occur that the absorbance vs. concentration line flattens out beyond certain absorbance values. If, however, the construction of our instrument permits one to meet the requirement that the halfwidth of the analytical band is at least 5 to 10 times larger than the spectral bandwidth, this source of error can be eliminated.

As shown in Section 2.J, stray light may also cause absorbance to deviate from linearity. However, in the wavelength region above 220 nm generally employed in quantitative analysis, the validity of Beer's Law can usually be satisfied for absorbances lower than 1 to 1.5 even with the use of simple spectrophotometers.

As to the chemical reasons for the deviation, it should be borne in mind that the Beer-Lambert Law can only hold if all of the substance to be investigated is in the same molecular state at different dilutions. Deviations can occur if the substance is in an associated state in relatively concentrated solutions and the association equilibrium shifts toward dissociation with dilution, and the spectra of the monomeric and associated species are different. However, such interactions occur mostly in concentrated solutions; in the concentration range of 10^{-4} to 10^{-5} mol/l generally applied in quantitative UV-VIS spectrophotometry this phenomenon is very rare with drug substances. As a typical example, in relatively concentrated solutions of chloramphenicol and sodium benzoate or salicylate the formation of molecular complexes can be detected by NMR, in dilute solutions as used for UV measurements no such interaction has been found.[68]

It is a more important problem that owing to the increase in the degree of dissociation or hydrolysis with dilution, the spectra of spectrophotometrically active organic acids or bases and their salts gradually shift with dilution between the protonated and deprotonated forms, and this may also cause deviation from the Beer-Lambert Law. As the degrees or dissociation of weak and intermediate acids and bases change significantly with concentration particularly in the concentration region for quantitative measurements, these changes should be suppressed by the appropriate choice of conditions. As shown in Section 3.P, this is readily achieved by performing the measurement in strongly acidic or alkaline media or in buttered solutions.

This is illustrated by the determination of 2,4,6-trichlorophenol (pK_a = 6.00), which can be performed in 0.01 M hydrochloric acid containing 10% of ethanol at 286 nm (ε = 2.210) or in 0.01 M sodium hydroxide ($\varepsilon_{245\ nm}$ = 9.260; $\varepsilon_{312\ nm}$ = 4.840). If the measurement is done without the addition of acid or base in aqueous medium containing 10% of ethanol at 245 nm, the calibration

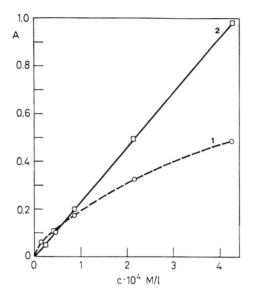

FIGURE 4.2. Deviation from Beer's Law due to acidic dissociation. The absorbance vs. concentration curves of 2,4,6-trichlorophenol. Solvent: water with 10% ethanol. 1: 245 nm; 2: 293 nm (isoabsorption point).

curve shown as Curve 1 of Figure 4.2 is obtained. The reason for the deviation from the straight line corresponding to Beer's Law is that whereas in the first measurement point the dissociation degree of trichlorophenol is ca. 0.21, this degree is reduced to ca. 0.05 in the case of the highest concentration measured. It can be seen from Curve 2 of Figure 4.2 that in the analysis of even this solution a linear calibration can be obtained if the measurement is performed at the iso-absorption point (at 293 nm) of the dissociated and undissociated forms.

The cases discussed so far suggest the conclusion that it is generally possible to eliminate deviations of absorbance vs. concentration curve from linearity, by optimizing the optical and/or chemical parameters of the measurement, particularly if the 1.5 to 4.0 absorbance region, which is available on the most expensive instruments only and is dubious for exact quantitative analysis, is not used. We feel it is important to emphasize this, since analysis can also be performed on the basis of non-linear calibration curves; up-to-date instruments with microcomputer support have programs for this purpose. However, if possible, one should avoid this situation, and base quantitative measurements on the more reproducible and reliable linear calibration.

C. BACKGROUND CORRECTION METHODS

1. Significance of Spectrum Background and of its Correction

It often occurs in the spectrophotometric analysis of drugs that the measurement of the absorbance of the component to be determined is subject to interferences by absorptions from other sources, and thus the determination cannot be carried out directly on the basis of the principles discussed so far.

This may be due to one or more other well defined, spectrophotometrically active components, the spectra of which more or less overlap with that of the analyte. The analysis of such binary or multicomponent systems is discussed in the next section (Section 4.D). This section deals with the equally important case when the spectrum of the analyte is readily recognizable in the spectrum of the sample, but it is more or less distorted, being superimposed on a featureless, unstructured spectrum, called background. This background is usually characterized by descending absorbance intensity in the direction of increasing wavelength. Over short wavelength ranges absorbance is an approximately linear function of wavelength, but over wider ranges it has a curvature (usually concave).

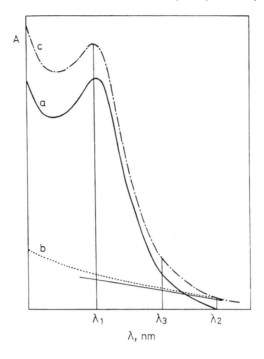

FIGURE 4.3.   Model spectra to illustrate background effect. a: component to be determined; b: background; c: spectrum distorted by background.

The presence of background may be due to different reasons. With samples of plant or animal origin, the undefinable accompanying substances of the biological sample form a background for the spectrum of the active substance. A similar situation may occur in the analysis of drugs in fermentation broths, mother liquors or raw products. With pharmaceutical preparations, the substances extracted from the excipients of tablet or ointment preparation and the oil component of injection products may be sources of background. The problem of background may also emerge when a small amount of spectrophotometrically active component (contamination, active principle of a pharmaceutical preparation) with a spectrum of relatively low intensity is to be determined in the presence of a large excess of a well defined component. The interferences due to background absorption can be eliminated by advanced techniques deserving separate discussion, like derivative spectroscopy (Section 4.F) and difference spectroscopy (Chapter 8). In addition, there are also simpler techniques available, which have the common feature that measurements are performed at other wavelengths as well as the absorption maximum, and background correction is done mathematically. It should be noted in advance that these methods are less accurate than the more advanced techniques, particularly those combined with chromatographic separation, and thus these simpler techniques are preferred only if background absorbance at the band maximum does not exceed 10 to 20% of the absorbance of the analyte. In certain cases (e.g., in the investigation of biological samples demanding lower accuracy) this restriction does not hold.

2.   Baseline Method and Other Simple Algebraic-Geometric Methods

Figure 4.3 shows a spectrum series that can be taken as typical. Curve "a" is the spectrum of the pure analyte, "b" is the background spectrum, which can usually not be recorded directly, and "c" is the spectrum distorted by the background. The simplest, but at the same time least effective method is to subtract from the absorbance measured at the wavelength of absorption maximum, λ_1, the absorbance measured at the wavelength, λ_2, where the absorption of pure analyte is zero.*

---

* As described in Section 4.G, this can be performed directly by means of a dual wavelength spectrophotometer.

For the calculation the specific absorbance of the pure analyte measured at wavelength λ_1 is used. The weakness of this method is that it takes no account of the variation of background in the $\lambda_1–\lambda_2$ interval. The associated error can be reduced by decreasing the range of this wavelength interval.[57] This is done preferably by looking for a wavelength on the descending branch of the spectrum (λ_3) of relatively low absorbance. If the absorbance difference between these two wavelengths is taken, the error arising from the neglect of wavelength dependence of background absorbance is reduced. Of course, in this case the difference between the specific absorption coefficients measured at λ_1 and λ_3 is used in the calculation. The application of this method is illustrated by several examples (including examples from various pharmacopoeias) in Chapter 10.

As also shown in Figure 4.3, a further correction method (assuming the recording of the complete spectrum) is to take the section of the spectrum in which the analyte has no absorption, i.e., the end of the spectrum of "pure" background (beyond λ_2). By extrapolating this section to wavelength λ_1, the correction term can be obtained directly.[118] An advantage of this method is that it attempts to take into account the change of background up to the maximum, but this cannot always be done exactly by extrapolation, and, in drawing a straight line, subjective elements can hardly be avoided.

Particularly after the color reactions of biological samples, the so-called Allen correction[5] is applied. The essence of this correction is that in addition to the absorption maximum, measurements are done at wavelengths equally spaced (Δ) to the left and to the right of the maximum, and the mean of the absorbances measured at the latter two wavelengths is subtracted from the absorbance measured at the maximum:

$$A_{corrected} = A - \frac{A_{\lambda+\Delta} + A_{\lambda-\Delta}}{2} \tag{4.3}$$

Of course, the calculation is performed with a specific absorbance corrected in the same manner. By this method constant background is fully corrected, and sloping background is corrected to sufficient accuracy if the band is symmetrical and the background is a linear function of wavelength in the $(\lambda - \Delta)$ to $(\lambda + \Delta)$ section. In this respect it is advantageous to choose as small Δ as possible, but in this case the reduction of corrected absorbance used in the calculations has adverse effects on the precision of analysis.

The base-line method is based on a similar principle as the Allen correction, but it is much more exact. The method was introduced by Morton and Stubbs for the determination of Vitamin A in cod-liver oil,[94,95] and then applied to the solution of several other problems of drug analysis.[10,76,80,96] An example of the latter is the determination of mestranol impurity and mestranol as tablet component in the company of large excess of 3-ketosteroids as shown by Legrand et al.[80] and Bastow.[10] The determination is based on the long-wavelength band of the phenol ether type spectrum of mestranol. The ketosteroids are reduced by potassium borohydride, since, however, they are present in an extremely large (100 to 1000 times) excess, their background spectrum must be corrected even after chemical reduction.

The determination is based on the following principle. In the spectrum of pure mestranol (Curve "a" of Figure 4.4) one wavelength each is selected on both sides of the 278 nm maximum, where the absorbances are identical (273 and 288 nm), then the ratio of absorbances measured at the higher wavelength and at the maximum is determined with high accuracy:

$$F = \frac{A_{288(St)}}{A_{278(St)}} = 0.76 \tag{4.4}$$

Then, the absorbances of the sample reduced with potassium borohydride (or, more preferably, sodium borohydride[56]) are determined at the same three wavelengths (Figure 4.4, Curve "b"). Due to the background, all the three values will be higher than the absorbances of mestranol at the same wavelengths. The correction at the absorption maximum consists of two terms: y is the value pertaining to the complete wavelength region, i.e., to 288 nm, too, and x is due to the rise of background between 288 and 278 nm. This value can be calculated from the difference between the absorbances measured at 273 and 288 nm on the basis of the relationship between the sides of similar triangles:

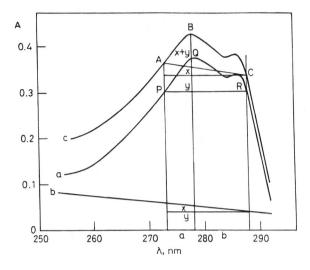

FIGURE 4.4. Determination of mestranol in the presence of a large excess of norethynodrel by Bastow's method,[10] using the baseline technique. Solvent: methanol. a: spectrum of mestranol (c = 3.86 mg/100 ml); b: background; c: spectrum of mixture. (From R.A. Bastow. *J. Pharm. Pharmacol.* 1967, *19*, 41.[10])

$$x = (A_{273} - A_{288}) \frac{288 - 278}{288 - 273} = 0.667(A_{273} - A_{288}) \qquad (4.5)$$

If x and y are taken into correction, Equation 4.4 will also hold for the absorbances of the sample with background absorption:

$$F = \frac{A_{288} - y}{A_{278} - (x + y)} = 0.76 \qquad (4.6)$$

Substituting x from Equation 4.4 and expressing y we obtain:

$$y = 2.055 A_{288} - 3.167 A_{278} + 2.111 A_{273} \qquad (4.7)$$

wherefrom the corrected absorbance at the maximum is:

$$A_{corrected} = A_{278} - (x + y) = 4.164 A_{278} - 2.776 A_{273} - 1.387 A_{288} \qquad (4.8)$$

From the corrected absorbance, by means of the specific absorbance of mestranol, the concentration of mestranol can be determined.

Attention is drawn to the paper of Traveset et al.,[124] in which the authors compare the baseline method and derivative spectroscopy (see Section 4.F) concluding that for several cases (e.g., the determination of phenobarbital in suppositories) the classical method yields results equivalent to the significantly more up-to-date and powerful technique.

3. The Method of Orthogonal Polynomials[83]

It is a common limitation of all background correction methods discussed so far that they take no account of the curvature of the background in the given wavelength region. This may cause significant error in the absorbance measurement of very dilute solutions with relatively strong background, such as, for instance, in the determination of the rate of the release of drugs incorporated into polymer films.[129] If the calculation of correction were based not only on the absorbances measured at three wavelengths but on a large number of absorbance values, for example at 2 to 4 nm intervals in the complete spectrum, then the correction could be made completely exact.

Obviously, with the increase in the number of data points, calculations become increasingly complicated, but these difficulties are easy to overcome by means of relatively simple computer programs.

With background correction methods based on several wavelengths it is generally assumed that the given section of the spectrum is a higher order function of the wavelength than the background. The classical case of calculations based on this principle is the determination of griseofulvin in fermentation broth using the method of Ashton, Brown and Tootil[7,8] and Daly,[27] in which it was assumed that background is a quadratic, whereas the spectrum is a cubic function of wavelength.

A much more widely applicable method, the method of orthogonal polynomials, has been introduced by Glenn and applied mainly by Egyptian authors for the solution of a number of practical problems.[53,132,137,138] Without going into the mathematical foundations and technical details of the method, the main features of the technique can be summarized as follows. The method is based on the fact that any function, thus the absorption spectrum, too, can be approximated in the vicinity of a given point as a sum of so-called orthogonal polynomials:

$$f(\lambda) = p_0 P_0 + p_1 P_1 + p_2 P_2 + \cdots + p_j P_j \tag{4.9}$$

where $f(\lambda)$ is the absorbance along the band, and the wavelengths are taken at equidistant (2 to 8 nm) intervals, although the method can be applied to non-equidistant intervals as well.[76] P_0, P_1, ... P_j are the orthogonal polynomials (constant, linear, quadratic, cubic, etc. functions of λ, generally up to P_5), and p_0, p_1, etc. are the coefficients of these functions, which depend on the concentration of the species to be determined. The latter are increased to various extents by background absorbance. As in the Ashton-Tootil method, a higher order polynomial should be taken for the spectrum of the analyte than for that of the background, the latter being at most quadratic. The system of orthogonal polynomials which gives the best description of the spectrum is selected empirically from the possible variants, the constants used in the calculations are taken from tables. The calculations, particularly with the use of computers, are rather straightforward, and thus the method is readily applicable for quantitative analyses performed on the basis of spectra with non-linear backgrounds, or even, in certain cases, for the simultaneous determination of two components with more or less overlapping spectra. In the latter case greater accuracy can be achieved than by measuring at two wavelengths and solving simultaneous equations with two unknowns.

The technique is illustrated by the method of Bedair et al.[13] used for the determination of diazepam in drug preparations, primarily in the presence of its hydrolytic decomposition product (2-methylamino-5-chloro-benzophenone, see p. 183) As can be seen in Figure 4.5, the hydrolytic product appears as background in the 272–360 nm region.

The coefficient of the combined polynomial is given in this case by Equation 4.10:

$$p_w = p_3 + 6 \, p_4 \tag{4.10}$$

The numerical value of p_w can be determined by means of Equation 4.11 from the absorbances measured at 8 nm intervals in the given region:

$$p_w = (1/36036)[A_0(+1122) + A_1(-750) + A_2(-1038) + A_3(-526)$$
$$+ A_4(-191) + A_5(+707) + A_6(+805) + A_7(+457) + A_8(-176) \tag{4.11}$$
$$+ A_9(-744) + A_{10}(-708) + A_{11}(+660)]$$

where A_0, A_1, A_2, etc. are the absorbances measured at 272, 280, 288 nm wavelengths, respectively. The regression equation used for the calculation of background-free concentration (g/100 ml) is

$$p_w = -0.0118 + 11.2050c \tag{4.12}$$

Of the large number of other applications, the simultaneous determination of phenol and adrenaline is cited as a classical application of the method.[137] Further applications include the measurement of salicylamide in the presence of chloroquine phosphate,[53] the determination of thiamine hydrochlo-

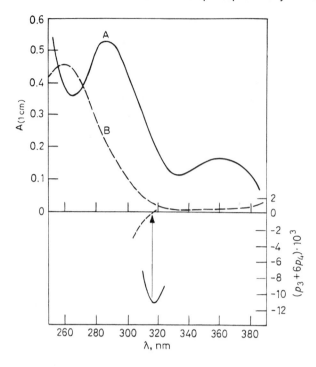

FIGURE 4.5. Spectra of diazepam (A) and its decomposition product (B) and the convolution curves (see text) calculated by the equation $p_3 + 6p_4$. Solvent: 0.1 M hydrochloric acid; $c = 0.001$ g/100 ml. (From M. Bedair; M.A. Korany; M.E. Abdel-Hamid. *Analyst* 1984, *109*, 1423.[13])

ride in the presence of its decomposition products,[135] the measurement of preservatives in Vitamin B_{12} injections,[52] the stability studies of methyl phenobarbitone,[136] and the investigation of atropine sulphate injection[131] and corticosteroid preparations.[6] A large number of further applications can be found in the various sections of Chapter 10. Special attention is merited by the paper of Bershtein and Lupashevskaya,[17] which compares the performances of various measurement and computer techniques based on the application of orthogonal polynomials in the analysis of antibiotics.

Background correction can be performed in basically the same approach, with the use of trigonometric functions instead of orthogonal polynomials, by means of Fourier series expansion. The two methods were compared for the determination of progesterone and testosterone in oily injection preparations by Wahbi et al.[133]

4. The Compensation Method

The compensation or curve inversion method used for the elimination of background effect is basically different in principle and in practice from the methods discussed so far.

The applicability of the method is essentially restricted to recording spectrophotometers although the application of single-beam instruments on the basis of the measurement at several wavelengths cannot be excluded. The method, introduced by Hiskey[66] and Tardiff[122] for drug analysis, is based on the following principle. The spectrum of the sample can be considered as the sum of the spectrum of the analyte with well defined maximum (maxima) and of a featureless background discussed above. If the spectrum of a solution of the sample is recorded against solutions of the analyte with known, gradually increasing concentrations, instead of the pure solvent as reference, then the spectrum of the analyte appears with gradually decreasing intensities until it vanishes and then, if the concentration of the analyte in the reference is increased further, the spectrum is inverted, i.e., an absorption minimum appears instead of the maximum. The concentration necessary in the reference solution to produce this inversion gives exactly the concentration of the analyte in the sample, and the spectrum observed at the inversion point is the pure background spectrum.

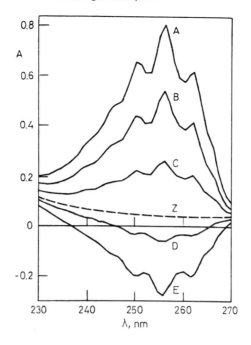

FIGURE 4.6. Determination of ephedrine hydrochloride in tablets by means of the compensation method. Solvent: 0.05 M sulphuric acid. A: the spectrum of the tablet extract; B to E: compensated and overcompensated spectra; Z: background. (From A.-A. M. Wahbi; M. Mahgoub; M. Barary. *Analyst* 1989, *114*, 505.[140])

The applicability of this ingenious and effective method has been obviously limited by its time consuming nature, since for the determination of the concentration at the inversion point a large series of dilutions should be prepared and measured. Almost the same amount of labor and time is required for the following alternative. Instead of a series of dilutions, one uses a variable path length cell or a flow cell with fixed thickness combined with a reservoir,[70] from which the concentrated reference solution can be fed between the individual spectroscopic measurements similarly to photometric titration.

The best solution, which could well lead to wider application of the method, would be a computer controlled version of the technique. An example of this has been described by Wahbi et al.[140] The example taken from their paper is shown in Figure 4.6.

The task was the determination of ephedrine hydrochloride in a tablet preparation after extraction with 0.05 M sulphuric acid. Curve A is the spectrum of the extract taken against 0.05 M sulphuric acid. The spectrum of a pure ephedrine hydrochloride solution of known concentration smaller than the expected concentration of the sample is also recorded and stored in the computer. The computer generates various versions of this spectrum by multiplying it by a gradually increasing factor and subtracts these versions from curve A, yielding curves B, C, D and E. The pure background spectrum corresponding to the inversion point is curve Z, which lies between curves C and D. The absorbance to be used for the calculation of ephedrine content can be determined either from the absorbance measured at the maximum of curve A using the corresponding value of curve Z as correction, or directly from the concentration of reference solution, taking into account the multiplication factor necessary to reach the inversion point.

The application of the method for derivative spectra[134,139] is also worth mentioning.

D. SPECTROPHOTOMETRIC ANALYSIS OF MULTICOMPONENT SYSTEMS

1. Determination by Means of Systems of Equations

A significant fraction of drug preparations contain more than one active principle to be determined. An important advantage of spectrophotometric determinations is, in contrast to the obviously more selective and sensitive separation methods, the simplicity of sample preparation and speed of measurement. It is, however, a disadvantage that even in the presence of two components a lengthy calculation is required and the simultaneous determination of several components requires complicated calculations which can be performed routinely only by means of computers.

The spectrophotometric measurement of multicomponent systems is principally possible since, to a good approximation, absorbance is an additive property in the concentration region usually applied in UV spectrophotometry. Thus, if in a system at a given wavelength several species absorb light, the measured absorbance is a sum of the absorbances of the individual species. In complicated systems, however, in which a large number of components is present or the concentrations of components are widely different, additivity should be checked by separate measurements and any deviations should be taken into account in the evaluation of results.

For the simultaneous determination of several components, the spectra of the pure components and their specific absorbances at all analytical wavelengths are always required. In many cases the accuracy of analysis can be increased by using the spectra of mixtures of known compositions.

In the analysis of binary mixtures the spectra of both pure components are recorded first, and two wavelengths are chosen for which the absorptions of the two components show maximum differences. In this selection preference is given to the band maxima of the components or to the regions in which only one of the components absorbs.* Then, at the selected wavelengths (λ_1 and λ_2) the specific absorbances of both components are determined. Since absorbances are additive, one may write:

$$A_1 = a_{1,\lambda_1}c_1b + a_{2,\lambda_1}c_2b \qquad (4.13)$$

and

$$A_2 = a_{1,\lambda_2}c_1b + a_{2,\lambda_2}c_2b \qquad (4.14)$$

where A_1 and A_2 are the absorbances of the mixture at wavelengths λ_1 and λ_2; a_{1,λ_1} and a_{1,λ_2} are the specific absorbances of component 1 at wavelengths λ_1 and λ_2; a_{2,λ_1} and a_{2,λ_2} are the specific absorbances of component 2 at wavelengths λ_1 and λ_2; c_1 and c_2 are concentrations of component 1 and 2, respectively; and b is cell thickness in cm.

In Equations 4.13 and 4.14 A_1 and A_2 are measured data, the specific absorption coefficients are constants, b is known from the experimental method, and c_1 and c_2 are the unknown concentrations to be determined.

It is theoretically possible in the analysis of binary systems that there are wavelengths in the spectra of both components where the other component does not absorb at all, and thus the concentrations of the two components can be determined independently by means of Equation 4.1. In practice, however, such a case never occurs.

In contrast, it is a frequent phenomenon that the spectra of the two components overlap only in part, i.e., there is a wavelength λ_1 where only one of the components has measurable absorption and the other does practically not absorb. Thus, the concentration of one of the components can be determined on the basis of the Beer-Lambert Law, by Equation 4.1, directly from the absorbance measured at wavelength λ_1.

This is the case with the determination of the drotaverine hydrochloride and nicotinic acid contents of Nicospan tablets. Since both components are very soluble in 0.1 M hydrochloric acid, the spectrophotometric measurement can be performed with a single solution.

As can be seen in Figure 4.7, drotaverine hydrochloride has three absorbance maxima in the UV region in the vicinity of 241, 303 and 354 nm. Nicotinic acid has a single maximum at ca. 260 nm (the maximum at ca. 215 nm is useless for analysis owing to the experimental errors occurring at low wavelengths). In the 354 nm region nicotinic acid has no appreciable absorption,

---

* Concerning the selection of optimum wavelengths in the analysis of multicomponent mixtures, attention is drawn to the paper of Bershtein.[16]

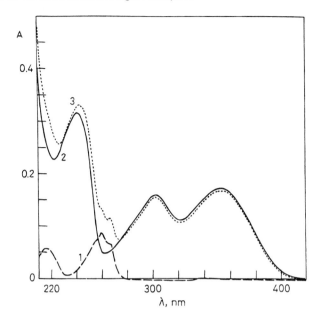

FIGURE 4.7. Spectra for the simultaneous determination of nicotinic acid (1) and drotaverine hydrochloride (2) in Nicospan® tablets (3). Solvent: 0.1 M hydrochloric acid. Concentrations: nicotinic acid 0.22 mg/100 ml; drotaverine hydrochloride 0.78 mg/100 ml; Nicospan® tablet: 2 mg/100 ml.

i.e., in Equation 4.13 $a_{2,\lambda1}$ is zero, therefore the concentration of drotaverine hydrochloride, c_1 can be determined from the absorbance measured at 354 nm:

$$c_1 = \frac{A_{354\ nm}}{a_{1,354\ nm}b} \tag{4.15}$$

In the possession of c_1, nicotinic acid content can be determined by means of Equation 4.16 from the absorbance measured at 260 nm:

$$c_2 = \frac{A_{260\ nm} - a_{1,260\ nm}c_1b}{a_{2,260\ nm}b} \tag{4.16}$$

The calculation is lengthier but still manually feasible if the spectra of the two components are completely overlapping, and at all measured wavelengths both components absorb. Then, concentration may be calculated from Equations 4.17 and 4.18:

$$c_1 = \frac{a_{2,\lambda_2}A_1 - a_{2,\lambda_1}A_2}{b(a_{1,\lambda_1}a_{2,\lambda_2} - a_{2,\lambda_1}a_{1,\lambda_2})} \tag{4.17}$$

$$c_2 = \frac{a_{1,\lambda_1}A_2 - a_{1,\lambda_2}A_1}{b(a_{1,\lambda_1}a_{2,\lambda_2} - a_{2,\lambda_1}a_{1,\lambda_2})} \tag{4.18}$$

The application of the system of equations with two unknowns is illustrated by the simultaneous determination of the two components of tablet Dolor®, aminopyrine and phenacetin.[55] In the

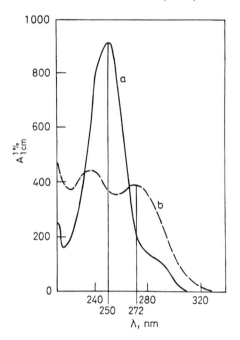

FIGURE 4.8. Spectra for the simultaneous determination of phenacetin (a) and aminopyrine (b). Solvent: ethanol.[54]

spectrum shown in Figure 4.8, the absorption maximum of phenacetin, 250 nm, and that of aminopyrine at 272 nm are chosen for the measurement. The specific absorption coefficients are, in ethanol solution:

	λ	A
phenacetin	250 nm	917
	272 nm	218
aminopyrine	250 nm	370
	272 nm	385

Substituting these values and the cell thickness (1 cm) into the above equations, the concentrations of the two components in the ethanol solution can be calculated as follows:

$$c \text{ (phenacetin)} = 0.00141 \cdot A_{250} - 0.00135 \cdot A_{272} \tag{4.19}$$

$$c \text{ (aminopyrine)} = 0.00337 \cdot A_{272} - 0.00080 \cdot A_{250} \tag{4.20}$$

It is noted that the third component of the tablet, ethylmorphine, does not interfere with the determination of the above two components, partly because of the relatively low content (phenacetine: 0.3 g/tablet, aminopyrine: 0.3 g/tablet, ethylmorphine: 0.02 g/tablet), and partly because of its much lower specific absorbance at the wavelengths of measurement. However, for this reason, ethylmorphine cannot be determined by direct spectrophotometric methods along with the other two components.

This method was described by Vierordt[130] back in 1873, and an extremely large number of applications can be found in the literature, both of the original method and of its more or less modified versions (see e.g., Refs. 12 and 13). Some of these applications are discussed in Chapter 10, including the method of Bayer[12] for the simultaneous determination of procaine and caffeine in injections, based on measurements at 273 and 290 nm. The analysis of another caffeine containing preparation, Acidotest I tablet, and of various sulphonamide mixtures was described by Milch et

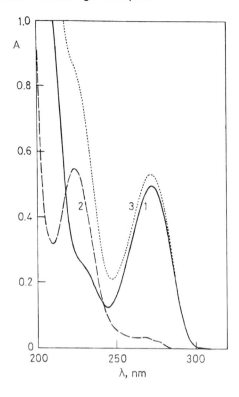

FIGURE 4.9. Spectra for the simultaneous determination of caffeine (1) and sodium benzoate (2) in Acidotest I tablets (3). Solvent: water. Concentrations: caffeine 1.00 mg/100 ml; sodium benzoate 0.994 mg/100 ml; tablet Acidotest I 2.00 mg/100 ml. (From Milch, Gy.; Csontos, A.; Borsai, M.; Vadon, E.; Mogács, I. *Acta Pharm. Hung.* 1966, *36*, 200.[88])

al.[87–89] As can be seen in Figure 4.9, the measurement at the absorption maximum of caffeine (273 nm) poses no technical problems, but at the maximum (224 nm) of the other component of the tablet, sodium benzoate, problems may occur since this falls on a steeply descending edge of the spectrum of the preparation. The resulting error can be reduced if, as suggested for single-component systems, the analysis is based on standards measured simultaneously instead of the specific absorbances determined prior to analysis.

It should finally be noted that by the extension of Vierordt's method to three wavelengths and the solution of a system of equations with three unknowns, the analysis of ternary systems can also be performed. As in the case of binary systems, as shown with the example of nicotinic acid–drotaverine system (Figure 4.7), otherwise tedious calculations can be greatly simplified if at the maximum of one of the three components (generally at the long wavelength end) the other two components have no absorption. Such a case is seen, for example, in the method for the simultaneous determination of Vitamins B_1, B_6 and B_{12} described by Machek.[82] Since the latter has a selective absorption at 361 nm, it can be determined separately here, and its absorption corrected at the maxima of Vitamins B_1, B_6 (235 and 292 nm). Then, the two vitamins can be determined by methods previously described in connection with binary systems.

Another similar example is the method of Sharma et al.[114] for the simultaneous determination of meclozine, caffeine and Vitamin B_6. The analytical wavelengths in 0.01 M sodium hydroxide solution are 230, 273 and 307 nm, the latter wavelength being specific for Vitamin B_6.

Systems where such simplifications cannot be made are more frequent in practice. The analysis of these three-component or more complex systems is generally done nowadays by computer techniques, thereby avoiding tedious calculations connected with the solution of multivariate systems of equations. This, in addition to reasons of convenience, significantly increases the selectivity and accuracy of the analysis. This statement also holds for the analysis of binary systems. These possibilities are discussed in the next section.

2. Analysis of Multicomponent Systems by Linear Combination

The possibility outlined in the previous section does not only mean that the complicated simultaneous equations occurring in the analysis of multicomponent systems are solved by computer. Modern mathematics and computer science offer much more prospects for the analyst. Even the task can be presented in a different way: the spectrum of a multicomponent system may be generated from the spectra of its pure components, or else, the spectra of the pure components could be deduced with high accuracy from the spectrum of the mixture.

From a mathematical point of view this is an approximation problem: a curve is to be generated as the sum of several curves such that deviations of the resulting curve from the measured one should be minimum. Since the additivity of absorbances holds for multicomponent systems, too, i.e., the spectrum of the mixture is a sum of the spectra of its components, the task is to find the concentration-proportional coefficients by which the individual spectra are multiplied before adding these spectra together and investigating the fit of the resulting sum to the experimental spectrum.

If the solution is based on linear combination, the values $c_1, c_2, \ldots c_n$ of the n components should be determined by solving the system:

$$A_1 = K_{1,1}c_1b + K_{2,1}c_2b + \cdots + K_{n,1}c_nb$$

$$\vdots \tag{4.21}$$

$$A_m = K_{1,m}c_1b + K_{2,m}c_2b + \cdots + K_{n,m}c_nb$$

The number of equations in the system is m, equal to the number of wavelengths for which absorbance values are taken. This must be at least equal to the number of components. The accuracy of the determination is increased if the measurement is performed using more points and in such cases the system is overdetermined. It will be shown later that up-to-date spectrophotometers equipped with computers enable a large number of data points to be recorded, even at nm intervals in the spectrum region investigated, and a very large system, consisting even of several hundred equations, can be solved by the computer.[19,98]

For the determination to be feasible, the following conditions must be met:

- the spectra of pure components must be available
- all components should obey the Beer-Lambert Law in the measured spectral range
- the spectrum of the multicomponent system must be representable as the sum of the spectra of the components
- spectral data should be available in digital form
- appropriate computer and program for the calculations should be available.

Concerning the spectra of pure components it is worth noting that the minimum number of standard spectra is equal to the number of components in the system, but the accuracy of analysis can be improved by increasing the number of standards. This can be done by taking the spectra of the same standard at different concentrations and in the neighborhood of the concentration of the analyte in the sample, or by preparing mixtures of known composition from the standards and measuring their spectra as well. The two methods may be combined, but the number of standards applicable is limited by the computer program. The more standards are used, the larger the dimensions of the matrices used by the program, and this requires larger computer memory and rapidly increasing running times.[108,142]

The application of the method is illustrated with the analysis of a three-component tablet, Quarelin. The tablet of 650 mg mass contains 400 mg of dipyrone, 60 mg of caffeine and 40 mg of drotaverine hydrochloride. The specific absorbances of the components in 0.1 M hydrochloric acid are:

	absorption maximum	
	λ_{nm}	$A_{1\ cm}^{1\%}$
dipyrone	259	270
caffeine	270	495
drotaverine HCl	354	235
	301	221
	241	435

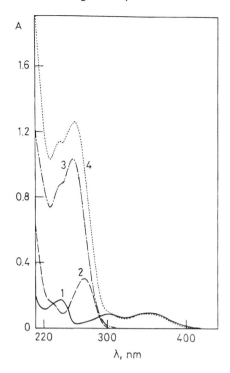

FIGURE 4.10. Spectra for the simultaneous determination of drotaverine hydrochloride (1), nicotinic acid (2) and dipyrone (3) in Quarelin® tablets (4) Solvent: 0.1 M hydrochloric acid. Concentrations: drotaverine hydrochloride 0.40 mg/100 ml; caffeine 0.60 mg/100 ml; dipyrone 4.00 mg/100 ml; Quarelin tablet 6.50 mg/100 ml.

As seen in Figure 4.10, the spectra overlap; the specific absorbances are similar in magnitude, and if the ratios of the components (10:1.5:1) are also taken into account, the conditions of analysis can be said to be poor.

It is quite evident from Figure 4.10 that only drotaverine hydrochloride has an absorption maximum suitable for independent determination, and below 310 nm the spectra of the three components overlap, and the spectrum of dipyrone practically covers the spectra of the other two components.

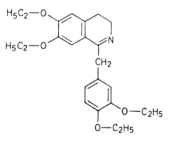

Drotaverine

First it should be investigated whether the Beer-Lambert Law is valid in the concentration range suitable for measurement. This can be done by linear regression analysis, which analyzes whether the measured points lie on the straight line described by the equation $A = a + bc$, or how far they deviate from it. The quality of the fit of measured points is most often investigated by the method of least squares fitting; the results of these calculations are shown in Table 4.1.

The concentrations of components vary to a certain extent in the process of drug production. This uncertainty is taken into account by the pharmacopoeia: a ±5% deviation from the declared drug content is usually permissible. On the other hand, it can be assumed that even with major fluctuations in technological parameters the deviation should not exceed double this permitted

TABLE 4.1. Linear Relation Between Absorbance and Concentration

Name of the substance	Absorbance maximum wavelength nm	Concentration μg/ml	Regression coefficients		Correlation coefficient r
			a	b	
Dipyrone	259	10–50	−0.053	0.026	0.9999
Caffeine	272	2.5–10.0	−0.060	0.048	0.9999
Drotaverine HCl	353	2.50–6.25	−0.042	0.023	0.9928
	303		−0.043	0.021	0.9873
	242		−0.060	0.042	0.9862

value, i.e., ±10%. Thus, the analyst should check the additivity of absorbances in a ±10% field, i.e., that there is no interaction between the components which would influence additivity.

For this investigation a series of experiments is performed in which the concentrations of the three components take the declared values, and in addition, 10% higher and lower values, respectively. The series of experiments contains all combinations of the possible concentrations (this is the so-called complete three-factor, two-level experimental plan, the three "factors" are the components, the two "levels" are the concentrations). By means of appropriate mathematical procedures the results of this series can be evaluated so that both the additivity of absorbances and the interactions between the components are revealed. There are several programs for this analysis.[32,113]

If the concentrations of the components are set to the ratios present in the Quarelin tablets, dipyrone to 30 μg/ml, caffeine to 4.5 μg/ml and drotaverine hydrochloride to 3.0 μg/ml, the absorbances fall into the measurable region. The concentrations of components in the individual experiments of the series are shown in Table 4.2.

In order to increase the reliability of the calculations, it is recommended that the measurement be performed on at least three solutions prepared independently.

Certain types of contemporary spectrophotometers are directly linked to a computer, and several programs are available for the necessary calculations from the spectra of standards and the measured data points.[19,108,142] For the analyst there is no need for special computer handling or programming knowledge. What is needed for the efficient use of the programs is an ability to exactly define the conditions of measurement and the kind of answer required from the calculations. We call the attention here to some simple but important points:

- when recording the spectra, data point density, i.e., the spacing between the data points should be chosen in a manner which makes the fine structure of the spectrum evident, but avoids the measurement of superfluous data points, since any increase in their number would increase calculation time. Spectra with sharp maxima require higher data point density then flatter spectra, but sometimes narrower region may suffice

TABLE 4.2. 2^3 Total Factor Plan

No. of experiment	Dipyrone μg/ml	Caffeine μg/ml	Drotaverine HCl μg/ml
0	30.0	4.50	3.00
1	27.0	4.05	2.70
2	27.0	4.05	3.30
3	27.0	4.95	2.70
4	27.0	4.95	3.30
5	33.0	4.05	2.70
6	33.0	4.05	3.30
7	33.0	4.95	2.70
8	33.0	4.95	3.30

TABLE 4.3. Determination of the Components of Quarelin Tablet

No. of experiment	Dipyrone concentration			Caffeine concentration			Drotaverine hydrochloride concentration		
	Set μg/ml	Measured μg/ml	Recovery %	Set μg/ml	Measured μg/ml	Recovery %	Set μg/ml	Measured μg/ml	Recovery %
0	30.00	30.573	101.91	4.50	4.357	96.82	3.00	3.082	102.73
1	27.00	27.510	101.89	4.05	3.871	95.58	2.70	2.746	101.70
2	27.00	27.477	101.77	4.05	3.871	95.58	3.30	3.406	103.21
3	27.00	27.527	101.95	4.95	4.825	97.47	2.70	2.748	101.78
4	27.00	27.547	102.03	4.95	4.778	96.53	3.30	3.351	101.55
5	33.00	33.470	101.42	4.05	3.953	97.60	2.70	2.801	103.47
6	33.00	33.520	101.58	4.05	3.834	94.67	3.30	3.377	102.33
7	33.00	33.713	102.16	4.95	4.750	95.96	2.70	2.784	103.11
8	33.00	33.750	102.27	4.95	4.750	95.96	3.30	3.376	102.30
Mean			101.89			96.24			102.46
Standard deviation			0.252			0.954			0.666
Variation coefficient %			0.25			0.99			0.65
Confidence interval (P = 0.95)			101.69–102.70			95.51–96.24			101.91–103.07

- different spectra in one series can be evaluated only if they have the same data point spacing
- the only spectral regions that can be used for the calculations are those in which both the standards and the multicomponent mixture absorb. For the selection of these regions trial calculations are recommended, to establish the region which gives the best fit. Some programs allow the spectral ranges containing little information to be skipped
- for the determination of components with widely different concentrations it is recommended to use the program option offering weight functions, which takes into account concentration ratios in calculating the results.

The operation and performance of one of the widely accepted programs will be described here emphasizing, however, that the application of a number of homemade and commercial programs can be found in the literature, and the task can be solved in several ways from the point of view of computing facilities.

Spectra can be recorded and stored in digital form, using commercially available programs, on computers coupled to Perkin-Elmer Lambda 5 UV-VIS spectrophotometer and the higher numbered instruments in this series. The program developed for the evaluation of multicomponent systems is capable of determining up to 10 components simultaneously. The program calculates the linear combinations of the spectra of the components, and fits the calculated spectrum to the measured one by a least squares method. The program looks for a linear combination of the spectra of the standards in which the deviation between the linear combination and the observed spectrum is minimum. The program calculates this error from the squared differences between the measured and calculated spectra summed over all data points (root mean square error).

If it is taken into account that the spectra of Quarelin are measured between 210 and 420 nm, and the resolution is the usual 1 nm applied in quantitative spectrophotometry, the program uses 211 data points which, as compared to the few data points used in the Vierordt method,[60] significantly increases the accuracy of the method. The calculation is refined further by the feature that the program uses a weight function which takes into account the ratio of components, thereby reducing the error due to the widely different concentrations.

The experimental results shown in Table 4.3 illustrate the accuracy to which the components could be quantitated in the mixtures of the series of known composition. As can be seen in the table, the recovery of dipyrone with the highest and drotaverine hydrochloride with the lowest concentration in the mixture is slightly higher whereas that of caffeine slightly lower than the true values. Nevertheless, the deviations fit into the accuracy limits described in the literature for the determination of multicomponent mixtures. The confidence intervals at 95% probability are very

TABLE 4.4. Evaluation of the Factor Plan by Three-way Analysis of Variance

Effect of the components	F-values calculated from the recovery		
	Dipyrone	Caffeine	Drotaverine HCl
Main effects			
×(1) Dipyrone	12319**	5.36*	0.91
×(2) Caffeine	0.85	2747**	2.36
×(3) Drotaverine HCl	9.48*	0.69	1845**
Interactions			
×(1)×(2)	0.53	4.18	1.65
×(1)×(3)	1.46	0.99	0.31
×(2)×(3)	0.99	0.99	0.20
×(1)×(2)×(3)	0.71	5.36*	0.96

$F_{(1,16)} = 4.49$; $P = 0.95$. * Significant, ** highly significant.

narrow for all components indicating that even if the concentrations deviate by $\pm 10\%$ from their declared value, the determination is sufficiently reproducible.

On the basis of the data shown in Table 4.3, the higher recovered concentrations of dipyrone and drotaverine hydrochloride and the lower concentration of caffeine appear to be a systematic error indicating an interaction, however weak, between the components. The extent of this interaction can be estimated by means of dispersion analysis. Programs for this analysis are available as integral parts of statistical program packages.

The analysis of the experimental plan by means of triple dispersion analysis yields the F-values which indicate whether the measured values of concentration are influenced significantly by the effects of components or the interactions between them. The program calculates the F-values which are well known statistical functions; the results of calculations are shown in Table 4.4.

By interpreting the data of the table it can be seen that the measured concentrations of the components are significantly influenced by their true concentrations, to an extent that the F-values characterizing this effect are three orders larger than those belonging to the concentrations of the other two components or to the interactions between the concentrations of components. It cannot be completely disregarded, however, that in addition to the above significant terms, the table contains other F-values above the significance level: the measured concentration of caffeine is affected significantly by the concentration of dipyrone and the interaction between the concentrations of the three components. The fact that the calculated caffeine concentration is somewhat lower than the true concentration may presumably be due to this effect. It is worth noting that the measured concentration of dipyrone is significantly affected by the concentration of drotaverine hydrochloride.

Consequently, it can be stated that the accuracy and reproducibility of calculation methods based on linear combinations and the confidence intervals of the results proved to be appropriate for the determination of the components of Quarelin tablet, although the measured concentrations of each component are slightly affected by the concentrations of other components.

The application of the method for the determination of Paniverin injection is also cited.[49] The injection contains, per milliliters, 40.0 mg of antipyrine, 32.1 mg of drotaverine hydrochloride and 8.8 mg of nicotinic acid, the ratio of components being ca., 10:8:2.2, i.e., more favorable than in Quarelin tablets. The spectra on which the analysis is based are shown in Figure 4.11. Of the components of Paniverin, the recovery of the least concentrated component, nicotinic acid, is better than that of the least concentrated component of Quarelin, caffeine (101.87% vs. 96.24%, see Table 4.5).

3. Other Computer Methods

In addition to the methods discussed so far, there are several computer techniques associated with expert systems[119] for the spectrophotometric analysis of binary and multicomponent systems. Thus,

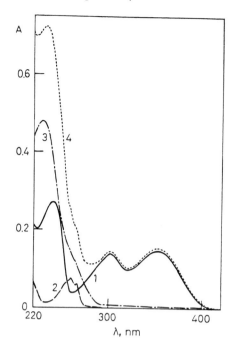

FIGURE 4.11. Spectra for the simultaneous determination of drotaverine hydrochloride (1), nicotinic acid (2) and antipyrine (3) in Paniverin injection (4). Concentrations: drotaverine hydrochloride 0.65 mg/100 ml; nicotinic acid 0.82 mg/100 ml; antipyrine 0.80 mg/100 ml. (From Géher, J.; Szabó, É. *J. Pharm. Biomed. Anal.* 1988, *6*, 757.[49])

TABLE 4.5. Determination of the Components of Paniverin Injection

Name of the component	Recovery %	Relative standard deviation %	Confidence interval (P = 0.95)
Antipyrine	101.61	0.47	101.5–101.7
Drotaverine HCl	98.20	0.90	97.6–98.7
Nicotinic acid	101.87	1.43	101.1–102.6

publications which compare the performance of these techniques are of particular interest. For example, the paper of Sala et al.[110] compares the performance of the multicomponent analysis of Hewlett-Packard, a program based on multiple regression and an iterative simplex optimization program, for the analysis of a 10:5:1 mixture of acetylsalicylic acid, acetaminophen and caffeine with overlapping spectra. The best results were obtained by the multiple regression program. It is particularly interesting in this comparison that the best result from an analytical point of view is not always provided by the calculation which gives the best fit. This, again, draws attention to the fact that computer techniques should not be used mechanically, their application always requires the consideration of analytical points of view and careful cross-checking.

The various spectrum convolution techniques,[141] in which complex spectra are constructed from simple curves (the spectra of the individual components, orthogonal and other polynomials, etc.) by means of algebraic-computer methods, have already been discussed in the earlier part of this section and in Section 4.C.

The opposite technique, deconvolution[20] applies when a complex spectrum is to be reduced into its components by computerized algebraic methods. Such a task is often encountered in the investigation of structure-spectrum relationships of molecules with complex chromophore systems, e.g., benzodiazepine derivatives,[54,91,93] for which the investigation is greatly facilitated if the spectrum is reduced into its components of Gaussian curve character. The method, as shown in Section 5.C, is important primarily for the investigation of peak homogeneity of substances separated

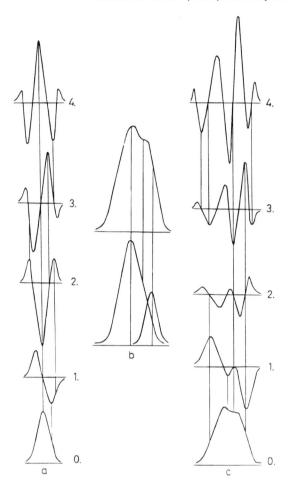

FIGURE 4.12. Derivative spectra. a: spectral band of Gaussian shape and its 1st to 4th derivatives; b: band arising from the overlap of two Gaussians of different intensities; c: composite band "b" and its 1st to 4th derivatives. (From G. Talsky; L. Mayring; H. Kreuzer. *Angew. Chem.* 1978, *17*, 785.[121])

by high performance liquid chromatography and in the quantitative evaluation of incompletely separated peaks.

Finally, digital filtering techniques,[71] primarily the application of Kalman filters,[48,61,63,103] is important for the treatment of background problems and the analysis of multicomponent mixtures. The discussion of these methods in full detail would exceed the scope of this book.

E. DERIVATIVE SPECTROSCOPY

1. Theoretical Background

For both background correction and the separation of overlapping bands for quantitative analysis, the most important development in the last decade was the introduction of derivative spectrophotometry into the practice of UV-VIS spectroscopy.

The principle of the method had been known for a relatively long time. Its recent propagation was made possible by the availability of spectrophotometers capable of recording the various derivatives generated either electronically or, more recently, by microcomputers as the spectrum is scanned.

In curves "a" of Figure 4.12 an idealized Gaussian absorption band (zero order derivative) is shown together with its first, second, third and fourth derivatives.

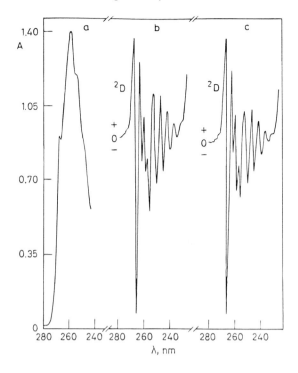

FIGURE 4.13. Identification of amphetamine analogs by derivative spectroscopy. Solvent: 0.1 M sodium hydroxide. a: common basic spectrum of amphetamine and phenylethylamine; b: second derivative spectrum of amphetamine; c: second derivative spectrum of phenylethyl amine (From A. H. Lawrence; J. D. McNeil, *Anal. Chem.* 1982, *54*, 2385.[79])

Since when derivatized the maxima and minima of the original function take zero values, and inflections are converted into maxima or minima, respectively, the derivative curves are much more structured than the original spectra. The first derivative, which is relatively the easiest to generate, has little practical significance; it hardly resembles the original spectrum: it takes zero value at the maximum of the original band and the band splits into two branches. The same applies to the other derivatives of odd order. The most widely used derivatives in practice are the second and fourth, though there are examples for the use of higher derivatives, too. As can be seen from the figure, the main peak is negative in the second derivative and positive in the fourth, and both derivatives contain a system of secondary peaks increasing in complexity with the order.

Excellent reviews of the theory and practical applications of derivative spectrophotometry have been published by Talsky et al.[121] and Fell.[41,44]

2. Derivative Spectrophotometry in Drug Analysis

Derivative spectrophotometry has led to advancements in UV/VIS spectroscopy compared with conventional spectrophotometry (often referred to as zero-order spectrophotometry in derivative spectroscopy publications) in four areas:

a. Qualitative Analysis. As already mentioned, derivative spectra are more structured than zero order spectra, enabling very tiny differences between the original spectra to be amplified. This holds particularly for substances containing isolated aromatic rings, the bands of which have fine structure. To illustrate this, not particularly widely used possibility, the spectra of amphetamine and phenylethylamine are shown in Figure 4.13 after Lawrence and McNeil.[79] Curve "a" shows the original spectra, the difference between them is so small that it cannot be seen in this plot. The second derivatives shown as curves "b" and "c" are also very similar, but in the intensity ratios of certain side maxima characteristic differences can be seen. The cited paper[79] shows fingerprint-like spectra for the identification of other amphetamine analog drugs, too. This technique was used successfully for the distinction and even for the quantitative analysis[78] of cannabinol and Δ^9-cannabinol which have very similar spectra. It is noted as a curiosity that the second derivative spectrum amplifies even the little fine structure of the weak n–π* band of the isolated keto group, which was utilized by Meal[86] for the identification of simple ketones.

Amphetamine   Methamphetamine

b. Removal of Monotonous Background Spectra in Quantitative Analysis.

One of the most important application fields of derivative spectrophotometry is background correction, which is in this case simpler and more effective than the methods discussed in Section 4.C. It follows from the rules of derivation that constant background is eliminated even in the first derivative:

$$\text{If A} = \text{const.,} \quad \frac{dA}{d\lambda} = 0$$

If background is a linear function of wavelength, the second derivative will vanish:

$$\text{If A} = a + b\lambda, \quad \frac{d^2A}{d\lambda^2} = 0 \tag{4.22}$$

In the case of fourth derivative, monotonous background curves corresponding to higher order functions may also be eliminated:

$$\text{If A} = a + b\lambda + c\lambda^2 + g\lambda^3, \quad \frac{d^4A}{d\lambda^4} = 0 \tag{4.23}$$

To solve relatively simple background problems, even the first derivative may prove sufficient, since with respect to short wavelength ranges background can be regarded as constant. This is the case e.g., in the determination of the active principle of spironolactone tablet (10 mg/tablet) as described by Uhlich.[125] Owing to excipients, the apparent active principle content calculated from the normal spectrum of a methanol extract (spectrum 4.14a) at the 242 nm maximum is 106.9%. As can be seen from curve "b" of Figure 4.14, the generation of the first derivative eliminates background practically fully in the neighborhood of the maximum and the active principle content calculated from the derivative spectrum is 100.8%.

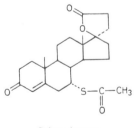

Spironolactone

A much more difficult background problem has been solved by means of the second derivative spectrum in the determination of flumecinol (Zixoryn) in an oily emulsion. Curves "a" and "b" of Figure 4.15 taken from Görög et al.[59] show the significant difference between the spectra of the preparation and the active ingredient. It can be seen from Figure 4.16 that the second derivative allows even this strong background to be eliminated, and quantitation can be based on this curve.

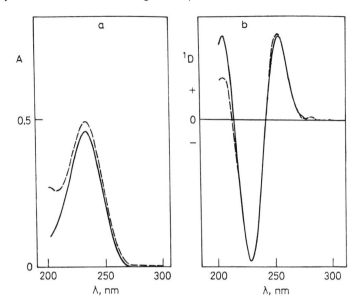

FIGURE 4.14. Determination of spironolactone in tablets by derivative spectrophotometric method. Solvent: methanol. a: basic spectra; —: spironolactone 1 mg/100 ml; – –: tablet extract; b: first derivatives of spectra "a". (From H. Uhlich. *Krankenhauspharmazie.* 1984 *5,* 167.[125])

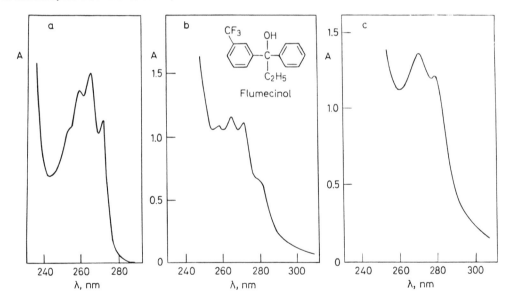

FIGURE 4.15. Determination of flumecinol in Zixoryn emulsion. Basic spectra. Solvent: 96% ethanol. a: Flumecinol (0.0376 g/100 ml); b: 1% Zixoryn emulsion diluted 50-fold; c: placebo emulsion diluted 25-fold. (From S. Görög; M. Rényei; B. Herényi. *J. Pharm. Biomed. Anal.* 1989, 7, 1527.[59])

The amplitudes on which the quantitative analysis can be based are also shown in the figure. It can be seen that either the amplitude taken from the baseline of the derivative spectrum, or the one taken from the baseline drawn between the two starting points of the band could be used. However, it is most generally the distance between a maximum and an adjacent minimum that is used. It is very important to note that although the amplitudes generally obey the Beer-Lambert Law, this should always be checked in any new problem. The signal on which quantitation is based has been denoted, after the suggestion of Fasanmade and Fell[36] by $^{n}D_{\lambda}$, with the upper left index giving the order of the derivative (1, 2, etc.), and the lower right index the wavelength where

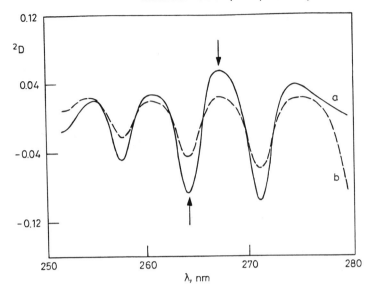

FIGURE 4.16. Determination of flumecinol in Zixoryn emulsion. Derivative spectra. a: 2nd derivative of curve a of Figure 4.15; b: 2nd derivative of curve b of Figure 4.15. (From S. Görög; M. Rényei; B. Herényi. *J. Pharm. Biomed. Anal.* 1989, 7, 1527.[59])

the signal was measured. If there is a single number here, it means that the signal is taken from the zero line at the given wavelength, if two numbers are given, the first is the wavelength of the maximum, the second is that of the adjacent minimum, and the quantitation is based on the distance between these points. Of the various possibilities offered by the spectrum in Figure 4.16, the amplitude between the 267 nm maximum and the 264 nm minimum in the second derivative ($^2D_{267,264}$) has been chosen for the quantitative analysis, since the slope of the response vs. concentration line is nearly the maximum and the intercept is the minimum for this value.

For the application of background correction in drug analysis, several further examples have been reported by Milch[90] and Uhlich.[125] The method of Hassan and Davidson[65] for the measurement of the active agent content of atropine tablets after precipitation with tetraphenyl borate and dissolution in dichloroethane is mentioned here. As can be seen in Figure 4.17, the fine structure of the low intensity bands of the isolated phenyl chromophores is hardly visible because of the background dominating the spectrum. However, the latter is completely eliminated in the second derivative spectrum, which is excellent for quantitation purposes.

This elegant possibility for background elimination can be utilized not only for light absorption. Background arising from light scattering can also be removed: derivative spectrophotometry is also applicable for the qualitative and quantitative analysis of opaque solutions.[40,99]

c. Elimination of an Overlapping Broad Band in the Quantitation of Substances with Narrow Bands.
A comparison of curve "c" with curves "a" and "b" in Figure 4.15 shows that in this case the background to be eliminated is due only in part to the background spectrum which is characterized by the descending edge, having a substantial contribution from a band with a maximum at 271 nm, which arises from sodium benzoate used as preservative. The successful elimination of the latter illustrates a not yet mentioned very important feature of derivative spectrophotometry: a possibility for the selective determination of a substance with a sharp absorption band in the presence of a substance which has a partly overlapping broad absorption band.

One of the most important basic equations of derivative spectrophotometry gives a relationship between the band intensity, A_n, of the nth derivative and the intensity (A_0) and band width (W) of the zero-order spectrum and the order, n:

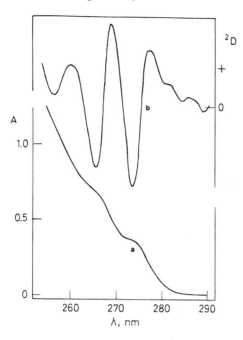

FIGURE 4.17. Determination of atropine in the form of its tetraphenylborate by derivative spectrophotometry. Solvent: dichloroethane; $c = 1.727 \cdot 10^{-4}$ M. a: basic spectrum; b: 2nd derivative spectrum. (From S. M. Hassan; A. G. Davidson. *J. Pharm. Pharmacol.* 1984, *36*, 7.[65])

$$A_n = k \frac{A_0}{W^n} \tag{4.24}$$

This relationship shows that the intensities of originally broad bands rapidly decrease with the order of the derivative, whereas the relative intensities of sharp bands increase. This effect is well demonstrated in Figure 4.18, where the zero-order spectrum (S) consists of two exactly coincident bands, X and Y with significantly different widths ($W_Y = 3W_X$). Performing the derivation of the resulting spectrum and the two components it can be seen that the intensity of the broad Y component decreases to barely 1/7 of its original intensity in the second derivative, and its fourth derivative merges almost completely into the baseline. Consequently, the derivatives of the X component and the resulting spectrum, S, coincide almost completely in the second order and completely in the fourth, i.e., not only continuous backgrounds but also broad bands can be eliminated by means of derivative spectrophotometry.

This provides an excellent opportunity for the qualitative and quantitative analysis of complex systems in which a sharp band or band system merges into a broad band since by generating an appropriately high order derivative, the broad band can be eliminated. It should be noted, however, that the generation of higher order derivatives has, as well as the advantages, several problems. The first is the reduction of signal-to-noise ratio, and another is that the side maxima may disturb the evaluation of adjacent peaks. It also follows from Equation 4.24 that derivative spectrophotometry cannot be used for the separation of overlapping bands with the same or closely similar band widths.

This method is significantly more applicable for the analysis of substances with spectra of narrow bands or band systems in the presence of substances with broad absorption bands even in large excess. It is particularly applicable for the determination of substances containing isolated phenyl chromophores. An example for this is presented in Figure 4.19, illustrating the method of Davidson and Elsheikh[29] for the determination of ψ-ephedrine in syrups in the presence of triprolidine hydrochloride and codeine phosphate. In curve "a" the α-band of the isolated aromatic ring of ψ-ephedrine, consisting of a low intensity system of sharp bands, is just discernible at the minimum of the spectrum. The fourth derivative shown as curve "b" is characteristic of ψ-ephedrine

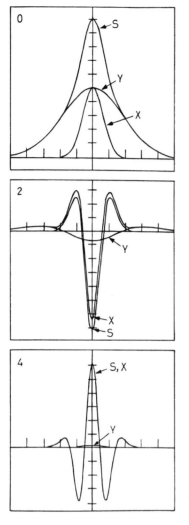

FIGURE 4.18. Elimination of a broad band (Y) during the measurement of a coincident narrow band (X) by means of the application of 2nd and 4th derivatives (S = X + Y). (From A. F. Fell, in *Amino Acid Analysis* J. M. Rattenbury, Ed.; Ellis-Horwood-John Wiley: New York, 1981; 86–118.[41])

only, the two intense and broad bands dominating the original spectrum could be eliminated. The relative standard deviation of this method is between 0.5 and 1%, indicating that the accuracy of derivative spectrophotometric method is similar to that of zero-order methods. However, in order to reach this accuracy, experimental conditions must be optimized carefully. It should be noted, for example, that the temperature dependence of derivative spectra is several times higher than those of zero-order.[28]

d. Analysis of Di- or Multicomponent Systems by the Combination of Derivative Spectrophotometry and Algebraic Methods.

Part "b" of Figure 4.12 shows the sum of two, partially overlapping curves of ideal, Gaussian shape, together with their first to fourth derivatives. It can be seen that the two curves can be separated in the second to fourth derivatives, which can be utilized in the quantitative analysis of the two components. This makes it possible to separate bands for quantitative analysis problems which could not be solved directly by the Vierordt method.

The bands on which practical analysis is based can generally not be regarded as strictly Gaussians, thus a pure band separation can scarcely be performed. The derivative spectra (particularly the higher order ones) have several maxima and minima arising from their structured nature. Thus it is often easy to find wavelength positions where the derivative of the signal of the analyte is maximum or closely maximum, whereas that of the other component is zero, and vice versa. These positions can be used for the selective determination of the individual components. Even if such

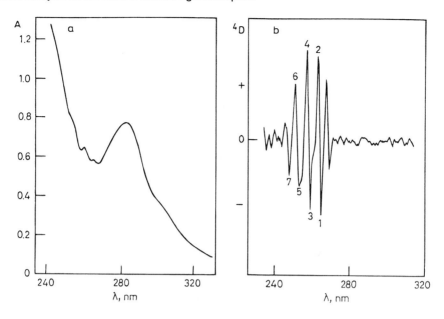

FIGURE 4.19. Determination of ψ-ephedrine in a syrup by means of derivative spectrophotometry. a: the basic spectrum of syrup containing ψ-ephedrine, codeine phosphate and triprolidine, in dilute hydrochloric acid, after purification by multiple extraction; b: fourth derivative spectrum. (From A. G. Davidson; H. Elsheikh. *Analyst.* 1982, *107*, 879.[29])

wavelength pairs cannot be found, the application of algebraic methods is still much more favorable with derivative spectra. For both techniques, characteristic examples will be given below.

The first, so-called zero-crossing method is illustrated by the analysis described by Berzas Nevado et al.[18] Figure 4.20a shows the spectra of sulphathiazole and sulphanilamide, Figure 4.20b shows the third, Figure 4.20c the fourth derivatives. As can be seen, amplitudes $^3D_{268}$ and $^4D_{278}$ are suitable for the determination of sulphanilamide, whereas $^3D_{297}$ and $^4D_{290.6}$ for the determination of sulphathiazole without the interference of the other component.

In connection with curves "b" and "c" of Figure 4.20 it is to be noted that they are, in apparent contrast to the above, not nearly as structured as could be expected for a 3rd or 4th derivative according to the principles discussed so far. The reason is that a very large $\Delta\lambda$ value was selected by the authors for the generation of the derivative $\Delta A/\Delta\lambda$ (28 nm for the third and 54 nm for the fourth derivative). The shape of the derivative functions can be affected by the value of $\Delta\lambda$ in a similar manner, but to a much greater extent, as by the effect of slitwidth on the original spectrum (see Section 2.K). The statements made so far referred to $\Delta\lambda = 1$ nm. The "smoothing" of derivative spectra with increasing $\Delta\lambda$ may be, in certain cases, of definite advantage from the point of view of quantitative analysis, but $\Delta\lambda$ is a generally important parameter in derivative spectroscopy, the optimization of which may significantly increase the selectivity of the method or the accuracy of quantitation.

To illustrate the application of algebraic methods, the method of Vetuschi and Ragno[126] for the simultaneous determination of atenolol and chlorothalidone is presented first. From the 4th derivatives of the very similar spectra shown in Figure 4.21a the concentration of atenolol can be determined selectively at 236 nm with a small intercept:

$$c_{atenolol} = 0.0160 \cdot {}^4D_{236} + 0.6575$$

$$\text{with } r = 0.99995 \tag{4.25}$$

The concentration of chlorothalidone can be determined from the amplitude between 229 and 218 nm by means of the equation:

$$c_{chlorothalidone} = 0.03903 \cdot {}^4D_{229,218} - 0.06604 \cdot {}^4D_{236} \text{ with } r = 0.99988 \tag{4.26}$$

Concentrations are expressed in mg/1, amplitudes ^4D are taken as mm read from the spectra.

The introduction of multiwavelength computer methods (see p. 108) may be a significant step forward in this field, too. To illustrate this, the method of Hoover et al.[67] for the simultaneous determination of ψ-ephedrine hydrochloride and chlorpheniramine maleate in tablets is shown. The direct analysis on the basis of the very similar original spectra shown in Figure 4.22a is impossible, but the first derivatives shown in Figure 4.22b are applicable if an appropriate computer program is used. Since the spectra were taken by a diode array instrument (see p. 22), this method, owing to its high speed, could also be used to study dissolution rates.

Other algebraic methods used for zero-order spectrophotometry, like the ratio method (p. 126), compensation method (p. 102), etc., may also be used in derivative spectroscopy; the combined application leads to a significant increase in selectivity.[84,100,134]

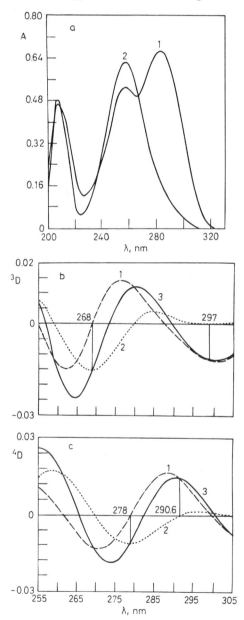

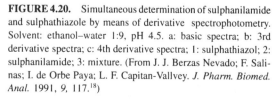

FIGURE 4.20. Simultaneous determination of sulphanilamide and sulphathiazole by means of derivative spectrophotometry. Solvent: ethanol–water 1:9, pH 4.5. a: basic spectra; b: 3rd derivative spectra; c: 4th derivative spectra; 1: sulphathiazol; 2: sulphanilamide; 3: mixture. (From J. J. Berzas Nevado; F. Salinas; I. de Orbe Paya; L. F. Capitan-Vallvey. *J. Pharm. Biomed. Anal.* 1991, 9, 117.[18])

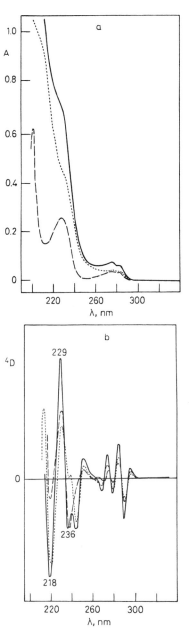

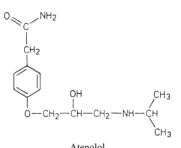

FIGURE 4.21. Simultaneous determination of atenolol and chlorthaildone by means of derivative spectrophotometry. Solvent: 95% ethanol. a: basic spectra; b: 4th derivative spectra; ----: atenolol; —: chlorthalidone; —: 1:1 mixture. (From C. Vetuschi; G. Ragno. *Int. J. Pharm.* 1990, *65*,177.[125])

Atenolol

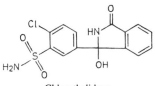

Chlorothalidone

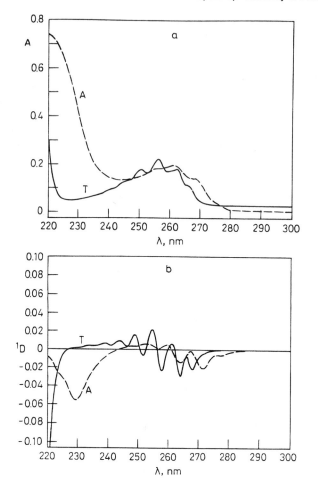

FIGURE 4.22. Simultaneous determination of ψ-ephedrine hydrochloride and chlorpheniramine maleate in drug preparations by means of derivative spectrophotometry. Solvent: water. a: basic spectra; b: 1st derivative spectra; ――: chlorpheniramine
maleate; ——: ψ-ephedrine hydrochloride. (From J. M. Hoover; R. A. Soltero; P. C. Bansal. *J. Pharm. Sci.* 1987, *76*, 242.[67])

3.   Difference Derivative Spectrophotometry*

Of the combined techniques, difference derivative spectrophotometry is discussed in this section. In
this method the derivatives of the difference spectra based on chemical changes, discussed in Chapter
8, are generated to significantly increase selectivity. The textbook example for the technique is the
method of Davidson and Mkoji[31] for the simultaneous determination of triprolidine, ψ-ephedrine
and dextromethorphan in drug preparations. The basic spectra show significant overlap, the 0.1 M
sulphuric acid solution taken against the 0.1 M sodium hydroxide solution as reference shows intense
difference spectrum only in the case of triprolidine which has a pyridine chromophore: the quantitative
analysis of the latter can be carried out at 301 nm. With the two other compounds, in which the
protonated nitrogen is far from the aromatic ring, the intensity of the difference spectra is low. However, due to their highly structured nature, they are excellent subjects for derivative spectrophotometry.
The contribution of triprolidine with a broad band difference spectrum is eliminated in the second
and particularly in the fourth derivative spectra, and thus it is easy to find wavelength pairs at which
the simultaneous determination of ψ-ephedrine and dextromethorphan is possible with the complete
exclusion of mutual interference. The method is illustrated in Figures 4.23–4.26.

---

* Owing to its great significance, and since it is based on the chemical reactions discussed in Chapters 6 and 7, difference
spectrophotometry as such will be dealt with separately, in Chapter 8.

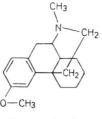

Dextromethorphan

Finally, the paper of Fell[42] should be mentioned, which presents the difference derivative spectrophotometric version of the sodium borohydride difference spectrophotometric determination of the active ingredient of norgestrel tablets.

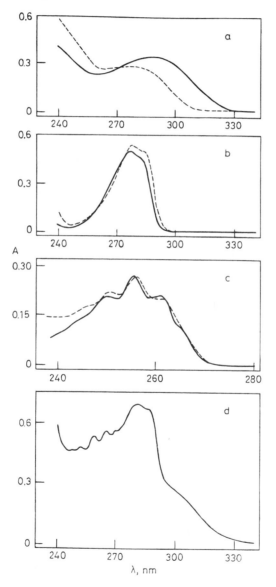

FIGURE 4.23. Determination of triprolidine, ψ-ephedrine and dextromethorphan by difference-derivative spectrophotometry. Basic spectra. a: triprolidine hydrochloride, 12.5 mg/l; b: dextromethorphan hydrobromide 100 mg/l; c: ψ-ephedrine hydrochloride 300 mg/l; d: combined preparation; —: 0.1 M sulphuric acid solution; – –: 0.1 M sodium hydroxide solution. (From A. G. Davidson; M. M. Mkoji. *J. Pharm. Biomed. Anal.* 1988, *6*, 449.[31])

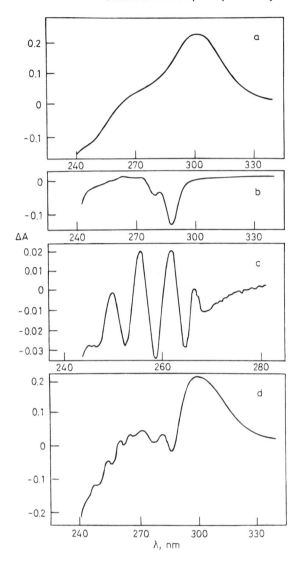

FIGURE 4.24. Determination of triprolidine, ψ-ephedrine and dextromethorphan by difference-derivative spectrophotometry. The difference spectra (0.1 M sulphuric acid solutions recorded against 0.1 M sodium hydroxide solutions). For the notation of curves and literature source, see Figure 4.23.

4. Practical Applications

In addition to the above examples, derivative spectrophotometry has been used for the solution of several other problems of pharmaceutical analysis. In this respect, the investigations of Fell can be regarded as pioneering. He applied this method, for instance, for the determination of amino acids in proteins[41] and enzymes,[39] for the determination of hyoscine hydrobromide, pilocarpine, etc.[37] in drug preparations containing conserving agents and other interfering components, as well as for the investigation of phenol and aromatic alcohols[38] and phenothiazine derivatives.[36]

The successful applications of this method in clinical and toxicological analysis[25,45,47,51,99,104] are also worth noting. For example, paraquat could be determined in biological samples in the ng/ml concentration range by taking the second and fourth derivative spectra after ion pair extraction and dithionite reduction.[45]

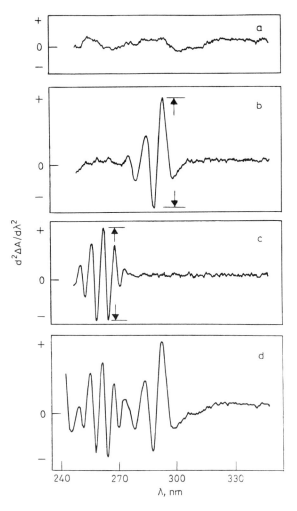

FIGURE 4.25. Determination of triprolidine, ψ-ephedrine and dextromethorphan by difference-derivative spectrophotometry. The 2nd derivatives of difference spectra. For the notation of curves and literature source, see Figure 4.23.

It is interesting to note some examples where this technique has been used for the analysis of impurities, thus, for instance, for the determination of 0.05% benzaldehyde as an impurity in benzyl alcohol.[33] A similar example is the selective determination of decomposition products in a matrix of undecomposed drugs. Of these, the determination of the oxidative and hydrolytic decomposition products of phenylbutazone,[106] the selective measurement of the photochemical decomposition product of nifedipine,[22] the determination of salicylic acid in acetylsalicylic acid,[72,85] the measurement of 2-aminopyridine in piroxicam,[23] and the determination of 5-fluorouracyl in 5-fluorocytosine[24,107] are examples. In other cases, derivative spectrophotometric methods have been applied in stability tests by ensuring the selective measurement of undecomposed substance in the presence of decomposition products, such as diazepam, oxazepam, other benzodiazepines,[26] paracetamol,[77] phenacetin,[77] glibenclamide,[14] mebeverine,[14] clopamide,[14] isothipendyl,[12] dimethothiazine[2] and 5-fluorocytosine.[24]

The main field of application of derivative spectrophotometry is the determination of the active ingredient content of drug preparations. In addition to the numerous examples discussed so far, one may mention the determination of several drugs containing phenyl chromophore[3,30] (cinnarizine, fenfluramine, phenyltoloxamine, dextropropoxyphene, tetrahydrozoline, tilidine, xylometazoline,

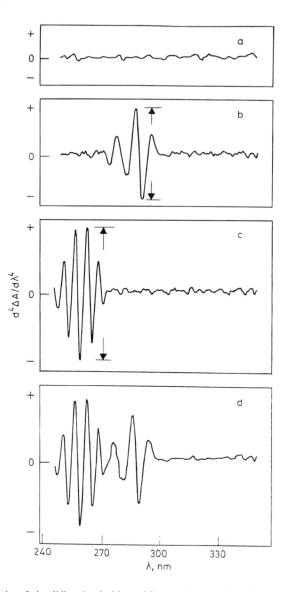

FIGURE 4.26. Determination of triprolidine, ψ-ephedrine and dextromethorphan by difference-derivative spectrophotometry. The 4th derivatives of difference spectra. For the notation of curves and literature source, see Figure 4.23.

benzatropine, bretylium tosylate, scopolamine, etc.), and the determination of acetaminophen and sodium salicylate,[123] homatropine methylbromide,[75] sodium fusidate,[64] oxprenolol and chlorthalidone,[128] sodium salicylate, caffeine, metronizadole benzoate, diazepam, chloroquine phosphate, paracetamol, etc.,[34] thiamphenicol glycinate,[98] salbutamol,[62] cilistatin sodium,[46] methadone and orphenadrine[21] in various drug preparations. The main features of some recently published methods are shown in Table 4.6.

The paper of Fabre et al.[35] is the first attempt to review the special aspects of the validation of derivative spectrophotometric methods. Of these the matrix suppression, light scattering suppression and recovery tests merit special attention. A number of further derivative spectrophotometric methods will be shown or mentioned in the various sections of Chapter 10.

TABLE 4.6. Some Derivative Spectrophotometric Methods

Compound	Matrix	Derivative signal	First author	Year	Ref.
		Single component (+ background)			
Cloctofol	suppository	$^2D_{298,286}$ or $^3D_{289,301}$	Vetuschi	1988	127
Clemastine fumarate	tablet, syrup, injectable	$^2D_{272}$	Bedair	1988	15
Domperidone	tablet, drops, suppository, suspension	$^1D_{294}$	Mohamed	1989	92
Miconazole	cream (+ benzoic acid), powder	$^2D_{285.2,280.8'}$ $^2D_{280.8,282.8'}$ $^3D_{274}$	Cavrini	1989	25
Econazole		$^2D_{285.2,281.2'}$ $^3D_{282.8'}$ $^3D_{279.6}$			
Mestranol	impurity in norethisterone	$^2D_{291,287}$	Görög	1989	59
2,6-Diisopropyl phenol	oil-in-water emulsion	$^2D_{286}$	Bailey	1991	11
Homatropine HBr, atropine sulfate	eye drops	$^2D_{257,254}$	Leung	1991	81
		Two components (+ background)			
Naphazoline + diphenhydramine	nasal drops	$^2D_{288,282}$ $^2D_{249}$	Santoni	1989	111
Tranylcypromin sulfate + trifluoperazin.2HCl	tablet	$^4D_{271}, ^1D_{262}$	Knochen	1989	74
Rufloxacin + rufloxacin-1-oxide		$^4D_{251,257'}$ $^4D_{265,261}$	Quaglia	1991	105
Nortriptyline.HCl + perphenazine	tablet	$^4D_{239.6'}$ $^4D_{268.8}$	Atmaca	1991	9
Amoxicillin + dicloxacillin	capsule	$^1D_{234'}$ $^1D_{225}$*	Abdel-Moety	1991	4
Imipramine + amitriptyline	serum	$^2D_{254'}$ $^2D_{268}$*	Garcia-Fraga	1991	47
Chlordiazepoxide** oxepam	urine	$^2D_{309'}$ $^2D_{316.2}$	Corti	1991	26

*Simultaneous equations with two unknowns.
**Data on several other benzdiazepines and their degradation products.

F. OTHER METHODS

1. Ratio Method for the Analysis of Binary Systems

The measurement of binary systems at two wavelengths is often achieved by the ratio method.[101,102,143] Of the numerous variants of this method the one discussed here is that in which the absorbances of the model mixtures with varying compositions of the two analytes are determined at two wavelengths (generally at the band maxima), then the absorbances measured at the two wavelengths are divided for all mixtures and these ratios are plotted against the composition of the mixture. The ratio R of the absorbances A_1 and A_2 measured at the two wavelengths can be expressed by dividing Equations 4.13 and 4.14 and simplifying with path length:

$$R = \frac{A_1}{A_2} = \frac{a_{a,\lambda_1} \cdot c_1 + a_{2,\lambda_1} \cdot c_2}{a_{1,\lambda_2} \cdot c_1 + a_{2,\lambda_2} \cdot c_2} \qquad (4.27)$$

The percentage amounts of components 1 and 2 are (taking the sum of the two concentrations as 100%):

$$\%_1 = \frac{100c_1}{c_1 + c_2}; \quad \%_2 = \frac{100c_2}{c_1 + c_2} = 100 - \%_1 \tag{4.28}$$

Consequently, Equation 4.27 can be converted into the following form:

$$R = \frac{(a_{1,\lambda_1} - a_{2,\lambda_1})\%_1 + 100a_{2,\lambda_1}}{(a_{1,\lambda_2} - a_{2,\lambda_2})\%_1 + 100a_{2,\lambda_2}} \tag{4.29}$$

Equation 4.29 determines a curve which can also be constructed by using the four absorption coefficients. Either constructed this way or plotted empirically from the ratios measured for the model mixtures, the calibration curve can be used to determine the percentage ratio of the two components of an unknown mixture from the measured absorbance ratio. If the second wavelength is not the band maximum of the second component but the wavelength corresponding to the crossing point of the two spectra (isoabsorption point), i.e., $a_{1,\lambda_2} = a_{2,\lambda_2}$, Equation 4.29 can be simplified into the following form:

$$R = \frac{(a_{1,\lambda_1} - a_{2,\lambda_1})\%_1 + 100a_{2,\lambda_1}}{100a_{2,\lambda_2}} \tag{4.30}$$

which represents, instead of a curve, a straight calibration line. Either Equation 4.29 or 430 is used, the ratio method may be used for the analysis of binary mixtures as an alternative to the method based on the solution of a system of equations discussed in Section 4.D.

An advantage of the ratio method is that in cases when one of the components is in large excess over the other analyte (analysis of impurities), it gives more accurate results than the method based on the solution of the system of equations. Bayer[12] applied this method successfully for the determination of cyanocobalamine (Vitamin B_{12}) as an impurity in hydroxocobalamine of analogous structure. In pure hydroxocobalamine the ratio of absorbances measured at 361 and 351 nm is 0.61, and the formula for the determination of cyanocobalamine impurity exceeding 1% is:

$$\% = \frac{190 - 120R}{0.78R + 0.61} \tag{4.31}$$

This formula was obtained by expressing $\%_1$ from Equation 4.29.

Equations 4.29 and 4.30 do not contain the absolute concentrations of the analytes, resulting in the most important advantage of the ratio method: to determine the ratio of the two components there is no need for accurate weighing. This is particularly significant if the sum of the concentrations of the two components is constant and known (e.g., when a chemical reaction is followed). This makes it possible to rapidly monitor the progress of a number of reactions used in the pharmaceutical industry. For example, the oxidation of 4-ethyl pyridine (λ_{max} 252 nm) into pyridine-4-carboxylic acid (i.e., isonicotinic acid, λ_{max} 272 nm) is easy to follow by measuring the ratio of absorbances at these two wavelengths: the samples taken without weighing from the reaction mixture can be measured directly after dilution.[56] The measurement of the ratio of absorbances of 4-ene-3-keto-steroids at 240 nm (band maximum) and of 1,4-diene-3-keto derivatives, obtained from them by microbiological oxidation, at 266 nm (shoulder) offers an excellent method for the analytical monitoring of fermentation.[69,73,109] If in such cases absolute concentrations are also required, this is readily calculated from the absorbance measured at the isoabsorption point.

Some further examples for the application of ratio method: the simultaneous measurement of tetracaine and benzocaine hydrochloride on the basis of the ratio of absorbances measured in the absorption maxima at 311 and 285 nm,[102] the simultaneous determination of theophylline and phenobarbital on the basis of the ratio $A_{274.5}/A_{252.5}$, where the two wavelengths are the absorption maximum of theophylline and the isoabsorption point, respectively, in pH = 9.5 borate buffer,[143]

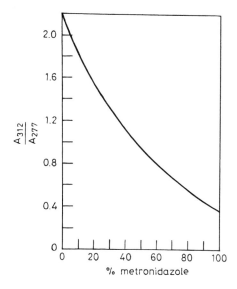

FIGURE 4.27. Calibration curve for the simultaneous determination of metronidazole and its 4-nitro isomer by the ratio method. Solvent: 0.03 M aqueous hydrochloric acid. (From Görög, S.; Csizér, É. *Proc. 3rd Anal. Conf. Budapest, Vol. 1,* Akadémiai Kiadó: Budapest, 1970; 65.[58])

the simultaneous determination of procaine and antipyrin in ear drops on the basis of the absorbance ratio of band maxima at 290 and 242 nm.[112]

Demanding problems like the simultaneous determination of metronidazole (1-hydroxyethyl-5-nitro-imidazole) and its 4-nitro isomer can also be solved by means of the ratio method. The spectra of the two isomers are essentially the same both as free base and in protonated form the only difference being that, owing to the different base constants (pK_b = 11.44 for metronidazole and 13.49 for the 4-nitro isomer), metronidazole is protonated at lower acidity than its isomer. Thus at appropriately selected acidity (0.03 M hydrochloric acid) the majority of metronidazole molecules are protonated, whereas those of the isomer are in the deprotonated state. If the absorbance of the isomeric mixture is measured at this acidity at the maxima of the protonated and deprotonated forms, the isomeric composition can be determined from the ratio by means of the calibration curve[58] shown in Figure 4.27.

2. Dual Wavelength Spectrophotometry

Spectrophotometers equipped with two monochromators are also commercially available. If the two monochromatic beams are passed alternately through the *same* cell, alternating in a way described for the double beam spectrophotometers, the detector output is directly the difference between the absorbances of the sample as measured at the two wavelengths.[44,115,116] Dual wavelength spectrophotometry offers advantages for several reasons. If the two wavelengths are chosen appropriately, e.g., one at the absorption maximum and the other on the long wavelength end of the spectrum where the sample has no absorption, very small absorbances can be determined in the presence of constant background or even of impurities causing opalescence. If it can be ensured that the absorbance of the solvent is the same at the two wavelengths concerned (colorimetry), no reference solution is required. It is interesting to note that if the two wavelengths are set close to one another and the spectrum is taken with this constant wavelength difference, the first derivative spectrum is obtained. (For other possibilities, see Section 4.E.E.)

Of the fairly scarce applications in drug analysis, the method of Shively and Simonelli[117] is noted for the determination of the dissolution rate of hydrocortisone microcapsules without the removal of the vehiculum of the capsules. The monochromators are set to 248 and 277 nm in the analysis (the absorption maximum and the cut-off wavelength of the spectrum of active ingredient). In their model experiments, the authors found the ΔA values of hydrocortisone acetate solutions constant with respect to time, in the concentration range of 0.05 to 0.6 μg/ml, even if a significant amount of active carbon settled from the disturbed solution.

3. Precision Spectrophotometry

Particularly for the investigation of strongly absorbing solutions, the accuracy of absorbance determination can be increased by taking a solution of the analyte of accurately known concentration but below the unknown concentration as reference, instead of the pure solvent. From the concentration of this reference and the now accurately measurable small absorbance, the concentration of the unknown is easy to calculate. A review of this technique together with other unusual techniques can be found in the book of Stern and Timmons.[120]

It should be noted here that of these spectrophotometric techniques difference spectrophotometry has particular significance in drug analysis. Therefore, this method will be discussed separately in Chapter 8.

REFERENCES

1. Abdel-Hamid, M. E.; Abuirjeie, M. A. *Analyst* 1988, *113*, 1443.
2. Abdel Hay, M. H.; Elsayed, M. A.; Barary, M. H.; Hassan, E. M. *J. Pharm. Belg.* 1991, *45*, 259.
3. Abdel-Khalek, M. M.; Abdel-Hamid, M. E.; Mahrous, M. S.; Abdel-Salam, M. A. *Anal. Lett.* 1985, *18*, 781.
4. Abdel-Moety, E. M. *J. Pharm. Biomed. Anal.* 1991, *9*, 187.
5. Allen, W. M. *J. Clin. Endocrinol.* 1950, *10*, 71.
6. Amer, A. M.; Hassan, S. M.; Kheir, A. A.; Mostafa, A. A. *Pharmazie* 1978, *33*, 344.
7. Ashton, G. C.; Brown, A. P. *Analyst* 1956, *81*, 220.
8. Ashton, G. C.; Tootil, J. P. R. *Analyst* 1956, *81*, 225, 232.
9. Atmaca, S.; Bilgic, Z.; Acikkol, M. *Pharmazie* 1991, *46*, 532.
10. Bastow, R. A. *J. Pharm. Pharmacol.* 1967, *19*, 41.
11. Bailey, L. C.; Tang, K. T.; Rogozinski, B. A. *J. Pharm. Biomed. Anal.* 1991, *9*, 501.
12. Bayer, J. *Magy. Kém. Lapja* 1969, *24*, 463.
13. Bedair, M.; Korany, M. A.; Abdel-Hamid, M. E. *Analyst* 1984, *109*, 1423.
14. Bedair, M. M.; Korany, M. A.; Abdel Hay, M. A.; Gazy, A. A. *Analyst* 1990, *115*, 449.
15. Bedair, M. M.; Korany, M. A.; Issa, A. S. *Analyst* 1988, *113*, 1137.
16. Bershtein, I. Y. *Z. Anal. Chem.* 1988, *332*, 227.
17. Bershtein, I. Y.; Lupashevskaya, D. P. *Z. Anal. Chem.* 1986, *323*, 117.
18. Berzas Nevado, J. J.; Salinas, F.; De Orbe Paya, I.; Capitan-Vallvey, L. F. *J. Pharm. Biomed. Anal.* 1991, *9*, 117.
19. Blackburn, J. A. *Anal. Chem.* 1965, *37*, 1000.
20. Blass, W. E.; Halsey, G. W. *Deconvolution of Absorption Spectra*, Academic Press: New York, 1981.
21. Bryant, S. L.; Neel, A. R.; Sewell, G. J. *J. Clin. Hosp. Pharm.* 1986, *11*, 327.
22. Carlucci, G.; Colanzi, A.; Mazzeo, P. *Farmaco* 1990, *45*, 751.
23. Carlucci, G.; Colanzi, A.; Mazzeo, P.; Quaglia, M. G. *Int. J. Pharm.* 1989, *53*, 257.
24. Cavrini, V.; Bonazzi, D.; Di Pietra, A. M. *J. Pharm. Biom. Anal.* 1991, *9*, 401.
25. Cavrini, V.; Di Pietra, A. M.; Gatti, R. *J. Pharm. Biomed. Anal.* 1989, *7*, 1535.
26. Corti, P.; Aprea, C.; Corbini, G.; Dreassi, E.; Celesti, L. *Pharm. Acta Helv.* 1991, *66*, 50.
27. Daly, C. *Analyst* 1961, *86*, 129.
28. Davidson, A. G. *Analyst* 1983, *108*, 728.
29. Davidson, A. G.; Elsheikh, H. *Analyst* 1982, *107*, 879.
30. Davidson, A. G.; Hassan, S. M. *J. Pharm. Sci.* 1984, *73*, 413.
31. Davidson, A. G.; Mkoji, L. M. M. *J. Pharm. Biomed. Anal.* 1988, *6*, 449.
32. Deming, S. N.; Morgan, S. L. *Experimental Design: A Chemometric Approach*, Elsevier: Amsterdam, 1987.
33. Di Pietra, A. M.; Cavrini, V.; Raggi, M. A. *Int. J. Pharm.* 1987, *35*, 13.
34. Eboka, C. J.; Adesanya, K. I. *Int. J. Pharm.* 1989, *51*, 263.
35. Fabre, H.; Karljikovic-Rajic, K.; Fell, A. F. *J. Pharm. Biomed. Anal.* 1995, in press.
36. Fasanmade, A. A.; Fell, A. F. *Analyst* 1985, *110*, 1117.
37. Fell, A. F. *Proc. Anal. Div. Chem. Soc.* 1978, *15*, 260.
38. Fell, A. F. *J. Pharm. Pharmacol.* 1978, *30*, 63P.
39. Fell, A. F. *UV Spectrometry Group Bull.* 1979, *7*, 5.
40. Fell, A. F. *UV Spectrometry Group Bull.* 1981, *8*, 5.
41. Fell, A. F. Derivative Spectroscopy in the Analysis of Aromatic Amino Acids, in *Amino Acid Analysis* J. M. Rattenbury, Ed.; Ellis Horwood: Chichester, 1981; 86–118.

42. Fell, A. F. The Quality Control of Steroids in Pharmaceutical Formulations by Higher Derivative Spectroscopy, in *Advances in Steroid Analysis,* S. Görög, Ed.; Elsevier: Amsterdam, 1982; 495–510.

43. Fell, A. F.; Chadburn, B. P.; Knowles, A. Derivative Spectroscopy, in *Practical Absorption Spectroscopy,* A. Knowles; C. Burgess, Eds.; Chapman and Hall: London; 1984, 179–184.

44. Fell, A. F.; Chadburn, B. P.; Knowles, A., Dual-wavelength Spectroscopy, in *Practical Absorption Spectroscopy,* A. Knowles; C. Burgess, Eds.; Chapman and Hall: London; 1984, 187–190.

45. Fell, A. F.; Jarvie, D. R.; Stewart, M. J. *Clin. Chem.* 1981, *27,* 286.

46. Forsyth, R. J.; Haynes, D.; Ip, D. *Pharm. Res.,* 1990, *7,* S26.

47. Garcia Fraga, J. M.; Jiménez Abizanda, A. I.; Jiménez Moreno, F.; Arias Leon, J. J. *J. Pharm. Biomed. Anal.* 1991, *9,* 109.

48. Gauglitz, G.; Mettler, M.; Weiss, S. *Trends Anal. Chem.* 1992, *11,* 203.

49. Géher, J.; Szabó, É. *J. Pharm. Biomed. Anal.* 1988, *6,* 757.

50. George, W. O.; Willis, H. A., Eds. *Computer Methods in UV, Visible and IR Spectroscopy* Royal Society of Chemistry: Cambridge, 1990.

51. Gill, R.; Bal, T. S.; Stewart, M. J. *J. Forensic Sci. Soc.* 1982, *22,* 165.

52. Girgis, E. H.; Negm, F.; Mahmoud, S. *J. Pharm. Pharmacol.* 1984, *36,* 840.

53. Glenn, A. L. *J. Pharm. Pharmacol.* 1963, *15,* 123T.

54. Gonzalez-Hierro, P.; Velázquez, M. M.; Cachaza, J. M.; Rodriguez, L. J. *J. Pharm. Biomed. Anal.* 1987, *5,* 395.

55. Görög, S., Ultraibolya és látható spektrofotometria (UV-visible Spectrophotometry), in *Gyógyszerészi Kémia (Pharmaceutical Chemistry),* Vol. 1; Gy. Szász Ed.; Medicina: Budapest, 1983, 68–120.

56. Görög, S. *Magy. Kém. Lapja* 1965, *20,* 503.

57. Görög, S. Csizér, É. *Acta Chim. Hung.,* 1970, *65,* 41.

58. Görög, S.; Csizér, É. *Proc. 3rd Anal. Conf. Budapest,* Akadémiai Kiadó: Budapest, 1970, Vol. 2, 65–70.

59. Görög, S.; Rényei, M.; Herényi, B. *J. Pharm. Biomed. Anal.* 1989, *7,* 1527.

60. Grizodub, A. I.; Levin, M. G.; Georgievskii, V. P. *Zhurn. Anal. Khim.* 1986, *41,* 1984; 1984, *39,* 1987.

61. Guan Luo; Jiaxue Qui; Yiming Wang; Zhengwei Yu, *J. Pharm. Biomed. Anal.* 1989, *7,* 507.

62. Hackmann, E. R. M.; Benetton, S. A.; Santoro, M. I. R. M. *J. Pharm. Pharmacol.* 1991, *43,* 285.

63. Hartwell, S. *International Laboratory,* March, 1988.

64. Hassan, S. M.; Amer, S. M.; Amer, M. M. *Analyst,* 1987, *112,* 1459.

65. Hassan, S. M.; Davidson, A. G. *J. Pharm. Pharmacol.* 1984, *36,* 7.

66. Hiskey, C. F. *Anal. Chem.* 1961, *33,* 927.

67. Hoover, J. M.; Soltero, R. A.; Bansal, P. C. *J. Pharm. Sci.* 1987, *76,* 242.

68. Hosia, L.; Peltonen, V.; Halmekoski, J. *Farm. Aikak.* 1972, *81,* 194.

69. Ivaskhiv, E. *Biotechn. and Bioeng.* 1971, *13,* 561.

70. Jones, J. H.; Clark, G. R.; Harrow, L. S. *J. Assoc. Agr. Chem.* 1951, *34,* 135.

71. Jones, R.; Leadley, B. S. *J. Pharm. Biomed. Anal.* 1986, *4,* 309.

72. Kitamura, K.; Majima, R. *Anal. Chem.* 1983, *55,* 54.

73. Knight, J. C., Analytical Control of Sterol Bioconversions, in *Steroid Analysis in the Pharmaceutical Industry,* S. Görög, Ed. Ellis Horwood: Chichester, 1989, 223–236.

74. Knochen, M.; Altesor, C.; Dol, I. *Analyst,* 1989, *114,* 1303.

75. Knochen, M.; Bardanca, M.; Piaggio, P. *Boll. Chim. Farm.* 1987, *126,* 294.

76. Korany, M.; Bedair, M.; El-'/azbi, F. A. *Analyst* 1986, *111,* 41.

77. Korany, M. A.; Bedair, M.; Mahgoub, H.; Elsayed, M. A. *J. Assoc. Off. Anal. Chem.* 1986, *69,* 608.

78. Lawrence, A. H.; Kovar, J. *Analyst* 1985, *110,* 827.

79. Lawrence, A. H.; MacNeil, J. D. *Anal. Chem.* 1982, *54,* 2385.

80. Legrand, M.; Delaroff, V.; Smolik, R. *J. Pharm. Pharmacol.* 1958, *10,* 683.

81. Leung, C. P.; Wong, K. C. C. *J. Pharm. Biomed. Anal.* 1991, *9,* 195.

82. Machek, G.; Lorenz, F. *Sci. Pharm.* 1963, *31,* 17.

83. Maddams, W. F., Numerical Methods in Data Analysis, in *Practical Absorption Spectrometry,* A. Knowles; C. Burgess, Eds., Chapman and Hall: London, New York, 1984, 160–177.

84. Mahgoub, H. *Drug. Dev. Ind. Pharm.* 1990, *16,* 2135.

85. Mazzeo, P.; Quaglia, M. G.; Segnalini, F. *J. Pharm. Pharmacol.* 1982, *34,* 470.

86. Meal, L. *Anal. Chem.* 1983, *55,* 2448.

87. Milch, Gy. *Acta Pharm. Hung.* 1962, *32,* 206.

88. Milch, Gy.; Csontos, A.; Borsai, M.; Vadon, E.; Mogács, I. *Acta Pharm. Hung.* 1966, *36,* 200.

89. Milch, Gy.; Borsai, M.; Mogács, I. *Acta Pharm. Hung.* 1967, *37,* 6.

90. Milch, Gy.; Szabó, É. *Analusis* 1988, *16*, 59.

91. Minguez, A. C.; Velázquez, M. M.; Rodriguez, L. J. *Farmaco, Ed. Prat.* 1987, *42*, 165.

92. Mohamed, M. E.; Al-Khames, H. A.; Al-Awadi, M. *Farmaco* 1989, *44*, 1045.

93. Moro, M. E.; Velázquez, M.M.; Rodriguez, L. J. *J. Pharm. Biomed. Anal.* 1988, *6*, 1013.

94. Morton, R. A.; Stubbs, A. L. *Analyst* 1946, *71*, 348.

95. Morton, R. A.; Stubbs, A. L. *Biochem. J.* 1947, *41*, 525.

96. Mulder, E.J.; Spruit, F. J.; Keuning, K. J. *Pharm. Weekblad.* 1963, *98*, 745.

97. Neuer, H. *Z. Anal. Chem.* 1971, *253*, 337.

98. Nobile, L.; Cavrini, V.; Raggi, M. A.; Di Pietra, A. M. *Int. J. Pharm.* 1987, *40*, 85.

99. O'Haver, T. C. *Clin. Chem.* 1979, *25*, 1548.

100. Parissi-Poulou, M.; Reizopoulou, V.; Koupparis, M.; Macheras, P. *Int. J. Pharm.* 1989, *51*, 169.

101. Pernarowski, M.; Knevel, A. M.; Christian, J. E. *J. Pharm. Sci.* 1961, *50*, 943.

102. Pernarowski, M.; Knevel, A. M.; Christian, J. E. *J. Pharm. Sci.* 1961, *50*, 946.

103. Poulisse, H. N. J. *Anal. Chim. Acta* 1979, *112*, 361.

104. Poulou, M.; Macheras, P. *Int. J. Pharm.* 1986, *34*, 29.

105. Quaglia, M. G.; Bossu, E.; Melchiorre, P.; Farina, A.; Salvatori, A. *Farmaco* 1991, *46*, 979.

106. Quaglia, M. G.; Carlucci, G.; Cavacchio, G.; Mazzeo, P. *J. Pharm. Biomed. Anal.* 1988, *6*, 421.

107. Quaglia, M. G.; Carlucci, G.; Maurizi, G.; Mazzeo, P. *Pharm. Acta Helv.* 1988, *63*, 347.

108. *QUEST–UV Applications Program*, Perkin-Elmer GmbH, Bodenseewerk, 1983.

109. Rössler, H.; Brückner, K. *Naturwissensch.* 1961, *48*, 695.

110. Sala, G.; Maspoch, S.; Iturriaga, H.; Blanco, M.; Cerda, V. *J. Pharm. Biomed. Anal.* 1988, *6*, 765.

111. Santoni, G.; Mura, P.; Pinzauti, S.; Gratteri, P.; La Porta, E. *Int. J. Pharm.* 1989, *50*, 75.

112. Santoni, G.; Mura, P.; Pinzauti, S.; Lombardo, E.; Gratteri, P. *Int. J. Pharm.* 1990, *64*, 235.

113. Scheffler, E.; *Statistische Methoden der Versuchsplanung und Auswertung*, Gustav Fischer Verlag: Leipzig, 1974.

114. Sharma, S. C.; Sharma, S. C.; Saxena, R. C.; Talwar, S.K. *J. Pharm. Biomed. Anal.* 1989, *7*, 321.

115. Shibata, S. *Angew. Chem. Ind. Ed.* 1976, *15*, 673.

116. Shibata, S.; Furukawa, M.; Goto, K. *Anal. Chim. Acta* 1969, *46*, 271.

117. Shively, M. L.; Simonelli, A. P. *Int. J. Pharm.* 1989, *50*, 39.

118. Shroff, A. P.; Grodsky, J. *J. Pharm. Sci.* 1967, *56*, 460.

119. Smyers Verbeke, J.; Detaevernier, M. R.; Massart, D. L. *Anal. Chim. Acta* 1986, *191*, 181.

120. Stern, E. S.; Timmons, C. J. *Gillam and Stern's Introduction to Electronic Absorption Spectroscopy in Organic Chemistry*, Edward Arnold: London, 1970, 202–219.

121. Talsky, G.; Mayring, L.; Kreuzer, H. *Angew. Chem. Int. Ed.* 1978, *17*, 785.

122. Tardiff, R. *J. Pharm. Sci.* 1961, *50*, 693.

123. Tobias, D. Y. *J. Assoc. Off. Anal. Chem.* 1983, *66*, 1450.

124. Traveset, J.; Such, V.; Gonzalo, R.; Gelpi, E. *J. Pharm. Sci.* 1980, *69*, 629.

125. Uhlich, H. *Krankenhauspharmazie* 1984, *5*, 167.

126. Vetuschi, C.; Ragno, G. *Int. J. Pharm.* 1990, *65*, 177.

127. Vetuschi, C.; Ragno, G.; Losacco, V. *Boll. Chim. Farm.* 1988, *127*, 193.

128. Vetuschi, C.; Ragno, G.; Mazzeo, P.; Mazzeo-Farina, A. *Farmaco Ed. Prat.* 1985, *40*, 215.

129. Vezin, W. R. *J. Pharm. Pharmacol.* 1979, *31*, 663.

130. Vierordt, K.; *Die Anwendung des Spektralapparates zur Photometrie der Absorptionsspektren zur quantitativen chemischen Analyse*, Verlag der Laupp'schen Buchhandlung: Tübingen, 1873.

131. Wahbi, A. M. *Pharmazie* 1971, *26*, 291.

132. Wahbi, A. M.; Abdine, H. *J. Pharm. Pharmacol.* 1973, *25*, 69.

133. Wahbi, A. M.; Abdine, H.; Korany, M. A. *Pharmazie* 1978, *33*, 278.

134. Wahbi, A. M.; Abounassif, M. A.; Al-Kahtani, M. G. *Analyst* 1986, *111*, 777.

135. Wahbi, A. M.; Belal, S.; Abdine, H.; Bedair, M. *Analyst* 1981, *106*, 960.

136. Wahbi, A. M.; Belal, S.; Abdine, H.; Bedair, M. *Talanta* 1982, *29*, 931.

137. Wahbi, A. M.; Ebel, S. *J. Pharm. Pharmacol.* 1974, *26*, 317.

138. Wahbi, A. M.; Ebel, S.; Steffens, U. *Z. Anal. Chem.* 1975, *273*, 183.

139. Wahbi, A. M.; El-Yazbi, F. A.; Barary, M. H.; Sabri, S.M. *Analyst* 1992, *117*, 785.

140. Wahbi, A. M.; Mahgoub, H.; Barary, M. H. *Analyst* 1989, *114*, 505.

141. Wahbi, A. M.; Unterhalt, B.; *Z. Anal. Chem.* 1976, *282*, 31.

142. Winfield, S. A.; Rhys Williams, A. *J. Pharm. Biomed. Anal.* 1984, *2*, 561.

143. Yokoyama, F.; Pernarowski, M. *J. Pharm. Sci.* 1961, *50*, 953.

Chapter 5

COMBINATION OF SPECTROPHOTOMETRIC METHODS WITH CHROMATOGRAPHIC TECHNIQUES

A. SIGNIFICANCE OF THE COMBINATION OF SPECTROPHOTOMETRIC AND CHROMATOGRAPHIC METHODS

As discussed in detail in Chapter 1 of this book, there has been only a modest progress in the last 30 to 40 years compared to the previous period in the methodology of UV-VIS spectrophotometry and its application to pharmaceutical analysis. In contrast, it was during the last 30 to 40 years which was the period of unprecedented progress and spread of chromatographic methods. Gas chromatography (GC), thin layer chromatography (TLC) and high performance column chromatography (HPLC) have become determinant methods in pharmaceutical analysis, too.

The fact that spectrophotometry, unlike other classical methods, has arrived into the 1990s without significant loss in importance is due mostly to its excellent combination possibilities with up-to-date chromatographic methods.

If this question is approached from the point of view of chromatography, it can be stated that this progress would have been impossible without the application of the methodology and approach of spectrophotometry. For instance, in most cases, UV spectrophotometers are applied to the detection and measurement of separated substances both in classical low-pressure and in modern high-pressure liquid chromatography. In a significant number of cases in TLC, in electrophoresis, capillary electrophoresis and related methods (isoelectric focusing, isotachophoresis, etc.) qualitative and quantitative analysis is based on the measurement of light absorption or reflection. It is worth noting that now more UV-VIS spectrophotometers are marketed as indispensable accessories of these techniques than as stand-alone instruments. In order to enhance the selectivity and sensitivity of detection and quantitative analysis, the results of spectrophotometry based on chemical reactions have been utilized, and new directions of this technique exploited to meet special demands.

If, on the other hand, the question is approached from the point of view of spectrophotometry, it can be seen that the application of chromatographic methods resulted in a breakthrough in an area which represents the greatest weakness of UV-VIS spectrophotometry, its poor selectivity. In a significant number of cases the efforts to increase selectivity discussed in Chapter 4 (optical methods) and Chapters 6, 7, 8 and 10 (application of chemical reactions) of this book are fruitless: the condition of sufficient selectivity is often the separation of the analyte from the matrix and/or other accompanying compounds with similar spectra. This separation can be done most simply and effectively by means of one of the chromatographic techniques.

According to the nature of coupling, combined methods can be divided into three groups:

1. Chromatography and spectrophotometric measurement are separated in time and space. This group includes, for example, the spectrophotometric measurement of substances separated by low-pressure column chromatography or eluted from thin-layer chromatograms. In these cases the spectrophotometric measurement itself can be done by methods discussed earlier.

2. Spectrophotometric measurement and chromatography are separated in time, but the former is not separated from the chromatogram in space. This group includes the *in situ* densitometric measurement (without elution) of plates produced by TLC or related techniques.
3. Chromatography and spectrophotometric measurement are practically not separated in time and space. This includes, for instance, HPLC separation with UV detection or, just mentioned as curiosity, GC separation with UV detection.

As can be seen from the examples, almost all important branches of chromatography are involved, and thus a detailed discussion would exceed the scope of this book.

B. SPECTROPHOTOMETRIC MEASUREMENT AFTER SEPARATION BY COLUMN CHROMATOGRAPHY

Column chromatographic separation of organic compounds is used primarily in preparative chemistry. Since this technique, depending on the size of the column, can be used for separation from the microgram to the ton range, separation on small columns has also been used for analytical purposes, including separation prior to spectrophotometric measurement, since the early development of the method. (It is noted that UV spectrophotometry has played a significant role in chromatographic separations on a preparative scale, since the easiest way of monitoring a separation is by means of the UV-spectrophotometric measurement of the effluent, either continuously using a flow cell or periodically on the basis of individual samples from fractions and observing the appearance of the UV-active components.) Column chromatographic separation for analytical purposes, although its significance has been reduced with the spread of TLC, is still an important method of pharmaceutical analysis, for the measurement of both drug preparations and biological samples. Of the huge literature on column chromatography, two books are recommended since they contain abundant examples from pharmaceutical analysis: the monographs edited by Deyl, Macek and Janák[11] and by Mikes.[47]

Column chromatographic separations for analytical purpose are usually performed on glass columns 15 to 30 cm in length and 1 to 3 cm in diameter. The columns may be filled with almost the full selection of chromatographic stationary phases ranging from silica gel or alumina as applied in classical adsorption chromatography through ion exchange resins[39] to chemically modified polydextran gels used primarily for the analysis of biological samples. Liquid-liquid partition column chromatographic methods, much widespread in the practice of pharmaceutical analysis, are worth mentioning separately. In this technique an inert carrier (Celite, Kieselguhr) is impregnated with an aqueous buffer or another polar solvent (dimethyl formamide, dimethyl sulfoxide, acetonitrile, etc.), and non-polar solvents are used as eluent combined occasionally with ion pair forming reagents. The acceptance of this method is demonstrated by the fact that it is represented by 126 references in the review of Doyle and Levine[12] dealing with the application of the technique to drug preparations. Concerning the choice of eluents, it is a general rule that if the eluted component is to be measured in the UV region, benzene, acetone or, depending on the wavelength of the measurement, other UV active solvents (see p. 25) cannot be used. If color reactions are employed, this limitation is considerably reduced.

For the management of eluent and the detection and measurement of the analytes in the effluent, the literature contains a wide variety of methods ranging from the simplest (manual control of the eluent of any gradients and of the effluent) to the fully automated ones (programmed gradient, automated fraction collectors). Since the aims of the separation steps employed for the solution of the problems of pharmaceutical analysis are usually quite modest (separation of an active ingredient from the vehiculum or another, chromatographically not problematic active ingredient or decomposition product), the simplest technical solutions usually suffice.

As an example, let us present the system developed by Graham[30,31] for the separation of testosterone esters from the oily vehicle used for the preparation of injections and from free testosterone obtained by decomposition.[30] The packing is filled into a 25×2.2 cm glass column narrowed down to 5×0.4 cm at the lower end. A slurry obtained by mixing 4 g of Celite 545 and 4 ml of acetonitrile is introduced into the column after closing its bottom with glass wool, and then the resulting packing is compressed by means of an aluminum rod.

The tablets containing the testosterone esters are extracted with methanol, the injections are diluted with chloroform. Portions of these stock solutions containing 1.25 mg of steroid are evaporated to dryness, dissolved in 1.5 ml of heptane and 1.5 ml of acetonitrile, respectively, the two solutions are mixed with 3 g of Celite and the resulting slurry is introduced to the top of the column. Elution is performed with 250 ml of heptane. The first 60 ml fraction is collected separately; this is used to determine testosterone esters, the second, 190 ml fraction is used to determine the free testosterone decomposition product. The analysis may be based on the direct measurement of the unsaturated keto group at 240 nm or on the measurement of the same group in the visible region after a reaction with isonicotinic acid hydrazide (see p. 326). In addition to separating testosterone from its esters, chromatographic separation removes both components of the oil which interfere with the spectrophotometric measurement. It does not remove, however, benzyl alcohol used as a preservative: it is eluted together with the testosterone esters in the first fraction. Consequently, in such cases only the isonicotinic acid hydrazide method may be used.

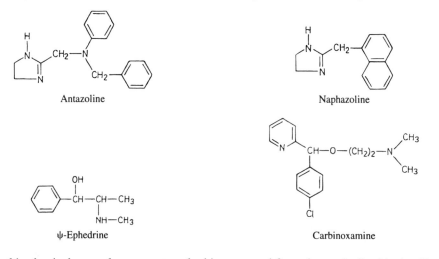

Antazoline

Naphazoline

ψ-Ephedrine

Carbinoxamine

After this classical example, a recent method is presented from the gradually thinning literature of the technique as described by Lopez Silva et al.[43] for the determination of the components of two complex drug preparations. Separation is performed on a column 14.5 cm in length and 3.2 cm in diameter, filled with Kieselguhr impregnated with phosphate buffer. Separation is based on the different retentions of the components with different basicities on the column at the optimized pH of the buffer. The optimum pH of the buffer is 6.0 for the combination of antazoline (pK_a = 10.1) and naphazoline (pK_a = 10.9) and 6.7 for the combination of ψ-ephedrine (pK_a = 9.9) and carbinoxamine (pK_a = 8.1). In the analysis of the former combination antazoline is eluted quantitatively after extraction with 150 ml of chloroform, whereas naphazoline is retained by the column. With the latter combination carbinoxamine can be eluted with 100 ml of chloroform, whereas ψ-ephedrine cannot. The retained components are eluted with 0.1 M hydrochloric acid, and these solutions can be measured directly, after appropriate dilution. The chloroform solutions are evaporated, then dissolved in 0.1 M hydrochloric acid for spectrophotometric measurement at the absorption maxima (naphazoline: 281 nm, antazoline: 241 nm, ψ-ephedrine: 256.5 nm, carbinoxamine: 264 nm).

C. HIGH-PERFORMANCE LIQUID CHROMATOGRAPHY

1. General Questions: Quantitative Analysis

The significance of classical column chromatography discussed in the former section has been greatly surpassed in analytical practice by that of contemporary high-performance or high-pressure liquid chromatography (HPLC), in which, by means of a high-pressure pump, the eluent can be passed through columns packed with absorbents of such small particle size which would make the column practically impenetrable at atmospheric pressure. Consequently, in a very short time a very efficient separation can be performed, and by the continuous detection of the eluate a chromatogram, also excellent for quantitative analysis is obtained. The brief discussion of this separation technique, which has in fact revolutionized contemporary pharmaceutical analysis,[58] is motivated by the fact that in the practice of pharmaceutical analysis the chromatograms are most frequently monitored by UV spectrophotometric detection. A UV detector is in fact a spectrophotometer, generally a double beam one, specially designed for recording chromatograms. Concerning its construction and operation, the main principles discussed in Sections 2.D and 2.F essentially apply here.

With the simplest and cheapest detectors a mercury lamp is used as source, which allows the monitoring of the chromatogram at 254 nm. The number of detection wavelengths can be increased by means of appropriate filters and phosphores, but for continuous wavelength detection, as with modern spectrophotometers, a deuterium lamp or, in the visible, tungsten lamp is generally required. The resolution of the grating monochromator applied in these detectors is, of course, significantly poorer than for stand-alone spectrophotometers. The measuring cell is attached directly to the chromatographic column. The cell should have a minimum volume in order to avoid peak broadening, but a possibly long effective path length for high sensitivity. The volume of modern cells does not exceed 5 to 10 μl, but micro cells with 1 to 2 μl volume are also available. Path length may reach 10 mm. The performance of a spectrophotometric detector is principally determined by its sensitivity and linear dynamic range, since the recording of small absorbance differences with acceptable signal-to-noise ratio is a much more frequent task in chromatographic detection than in conventional spectrophotometry. For trace analysis, for example, the most sensitive detectors can be used in a mode that the full scale of the recording unit corresponds to 0.005 absorbance units.

The broad linear dynamic range of UV photometric detectors (which is ensured, in addition to the appropriate parameters of the detector, by the validity of the Beer-Lambert Law for this special field of application, too) allows quantitative analysis of known components to be performed after proper calibration on the basis of peak area or, in the case of appropriately sharp peaks, on the basis of peak height. Just as in spectrophotometry, the sensitivity of analysis depends on the molar absorptivity of the analyte. Therefore, the detector permitting, it is recommended to choose the absorption maximum as the wavelength of chromatographic detection. When the sample contains several known, chromatographically separated components, it may occur that at the optimum wavelength for one of the components the others can only be detected with low sensitivity, if at all. If the task is the quantitative determination of all components, a wavelength should be selected at which all components have appropriate absorption, and the chromatogram is recorded at this wavelength. Another method is to record the chromatogram several times and always at the absorption maxima of the various components to be determined.

With certain UV detectors available on the market, it is possible to change the wavelength between the peaks to the optimum wavelength of the next component, and the required changes can be entered into the computer program controlling the detector. In this respect the most up-to-date solution is the use of diode array UV detectors, enabling a simultaneous detection to be performed at several wavelengths. This is illustrated by the determination of 9(11)-dehydro-mestranol impurity in mestranol (Figure 5.1). It can be seen that there is a significant increase in sensitivity if the chromatogram is detected at the absorption maximum of the impurity (261 nm) instead of that of the main component (280 nm).[28]

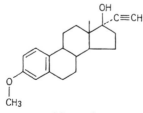

Mestranol

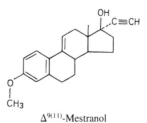

$\Delta^{9(11)}$-Mestranol

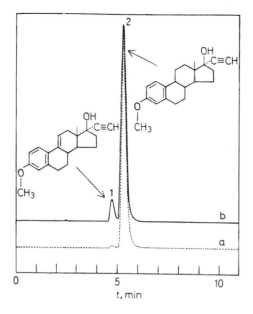

FIGURE 5.1. Liquid chromatogram of mestranol (Peak 2) containing 0.5% of 9(11)-dehydro-mestranol (Peak 1) as impurity. Column: 250 × 4 mm of 10 μm LiChrosorb RP-18; eluent: methanol-water 9:1; 1 ml/min. a: taken at 280 nm; b: taken at 261 nm. (From S. Görög; A. Laukó; B. Herényi. *J. Pharm. Biomed. Anal.* 1988, *6*, 697.[28])

The great majority of substances investigated in pharmaceutical analysis have an absorption which makes it possible to detect them at an appropriate wavelength. Even substances regarded as UV-inactive have absorption at the short wavelength permitting detection: in the vicinity of 200 nm all substances of interest in pharmaceutical analysis, with negligible exceptions, can be detected with more or less reasonable sensitivity. However, measurements at very short wavelength, as in spectrophotometry, pose a series of problems: only very high quality, well maintained detectors may be used, and the solvents used for elution, too, should meet very strict requirements. It is only water, simple alcohols, hydrocarbons and acetonitrile that can be used, and even then high purity is required ("HPLC grade").

If the sample contains unknown components, the wavelength(s) should be selected very carefully. If the chromatogram is taken at the absorption maximum of the main component only, it may often occur that a major secondary component, impurity or decomposition product, which does not absorb at this wavelength, is missed. In such cases it is desirable to take the chromatogram in the vicinity of 200 nm in addition to the optimum wavelength.

Modern instruments equipped with electronic integrators give the peak areas of the components and their percentage values related to the total area of the chromatogram. These values should only be taken, even as approximations, as representative of percentage concentration ratios of the components concerned if the absorptivities of all components are approximately the same at the monitoring wavelength. Such a case is shown in Figure 5.2 displaying the chromatogram of the ergolene-type alkaloids present in a fermentation broth.[41] Since the spectrophotometrically active lysergic acid component of all alkaloids is the same (see Curve b in Figure 3.23), and differences occur only in the spectrophotometrically inactive peptide part, the molar ratios of the components can be determined approximately well from the ratios of peak areas. However, for the majority of cases this condition does not hold; thus the values read from the integrator may not even provide an indication of the composition of the mixture, since the specific absorbances of the components at the wavelength of measurement could differ by several orders. Therefore, extreme care should be exercised when this method is applied for the determination of the impurity profiles of the bulk materials of pharmaceutical compositions, and the structures and spectra of not all components are known. As mentioned, it may occur that certain impurities are not detected at all, whereas others are overestimated in the chromatogram. Of course, the latter statement may also be worded so that the method provides an excellent chance for the very sensitive determination of certain

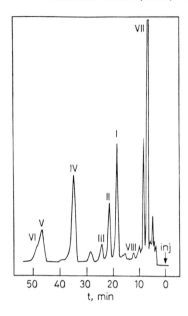

FIGURE 5.2. Liquid chromatogram of a fermentation broth obtained in the industrial production of ergocornine–ergocryptine. Column: 250 × 4 mm of 10 μm LiChrosorb RP-18; eluent: tetrahydrofuran–0.01 M aqueous ammonium acetate 4:6; 1.5 ml/min; 322 nm. I: ergocornine; II: α-ergocryptine; III: β-ergocryptine; IV: ergocorninine; V: α-ergocryptinine; VI: β-ergocryptinine; VII: ergometrine; VIII: ergometrinine. (From B. Herényi; S. Görög. *J. Chromatogr.* 1982, *238*, 250.[32])

impurities, decomposition products, etc. As an illustration, the determination of 4,5-dihydro- and $\Delta^{5(10)}$-derivatives as impurities in norethisterone is cited. These derivatives which have no conjugated bond system even in the percent concentration range produce no measurable signal at the absorption maximum (240 nm) of norethisterone which has a conjugated keto group.

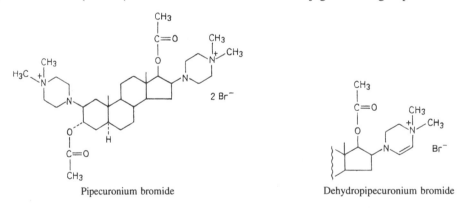

Pipecuronium bromide   Dehydropipecuronium bromide

As another illustration, the determination of dehydropipecuronium bromide impurity in pipecuronium bromide is shown in Figure 5.3. Since the main component has a completely saturated skeleton, the chromatogram should be detected in the short wavelength region (at 215 nm). Since, however, the specific absorbance of the main component is relatively low even at this wavelength, because the only chromophores are the acetoxy and tertiary amino groups, the amount of the impurity with an intensive spectrum is strongly overestimated at this wavelength if the peak intensities are taken into account directly. When detected at the absorption maximum (235 nm), the 1% impurity looks already as main component.[24]

Consequently, a reliable quantitative analysis can be done only by preparing and using an "impurity standard" for calibration. As it will be repeatedly emphasized in the next subsection, this standard has an important role in the final identification of components with structures suggested by combined chromatographic–spectroscopic methods.

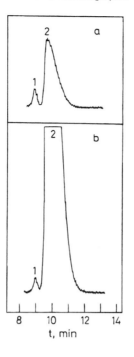

FIGURE 5.3. Liquid chromatographic determination of 2′ (3′)-dehydro-pipecuronium bromide (Peak 1) in pipecuronium bromide (Peak 2). Column: LiChrosorb SI-60 (5 μm), 250 × 2.6 mm. Eluent: 100 mM each of ammonium chloride and ammonium carbonate dissolved in a methanol–acetonitrile-conc. NH$_3$ 43:14 mixture.[55] a: at 235 nm; b: at 215 nm. (From Gazdag, M.; Görög, S., *unpublished results*)

2. Structure Elucidation of Separated Components

In addition to applications in quantitative analysis presented so far, there are significant application potentials of HPLC with UV detection in the identification and structure elucidation of the separated components. Even the cheapest UV detectors with variable wavelength make it possible to measure the approximate spectra of the separated components. This can be done in a way that the chromatogram is taken repeatedly by changing the wavelength in 5 to 10 nm steps carefully keeping the other parameters (injected volume, flow rate, etc.) constant. Then, for the components with the same retention times, the peak heights (or peak areas) are plotted against wavelength. Of course, with this extremely time and labor consuming technique, a spectrum of only very poor quality can be obtained, due partly to the large wavelength increment, which has to be chosen for practical reasons and partly to the poor resolution of the monochromator of the instrument. In the early days of liquid chromatography, instruments were marketed with recording units built into their monochromators. These instruments, now regarded as obsolete, stopped the flow of eluent at the maximum of the chromatographic peak, and the spectrum was recorded in a stationary system ("stopped flow" method).

The most convenient and advanced approach to date is the application of diode arrays as detector.[16,19,20,25] This detector operates according to the principles shown on pp. 18 and 22. The fact that the complete spectrum can be recorded within milliseconds allows the spectrum to be taken without stopping the chromatographic process. The computer interfaced with the detector stores the spectra, and at the end of the chromatographic separation all spectra may be retrieved, or the chromatogram can be presented as a 3-D (three dimensional) plot (absorbance plotted as a function of retention time and wavelength), which can be used to solve various practical problems.

As an example, the 3-D chromatograms of zimeldine and its main metabolites are shown in Figure 5.4 in two forms of presentation. Part "a" of the figure shows the 3-D chromatogram as seen from the left, at an angle of 35°. It is evident that the sections of the 3-D surface cut in parallel with the wavelength axis give the spectrum of the solution in the cell of the detector at the given instant, whereas sections parallel with the retention time axis are chromatograms at

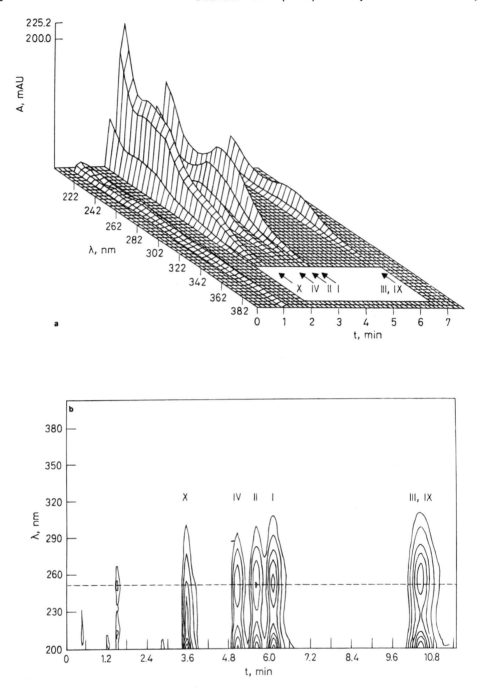

FIGURE 5.4. 3-D liquid chromatograms of Zimeldine and its metabolites. Column: 100 × 2.9 mm Nucleosil C-18 (5 μm). Eluent: acetonitrile—0.05 M sodium bisulphate, 6 × 10⁻⁴ M N, N-dimethyl-N-octylamine; pH = 1.9–2.0 1:9; 1 ml/ min.[53] I-X: see text. a: steric view; b: contour plot (contour intervals: 10, 25, 50, 100, 125, 150, 175, 200 mAU).

different fixed wavelength values. The other form of presentation (Figure 5.4b) is the so-called contour plot in which, similar to the height contours of maps, the points with the same absorbance values are joined. The asymmetry of contour lines generally indicates an inhomogeneity of peaks, as shown e.g., in the case of the overlapping peaks of metabolites denoted as III and IX.[17]

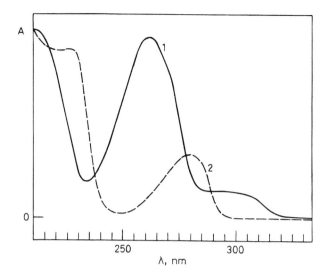

FIGURE 5.5. Spectra of 9(11)-dehydro-mestranol impurity (Curve 1) and mestranol (Curve 2) taken during chromatographic separation with diode array UV detector. (For chromatographic conditions and source, see Figure 5.1)

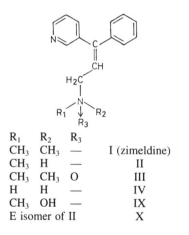

R₁	R₂	R₃	

R₁ R₂ R₃

CH₃	CH₃	—	I (zimeldine)
CH₃	H	—	II
CH₃	CH₃	O	III
H	H	—	IV
CH₃	OH	—	IX
E isomer of II			X

The identification of byproducts, impurities and decomposition products separated by liquid chromatography is a particularly important field of application. Although, as pointed out in Chapter 3, the potential of UV spectrophotometry in structure elucidation is much poorer than those of NMR, IR and mass spectrometry, in fortunate cases very useful data can be obtained quite readily. This occurs if the spectrum corresponding to the peak in question is significantly or characteristically different from the spectrum of the main component obtained under similar conditions. From this fact, knowing the history of the sample, it is often easy to deduce a structure for the substance, which should be, however, always confirmed by other spectroscopic methods and/or the synthesis of that substance followed by the comparison of the chromatographic retention times of the component in question and the reference substance.

Three examples are shown as illustrations. The structure elucidation of 9(11)-dehydro-mestranol, an impurity of mestranol, as mentioned above, was based to a great extent on the UV spectrum recorded by the diode array detector.[26,28] With a C_{18} column and a 9:1 methanol-water mixture eluent, the impurity has a relative retention of 0.89 with respect to the main peak. The spectra taken on-line with the chromatographic process are shown in Figure 5.5. It can be seen that, compared to the phenol ether type spectrum of mestranol, in the spectrum of the impurity the strong band is subjected to a bathochromic shift of ca. 32 nm and the long wavelength band to one of 20 nm. This indicates the presence of a double bond conjugated with the phenolic ring. Of

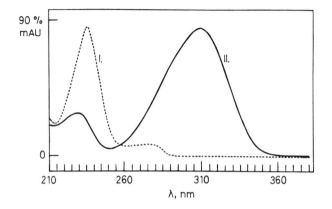

FIGURE 5.6.   Spectra taken with diode array UV detector during the chromatographic measurement of trans-(4-methoxy-phenyl)-oxiranecarboxylic acid. I: main component; II: by-product (α-chloro-4-methoxycinnamic acid methyl ester). (From S. Görög, M. Rényei and B. Herényi. *J. Pharm. Biomed. Anal.* 1989, 7, 1527.[29])

the three possible isomers (Δ^6, $\Delta^{8(9)}$ or $\Delta^{9(11)}$), the decision could be made in favor of the $\Delta^{9(11)}$ derivative by comparing the position and intensity ratio of the two bands with the corresponding data of analogs. This conclusion has been confirmed by both other spectroscopic studies and the agreement of retention times after the synthesis of the assumed substance.

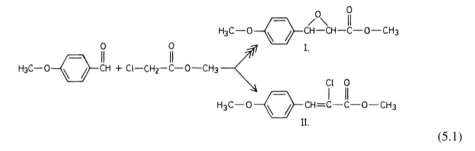

$$(5.1)$$

The subject of our second example is the structure elucidation of another phenolic type substance.[27,29] Scheme 5.1 shows a Darzens reaction between 4-methoxybenzaldehyde and chloroacetic acid methyl ester to yield substance I. A byproduct of this reaction is assumed to be substance II. Using a C_{18} column and a 1:1 acetonitrile-water mixture, the chromatographic separation leads to retentions of k' = 3.8 for the main component and k' = 10.6 for II. The diode array UV spectra are shown in Figure 5.6. Compared to the phenol ether type spectrum of I, the bathochromic shift in the spectrum of II is even stronger (73 nm) than in the previous example, as a result of which the long wavelength band is hidden under the main band indicating the conjugation of two double bonds with the aromatic ring. Taking into account the starting substances, this is possible for the substance shown as II in the reaction scheme. This tentative structure was confirmed by further spectroscopic studies.

The third example[26] illustrates that smaller but characteristic differences may also be of diagnostic value in the identification of impurities on the basis of their diode array UV spectra. The spectra

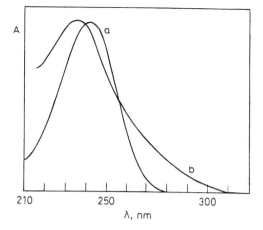

FIGURE 5.7.   Spectra taken with diode array UV detector during the chromatographic measurement of norgestrel. a: main component; b: by-product (8(14)-dehydro-norgestrel). (From S. Görög; B. Herényi. *J. Chromatogr.* 1987, *400*, 177.[26])

of norgestrel and its impurity are shown in Figure 5.7. The spectra were obtained after separation on a C_{18} column with a 7:3 methanol-water mixture, where the relative retention time of the impurity was 0.84. Both spectra are characteristic of a 4-ene-3-keto structure. The slight hypsochromic shift and band broadening observed in the spectrum of the impurity suggest the presence of a $\Delta^{8(14)}$ bond in the molecule, which makes the skeleton more rigid thereby indirectly affecting conjugation ("through-space conjugation"[33]). The 8(14)-dehydro hypothesis made on this basis has been supported and confirmed by other spectroscopic techniques and, ultimately, by the synthesis of the suggested structure.

The technique has an important role in the structure elucidation of metabolites.[8,17,19,38] Another important application of HPLC–diode array UV combination is the identification of drugs in complex multicomponent mixtures (like toxicological samples) by means of computer searches of spectrum libraries.[1,3,15,18,34,48,49,64]

3. Peak Homogeneity (Purity) Studies

Another very important application field of liquid chromatography–diode array UV detector combination is the investigation of the homogeneity of chromatographic peaks. This possibility is important since in those cases where components coelute, the poor separations manifest themselves only (sometimes) as observable distortion in peak shapes, and the presence of a non-separating component, due to its low relative concentration, often remains unnoticed. It may also occur that the peaks of two components with closely similar retention times and similar intensities merge into one another to yield an apparently symmetric peak, or, in the case of exactly identical retention times, a genuine single peak occurs. This phenomenon may cause a positive deviation (occasionally a very serious one) in the quantitative analysis of the components of multicomponent mixtures (e.g., biological samples, impurity profiles).

Therefore, in more demanding chromatographic analyses, peak homogeneity must be checked. The classical, but still most widespread method for this check is the absorbance ratio method, which is based on the fact that in the solution of a substance, the ratio of the absorbances measured at two selected wavelengths does not depend on concentration. Thus, if one measures the absorbance ratio at two characteristic wavelengths at the ascending and descending edges and at the maximum of a chromatographic peak, in the case of a homogeneous peak a constant ratio, which does not depend on the retention time, is obtained. Any deviation from this constant indicates inhomogeneity. (Of course, a condition of the applicability of this method is that the spectra of the two, co-eluting components are sufficiently different.[52]) The measurement of such absorbance ratios was also possible before the introduction of diode array UV detectors, e.g., by applying the "stopped flow" technique[67] or by means of two parallel detectors set at different wavelengths.[65]

Obviously, with the application of diode array detectors, the measurement and continuous recording of arbitrary absorbance ratios has become a simple routine task. Most generally, the method is applied almost obligatorily in the validation of HPLC methods, and in the case of properly optimized systems the uniqueness of the peak is established. If inhomogeneity is found, the spectra taken at the maximum, the head and the tail of the peak yield data from which the reasons for inhomogeneity and the nature of the non-separating component can be derived. Further assistance comes from the derivatives, namely the derivative of chromatograms with respect to time and that of spectra with respect to wavelength.[8,19,20] In the case of gradient elution, the variation of the spectra with the continuously changing solvent composition must also be accounted for.[13] An interesting example for the application of absorbance ratio method has been reported by Szepesi, Gazdag and Mihályfi[59] concerning the validation of the impurity analysis of pipecuronium bromide already mentioned in this section.

Using more advanced processing of the huge wealth of data collected by the computer of a diode array detector, more sensitive solutions providing more pieces of information than the simple absorbance ratio method have been developed for the investigation of peak homogeneity. The "purity parameter" introduced by Alfredson et al.[1] is one of these solutions. This parameter is the mean of the wavelengths in nm weighted by the absorbances in an appropriately chosen wavelength region. Of course, in the case of a homogeneous peak, this parameter is constant along the full breadth of the peak. The sensitivity of small deviations for the detection of inhomogeneity has been shown by Chen and Carr:[7] in the study of temazepam and lormetazepam (for structures, see p. 289) with very similar spectra and retention times (3.32 min for temazepam and 3.29 for lormetazepam) an 0.5% impurity of the latter could be detected in the former compound.

If the methods of qualitative character discussed so far indicate an inhomogeneity of peaks, i.e., direct quantitative analysis is impossible, an improvement of chromatographic separation should be attempted first by changing chromatographic parameters. If this is impossible for any reason, then the so-called spectral suppression methods may be applied to separate partially or fully overlapping peaks by algebraic-deconvolution methods, enabling quantitative analysis to be performed in the next step.

The simplest approach of this kind is the method of Carter et al.[5] in which absorbance ratio is converted into the linear combination of chromatograms:

$$\frac{A_{\lambda_1}}{A_{\lambda_2}} = k_{1,2} \tag{5.2}$$

where from

$$A_{\lambda_1} = k_{1,2} A_{\lambda_2} \tag{5.3}$$

where $k_{1,2}$ can be determined from the spectrum of the component to be suppressed. In the simplest case two isoabsorption wavelengths are chosen for which $k_{1,2} = 1$, and thus $A_{\lambda_1} = A_{\lambda_2}$, and their difference is zero. As an example, the spectra of carbamazepine (see p. 78) and its impurity, acridone, are shown in Figure 5.8. The two wavelengths for the detection of the two perfectly co-eluting peaks were 256 and 307 nm.

The result can be seen in Figure 5.9. Chromatogram A shows the two overlapping peaks detected at 254 nm. Chromatogram C is the curve of pure carbamazepine, indicating the full suppression of the peak by recording the difference of the absorbances measured at the above two wavelengths. Curves a, b and c of chromatogram B indicate that after a suppression of the main component an impurity in the 0.1% concentration region is easy to measure.

The spectral suppression method has been extended by Fell et al. from two wavelengths to several wavelengths, by using practically all absorbance data stored in the computer for the more sensitive and more accurate solution of problems exceeding in difficulty those discussed so far.[6,17,21,50,51] Although the mathematical details of this method are beyond the scope of our book, it is noted that Marr et al.[46] reported the extension of the method developed originally for two overlapping peaks to multicomponent systems. The application of this method was illustrated by

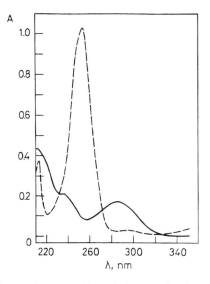

FIGURE 5.8. UV spectra. — carbamazepine; – – acridone. Solvent: methanol–water 55:45; c = 10 mg/l. (From G.T. Carter; R.E. Schiesswohl; H. Burke; R. Yang. *J. Pharm. Sci.* 1982, *71*, 317.[5])

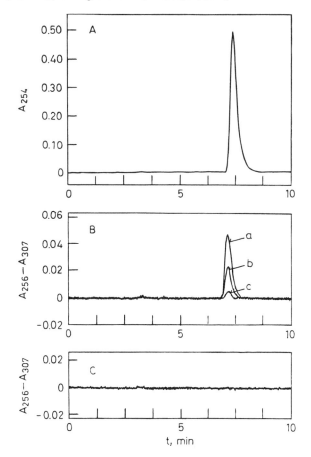

FIGURE 5.9. Liquid chromatograms of carbamazepine with acridone impurity. Column: 300 × 3.9 mm reversed phase column. Eluent: methanol–water 55:45, 1 ml/min. Injected sample: ca. 5 μg of carbamazepine. Wavelengths and wavelength combinations: see figure. A. Carbamazepine containing 1% acridone B. a: Carbamazepine containing 1% acridone; b: Carbamazepine containing 0.5% acridone; c: Carbamazepine containing 0.1% acridone; C. Carbamazepine without acridone impurity. (From G.T. Carter; R.E. Schiesswohl; H. Burke; R. Yang. *J. Pharm. Sci.* 1982, *71*, 317.[5])

the quantitative determination of the individual components of ajmalicine, vincristine and catharantine in the HPLC analysis of *Catharanthus roseus* extract after the spectral suppression of the other two components.

4. Derivatization Reactions

It is noted finally that, usually to enhance the sensitivity and selectivity of detection, chemical reactions are often applied to increase the intensity of the UV-VIS absorption of analytes or to shift absorptions toward the longer wavelengths. These reactions are usually performed prior to injection into the chromatograph, and thus it is the derivatives which are separated. The reactions are essentially the same as those used for spectrophotometric purposes, and thus we refer here to Chapter 7 and to the publications dealing with the preparation of derivatives for chromatographic purposes.[4,23,41,42] In other cases underivatized substances are separated, and derivatization is done in a reactor inserted between the chromatograph and the detector. This derivatization may range from an instantaneous and reversible ionization effected by a simple change of pH (e.g., the method based on the bathochromic shift of the spectra of barbiturates upon basification[9,49,53]) to complicated reactions for which several examples can be found in the publications cited.

D. SPECTROPHOTOMETRIC MEASUREMENT AFTER ELUTION FROM THIN-LAYER CHROMATOGRAMS

Thin-layer chromatography appeared in pharmaceutical analysis at the end of the 1950s, and in the next 10–15 years it had almost completely replaced paper chromatography. Since the appearance of this new technique, there have been attempts to apply it for the quantitative analysis of separated compounds, including a large number of drugs. Two quantitative approaches have developed in parallel: the elution and the *in situ* method, of which the former is discussed in this section.

Even a summary of the fundamentals and techniques of thin-layer chromatography would exceed the scope of this book. Of the several publications and handbooks the fundamental work of Stahl,[57] the work of Touchstone and Dobbins,[60] as well as the recently published books of Treiber[63] and Sherma and Fried[56] are mentioned. Special applications in pharmaceutical analysis are discussed in the book of Macek.[44]

Those important factors which primarily determine the efficiency of chromatographic separation also significantly affect the success of quantitative measurements since a quantitative measurement based on poorly separated spots must obviously be inaccurate. Thus, in addition to the judicious choice of sorbent and eluent system, the method of applying the sample to the layer is also important. An accurate volume of the sample solution in an appropriately volatile solvent should be applied to the layer in a minimum surface area without damaging the layer. This can be done in the form of spots or strips. An advantage of the latter is that larger quantities can be applied to the plate without adversely affecting separation.

After running the chromatogram and removing the solvent, the first step which determines the success of quantitative analysis is the marking of the zone to be eluted. With colored substances or those displaying fluorescence in long-wavelength UV light (366 nm) this is relatively easy. The fluorescence quenching of substances absorbing UV light in the short-wavelength UV region may also be utilized if a layer impregnated with a fluorescent substance is used for separation. If the substances can be made visible by chemical reaction only, the reaction is run on the two edges of the plate, and the zone to be eluted is marked here. However, the non-uniform retention across the plate may cause errors in this case. This problem may be overcome if the plate is developed with iodine vapor: by this method most of the organic substances can be detected with sufficient sensitivity, and after evaporating iodine from the plate, extraction and quantitative analysis become possible in the majority of cases. If the analysis is based on the measurement of a color produced by chemical reaction instead of the native absorption of the analyte, it is advantageous to develop the spot with the same reagent as used after elution to develop the color (a practical example for this method will be shown at the end of the section).

Owing to the potential for decomposition of certain compounds on the active layer, the time spent by the sample on the layer at the starting spot before separation and in the separated spot before elution should be minimized, and, in addition, the plates should be protected from strong illumination. For this reason, one should also be careful with detection by means of UV illumination.

For the extraction of the zone containing the analyte, several techniques are known. One is to remove the sorbent containing this fraction from the glass plate by means of a spatula, and simply put the removed substance into an extraction flask. However, vacuum cleaner-like glass devices are also available for the removal of the zone, which collect the removed substance on glass filters, from which the analyte can be eluted by an appropriate solvent. The extraction of layers on aluminum plates can be done simply by cutting out the zone with scissors, then cutting it to small pieces and extracting the layer together with the aluminum support. Devices are also available for the direct extraction of zones from the plate with minimum amounts of solvent, without removal from the support (e.g., Camag Eluchrom).

Depending on the solubility of the substance to be eluted, a wide choice of solvents (primarily of medium polarity) can be used. Depending on the strength of binding to the layer, in some cases simple shaking may be sufficient for quantitative extraction, in other cases an intense treatment with the solvent (e.g., Soxhlet-extraction) may prove necessary. Based on experience gained with a wide variety of drugs, it may be stated that full quantitative recovery is very hard to achieve: one must often reckon with a few percent of irreversible adsorption. If the extent of this irreversible adsorption does indeed not exceed some percent, the resulting negative error can be eliminated by applying standards to the same plate along with the spot of the unknown in similar concentration, and taking into account in the evaluation the response obtained from the standard. It is also worth noting that if the material of the plate is separated from the extract by filtering, the colloidal particles of the sorbent may pass the filter and cause error in the spectrophotometric measurement owing to light scattering. This can be avoided by centrifuging instead of filtering.

The analyte can be determined by various analytical methods in the solution prepared by the above method. The spectrophotometric method is based most often on native absorption or on a chemical reaction. The literature of spectrophotometric methods based on elution is very rich; several references will be given in Chapter 10. Here only two book chapters are mentioned: the chapter by Court[10] in the fundamental book of Shellard reviews the methodological problems and early literature of the field. There is a valuable chapter dealing with this field also in the book of Touchstone and Dobbins.[60]

If the measurement is based on natural UV absorption, the impurities dissolved from the layer cause several problems, as they may have a typical background spectrum. This effect can be suppressed by the solvent extraction of the layer or by the application of a development system to the layer prior to separation. Measurement in the short-wavelength region (below 240 nm) should be avoided anyway, if possible. As to the solution of background problems in general, we refer to Section 4.C. In this case, however, the simplest way of eliminating background effects is to use a solution obtained from the empty zone of the chromatogram as reference for the measurement by taking the same area and treating it in the same manner as the zone containing the analyte.

Of the numerous examples found in the literature, the method of Macioci and Fiotek[45] for the determination of chloramphenicol, hydrocortisone and methyl paraben in ointments is cited. A methanol extract of the preparation is applied to a silica gel plate together with the appropriate standards. The spots are marked on the basis of fluorescence quenching after running the chromatogram in a 254:30:14 mixture of dichloromethane, n-butanol and acetone. The spots are scraped from the layer, shaken for 10 minutes with methanol, and then the slurry is centrifuged. The three components (in the order given above) are determined by measuring their absorbances at 274, 242 and 255 nm, respectively. The extract of a blank zone of the layer prepared under the same conditions is used as reference solution. Recoveries were 98.9, 100.9 and 96.4%, the relative standard deviation did not exceed 0.6% in any of the cases. (The relative standard deviation of the large number of methods collected by Court[10] is higher than that, but with methods regarded as modern this deviation does not exceed 2 to 3%.)

Analysis is more often based on the measurement of color produced by chemical reactions than on natural absorption. In this case the light absorption of substances eluted from the layer causes no problem, and compounds with weak spectrophotometric activity can also be analyzed. The analysis can be based on the same reactions as those applied without chromatographic separation (Chapter 7), and there are no significant differences in the practical realization of the reaction and the spectrophotometric measurement, either.

As an example, the method of Horváth[35] is shown for the determination of digoxin and of the components of *Digitalis lanata* convertible into digoxin by enzymatic and then sodium methoxide treatment. Enzymatic cleavage is followed by ethyl acetate extraction, evaporation to dryness and then deacetylation with sodium methoxide. A methanol-chloroform solution is applied to the silica gel plate together with the appropriate standard solutions. The chromatogram is run with a 85:13.5:1.5 mixture of dichloromethane, methanol and water. The spots are marked after spraying the plate with xanthydrol reagent (0.1% solution in a 96:3:1 mixture of glacial acetic acid, sulphuric acid and hydrochloric acid), and heating to 50°C. The corresponding zones of the sample and the standards are cut from the aluminum-supported plate, and then extracted for 15 min with 5 ml of 80°C glacial acetic acid. The solution is reacted for 2 min at 80°C with the xanthydrol reagent mentioned above. Absorbance is read at 530 nm against the reagent as reference. The relative standard deviation of the method is 2.5%, which can be regarded as characteristic of the precision of similar methods. Since, as shown, elution methods have various sources of error, this performance can be achieved only by very careful and precise work after sufficient experience. This fact and the time- and labor-consuming nature of the technique have caused the method to be avoided in laboratories where the separation problems can be solved by thin layer densitometric (see next section) or HPLC methods. At the same time, the simplicity and the applicability in even poorly equipped laboratories ensure the continued popularity of this technique.

E. THIN-LAYER DENSITOMETRY (also see Section 2.M)

The most problematic step of the elution method discussed in the previous section is the recovery of separated substances for analysis. This is the source of most of the errors; this limits the sensitivity of the method; and this is the reason for its labor- and time-consuming nature. Thus, even in the early period of thin-layer chromatography there was an endeavor to perform the spectrophotometric quantitative analysis of separated substances *in situ*, i.e., without elution from the plate. For instance, in the book mentioned in the previous section, published in 1968, a separate chapter dealt with the possible application of densitometry in quantitative TLC analysis.[55]

The early methods, in which the measurement was based primarily on the relationship between the area of the spot measured planimetrically and the concentration of the substance in it, involved several factors of uncertainty: due to the poor sensitivities of instruments, relatively large amounts of sample, 10 to 100 µg, had to be introduced to the plate, usually yielding non-linear calibration curves. The area of the measured spot was strongly affected by the size of the sample spot, the value of R_f, etc.

The progress in densitometry has brought significant changes in the last 10 to 20 years to the potentials of *in situ* quantitative thin-layer densitometry, too. This started a major development of applications; in several fields, including pharmaceutical analysis, thin-layer densitometry has become a method of equivalent significance to HPLC and GC. This is shown by the publication of special monographs,[2,40,62] and by the fact that a number of books on TLC also discuss the theory, methodology and practical application of densitometry.[22,44,56,57,61,63,68]

The theoretical and practical foundations of the measurement of light reflection were discussed in Section 2.M. The advantages of modern thin-layer densitometry can be summarized as follows.

1. Excellent sensitivity. With reflection and transmission techniques the optimum sensitivity range is 0.5 to 5 µg/spot. The application of fluorimetric techniques or the above techniques with micro optics reduces this range to 10 to 100 ng/spot. This sensitivity is 2 to 3 orders higher than that attainable by the elution technique, and corresponds roughly to the sensitivity of GC with flame ionization detector (FID) or HPLC with UV detection.
2. Linear calibration. As pointed out in Section 2.M, for these low concentrations linear calibration curves can be obtained with systems which show non-linear behavior at higher concentrations. (This means that in the densitogram, in which a peak represents a chromatographic spot on the plate, peak height, or even more so the peak area determined by an integrator, is a linear function of the amount of sample in the spot.)
3. Applicability of the HPTLC technique. It is obvious from the above that densitometry can

be combined excellently with high performance thin-layer chromatography (HPTLC).[2,40] Thus, not only the sensitivity but also the resolution of the method approaches or, in some cases, surpasses that of GC and HPLC.

4. Possibility for measurement in the UV region. Since the advanced densitometers used nowadays are equipped with UV sources and detectors as well, it is possible to analyze several compounds on the basis of native absorption, without derivatization.

5. Possibility for reflection spectroscopy. In the investigation of unknown mixtures, the ability of modern densitometers to take reflection spectra is a great advantage since they are usually similar to the absorption spectra taken in solution phase. Such a spectrum is presented in Figure 2.19.

6. Great measurement capacity and flexibility. Since the optimization of a chromatographic system can be carried out independently of the instrument, and since the recording of the densitogram takes less time than the other steps (sample preparation, transfer to the plate, running of chromatogram), one densitometer may be used to evaluate the chromatograms taken by several analysts working simultaneously.

7. Versatile measurement possibilities. The method is applicable for measurements based on both natural absorption and chemical reactions. Chemical reactions may be applied prior to chromatographic separation (in this case the species measurable by densitometry is separated), or after chromatographic run (when the separated spot is made visible and thus measurable by densitometry by means of spraying with or dipping into a reagent). In the latter case a condition of measurement is that the excess of reagent does not cause interference. The chemical reactions used for this purpose are essentially the same as those used for non-chromatographic spectrophotometric analyses based on chemical reactions (see Chapter 7).

On the other hand, new possibilities of measurement techniques are also available. It is possible to perform the measurement in transmission, reflection or fluorescence modes. The latter mode, in which the presence of an appropriate fluorophore group or the possibility of forming such a group in the molecule may increase the sensitivity of the method into the picogram concentration range, is beyond the scope of this book. Transmission mode can be used only to measure colored spots in layers supported by glass plates, and thus reflection mode is the most commonly applied.

For both modes single- and double-beam instruments may be used. In the latter, the empty section of the layer is also scanned continuously, thereby compensating for distortions due to the unevenness of layer, impurities of the solvent and the layer, etc. In double beam mode, signal-to-noise ratio and the smoothness of baseline are also improved. A similar effect can be obtained by double monochromator instruments scanning simultaneously at two different wavelengths and recording the difference between the signals obtained at the two wavelengths.[14,66] (For details of the measurement at two wavelengths, see Section 4.G.)

With most instruments, the beam scanning the chromatogram is narrow for good chromatographic resolution, but its width makes it possible to monitor the full breadth of the spot simultaneously. However, so called "flying spot" densitometers are also available, in which a nearly point-like beam scans the chromatogram not only along its development axis (or perpendicularly to it) but alternately in two directions, in "zig-zag", or another fashion, to collect and average information from all points of the chromatographic spots.

The wide choice of instrumental techniques and measurement methods leads to an extreme versatility in selecting the best solution for the given analytical task. Therefore, thin-layer densitometry is used for the solution of the most diverse problems of pharmaceutical analysis (including the most delicate ones): medicinal plant analysis, impurity profile studies, determination of the active principle content of pharmaceutical preparations, stability studies and even investigations of biological samples. The books cited discuss several analytical methods, and the number of papers published in this field is enormous; even a review of this literature would exceed the scope of this book.

As an illustration, only one example is given: the method of Horváth and Iványi[36] for the determination of vinblastine in the extract of *Catharanthus roseus*. The determination of the active principle present in 0.02% in the plant accompanied by other alkaloids in larger (in some cases an order larger) amounts is, indeed, a very delicate task. The determination is carried out from 10 g of dried herb. After an ethyl acetate extraction (no details given here), evaporation of the extract to dryness, and dissolution of the extract in 10 ml of 1:1 methanol-chloroform mixture, 6 µl of the resulting solution are applied, accompanied by an appropriate amount of vinblastine standard solution, to the plate in the form of a strip 5 mm in length. As plate, a 20 × 20 cm silica gel 60 layer on aluminum foil support is used. After 2-D run (I: ethyl acetate–benzene–ethanol–10% ammonia, 100:5:23:2.3; II: dichloromethane–methanol–water 100:7:0.7), the spots are marked under UV light at 254 nm, and then measured by densitometry. The measurement is performed in reflection mode at the absorption maximum of vinblastine (see Figure 3.24), 289 nm; the slit applied is 10 mm in length and 0.2 mm in width. Vinblastine content is calculated from the area under the sample peak and the area under the peak of the standard. By using appropriate standards, other alkaloids can also be determined in a similar manner from the same chromatogram. Relative standard deviation is 7.5%, but in the case of simpler applications this can be reduced to 1 to 2%.

REFERENCES

1. Alfredson, T.; Sheehan, T.; Lenert, T.; Aamodt, S.; Correia, L. *J. Chromatogr.* 1987, *385*, 213.
2. Bertsch, W.; Hara, S.; Kaiser, R. E.; Zlatkis, A., Eds. *Instrumental HPTLC*, Hüthig Verlag: Heidelberg, 1980.
3. Binder, S. R.; Adams, A. K.; Regalia, M.; Essien, H.; Rosenblum, R. *J. Chromatogr.* 1991, *449*, 550.
4. Blau, K.; King, G. S. *Handbook of Derivatives for Chromatography*, Heyden: London, 1978.
5. Carter, G. T.; Schiesswohl, R. E.; Burke, H.; Yang, R. *J. Pharm. Sci.* 1982, *71*, 317.
6. Casteldine, J. B.; Fell, A. F.; Modin, R.; Sellberg, B. *J. Pharm. Biomed. Anal.* 1991, *9*, 619.
7. Chan, H. K.; Carr, G. P. *J. Pharm. Biomed. Anal.* 1990, *8*, 271.
8. Clark, B. J.; Fell, A. F.; Scott, H. P.; Westerlund, D. *J. Chromatogr.* 1984, *286*, 261.
9. Clark, C. R.; Chan, J.-L. *Anal. Chem.* 1978, *50*, 635.
10. Court, W. E.; Quantitative Thin-Layer Chromatography Using Elution Techniques, in *Quantitative Paper and Thin-Layer Chromatography*, E. J. Shellard, Ed.; Academic Press: London-New York, 1968; 29–49.
11. Deyl, Z.; Macek, K.; Janák, J. *Liquid Column Chromatography*, Elsevier: Amsterdam, 1975.
12. Doyle, T. D.; Levine, J., J. *J. Assoc. Off. Anal. Chem.* 1978, *61*, 172.
13. Drouen, A. C. J. H.; Billiet, H. A. H.; De Galan, L. *Anal. Chem.* 1984, *56*, 971.
14. Ebel, S.; Herold, G. *Arch. Pharm.* 1976, *309*, 660.
15. Ebel, S.; Mueck, W. *Z. Anal. Chem.* 1988, *331*, 359.
16. Fell, A. F. *Anal. Proc.* 1983, *20*, 356.
17. Fell, A. F.; Clark, B. J.; Scott, H. P. *J. Pharm. Biomed. Anal.* 1983, *1*, 557.
18. Fell, A. F.; Clark, B. J.; Scott, H. P. *J. Chromatogr.* 1984, *316*, 423.
19. Fell, A. F.; Scott, H. P. *J. Chromatogr. Biomed. Appl.* 1983, *273*, 3.
20. Fell, A. F.; Scott, H. P.; Gill, R.; Moffat, A. C. *Chromatographia* 1982, *16*, 69.
21. Fell, A. F.; Scott, H. P.; Gill, R.; Moffat, A. C. *J. Chromatogr.* 1983, *282*, 123.
22. Frei, R. W. Reflectance Spectroscopy in Thin-Layer Chromatography, in *Progress in Thin-Layer Chromatography and Related Methods*, A. Niederwieser; G. Pataki, Eds. Vol. II, Ann Arbor Sci. Publ.: Ann Arbor, 1971, 1–61.
23. Frei, R. W.; Lawrence, J. F. *Chemical Derivatization in Analytical Chemistry*, Vol. 1–2, Plenum Press: New York, London, 1982.
24. Gazdag, M.; Görög, S., unpublished results.
25. George, S. A.; Maute, A. *Chromatographia* 1982, *15*, 419.
26. Görög, S.; Herényi, B. *J. Chromatogr.* 1987, *400*, 177.
27. Görög, S.; Herényi, B.; Rényei, M.; Georgakis, A.; Balogh, G.; Csehi, A.; Gizur, T. *Magy. Kém. Foly.* 1989, *95*, 504.
28. Görög, S.; Laukó, A.; Herényi, B. *J. Pharm. Biomed. Anal.* 1988, *6*, 697.
29. Görög, S.; Rényei, M.; Herényi, B. *J. Pharm. Biomed. Anal.* 1989, *7*, 1527.
30. Graham, R. E.; Biehl, E. R.; Kenner, C. T. *J. Pharm. Sci.* 1979, *68*, 871.
31. Graham, R. E.; Kenner, C. T. *J. Pharm. Sci.* 1973, *62*, 1845.
32. Herényi, B.; Görög, S. *J. Chromatogr.* 1982, *238*, 250.
33. Herrmann, E. C.; Hoyer, G.-A. *Chem. Ber.* 1979, *112*, 3748.
34. Hill, D. W.; Kelley, T. R.; Langer, K. J. *Anal. Chem.* 1987, *59*, 350.
35. Horváth, P. *Acta Pharm. Hung.* 1982, *52*, 133.
36. Horváth, P.; Iványi, G. *Acta Pharm. Hung.* 1982, *52*, 150.
37. Huber, J. F. K. Ed. *Instrumentation for High-Performance Liquid Chromatography*, Elsevier: Amsterdam, 1978.
38. Huber, L.; Zech, K. *J. Pharm. Biomed. Anal.* 1988, *6*, 1039.

39. Inczédy, J. *Analytical Applications of Ion Exchangers,* Pergamon Press: Oxford, 1967.

40. Kaiser, R. E., Ed. *Instrumental High-Performance Liquid Chromatography,* Institute for Chromatography: Bad Dürkheim, 1982.

41. Knapp, D. R. *Handbook of Analytical Derivatization Reactions,* Wiley: New York, 1979.

42. Lawrence, J. F.; Frei, R. W. *Chemical Derivatization in Liquid Chromatography,* Elsevier: Amsterdam, 1976.

43. Lopez Silva, F.; Bocic Vildosola, R.; Vallejos Ramos, C.; Alvarez Lueje, A. *An. Real. Acad,. Farm.* 1990, *56,* 19.

44. Macek, K., Ed. *Pharmaceutical Applications of Thin-Layer and Paper Chromatography,* Elsevier: Amsterdam, 1972.

45. Macioci, F.; Fiotek, C. *Boll. Chim. Farm.* 1975, *114,* 468.

46. Marr, J. G. D.; Horváth, P.; Clark, B. J.; Fell, A. F. *Anal. Proc.* 1986, *23,* 254.

47. Mikes, O., Ed. *Laboratory Handbook of Chromatographic and Allied Methods,* Ellis Horwood: Chichester, 1979.

48. Minder, E. I.; Schaubhut, R.; Vonderschmitt, D. J. *J. Chromatogr.* 1987, *419,* 135.

49. Minder, E. I.; Schaubhut, R.; Vonderschmitt, D. J. *J. Chromatogr. Biomed. Appl.* 1988, *428,* 369.

50. Owino, E.; Clark, B. J.; Fell, A. F. *J. Chromatogr. Sci.* 1991, *29,* 298.

51. Owino, E.; Clark, B. J.; Fell, A. F. *J. Chromatogr. Sci.* 1991, *29,* 450.

52. Schieffer, G. W. *J. Chromatogr.* 1985, *319,* 387.

53. Scott, E. P. *J. Pharm. Sci.* 1983, *72,* 1089.

54. Scott, R. P. W. *Liquid Chromatography Detectors,* Elsevier: Amsterdam, 1977.

55. Shellard, E. J. Quantitative Thin-Layer Chromatography Using Densitometry, in *Quantitative Paper and Thin-Layer Chromatography,* E. J. Shellard, Ed., Academic Press: London, New York, 1978.

56. Sherma, J.; Fried, B., Eds. *Handbook of Thin-Layer Chromatography,* Marcel Dekker: New York, 1991.

57. Stahl, E., *Die Dünnschicht-Chromatographie,* Springer Verlag: Berlin, Göttingen, Heidelberg, 1965.

58. Szepesi, G. *HPLC in Pharmaceutical Analysis,* Vol. I, II, CRC Press: Boca Raton, 1990–1991.

59. Szepesi, G.; Gazdag, M.; Mihályfi, K. *J. Chromatogr.* 1989, *464,* 265.

60. Touchstone, J. C.; Dobbins, M. F. *Practice of Thin-layer Chromatography,* Wiley: New York, 1978.

61. Touchstone, J. C.; Rogers, D., Eds. *Thin-Layer Chromatography: Quantitative Environmental and Clinical Applications,* Wiley: New York, 1980.

62. Touchstone, J. C.; Sherma, J., Eds. *Densitometry in Thin-Layer Chromatography,* Wiley: New York, 1979.

63. Treiber, L. R., Ed., *Quantitative Thin-Layer Chromatography and its Industrial Applications,* Marcel Dekker: New York, 1987.

64. Turcant, A.; Premel-Cabic, A.; Cailleux, A.; Allain, P. *Clin. Chem.,* 1991, *37,* 1210.

65. White, P. C. *J. Chromatogr.* 1980, *200,* 271.

66. Yamamoto, H.; Kurita, T.; Suzuki, J.; Hira, R.; Nakano, K.; Makabe, H.; Shibata, K. *J. Chromatogr.* 1976, *116,* 29.

67. Yost, R.; Stoveken, J.; MacLean, W. *J. Chromatogr.* 1977, *134,* 73.

68. Zlatkis, A.; Kaiser, R. E., Eds., *HPTLC—High-Performance Thin-Layer Chromatography,* Elsevier: Amsterdam—Institute of Chromatography: Bad Dürkheim, 1977.

Chapter 6

SPECTROPHOTOMETRIC METHODS
BASED ON CHEMICAL REACTIONS

A. FACTORS NECESSITATING THE APPLICATION OF CHEMICAL REACTIONS IN SPECTROPHOTOMETRY

The previous chapters were mainly concerned with those applications of spectrophotometric drug analysis where the investigation can be based on the natural ultraviolet-visible absorption of the analyte. The limitations of this approach have been pointed out several times from the points of view of the field of applicability, sensitivity, selectivity and effectiveness of the measurement. In this chapter, one of the most important methods to extend the possibilities of spectrophotometric analysis is discussed, namely the combination of the spectrophotometric measurement with a preliminary chemical reaction where the reaction product is the subject of the spectrophotometric measurement. The chemical reaction is selected in such a way that the absorption spectrum of the reaction product should be shifted toward the long wavelengths with occasional increase of its intensity as compared with that of the parent compound. By applying this approach, the scope of the spectrophotometric analysis of drugs can be expanded in the following areas:

1. Determination of spectrophotometrically inactive compounds. As it has already been discussed in the previous chapters, substances with absorption maxima below 220 nm are usually considered to be spectrophotometrically inactive; the possibilities of the direct spectrophotometric determination of this class of compounds is very difficult and in the majority of cases even impossible. The structural basis of spectrophotometric inactivity is discussed in Sections 3.C and 3.D, where several examples are also shown to characterize this group of drug materials and related substances. These materials usually contain functional groups (hydroxy, oxo groups, etc.) which react with spectrophotometrically active reagents leading to derivatives with absorption maxima in the long-wavelength ultraviolet or visible spectrum range. Applying this principle the indirect determination of the majority of spectrophotometrically inactive compounds becomes possible.
2. Increasing the selectivity of the measurement. Preliminary chemical reactions are often applied in those instances also when the compound to be determined does have its own spectrum in the spectrophotometrically accessible wavelength range but this spectrum is not selective enough to allow its selective determination. This is the case e.g., with the simultaneous selective determination of phenols and their ether derivatives or spectrophotometrically active alcohols and their acyl derivatives or carboxylic acids and their esters. Thus, in these cases the spectra of the two compounds are too similar to allow the possibility of the selective spectrophotometric determination of either of them. In such instances by means of a suitable reaction the spectrum of one of them is selectively shifted to such a spectral range that the selective determination can be carried out. In addition to the possibilities described in Section 4.C this principle is also suitable to solve background problems.
3. Increasing the sensitivity of the measurement. The natural absorption of spectrophotometrically active compounds is in many cases sufficient for the solution of various problems. Sometimes, however, when high sensitivity would be necessary for the solution of the

problem the natural absorption is too weak for the required purpose. In these cases a suitable reaction leading to a product with considerably higher molar absorptivity can solve the problem.

A typical example is the determination of oestrogenic steroids. The α-band of these phenol-type compounds at about 280 nm with its medium intensity ($\varepsilon = 2,000$) enables their determination in several pharmaceutical formulations. This intensity is, however, insufficient for their determination in contraceptive pills and especially in biological samples where their concentration is very low. In such instances color reactions are widely used leading to reaction products with molar absorptivities in the order of 30–40,000.

4. Another purpose for using chemical reactions in spectrophotometric drug analysis is to enable certain instruments (colorimeters) to be used. As is discussed in Section 1.A, in the early period of spectrophotometry for the determination of colorless compounds, this was the only possibility because in that time the overwhelming majority of the instruments available for drug analysts worked in the visible spectral range only. Updated versions of these classical colorimeters are still in use, especially for clinical drug analysis. For this reason, the application of color reactions is still of considerable importance in this field. It should also be mentioned here that the majority of automatic analyzers use visible spectrophotometers or filter photometers (see Section 9.J). Important parts of these analyzers are the reagent pumps and the reaction coil where the color reaction takes place.

B. CONSIDERATIONS FOR SELECTION OF THE REACTION

In the preceding section the main factors requiring the use of chemical reactions prior to the spectrophotometric measurement were briefly outlined. If the primary goal for the application of a chemical reaction is to increase the selectivity of the measurement it is necessary to decide first which functional group in the molecule is transformable by means of an appropriate reaction to a chromophoric system suitable for the selective measurement. As an example, the selectivity problems in the assay of corticosteroid preparations (e.g., hydrocortisone) are briefly described. The non-selective but very simple method for their assay is to measure the absorbance of the intense band of the unsaturated 3-keto group at 240 nm. If excipients or other active ingredients of the formulation interfere with this assay, the selectivity can be increased by using e.g., isonicotinoyl hydrazide to react with the 3-keto group, shifting its absorption maximum to about 380–400 nm (see Section 10.M.2).

This reaction is suitable to solve the selectivity problem originating from foreign absorbances but it is by no means suitable to serve as the basis for stability assays. This is because both the oxidative and the hydrolytic decomposition of the corticosteroids takes place at the α-ketol side chain.

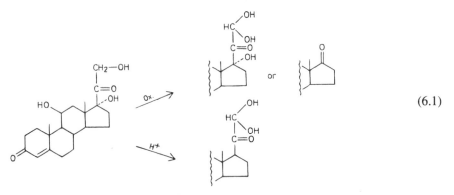

$$(6.1)$$

As seen in Equation 6.1, the degradation products contain the 4-ene-3-keto group and for this reason simple absorbance measurement at 240 nm is not stability-indicating. The application of the above mentioned isonicotinoyl hydrazide method which is based on the same group does not solve the problem of the stability assay either: the degradation products also give positive reactions. This applies to several other reactions based on the hydroxyl and ketone groups in the molecule.

Only those reactions which are based on the reducing properties of the α-ketol type C-17 side chain are suitable for the stability-indicating assay because this property is lost in the course of any kind of the degradation pathways in Equation 6.1. Accordingly, the tetrazolium reaction, by which the undecomposed corticosteroids reduce the colorless tetrazolium reagents to colored formazans while the decomposition products do not, is successfully applied to their stability assay (see Section 10.M.2).

However, if the analytical task is the selective determination of free hydrocortisone in the presence of its 21-ester derivatives, e.g., as an impurity in the latter not even the tetrazolium method or similar procedures can be successfully applied. The reason for this is that in the strongly alkaline reaction medium of the tetrazolium reaction the 21-acyloxy group hydrolyzes and hence the fact that they do not react directly with the tetrazolium reagents cannot be exploited for the selective determination of the 21-hydroxy derivatives. A suitable procedure for this purpose is the 2-step reaction described on p. 332 (see reaction scheme 10.19). In this procedure the 21-hydroxy group is first oxidized to aldehyde (the 21-ester derivatives do not react) followed by the condensation of the 20-keto-21-aldehyde system with 4,5-dimethyl-o-phenylenediamine reagent to form quinoxaline derivatives which are measurable around 350 nm[3].

In addition to selectivity another important factor that has to be taken into consideration in the selection of the chemical reaction prior to the spectrophotometric measurement is the sensitivity of the latter. In the majority of cases of application of those reactions it is the aim of the selection to afford at least 5,000 as the molar absorptivity of the reaction product. Sensitivity which can be characterized by molar absorptivities in the range of 10–20,000 is generally sufficient to solve the majority of problems in drug analysis. When higher sensitivity is necessary (analysis of very low-dose pharmaceutical preparations, determination of drugs and metabolites in biological samples, etc.) derivatives with molar absorptivities around or above 50,000 may be necessary.

Further factors to be taken into account in the course of selecting chemical reactions for spectrophotometric assays include the requirement for well defined chemistry of the reaction. In the early period of photometry the restrictions in the instrumentation made it necessary to transform the usually colorless analyte to colored derivatives. In many cases the chemical background of these procedures was not clear, but subsequent research often revealed that the reaction leading to the colored products had not been unidirectional: the yields were not quantitative or, in many cases, definitely poor. The reproducibility and the reliability of these methods are usually insufficient. The guiding principle in this respect in modern drug analysis is that the description of a new method is only acceptable if it is proved that the chemistry of the reaction leading to the chromophoric product is well established, unidirectional, and quantitative, or at least almost quantitative. In the case of old, classical reactions exceptions sometimes have to be made: even the pharmacopoeias make concessions in some cases by applying spectrophotometric methods with dubious chemical background. In such cases the careful optimization of the method is especially important and the optimal conditions should be strictly adhered to.

Another important point is the time and labor consumption of the procedure. Long reaction time is a disadvantage just as the necessity of separating the reaction product from the excess of the reagent by extraction or chromatography. Important factors are the stability of the reaction product under the conditions of the reaction and the measurement as well as the capability for automation of the reaction and the subsequent measurement.

When selection has to be made among virtually equivalent methods, the analytical accuracy and precision attainable with the reaction is also an important factor. Great caution is necessary if this comparison is not based on the analyst's own experimental results but on data taken from the literature: the statistical data characterizing the accuracy and precision of a method are often of very variable standards.

Similarly, one must be very cautious before adopting a method from the literature described specifically for the determination of a drug in biological samples for the analysis of pharmaceutical products. The demands regarding accuracy and precision are usually lower in biomedical analysis and often do not meet the stricter requirements of pharmaceutical analysis.

A general problem with the spectrophotometric methods based on chemical reactions is that in many of the methods published in the literature, it is not clear what the authors' aim was when they decided to develop a new method, whether they carefully studied existing methods for the solution of a certain drug analytical problem before starting to develop their own. In many cases not even the conclusions of the papers contain convincing evidence about the superiority of the published new method to the existing ones. The "confusion of abundance" which is evident in the sections of Chapter 10 certainly means that many of the published methods are unnecessary, "l'art pour l'art" procedures. Some of these methods have been developed at a sufficiently high level and more or less carefully validated, many others, however, are not. For this reason it is not an easy task for an analyst (especially for beginners) to make the decision after studying the literature available for the solution of a certain problem as to which of the existing methods is suitable for solving the given problem or whether it is reasonable to develop a new method by selecting a new (or at least from the analytical point of view, new) reaction for the measurement. Some more aspects of this question are discussed in Section 10.A.

C. ASPECTS OF THE OPTIMIZATION OF THE METHOD

If the most apparently suitable reaction has been selected for the solution of a given problem, the method has to be optimized taking into consideration the following points.

1. Solvent, Composition and Dilution of the Reaction Mixture

In advantageous cases the solvent for the reaction and for the spectrophotometric measurement (as well as for the extraction of the drug to be determined from various matrices) is the same. Very often, however, the situation is not so simple: the solvent for the extraction or for the reaction is not suitable for the spectrophotometric measurement or the ideal solvent for the measurement is unsuitable for the reaction or the extraction. In such cases either solvent-solvent extraction or evaporation to dryness and redissolution have to be applied.

The above problem can often be solved in such a way that the reaction is allowed to run in relatively concentrated solution which is considerably diluted with a suitable solvent prior to the spectrophotometric measurement. This can be achieved if large quantities of samples are available for the analysis and 50–100 mg weighings can be applied. If the sample size is limited, a few milligrams are weighed using a micro balance and the reaction is run at the same concentration as that used for the spectrophotometric measurement.

The same applies to the analysis of low-dosed pharmaceuticals and biological samples. The greatest sensitivity is attainable only if the reagents are added to the sample dissolved in a minimum volume of the solvent and no further dilution is made. In this case the only requirement regarding the volume of the solution is that it should be enough to fill the cell. This means about 5–10 ml in the case of the usual cells and about 1 ml in the case of micro cells. In this way the lowest measurable quantity is in the order of 10 μg (although in the case of the application of very sensitive methods and/or micro cells it is in the order of 1 μg). This high sensitivity is usually necessary only in the case of the analysis of biological samples. In the analysis of pharmaceutical products the test solution is usually prepared by diluting it to the final volume in a volumetric flask, thus ensuring higher accuracy and precision.

2. Reagent Concentration

It is an important step for the optimization of a procedure to find the optimum reagent concentration which enables the reaction to run at an appropriate rate and, if possible, without side reactions and equilibria. From this point of view, of course, high reagent concentrations would be most advantageous. However, high reagent concentrations can only be used if the absorbance of the reagent is low compared with that of the reaction product at the absorption maximum of the latter. If this requirement is met, then this low absorbance does not interfere since it is also present in the reference solution. Excessively high light absorption of the reference solution is still to be avoided since it reduces the spectral purity of the measurement and decreases its sensitivity as well. The reason for this decrease is the following. If the absorbances of the reaction product and the excess reagent are commensurable values, the reaction results, on the one hand, in the increase of the absorbance of the test solution (as a consequence of the absorbance of the reaction product). On the other hand, however, a decrease will take place (due to the consumption of the absorbing reagent during the reaction). The measured absorbance is the difference of the two values. This is why the apparent molar absorptivity thus obtained may be considerably lower than that of the isolated reaction product. The use of reagents with the above described spectral characteristics should be avoided as far as possible or the reaction product and excess reagent have to be separated from each other by extraction or chromatography.

Taking the above points into consideration the reagent concentration–absorbance plot has to be established at the maximum of the spectrum under otherwise optimized conditions at constant analyte concentration. At the low concentration range of the reagent, the absorbance will naturally increase with increasing reagent concentration and then, in ideal cases, it remains constant upon further increase of the concentration; (sometimes it decreases). It is quite inadvisable to select the reagent concentration from the first concentration range, since in this case increasing the concentration of the analyte would lead to negative deviation from Beer's Law. For the same reason it is expedient to select such a (constant) concentration for the analyte during the investigation of the reagent concentration–absorbance correlation which corresponds to the maximum concentration where the method is intended to be used. In practice it is usual to select a reagent concentration which is higher than the value where the reagent concentration–absorbance plot reaches its plateau by about 50–100%

3. The pH of the Solution

In the majority of spectrophotometric methods the proper selection of the pH is an important factor of method optimization. This problem is often divided into two parts: the optimization of the pH of the reaction mixture and of the spectrophotometric measurement after the completion of the reaction have to be treated separately.

As for the optimum pH of the reaction, this question has to be investigated from two points of view. The pH of the reaction mixture may influence the rate of the reaction: from this point of view, of course, the goal is to increase the reaction rate. In the case of reactions leading to an equilibrium, the pH may influence the quantitative running of the reaction. Both points of view have to be taken into account when selecting the optimum pH for the reaction mixture. If this value is between 2 and 12, buffers have to be used which should be selected in such a way that they must not interfere with the reaction and the spectrophotometric measurement.

The optimization of the pH of the test solution in the spectrophotometric measurement is discussed in detail in Section 3.P. The most important point is that in the case of acidic or basic analytes where the spectrum depends on the pH, such a pH has to be selected where they are present either in fully protonated or deprotonated form. The question whether the protonated or the deprotonated form is more suitable for the measurement can be decided on the basis of the attainable sensitivity or selectivity taking into account also the influence of the pH on the stability or the reaction product.

Since the optimum pH of the reaction mixture and of the spectrophotometric measurement are not necessarily the same, the pH has often to be changed before the spectrophotometric measurement is carried out.

4. Temperature

In the overwhelming majority of cases the spectrophotometric measurement is carried out at room temperature but the reaction is often run at different temperatures. The reaction mixture is heated if the reaction needs to be accelerated and cooling is applied if side reactions have to be avoided. Care should be exercised with the application of elevated temperatures: since this could result in side reactions, the rates of which may be negligible at ambient temperature.

The term "room (ambient) temperature" is also to be used cautiously since fluctuations up to 10°C may occur. Since the rates of average chemical reactions are increased two- or three-fold by a 10°C increase of the temperature, this can be the source of serious errors.

In general it can be stated that whether the reaction runs at room temperature or under different conditions, the temperature should be given (and maintained) with an accuracy of 1°C and the reaction temperature should be chosen in such a way that a change of 1–2°C does not influence the absorbance.

5. Reaction Time: Stability of the Reaction Product

These are among the most important parameters of the optimization of spectrophotometric methods. Under otherwise optimum conditions, the reaction time—absorbance curve has to be established. In ideal cases the curve can be characterized by an ascending time vs. absorbance section which gradually changes to plateau. The constancy of the absorbance means that the spectrophotometrically active reaction product is stable under the conditions of the reaction. In such cases it is prudent to select a reaction time which is 1.5 to 2-times the time needed to reach constant absorbance. Because of practical reasons prolonged reaction time (several hours) is not expedient. The reduction of the reaction time can be achieved by increasing the concentration of the reagent, changing the pH and mainly by increasing the reaction temperature.

Much more problematic are the reactions where the instability of the reaction product results in time-absorbance plots which can be characterized as curves with maxima. If the rate of the decrease of the absorbance after the maximum is not too high (not exceeding 5–10% per hour) the measurement can be done provided that the reaction conditions are strictly followed. Color reactions with sharp maxima in the time-absorbance plot (characteristic of many of the old methods mainly in the biochemical literature) should be avoided.

In these cases considerable decomposition of the reaction product takes place before reaching the maximum and the shape of the curve and the position of the maximum are influenced by barely controllable factors; this can be a source of very serious errors, even if the calculation is based on simultaneously run standards.

It is to be noted that in those cases when the composition of the reaction mixture and of the test solution differs considerably, the stability of the reaction product should be checked not only in the reaction mixture but also in the test solution.

6. The Quantitative Characteristics of the Reaction

A spectrophotometric method based on chemical reaction can be considered to be well developed if it follows a strict stoichiometry and the reaction leads quantitatively to the chromophoric product. In evaluating the fulfillment of the latter demand it is not enough to establish the reaction time required for the absorbance to become constant, since in many cases this reaction time is only the time necessary to reach an equilibrium (possibly with low yield). Even more characteristically, the yield of the reaction could be reduced by simultaneous side reactions. It is part of the optimization of the method to find the parameters (reagent concentration, solvent, pH, etc.) which influence the extent of the reaction.

Generally speaking it is not an easy task to determine the absorbance which corresponds to 100% completion of the reaction. The ideal solution to the problem is to prepare the chromophoric reaction product in pure form and determine its molar absorptivity at the analytical wavelength. If this is possible, with this data the course of the reaction can easily be measured from samples taken from the reaction mixture. Reactions where this can be done are preferred to reactions with an obscure chemical background.

7. Reagent Blanks

The success of a spectrophotometric method based on chemical reaction requires the consistent application of the principle that the solution in the reference cell (reagent blank) should contain all components (except the analyte) at the same concentration as they are present in the sample solution (solvents, reagent, buffer, etc.). It is also very important that the sample and reference solutions should undergo identical treatment (equal reaction time, temperature, etc.). In this way a considerable proportion of background absorption can be avoided.

8. The Validity of Beer's Law

In the course of the development of a method based on a chemical reaction it is important to check the validity of Beer's Law, i.e., to investigate if there is a linear relationship between the concentration of the analyte and the absorbance at a characteristic wavelength of the spectrum of the chromophoric reaction product (usually at the absorption maximum). In addition to the factors described in Section 4.B deviations from the linear relationship can be due to the improper optimization of the reagent concentration, pH and other factors discussed in the preceding paragraphs. In this way, the linearity check can be considered to be a control of the optimization. Positive or negative intercepts of the calibration graph usually indicate that the reference solution has been incorrectly prepared. Apart from absolutely obscure reactions of unknown chemical basis, a linear calibration graph with negligible intercept is usually attainable by the systematic optimization of the parameters. The regression equation of the calibration graph and the concentration range of the validity of Beer's Law are indispensable parts of the documentation of a new spectrophotometric method. It has to be noted that—due to the limited excess of the reagent—in the case of methods based on chemical reactions this range is usually narrower than in the case of methods based on natural absorption. It is usually considered to be satisfactory if Beer's Law is obeyed in the 0–1.5 absorbance range. To make calculations from absorbances above the linearity range may be a source of serious errors.

9. The Application of Standards

It can be concluded from the points discussed so far that the development and optimization of spectrophotometric methods based on chemical reactions are more delicate tasks with much more potential sources of error than is the case with methods based on natural absorption. Even if all the above discussed hints are strictly followed, it is indispensable to reduce the systematic error of the method by simultaneously running a reaction under exactly identical circumstances with a reference standard. After having already checked the linearity of the concentration—absorbance graph it is sufficient to adopt only one or two standardization points within the same concentration range as that expected for the test solution. The calculation is then based on the absorbance value(s) obtained for the standard. If the measurement and the calculation are not based on *simultaneously* run standards, but on a specific absorbance value determined in the same laboratory using the same instrument, minor errors may occur, relying, however, on specific absorbance values taken from the literature can be the source of very serious errors especially in the case of insufficiently reliable literary data.

Measuring according to the above outlined principles increases the accuracy of the method. One disadvantage, however, is that a standard sample of high purity is necessary. In the case of serial measurements, so called "working standards" can also be used which are somewhat less pure than the reference standard; their assay is carried out with great care using the same method and the purity obtained (expressed in percentage) is taken into account in the calculations.

To avoid the use of standards, in many cases even at the compendial level the description of the analytical procedure contains the specific absorbance which is the basis of the calculations. This approach is characteristic of many of the monographs of the British Pharmacopoeia, mainly of those which are based on natural absorption but in some cases no use of standard is prescribed for the spectrophotometric methods based on chemical reactions either. In these cases we have good reasons to assume that those experts who developed the monographs in question carefully optimized the method and determined the given specific absorbance value with the highest possible accuracy. If the users of the method reproduce it under carefully validated experimental conditions the error caused by the omission of the standard can be minimized. This naturally does not apply to specific absorbances originating from unreliable sources. In general the approach of the United States Pharmacopoeia presents greater safety to the analyst by prescribing all spectrophotometric measurements (based on natural absorption and chemical reactions) with simultaneously run reference standards.

When the aim is to reduce the systematic error of the measurement in the course of the determination of drug substances in various matrices by using standards, a further possibility is not to investigate the standard separately but to introduce it into the matrix (placebo tablet or injection, blood or urine sample not containing the drug to be determined). The standard addition method which is generally used in electroanalytical chemistry is only seldom applied in spectrophotometric drug analysis. Here the standard is added in measured concentration to the sample to be investigated and the calculation is based on the absorbances measured with and without the addition of the standard.

10. Statistical Evaluation of the Results

The most important criterion of the applicability of an analytical method to a given problem is the attainable accuracy and precision. The possibility of decreasing the systematic error (i.e., increasing the accuracy) was discussed in the preceding section. The question of the precision of the method has also to be carefully investigated. In the case of working out a new method or adopting one from the literature the mean error of the method has to be determined. If this is higher than 1–1.5% which is generally considered to be acceptable then the influence of the parameters discussed in the previous sections on the precision of the method has to be checked in order to be able to decrease it. If these efforts are unsuccessful then this usually means that the chemical basis of the method is problematic. If the attainable precision does not meet the established requirements of the method to solve the given problem, the method should be considered as unsuitable for this purpose.

D. A PRACTICAL EXAMPLE: SELECTION AND OPTIMIZATION OF THE REACTION FOR THE DETERMINATION OF PREDNISOLONE-21-MESYLATE; APPLICATION OF THE METHOD FOR THE DETERMINATION OF THE LATTER AS AN IMPURITY IN MAZIPREDONE

1. The Analytical Task

Mazipredone®, the water-soluble prednisolone derivative of the Chemical Works of Gedeon Richter Ltd. (Budapest), the active ingredient of the Depersolone® products is prepared from prednisolone.[9] As is seen in the reaction scheme 6.2[2] the key step of the synthesis is the reaction of prednisolone 21-mesylate with N-methylpiperazine. It was therefore an important task to develop a selective method for the determination of the key intermediate prednisolone-21-mesylate (PM) with a view to the determination of PM in mazipredone. It was a requirement of the method to be developed that it should enable the selective determination of PM in the presence of the starting material of the synthesis (prednisolone) and the final product (mazipredone), and it should additionally allow its determination in the latter at the 0.1% level.

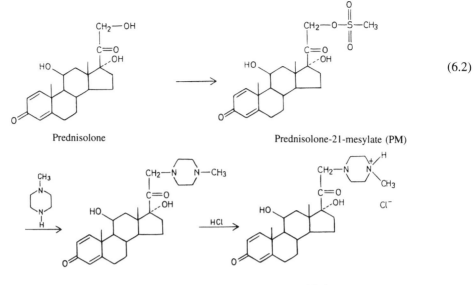

(6.2)

Prednisolone

Prednisolone-21-mesylate (PM)

Mazipredone

2. Aspects of the Selection of the Reaction for the Determination of PM

As seen in reaction scheme 6.2, prednisolone, PM and mazipredone possess the same chromophoric group (1,4-diene-3-ketone) and hence no method based on natural absorption could come into question for the selective determination of PM. Similarly, no methods could be considered for this purpose which were based on the reactions of the common functional groups of the three substances. The difference is only at the substituent in the 21-position and for this reason the method has to be based on this. The methods described in Section 10.M. for the determination of corticosteroids are not suitable for the selective determination since these depend on the reducing properties of the α-ketol-type side chain. Prednisolone and mazipredone are strong reducing agents giving e.g., positive tetrazolium reaction[3] the very reactive, active-ester type PM in turn hydrolyses to prednisolone under the alkaline conditions of the tetrazolium reaction and gives therefore also positive reaction.

The final selection of the reaction meeting the outlined requirements was based on the great reactivity of the 21-active ester group of PM. The great reactivity does not only mean strong tendency to hydrolyze but it is reactive toward various nucleophilic agents, too. Of the nucleophilic reagents, pyridine was chosen because of the following reasons:

1. It is inexpensive and easily available.
2. The reaction product, 21-deoxy-21-(1-pyridinium)-prednisolone mesylate (or, in short pre-dnisolone pyridinium mesylate, PPM) (see reaction scheme 6.3) is the isolated byproduct of the synthesis of mazipredone[2] which was available in crystalline form. Its chemical and spectroscopic properties were well known and hence it was possible to fulfill the requirements described in Section 6.C.5.

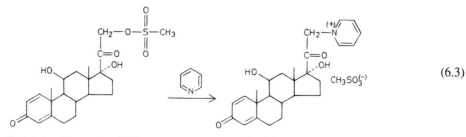

(6.3)

Prednisolone-21-mesylate (PM) Prednisolone-pyridinium-mesylate (PPM)

3. PPM formed in the reaction (6.3) undergoes intramolecular rearrangement in alkaline medium to form the zwitterion-like prednisolone pyridinium enolate (PPE).[2,5] This derivative was also available in crystalline form. This rearrangement (see reaction scheme 6.4) leads to an equilibrium depending on the pH of the reaction mixture. After having found evidence for the reversibility of the reaction (by repeated acidification and basification of the reaction mixture) the equilibrium constant was determined on the basis of the absorbance vs. pH plot at the maximum of the enolate form. It was found to be $K = 5.43 \cdot 10^{-12}$ in aqueous medium. This means that in strongly alkaline medium PPM can be quantitatively transformed to PPE.

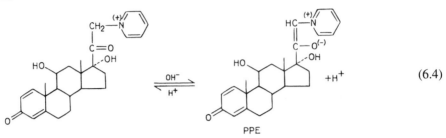

$$(6.4)$$

4. The spectrophotometric properties of PPE are ideally suited to the solution of this analytical task. Under the selected experimental conditions to be described later the absorption maximum is at 414 nm. This means that the reaction product is colored (yellow). Neither the excess reagent nor prednisolone and mazipredone possess light absorption at this wavelength. Since neither of the latter reacts with pyridine, under the conditions of the reaction of PM, the selectivity of the reaction seemed to be ascertained. The molar absorptivity of PPE is high under the given experimental conditions ($\varepsilon = 21,900$) and hence the prospects were favorable even from the point of view of sensitivity.

3. Optimization of the Reaction Conditions

In this section the detailed description of the development and optimization of the method is described in the spirit of Section 6.5 and following the sequence of optimization parameters described there. At first the optimization of the reaction PM $\rightarrow$ PPM $\rightarrow$ PPE will be given[5] while the problems of the determination of PM as an impurity in mazipredone will be the subject of the subsequent sections.

Since, as has already been mentioned, pyridine was selected as the *reagent* for the reaction, it seemed to be expedient to use it at the same time as the reaction *solvent* (6.3), too. In such a way the reagent could be used in great excess, using the previously discussed fact that the excess reagent does not interfere with the measurement of the reaction product. Pyridine is a good solvent for PM and also for mazipredone at a sufficient concentration to form the basis of the determination of the PM impurity. Its high boiling point enables the reaction time to be shortened by increasing the temperature of the reaction mixture. For the preceding reasons, reaction (6.3) was carried out in all cases during the subsequent study by dissolving the samples in 10 ml of pyridine.

After the completion of reaction (6.3) the equilibrium (6.4) is shifted toward the right by basification. It is expedient to do this after ten-fold dilution of the reaction mixture from reaction 6.3. Ten percent of pyridine does not interfere with the establishment of the equilibrium and, for this reason, its removal is not necessary. As solvents for the dilution water and ethanol were considered. The spectra of PPE in Figure 6.1 were prepared in aqueous and ethanolic solutions containing 10% v/v pyridine in the presence of the optimal tetramethylammonium hydroxide concentration to be discussed later on.

As shown in Figure 6.1 a considerable bathochromic and hyperchromic shift is produced by ethanol as compared with the spectrum taken in the water–pyridine mixture. For this reason and because the 9:1 mixture of ethanol and pyridine is a much better solvent for the investigated steroids than the similar mixture of water and pyridine, ethanol was used to dilute the reaction mixture of reaction (6.3) for the subsequent spectrophotometric measurement. A further advantageous feature of ethanol is that it influences the equilibrium (6.4) favorably: ten-fold lower tetrameth-

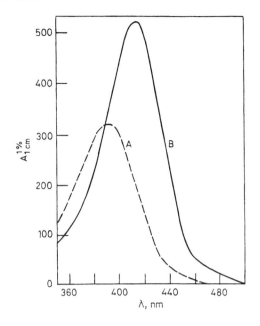

FIGURE 6.1. Spectra of 21-deoxy-21-(1-pyridinium)-prednisolone enolate (PPE). Solvent: A. 0.01 M tetramethylammonium hydroxide in water containing 10% v/v pyridine; B. 0.01 M tetramethylammonium hydroxide in ethanol containing 10% v/v pyridine. (From S. Görög; Z. Tuba, *Analyst* 1972, *97*, 523.[5])

ylammonium hydroxide concentration is necessary to shift the equilibrium toward the formation of PPE than is necessary when water is used for the dilution.

The excellent sensitivity of the method makes it possible that the above mentioned 10 ml of pyridine (which is subsequently diluted to 100 ml with ethanol) contain only 0.1–2 mg of PM (or a quantity of mazipredone containing this amount of PM) to obtain absorbances in the range of 0.05–1.

The approach of using the same material as the reagent and the solvent makes it unnecessary to deal with the effect of *excess reagent* but the investigation of the *effect of pH* is still very important. As is described in Section 6.C.3 the question of the optimization of the pH of the reaction mixture and of the test solution has to be dealt with separately. As for the reaction mixture it can be stated that although reaction 6.4 takes place in pyridine without the addition of any further reagent, in order to stabilize the forming PPM, 0.1 ml of 5 M hydrochloric acid was added to the reaction mixture before the reaction. The detailed reasons for this and the optimization of the quantity of hydrochloric acid will be discussed later in the section describing the stability of the reaction product.

The equilibrium 6.4 is shifted toward the formation of the yellow reaction product by adding a suitable quantity of 10% aqueous solution of tetramethyl ammonium hydroxide to the reaction mixture of reaction 6.3 after its completion. This reagent has the advantage that, unlike with sodium or potassium hydroxide, the solutions do not become opalescent after the addition of the reagent. The neutralization of the above mentioned 0.1 ml of 5 M hydrochloric acid requires about 0.4 ml of 10% tetramethylammonium hydroxide solution. When smaller volumes than this are added to the reaction mixture, the color naturally does not develop, in the ethanol–pyridine mixture, however, even a small excess of the base causes the equilibrium to shift towards the formation of PPE. This can be characterized by the following set of data. To the reaction mixture containing 0.868 mg of PM and 0.1 ml of 5 M hydrochloric acid dissolved in 10 ml of pyridine 0.3, 0.6, 1.0 and 1.4 ml aliquots of 10% solution of tetramethylammonium hydroxide were added after the completion of reaction (6.3). After diluting these mixtures to 100 ml with ethanol, the absorbances of the resulting solutions were 0.005, 0.426, 0.430, and 0.428. This indicates that 0.6 ml of tetramethylammonium hydroxide solution is enough to complete the formation of PPE, but the spectrum does not change even if 1.4 ml of the reagent is added. For this reason the addition of 1 ml of the tetramethylammonium hydroxide reagent was prescribed in the final version of the method.

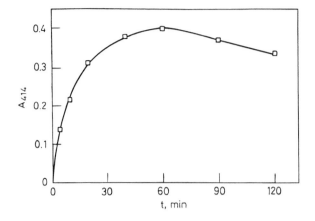

FIGURE 6.2. Time–absorbance curve of the reaction of prednisolone 21-mesylate (PM) with pyridine at 100°C. Solvent: pyridine; c_{PM} = 0.934 mg/10 ml. Measurement at 414 nm after the dilution of the reaction mixture + 1 ml of 10% tetramethylammonium hydroxide to 100 ml with ethanol.

In the course of the investigation of the effect of temperature on reaction (6.3) it was first stated that the rate of the reaction is very low at room temperature: A_{424nm} = 0.029 after a reaction time of 120 min. The rate of reaction rapidly increases by increasing the temperature but the reaction time is longer than 120 min even at 70°C. On the basis of practical considerations only two reaction conditions might come into consideration: reaction immersed in a boiling water bath, or at the boiling point or pyridine (115°C).

The time–absorbance curve of the reaction in pure pyridine (without hydrochloric acid) at 100°C is shown in Figure 6.2. As it is seen, the absorbance reaches a maximum value after a reaction time of 60 min and a 15% decrease is observable after heating the reaction mixture for another 60 min. As a consequence of the degradation of the reaction product, it follows that not even the absorbance measured after 60 min corresponds to the full completion of the reaction since the decomposition of the reaction product (PPM) must begin in the first period of the reaction before reaching the maximum absorbance. Calculating with the molar absorptivity of PPE (21,900) a conversion of 85% can be obtained from the absorbance measured after 60 min reaction time.

On the basis of what is described in Section 6.C.5, there was no reason for regarding the stability of the reaction product under the conditions of reaction 6.3 as satisfactory. The decomposition is probably due to the oxidation of the C-17 side chain by atmospheric oxygen principally to etianic acid derivatives (see Equation 6.5).

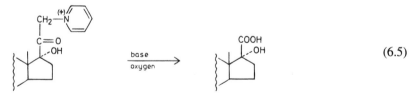

$$(6.5)$$

Since this reaction is catalyzed by bases, it seemed to be reasonable to try to stabilize the reaction product by adding hydrochloric acid to the reaction mixture in small quantities relative to the quantity of pyridine. Figure 6.3 shows the time—absorbance curves of solutions containing 0.1 ml of 5 M hydrochloric acid in 10 ml of pyridine at 100°C and at the boiling point of pyridine. From this figure the following conclusions could be drawn:

1. The stabilization of the reaction product was successful.
2. The reaction is completed within 60 min at 100°C and 30 min at the boiling point temperature.

To optimize the concentration of hydrochloric acid 0.1 ml volumes of water, 2.5, 5, and 7.5 M hydrochloric acid were added to 10 ml volumes of pyridine each containing 1 mg quantities of

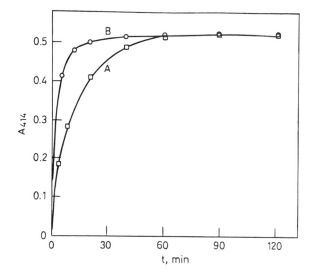

FIGURE 6.3. Time–absorbance curve of the reaction of prednisolone 21-mesylate (PM) with pyridine in the presence of hydrochloric acid. Solvent: mixture of 10 ml of pyridine and 0.1 ml of 5 M hydrochloric acid; c_{PM} = 1.061 mg/10 ml. Temperature: A: 100°C; B: boiling point. Measurement: as in Figure 6.2.

PM. The reaction was carried out using the optimized reaction conditions and the following conversions were obtained in the above sequence using the molar absorptivity of 21,900 for the calculation: 78.1, 95.0, 97.5, and 96.2%. On this basis, 0.1 ml of 5 M hydrochloric acid was selected for being added to the reaction mixture prior to the reaction.

After having solved the problem of the stabilization of the precursor of the colored product under the conditions of the reaction leading to its formation, along the lines of Section 6.C.5 the stability of the colored product in the test solution to be measured spectrophotometrically also had to be investigated. This is a problematic point of the method since the decomposition reaction 6.5, leading to the decolorization of PPE, is catalyzed by bases and in the test solution, strongly basic medium has to be applied in order to ensure the exclusive presence of PPE.

Investigating the absorbance of the test solution in the cell as a function of time elapsed after the basification of the solution and filling it to the mark it was found that after 30, 60, 90 and 120 min, the absorbance had decreased by 1.7, 3.4, 6.0, and 7.4% relative to the absorbance measured immediately after filling the solution to the mark. From these data it was concluded that the absorbance should be measured within 5 min. If this is done the error originating from the instability of the reaction product is negligible.

4. Determination of Prednisolone-21-mesylate (PM)

On the basis of the optimization described in the preceding section the following method was finally described for the determination of prednisolone-21-mesylate.[5]

The accurately weighed sample containing 0.5–1 mg of PM is dissolved in 10 ml of pyridine or an aliquot volume of a stock solution in pyridine containing the same is diluted to 10 ml. 0.1 ml of 5 M hydrochloric acid is added and the solution is refluxed for 60 min. After cooling, the mixture is transferred to a 100 ml volumetric flask with the aid of ethanol, 1 ml of 10% tetramethylammonium hydroxide is added and the flask is made to volume with ethanol. The absorbance of the solution is measured within 5 min at 414 nm against a reagent blank prepared in the same manner but without the sample. The absorbance of standard PM is determined simultaneously after the same treatment. The PM content of the sample is calculated from the two absorbances.

The spectrum of the test solution is completely identical with that of standard PE in the spectral range of 350–500 nm, shown in Figure 6.1B. This indicates that the measurement is not disturbed by foreign absorbances. Strict adherence to Beer's Law was found in the 0–1.5 absorbance range. The regression equation is given in Equation 6.6.

$$A_{414nm} = -0.0012 + 0.483 \cdot c \quad r = 0.9997 \tag{6.6}$$

where c is the concentration of PM in the test solution expressed in mg/100 ml. As seen, the intercept of the regression line is negligible and the regression coefficient characterizes a method suitable for the solution of practical analytical problems. A molar absorptivity value of 21,150 was calculated from the slope of the regression line. If this value is compared with the molar absorptivity of PPE (see Equation 6.4) measured by using crystalline PPE (21,900) it can be stated that the over-all conversion of the sequence of reactions PM → PPM → PPE is 96.6%, indicating that the reactions run practically quantitatively according to the reaction schemes (6.3) and (6.4). This means that the method meets the requirements for a modern spectrophotometric method from this respect, too.

The applicability of the method to practical purposes greatly depends on how much the method is selective to the compound to be determined. Thorough investigations revealed that steroids not containing the C-17 side chain do not give positive reactions. No positive reaction was obtained with 21-unsubstituted 20-keto pregnanes, 21-hydroxy corticosteroids and their carboxylic acid ester derivatives and 21-aminocorticoids, either. This enables the reaction to be used for the solution of the analytical tasks outlined in Section 6.D.1. It has to be noted that 21-chloro and bromo derivatives as well as active esters of 21-hydroxy corticosteroids give positive reaction. Steroid derivatives successfully determined by the described method include in addition to prednisolone 21-mesylate discussed so far prednisolone 21-tosylate, 21-tosylate of Reichstein's substance S, 21-deoxy-21-bromoprednisolone, 21-chloroprogesterone, 11β-hydroxy-21-chloroprogesterone, etc. The absorption maxima of these derivatives after the above described treatment range between 413 and 417 nm and the molar absorptivities are also similar to that of prednisolone 21-mesylate (20, 240–21, 900).[5]

5. Determination of Prednisolone-21-mesylate Impurity in Mazipredone

After the successful development of the method and the promising findings regarding its selectivity the applicability of the method was investigated for the determination of PM in the presence of 100- to 500-fold quantities of mazipredone, i.e., its determination as an impurity in the latter. It was stated that the optimum values of the various parameters described in Sections 6.D.3 and 6.D.4 apply to this analytical task, too, with the exception of hydrochloric acid added to the reaction mixture in order to stabilize the reaction product. Although, as already mentioned, mazipredone does not give a positive reaction, but it decomposes in strongly acidic solution and the decomposition products do interfere with the assay causing a negative error. It was found, however, that the quantity of hydrochloric acid introduced into the solution in pyridine with the investigated mazipredone hydrochloride (about 30–40 mg) is sufficient for the stabilization of PPM, and for this reason, no hydrochloride was added to the reaction mixture when the method was applied for the determination of PM impurity in mazipredone. Using this modification, the decomposition of mazipredone leading to the above described problems could be suppressed without significantly decreasing the stability of PPM under the conditions of boiling in pyridine solution.

The following regression expression, Equation 6.7 was obtained when the concentration of PM was varied in the presence of constant concentration of mazipredone (35 mg/10 ml pyridine) and omitting the addition of hydrochloric acid:

$$A_{414nm} = 0.0021 + 0.460 \cdot c \quad r = 0.9995 \tag{6.7}$$

As can be seen, the values of the slope and the regression coefficient are somewhat lower than in the case of Equation 6.6 but the method was still found suitable for the determination of prednisolone 21-mesylate impurity in mazipredone above the level of 0.2%. For example, in the case of a mazipredone sample spiked with 0.51% of prednisolone 21-mesylate, the result (mean

± SD; n = 7) of the assay was 0.50 ± 0.012%. The recovery of 98.0% and the relative standard deviation of ±2.4% indicate the suitability of the method to solve the problem.

E.   VALIDATION OF SPECTROPHOTOMETRIC MEASUREMENTS AND METHODS

1.   Introduction

In the last decade the requirements toward the quality of analytical measurements have rapidly and greatly increased. The harmonization of requirements was a most important factor for enabling an examiner to judge the quality of an analytical method on the basis of objective criteria. This has now led to the development of the principles and practice of the validation of analytical measurements and methods. In the field of pharmaceutical analysis the aspects of validation have already been set out at the pharmacopoeia level.[8]

Of course spectrophotometric methods must also be validated and the presentation of the validation data is part of the documentation of a spectrophotometric method for it to be considered at an acceptable level.

In this section the most important aspects of the validation of spectrophotometric methods are summarized. In spite of the great importance of the problem, detailed discussion is not necessary since the majority of these aspects are discussed in various chapters of this volume. For this reason, only an outlined presentation will be given here, calling the reader's attention to various sections of the book where a more detailed discussion of the aspects in question can be found.

The necessity for the validation of methods is not restricted to those methods based on chemical reactions but it is indispensable in the case of methods based on natural absorption, too, for the sake of uniform treatment the section dealing with validation is inserted towards the end of Chapter 6 dealing with the application of chemical reactions in the spectrophotometric drug analysis.

2.   Validation of the Measurement of the Absorbance

The fundamental prerequisite of the development and description of a well-validated spectrophotometric method is that the spectrophotometer and its accessories should enable the accurate determination of the absorbance in such a manner that the measurement is also reproducible in other laboratories using other instruments.

As it has been set out in detail in Sections 2.I and 2.J the condition of a spectrophotometer can easily be judged by checking the slitwidth-absorbance and stray light conditions. By the standardization of these, a firm basis can be set for the measurement of absorbance. Important aspects of the validation of the measurement of absorbance are the calibration of the wavelength and absorbance scales of the spectrophotometer from time to time and the check of the calibration (see Sections 2.L.1 and 2.L.2). When the aim of the measurement of the absorbance is quantitative analysis and the determination and calculation are based on a simultaneously run reference standard any errors originating from an out-of-calibration instrument can be diminished but the consequences of this can by no means be completely eliminated by using the reference standard. For this reason the validation of the measurement of the absorbance still remains an important task.

Matching the cells, i.e., checking their equivalence from the point of view of the spectrophotometric measurement (see Section 2.D.3) as well as ensuring and checking their cleanness (2.H) are also important aspects of the validation of the measurement of the absorbance.

The absorbance is usually determined in very dilute solutions. For this reason in addition to the factors discussed so far related to the spectrophotometer and its accessories, the accuracy of the measurement is influenced at an equally important level by the accuracy of the volumes of the volumetric flasks and pipettes used in the dilutions. Calibration of this glassware is also part of the validation procedure.

3. Validation of the Spectrophotometric Methods

In this section the special aspects of quantitative spectrophotometric analysis of drugs are briefly summarized following the sequence of parameters in the general discussion of the validation of quantitative drug analysis in the United States Pharmacopoeia.[8]

a. Precision.   This parameter characterizes the reproducibility of the measurement by determining the scatter of the measurement about the mean value ($\overline{X}$). To determine the standard deviation (SD or s) a set of at least 8–10 samplings taken from a homogeneous sample is analyzed.

$$SD = \sqrt{\frac{\sum\limits_{i=1}^{n} (X_i - \overline{X})^2}{n - 1}} \tag{6.8}$$

where n is number of the measurements and X_i are the values of the individual results within the set.

   More frequently the relative standard deviation (coefficient of variation) is used which is the standard deviation expressed as the percentage of the mean:

$$RSD = \frac{100 \cdot SD}{\overline{X}} \% \tag{6.9}$$

   It has to be emphasized that in the course of method validation and its documentation all samples taken from the homogeneous sample should undergo the full analytical procedure beginning with the sampling and up to the measurement of the absorbance. The RSD values thus obtained are naturally higher than that of the simple absorbance reading measurable by repeated filling and emptying of the cell.

   The accuracy of the measurement of absorbance is dealt with in Section 2.G. Under normal circumstances the RSD can be decreased to about 0.3–0.5%. This precision can be approached in the case of methods based on natural absorbance requiring simple sample pretreatment only. In the case of methods requiring more complicated sample pretreatment procedures (extraction, etc.) and/or the use of a chemical reaction, this value can increase up to 1–2% even if the optimization has been carried out carefully.

   If this question is approached from the point of view of the demands towards the precision of the analytical methods to be used in pharmaceutical analysis the following can be stated. In the course of the determination of the active ingredient content of bulk drugs the RSD must not be greater than 0.5–1%. When formulations are assayed the requirement is somewhat more relaxed (1–2% max.). In other cases, e.g., direct determination of impurities, determination of drugs and metabolites in biological samples even higher RSD values may be acceptable depending on the task.

   Finally it is mentioned that, similar to other fields of analytical chemistry, in spectrophotometric drug analysis, too, the precision of the method is often expressed by the confidence limits:

$$\text{confidence limits} = \overline{X} \pm \frac{t \cdot SD}{n} \tag{6.10}$$

where t is a table value characterizing that a result will fall within a certain (e.g., 95%) probability between the calculated confidence limits.

b. Accuracy.   The accuracy of a method characterizes its exactness, i.e., the systematic error of the method. It is defined as the difference between the mean of the set of results and the true value. It is often expressed as the percentage of the true value and is termed as percent recovery.

   For its determination the possession of the "void" matrix in which the analyte is to be determined is necessary. When an impurity or degradation product has to be determined in a bulk drug material then this is the pure drug itself. In the case of the determination of the drug in pharmaceutical formulations or biological samples the placebo and analyte-free blood, urine, etc; samples, respectively, are necessary. Adding known amounts of the analyte to these and carrying out the determination according to the method to be tested the recovery is directly obtained by the comparison of the result with the amount of the introduced material. More exact values can be obtained if the determination of the accuracy is not based on a single point but the void matrix is spiked with the analyte at several points below and above the expected value i.e., at 60, 80, 100, 120, 140% relative to the expected value (e.g., label claim in the case formulations). Plotting the concentrations thus obtained vs. the concentration of the added material a calibration line is obtained whose parameters characterize the accuracy of the method: in the ideal case, the intercept is zero and the slope equals unity. (It is noteworthy that in publications dealing with the assay of pharmaceutical formulations many authors find it satisfactory to calculate the accuracy by directly comparing the result of the assay with the label claim. Needless to say, this approach is much less informative regarding the accuracy of the method than the above described procedure.)

c. Limit of Detection.   This parameter is in use in pharmaceutical analysis primarily to characterize limit tests. This is the lowest concentration of e.g., an impurity which can be detected but not quantitated.

In quantitative terms the limit of detection is usually defined as the concentration of the analyte where the signal-to-noise ratio at the analytical wavelength is at least 3. It is often determined by measuring the standard deviation of the absorbance of solutions containing only the "void" matrix at the analytical wavelength. This value is then multiplied by a factor of 3 to get the value for the calculation. Since, in this very low absorbance range, selectivity problems often occur this parameter is of limited value in spectrophotometric drug analysis. More important is the question of the detection limit of limit tests based on observation or visual comparison in the visible spectrum range. The limit of detection can be estimated here with the aid of the visual inspection of solutions containing the matrix and a series of known concentrations of the analyte.

d. Limit of Quantitation.   The considerations described in the preceding paragraph (c) apply to the limit of quantitation (i.e., the lowest measurable concentration), too, the difference being that in this case a signal-to-noise value of 10 is usually required. This concentration is connected with the sensitivity of the method; some more details can be found in this section in Paragraph f, dealing with linearity and and the range of the methods.

e. Selectivity (Specificity).   The term selectivity expresses how much the results obtained by the method for a given analyte are influenced by the presence of foreign substances (other component of a combined formulation, impurities, degradation products, excipients, endogenous components of biological samples, undefined materials causing background absorption).

Questions related to the selectivity of spectrophotometric methods are among the most problematic areas in the application of this technique to pharmaceutical analysis. The considerations regarding this matter are the subject of Sections 4.A.2, 9.B.1 and 9.B.2 in the present volume. Of the possibilities to solve the selectivity problems the following are discussed (with the number of the respective sections in parentheses): algebraic background correction methods (4.C), analysis of multicomponent systems by algebraic (4.D), derivative spectrophotometric (4.E) and other (4.F) methods, by means of preliminary chromatographic separation (Chapter 5), chemical reactions (Chapters 6 and 7) with special respect to difference spectrophotometric methods (Chapter 8).

The identity (or at least close similarity) of spectra which prove that the determination can be selectively carried out at the analytical wavelength should be attached to the validation documentation of spectrophotometric methods. It is equally important to present the list of substances which are likely to be present in the sample beside the analyte together with their highest concentration relative to the analyte in which they do not influence the absorbance at the end of the assay.

In the case of those analytical problems where the analyst does not have sufficient information about the spectrophotometric characteristics of the matrix, thus excluding the possibility of applying the above described principles, the validation is carried out in such a way that the recovery obtained with the spectrophotometric method is compared by means of the methods of modern mathematical statistics with the results of a more selective (usually chromatographic) method.

f. Linearity and Range.   As is described in Section 4.A.1 the calibration line derivable from the Beer's Law can be characterized by the regression parameters: intercept, slope, regression coefficient. These as well as the presentation of the F-test characterizing the linearity are among the most important validation data.

The lower limit of the range is the lowest measurable concentration (discussed in Paragraph d). The upper limit is the highest concentration where the linear relationship between concentration and absorbance holds. The latter is not a very important parameter since solutions which are too concentrated can easily be diluted to fall into the range.

Of the regression parameters the slope is of prominent importance namely this determines the sensitivity of the method. This is usually given by means of one of the two data in Section 2.C the specific absorbance or the molar absorptivity but Sandell's sensitivity index is also in use. Its dimension is $\mu g/cm^2$ and the value is equal to the concentration (in $\mu g/ml$) of a solution the absorbance of which is 0.001 at a cell length of 10 mm.

g. Ruggedness. This validation parameter expresses how the reproducibility of the method is influenced by systematically changing each of the parameters of the method while keeping the others constant. The parameters most often checked for the ruggedness test are: the spectrophotometer, the laboratory where the analysis is performed, including the analyst who performs the analysis, origin (moreover the lot) of the reagents and solvents, the day of the analysis (day-to-day variation tests). Among these parameters, in particular, the effects of the transfer of the method from one laboratory to another, were thoroughly investigated.[6]

4. Special Aspects of the Validation of Spectrophotometric Methods Based on Chemical Reactions

The validation parameters summarized in the preceding section apply to all kinds of spectrophotometric methods whether they are based on natural absorption or on chemical reactions. However, the validation documents of the latter should contain the validation of the conditions of the reaction, too. The aspects of this are summarized in Section 6.C and, as an example, the validation of a method based on chemical reaction is presented in Section 6.D.3. The most important point is the demonstration by sets of data that minor changes in e.g., reagent concentration, temperature, reaction time, pH, etc., do not markedly influence the results obtained with the method. Finally two papers are mentioned dealing specifically with the validation of spectrophotometric methods in drug analysis.[1,7]

REFERENCES

1. Gemperline, P. J.; Salt, A. *J. Chemometrics* 1989, *3*, 343.
2. Görög, S., Ed. *Steroid Analysis in the Pharmaceutical Industry* Ellis Horwood: Chichester, 1989, 196–197.
3. Görög, S.; Szepesi, G. *Analyst* 1972, *97*, 519.
4. Görög, S.; Szepesi, G. *Anal. Chem.* 1972, *44*, 1079.
5. Görög, S.; Tuba, Z. *Analyst* 1972, *97*, 523.
6. Kivalo, P. *Standardization within Analytical Chemistry,* Akadémiai Kiadó: Budapest, 1989.
7. Setnikar, I.; Senin, P.; Arigoni, R. *Boll. Chim. Farm.* 1984, *123*, 263.
8. *The United States Pharmacopoeia XXII,* USP Convention Inc.: Rockville, 1990, 1710–1712.
9. Tuba, Z.; Szporny, L.; Tóth, J. *Hungarian Patent* 150, 350; *U.S. Patent* 1,123,598.

Chapter 7

CHEMICAL AND SPECTROSCOPIC CHARACTERIZATION OF DERIVATIZATION REACTIONS*

A. INTRODUCTION

This chapter deals with the chemical reactions most often used in spectrophotometric pharmaceutical analysis, arranged according to reaction type. The conditions for the quantitative conversion of the various functional groups of the molecule will be shown, the type of spectrum expected from the reaction product will also be discussed. This background is absolutely necessary for the analyst to be able to select, on the basis of the principles given in Chapter 6, the reaction most suited to the task for both selectivity and sensitivity.

Of course, only the most important reaction types can be discussed, and thus, like in Chapter 10 dealing in detail with the application of reactions, we endeavored to emphasize the reactions which are still applied and the chemistry of which, in accordance with Section 6.B, is known.

B. REVERSIBLE ACID–BASE REACTIONS

Since most compounds of pharmaceutical interest have acidic or basic functional groups, their simplest and most general reaction is the deprotonation of acids and protonation of bases. These reversible processes and the spectroscopic changes involved were discussed at length in Section 3.Q. From the principles given there it can be derived whether for a given substance the increase or the decrease of pH leads to a bathochromic or hyperchromic shift, allowing a more selective or sensitive analysis to be done. Thus, this problem is not discussed here, stressing only that for a quantitative analysis the value of pH should be selected so that (either by calculation from the dissociation constant or, if this is not known, by an empirical method) the analyte is in a uniformly protonated or deprotonated form.

The methods based on acid-base equilibria have exceptional importance in difference spectrophotometry; thus these methods will be discussed separately in Section 8.B.

C. ACYLATION REACTIONS

The general reaction given in Scheme 7.1 is the basis of several spectrophotometric methods:

$$R - X(H) + R'-\overset{\overset{\text{O}}{\|}}{C}-Y \longrightarrow R-X-\overset{\overset{\text{O}}{\|}}{C}-R' + HY \tag{7.1}$$

where X = –O–, –S–, –NH– or –N-alkyl, etc., and Y = Hlg, –O–CO–R', etc.

---

* See References 77, 82, 92, 129, 130, 133, and 157.

As acylating agent, generally acid anhydrides or acid chlorides are used, and the reaction is carried out in non-protic solvent. Hydroxy groups are generally less reactive, and reactivity decreases in the sequence of primary, secondary, tertiary. Their quantitative acylation usually requires catalysts. If the measurement is performed in the VIS region, generally pyridine is used; if, however, the UV spectrum is measured, a spectrophotometrically inactive catalyst, triethylamine, should be adopted.

Amines are usually more reactive than alcohols, although steric hindrance causes significant differences here, particularly with secondary amines. If acylation is carried out with an acid chloride, it is worth running the reaction in the presence of an acid binding tertiary amine to make it quantitative.

Spectrophotometrically inactive acylating agents (like acetic anhydride or chloride) generally cause no significant change in the spectra. An exception is the acylation of aromatic amines, for which, as shown on p. 59 on the example of aniline and acetanilide, acylation results in a significant bathochromic and hyperchromic. Such reagents are generally not used for spectrophotometric purposes. Spectrophotometrically active acylating agents, of course, result in derivatives with appropriate spectrophotometric activity. For instance, benzoyl derivatives with a strong band at ca. 230 nm can be produced with benzoyl chloride. The direct application of this derivative in spectrophotometric practice is, however, problematic, since the excess of reagent has an absorption maximum close in wavelength to that of the derivative, and their separation by chromatography or selective extraction is complicated. On the other hand, benzoyl chloride is widely used as reagent in HPLC analysis with UV detection to convert spectrophotometrically inactive alcohols into readily detectable benzoyl derivatives.[14,34,93]

Owing to their more favorable spectroscopic properties, the nitro derivatives of benzoyl chloride are preferred to benzoyl chloride itself. 4-Nitrobenzoyl chloride was, for example, applied successfully to improve the HPLC detection of polyhydroxy derivatives. The resulting 4-nitrobenzoyl esters absorb at 260 nm with a molar absorptivity of 14,800 per OH unit, ensuring high sensitivity in detection.[119] In spectrophotometric analysis 3,5-dinitrobenzoyl chloride is the most widely used reagent. As can be seen in Scheme 7.2, the 3,5-dinitrobenzoyl ester (and amide) derivatives assume quinoidal anion structure in alkaline media. The stability of the red product is satisfactory, making it applicable for the determination of primary and secondary alcohols, amines and thiols.

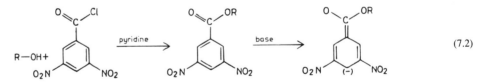

$$(7.2)$$

If sodium hydroxide is used as a base (λ_{max} 525 nm), the product and the excess of reagent should be separated.[75] Similarly with the variant of the method in which the base is propylene diamine,[75] if, however, ammonia[175] or piperazine[176] is used for basification (λ_{max} 555 and 510 nm, respectively), separation is unnecessary, since in this case the excess of reagent does not convert into a colored product.

Reaction 7.1 allows the determination of not only the acylated hydroxy, amino and thiol groups, but of the acylating agent or compounds convertible into the acylating agent, too. For example, acid chlorides or anhydrides can be determined by acylating 2-nitrophenylhydrazine with them, and selectively rearranging the resulting hydrazide in alkaline medium into a quinoidal anion.[117] The method is also applicable for the determination of free carboxylic acids if they are converted prior to analysis into active acylating agents e.g., by forming mixed anhydrides with dicyclohexylcarbodiimide[118] or diethyl chlorophosphite.[48,50] Thus a very stable product can be obtained (see Scheme 7.3) with a medium absorptivity (ε = 5,000 to 7,000) at about 550 nm. If aniline is used as reaction partner instead of 2-nitrophenylhydrazine and the spectrum is measured in acidic medium, the maximum appears at 243 nm, but with 2 to 3 times higher intensity[49] (see the spectral data of acetanilide on p. 57).

D. CONDENSATION

The various types of condensation reactions play a most important part in the spectrophotometric analysis of pharmaceutical substances based on chemical reactions (in fact, the acylation reactions discussed separately in the previous section can also be included in this group).

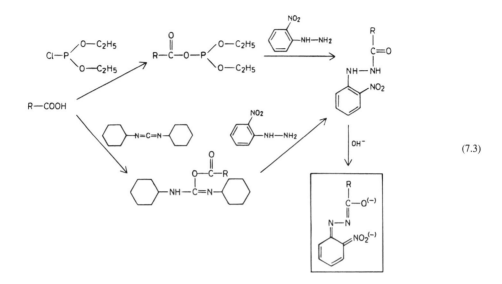

(7.3)

This section deals first with the reactions characterized in general by Scheme 7.4, in which the common feature is the formation of a C=N bond with water elimination.

$$\begin{matrix} R_1 \\ \diagdown \\ C=O \\ \diagup \\ R_2 \end{matrix} \quad + \quad HN-X \quad \xrightarrow{H^+} \quad \begin{matrix} R_1 \\ \diagdown \\ C=N-X \\ \diagup \\ R_2 \end{matrix} \quad + \quad H_2O$$

(7.4)

If this reaction is used for the determination of aldehydes or ketones, the following types of reagent H_2N-X can be used:

1. Aromatic amines (Schiff's base formation). Usual reagents are 4-dimethylaminoaniline[68,128] or 4-aminoantipyrine.[152] The absorption maximum of the Schiff's base formed in the acid catalyzed reaction is in the 350 to 450 nm region with medium intensity ($\varepsilon \approx 10,000$), depending on the reagent and the neighborhood of the carbonyl group. An advantage of these reagents is that they have no significant absorption at the maximum of the product.
2. Hydrazine derivatives. Condensation with free hydrazine does not increase the spectrophotometric activities of aldehydes and ketones significantly, however, hydrazines substituted with aromatic rings are among the most widely used reagents. Phenylhydrazine itself is used mostly for the determination of glucocorticoids,[136] as discussed in detail in Section 10.M, whereas 4-nitrophenylhydrazine[120,127] and particularly 2,4-dinitrophenylhydrazine are generally used. The various derivatives of oxo-compounds with the latter have absorption maxima in the 350 to 400 nm region ($\varepsilon = 20,000$ to $40,000$), depending on the neighborhood of the oxo-group.
   A common problem of the methods based on these spectrophotometrically very advantageous compounds is that the excess reagent also has significant absorption in this spectral region. This problem can be overcome by the selective extraction[96] or chromatographic separation[61] of the hydrazone product, or by the application of glacial acetic acid as solvent, in which the excess of reagent is colorless.[126] Even more generally, the red quinoidal structure that 2,4-dinitrophenylhydrazones assume in strongly alkaline media is utilized; the corresponding 50 to 100 nm bathochromic shift improves the selectivity of determination significantly.[76]

3.　Acid hydrazides. Of the carboxylic hydrazides, isonicotinic acid hydrazide is the most important reagent for unsaturated ketosteroids[174] (reaction schemes and details are given on p. 328). Thiosemicarbazide, as an equally selective and sensitive reagent primarily of unsaturated ketones is also worth noting.

If the possibilities given by Scheme 7.4 are utilized for the determination of amines (of course, always primary amines) or hydrazine derivatives, aromatic oxo-compounds are preferably used as reagents for the formation of Schiff's bases or hydrazones. The most widely used reagents for this purpose are salicylic aldehyde,[111] 4-dimethylamino-benzaldehyde[185] or 4-dimethylamino-cinnamic aldehyde.[127]

Other important methods, like the determination of amino acids by the ninhydrin or o-phthalic aldehyde methods, are also based on the condensation of the amino group and oxo-reagents. They are discussed in Section 10.M.

In addition to the methods discussed so far, based on the formation of a C=N double bond, methods based on the formation of C–N single bonds are also used in the analysis of amines. The reagent (usually a halogen derivative) should be reactive enough for the method based on Scheme 7.5 to be sufficiently effective.

$$\begin{array}{c} R_1 \\ \diagdown \\ NH \\ \diagup \\ R_2 \end{array} + HlgX \longrightarrow \begin{array}{c} R_1 \\ \diagdown \\ N-X \\ \diagup \\ R_2 \end{array} + H-Hlg \tag{7.5}$$

Typical Hlg-X reagents are, for instance, 2,4-dinitro-fluorobenzene[83,145] and 9-chloroacridine,[162] which can be replaced, with sufficiently reactive partners, by 9-methoxy-acridine[161] (in this case methanol is eliminated instead of hydrochloric acid).

Spectrophotometrically active reagents with sufficiently mobile halogen can also be used to derivitize functional groups other than amines. Thus, for instance, carboxylic acids can be converted into spectrophotometrically active phenacyl esters with phenacyl bromide (Scheme 7.6). Although with the use of the 4-nitro-derivative of the reagent this reaction can be used in direct spectrophotometry,[10] too, it is generally applied in HPLC with UV detection as a standard derivatization technique of fatty acids, bile acids, etc.[1]

$$R-COOH + \underset{\displaystyle}{\boxed{}}\overset{\displaystyle O}{\overset{\displaystyle \|}{C}}-CH_2-Br \longrightarrow \underset{\displaystyle}{\boxed{}}\overset{\displaystyle O}{\overset{\displaystyle \|}{C}}-CH_2-O-\overset{\displaystyle O}{\overset{\displaystyle \|}{C}}-R \tag{7.6}$$

In addition to active halogen derivatives, the formation of C–N bonds can also be achieved with other reagents. Of these reagents, 1,2-naphthoquinone-4-sulphonic acid[15] can be widely used in the determination of primary and secondary amines. As shown by Schemes 7.7 and 7.8, the reaction mechanisms are different for the different amines,[166] but the bands appearing in the 460 to 480 nm region with a molar absorptivity of ca. 4000 allow a wide range of amines to be determined.

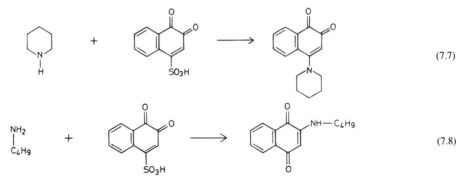

(7.7)

(7.8)

One of the classical and still widely used methods of colorimetric pharmaceutical analysis, the coupling reaction between diazotized aromatic amines and phenols or aromatic amines[16,129a] can also be discussed in the group of condensation reactions. This is illustrated by Scheme 7.9 on the example of β-naphthol or N-(1-naphthyl)-ethylenediamine and diazotized sulphanilic acid:

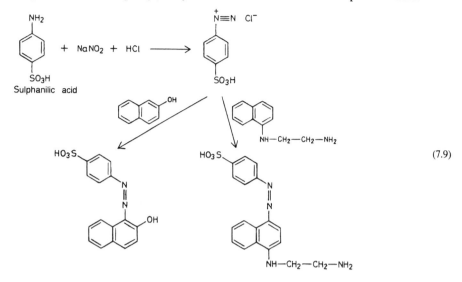

(7.9)

The reaction can be applied for quantitative analysis in the following cases:

1. Determination of primary aromatic amines. In this case the diazonium salt is formed by nitrous acid released *in situ* from sodium nitrite. This is coupled with an appropriate phenol or (generally secondary or tertiary) aromatic amine.
2. Determination of secondary or tertiary aromatic amines and phenols. In this case the reagent is either an *in situ* diazotized primary aromatic amine or another, stable diazonium salt, e.g., 4-nitrophenyldiazonium fluoroborate, available in crystalline form.
3. Indirect determination of substances which produce nitrous acid in certain reactions (e.g., oxidative cleavage of nitro compounds). In this case the diazonium salt (coupled e.g., with phenol) is formed with an excess of primary aromatic amine in the second step.

It is to be noted generally that the diazonium reagent reacts with phenols or aromatic amines in which the para or ortho position is free. Usually, the rates of both reactions make it possible to run the process at room temperature. The color of the products varies from yellow to red (λ_{max} between 400 and 600 nm). Sensitivity is also excellent: molar absorptivities vary usually between 10,000 and 40,000.

As a curiosity, the intramolecular variant of the reaction is shown in Scheme 7.10 for the determination of hydralazine hydrochloride.

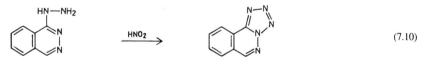

(7.10)

Finally, it is mentioned that condensation reactions involving the formation of C=C bonds may also play certain role in spectrophotometric pharmaceutical analysis. With the use of appropriate reagents condensation proceeds on the active methylene group adjacent to carbonyl groups. Such reagents are, for example, diethyl oxalate[45,53,168] and 1,3-dinitrobenzene.[184] Reaction schemes are shown on p. 213 and 329.

E. REDOX REACTIONS

Of the several redox reactions used in pharmaceutical analysis, the ones shown in first belong, in fact, into the group of condensation reactions discussed in the previous section, however, the substance to be analyzed is converted by means of a redox reaction into a form suitable for reaction with the condensation reagent. These methods include, for example, the conversion of secondary alcohols into ketones by chromic acid oxidation, followed by condensation with e.g., 4-nitrophenyl-hydrazine[120] or the determination of natulan by cerium (IV) oxidation and the condensation of the product with 2,4-dinitrophenylhydrazine (Scheme 7.11).[62]

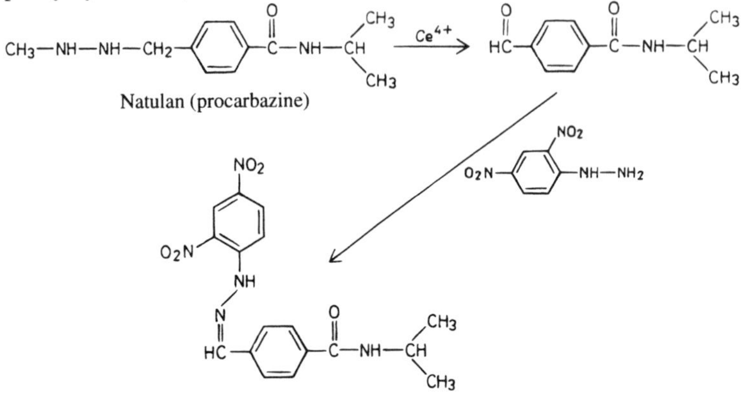

(7.11)

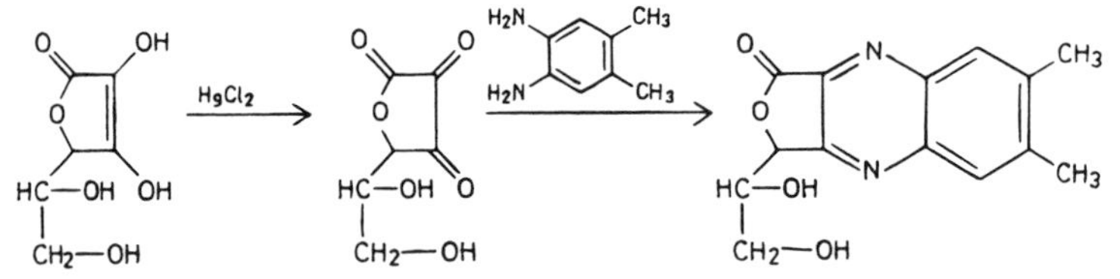

(7.12)

Ascorbic acid Dehydroascorbic acid

This method is the basis of the indirect determination of ascorbic acid by its oxidation into dehydro-ascorbic acid with mercury (II) chloride, and the conversion of the product into a quinoxaline derivative by ring closure with 4,5-dimethyl-o-phenylenediamine (Scheme 7.12).[167] Similarly, the 21-hydroxy group of corticosteroids or the amino group of 21-aminosteroids can be oxidised selectively by copper (II) acetate[54] and mercury (II) acetate,[55] respectively, into 21-aldehydes, which can be converted, again, into a quinoxaline derivative by condensation with 4,5-dimethyl-o-phenylene-diamine (for the reaction scheme, see p. 332).

The application of 4-dimethylaminobenzaldehyde, mentioned previously as a reagent of primary amines, for the determination of indol derivatives (van Urk reaction) can also be mentioned in this section. The structure[131] of the blue reaction product (λ_{max} 580 to 610 nm) obtained in the case of 3-substituted indole derivatives (ergot alkaloids) makes it obvious that a key role is played by the oxidant (usually iron (III) chloride) in a reaction between the reagent and the alkaloid. It is worth noting that even if position 2 of the indole derivative, which takes part in this condensation reaction, is occupied (such as in the case of reserpine, yohimbine, indomethacine, etc.), a positive reaction is obtained, only the maximum is shifted by 30 to 40 nm toward the short wavelengths. In these cases the condensation involves the benzene ring of the indol skeleton.[132]

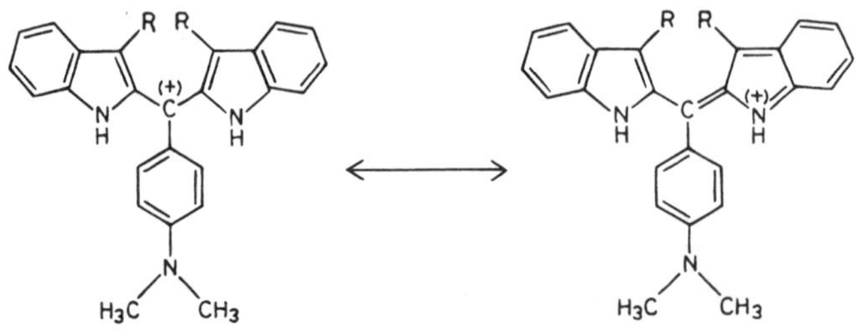

(7.13)

The limiting structures of the van Urk reaction products of 3-substituted indol derivatives.

The oxidative condensation of formaldehyde with chromotropic acid (1,8-dihydroxynaphthalene-3,6-disulphonic acid) leads to a very similar product. The role of the oxidant is played here by concentrated sulphuric acid which is also used as solvent; the methine bridge arising from formaldehyde links the two chromotropic acid units at position 4. One of the units assumes quinoidal structure (λ_{max} 540 nm, $\varepsilon = 18,200$).[17]

The significance of this reaction is greater than for the determination of formaldehyde itself: it also allows the indirect determination of all compounds producing formaldehyde on oxidative cleavage, and even of those producing methanol by acidic cleavage, since the latter can be converted into formaldehyde with potassium dichromate.

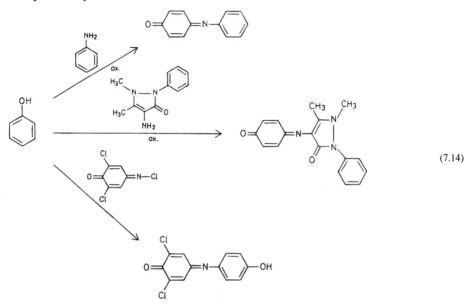

(7.14)

Of the photometric methods based on oxidative condensation, one of the most widely used is the analytical application of indophenol reaction. As can be seen in Scheme 7.14, 4-unsubstituted phenols are condensed in the presence of oxidants into blue indophenol derivatives with primary aromatic amines, or they give a red substance of similar, quinoneimine-type structure with 4-aminoantipyrine. The former reaction is widely applied for the determination of phenols and aromatic amines, the latter (the mechanism of which has been subjected to extensive studies[165]) for that of phenols only. Scheme 7.14 also shows the reaction of 2,6-dichloro-quinonechloroimide with phenols: this reagent yields substances of similar structure without the use of oxidants. For the application of all the three reactions, several examples can be found in Chapter 10.

Several methods based on direct oxidation are also applied in the practice of spectrophotometric pharmaceutical analysis. Of these, the ones based on unambiguous stoichiometry are particularly worth attention. Thus, for example, the determination of ephedrine and a number of other phenylalkanolamines is possible on the basis of the determination of their benzaldehyde oxidation product with an intense absorption maximum at 237 nm.[23]

$$
\text{Ephedrine} + IO_4^- = \text{Benzaldehyde} + CH_3{-}CHO
$$

(7.15)

Ephedrine   Benzaldehyde

The determination of spectrophotometrically inactive steroid 2- or 3-enes by converting them into active α,β-unsaturated ketones with chromic acid[57] can also be mentioned in this section.

Several methods based on interaction with strong oxidants, e.g., the determination of phenytoin and carbamazepine after potassium permanganate oxidation[74] or that of phenothiazine derivatives after N-bromosuccinimide oxidation,[170] etc., do not always have unambiguous stoichiometry.

It has been a common feature of all reactions discussed so far that the spectrophotometric activity of the molecule to be determined increased upon the reaction. However, there are important spectrophotometric methods in which the activity of the molecule to be analyzed does not change remarkably; the measurement is based in such cases on a spectrophotometrically active substance being formed from the inactive oxidant in a reduction process caused by the substance to be determined. Of these indirect methods, the methods involving the use of tetrazolium are of particular significance. These colorless cyclic reagents open into colored formazans in strongly basic media in the presence of reducing agents. As shown in Scheme 7.16 in the case of triphenyltetrazolium chloride, the reduction is a two-electron process.

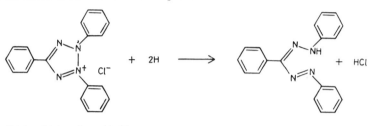

Triphenyl tetrazolium chloride   (7.16)

The absorption maximum of the resulting formazan is at ca. 485 nm, and the molar absorptivity is 16,200. Tetrazolium blue, containing two tetrazolium rings, i.e., 3,3'- (3,3'-di-methoxy-4,4'-biphenylidene)-bis-(2,5-diphenyl-2H-tetrazolium chloride), can be reduced in two steps: the absorption maximum of monoformazan is at 525 nm (ε = 24,000) whereas that of the diformazan at 590 nm (ε = 42,000). If the reagent is applied in sufficient excess, under analytical conditions, practically only monoformazan is formed.[47] Although the tetrazolium methods allow several other reductants to be determined,[56] after the publication of the first analytical application,[101] they were widely applied primarily in the analysis of corticosteroids containing reducing side chains (see Section 10.M).

Of the reagents reducible into colored products, one should mention phospho-molybdic acid or paramolybdate,[72] the Folin-Ciocalteu reagent[35] (phospho-molybdo-tungstic acid), used primarily for the determination of phenols on the basis of the spectrophotometric measurement of the resulting blue complex,[58,153] and o-dinitrobenzene,[33,99] the reagent for corticosteroids, which is converted into di-nitronic acid anion on reaction with reducing agents. Potassium ferricyanide is a similar reagent, which is reduced into ferrocyanide by phenolic type reductants (e.g., butorphanol[176]), and the ferrocyanide can be converted, using iron(III) reagent, into Berlin Blue with an absorption maximum at 770 nm, forming the basis of the indirect determination of phenols. If iron(III) salts are used as oxidant, the measurement can be based on a reaction of the selective photometric reagents (2,2'-dipyridyl, o-phenanthroline, etc.) with iron(II) ions.[73]

In this section, dealing with the application of redox reactions, we have been concerned so far with oxidants as reagent. Reductants have much less significance, since in most cases reduction decreases, instead of increasing, spectrophotometric activity. It should be noted, however, that the sodium borohydride difference spectrophotometric method,[44] discussed in Section 8.D, is based just on this effect. One may mention here the indirect method based on the measurement of iodine produced from iodine ions by oxidant type analytes. This is the principle for the determination, for example, of ACTH containing sulphoxide impurity on the methionine unit.[143] Amides and imides are converted into N-bromoderivatives by elementary bromine, which also release iodine from iodide ions, permitting their analysis. This principle is used in the determination of nicotinamide, alloxane, saccharine, barbital and theobromine.[80]

F. ACID CATALYZED REARRANGEMENT, ELIMINATION AND CLEAVAGE REACTIONS

In addition to the methods based on reversible acid-base equilibria, discussed in Section 7.B, there are several cases in pharmaceutical analysis in which an irreversible rearrangement or elimination reaction or even the cleavage of the molecule takes place by acid catalysis, and this reaction involves an increase in spectrophotometric activity. A common advantage of these methods is that the strong acid reagent is spectrophotometrically inactive. Consequently, the methods based on this reaction are excellent for difference spectroscopy, and thus several examples will be given in Section 8.C.

The determination of spectrophotometrically inactive $\Delta^{5(10)}$-3-keto-steroids after an acid- or base-catalyzed rearrangement of the 5(10)-double bond into position 4 (Δ^4-3-keto structure) will also be shown in Section 8.C.

In addition to the examples for the application of elimination reactions shown in Section 8.C, two further examples are presented here from the field of α-ethylbenzhydrols. As can be seen from Figure 7.1, flumecinol with isolated aromatic rings and tertiary alcoholic hydroxy group has poor spectrophotometric activity, whereas its conjugated, styrene-type product formed upon hydrochloric acid treatment (Scheme 7.17) is readily measured using the absorption maximum at 246 nm.[51]

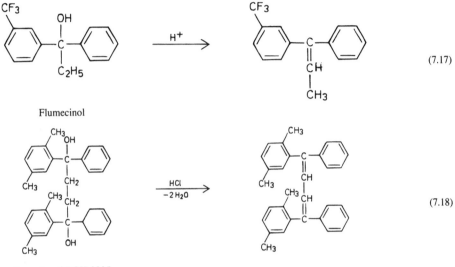

(7.17)

Flumecinol

Impurity of RGH-3395

(7.18)

In fortunate cases, such elimination reactions may lead to a remarkable selectivity. A similar benzhydrol derivative, 2,5-dimethyl-α-ethylbenzhydrol (RGH-3395) can be converted into a readily measured product (at 247 nm) in an analogous reaction (Scheme 7.18). The main contaminant of the drug, the dimeric 1,4-di-(2,5-dimethylphenyl)-1,4-diphenylbutane-1,4-diol, loses two water molecules in an analogous reaction. The resulting tetraphenylbutadiene derivative has an intensive absorption band at 332 nm (ε = 24,000), allowing this contaminant to be measured selectively after hydrochloric acid treatment.[51]

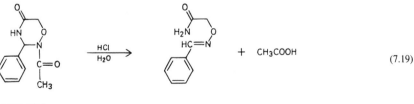

(7.19)

RGH-4615

Unlike the water elimination reactions discussed so far, deacetylation and subsequent ring opening is the basis of the determination of 2-acetyl-3-phenyl-tetrahydro-1,2,4-oxadiazin-5-one (RGH-4615). As shown in Figure 7.2, the substance of poor spectrophotometric activity is converted into a benzaldehyde oxime with a very intense spectrum during the reaction shown in Scheme 7.19.[51] The elimination of acetic acid is also the basis of the determination of ethynodiol diacetate discussed in Section 8.C (Scheme 8.1).

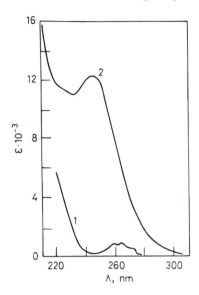

FIGURE 7.1. Spectra of flumecinol (Curve 1) and of its dehydrated (see Scheme 7.17) product (Curve 2). Solvent: ethanol. (From Görög, S.; Rényei, M.; Laukó, A. *J. Pharm. Biomed. Anal.* 1983, *1*, 39–51.)

The selectivity attainable in certain cases by the appropriate choice of the conditions of acidic cleavage is illustrated by the set of reactions shown in Scheme 7.20.

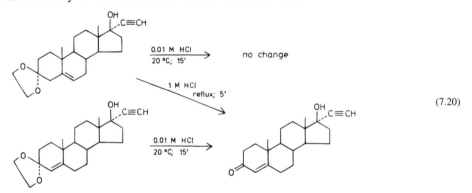

(7.20)

Thus, in a mixture of 3-ethylenedioxy-4-ene and 3-ethylenedioxy-5-ene isomers, the 4-ene isomer can be determined selectively if the mixture is subjected to very mild acid treatment. Under these conditions only one of the spectrophotometrically inactive isomers undergoes ethylene glycol cleavage to yield the spectrophotometrically active 3-one-4-ene derivative. Under more severe conditions, the 5-ene isomer is also converted, and thus the total amount of the two isomers can be determined.[48]

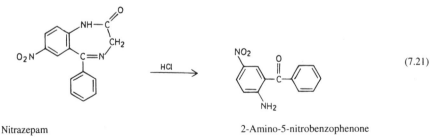

Nitrazepam   2-Amino-5-nitrobenzophenone

(7.21)

An acid catalyzed ring opening and subsequent amino acid cleavage takes place in the case of benzodiazepines, and this reaction, shown in Scheme 7.21, is the basis of the methods utilizing

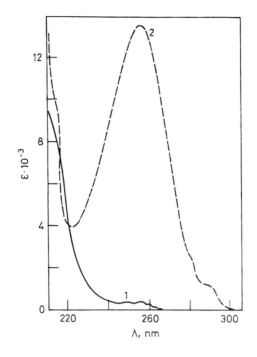

FIGURE 7.2. Spectra of RGH-4615 (Curve 1) and of the reaction product (Curve 2) of its hydrochloric acid treatment (Scheme 7.10). Solvent: ethanol. (From Görög, S.; Rényei, M.; Laukó, A. *J. Pharm. Biomed. Anal.* 1983, *1*, 39.[51].)

the fact that the direct spectrophotometric determination of the resulting amino-benzophenone derivatives is more advantageous than that of the parent benzodiazepine derivatives.[8,29,156] The reaction is illustrated by the example of nitrazepam.

Nitrazepam		2-amino-5-nitrobenzophenone	
λ_{max},nm	ε	λ_{max},nm	ε
218	24,600	235	17,470
257	15,700	358	18,590
308	10,300		

Solvent: ethanol

G.   THE FORMATION OF METAL COMPLEXES

The formation of colored metal complexes is the principle of several spectrophotometric determinations from the early days of pharmaceutical analysis and which are still used. For the analysis of metal salts, which was extremely important in the early periods of pharmaceutical chemistry, and, though gradually decreasing in significance, is still used, photometric analysis based on complex formation was particularly significant. Even a review of the classical methods would exceed the limits of this book, and thus only attention is drawn here to some important publications.[21,104,149,159] Although the importance of complexing reagents has been significantly reduced by the appearance of modern atomic spectroscopic and other methods, there are still publications on the introduction of new complexing agents permitting a more selective and sensitive determination of certain substances, which sometimes find application in pharmaceutical analysis. As an example, the determination of mercury-containing drugs is mentioned. The classical reagent of mercury(II) ions obtained after a permanganate-nitric acid digestion of the organic substance is dithizone,[110] but a very selective and sensitive determination can be based on the 2:1 Hg complex of 1-salicylidene-5-(2-pyridylmethylene)-isothiocarbonohydrazide of yellow color (λ_{max} 400 nm, $\varepsilon = 64{,}000$).[144] The spectrophotometric analysis of metal-containing pharmaceutical preparations will be discussed in Section 10.P.

In contemporary pharmaceutical analysis, the methods in which metal salts are the reagents and the colored products formed upon complexation are measured to analyze ligand type pharmaceuticals, are much more important. The most widely used ion is iron(III), which is the classical reagent of salicylic acid[121] (see Section 10.C). Since a condition for the appearance of the violet color of the complex is the simultaneous presence of free carboxylic and phenolic hydroxy groups, the method is widely used for the selective determination of free salicylic acid impurity or decomposition product in acetylsalicylic acid.[18,163,172] The method can also be applied for the determination of salicylic acid in biological samples;[173] an automated version of this analysis has also been described[91] (see Section 9.H).

Another important method in which iron(III) ions are used as reagent is the determination of esters convertible into hydroxamic acid derivatives in the form of an iron(III)–hydroxamic acid complex. This method was used, among others, for the determination of steroid esters,[52] piracetam,[123] the polyvinyl acetate content of polyvinyl alcohol,[6] and of several other ester, carboxylic amide and carbamoyl derivatives, such as carbachol.[30] Chapter 10 contains a number of further examples for the application of the method. It is worth noting that the method has also been extended for the analysis of substances with free carboxyl groups (e.g., ibuprofen, mefenamic acid). In these analyses the carboxyl group is coupled to hydroxylamine by dicyclohexylcarbodiimide[107] (Scheme 7.22).

(7.22)

Of the other pharmaceuticals determined in the form of their iron(III) complexes, furosemide,[186] alprenolol[140] and penbutolol[139] are mentioned. These methods are based on the chloroform extraction of the complex; with the latter two substances, a ternary complex formed with thiocyanate ions is extracted and measured. The formation of a ternary complex is also the basis of the determination of pyridone carboxylic acids with iron(III) and o-hydroxy-hydroquinonephthaleine. The method is extraordinarily sensitive: the molar absorptivity for nalidixic acid[38] is, for instance, 130,000.

Some examples based on the formation of complexes with other metals will now be discussed. Cobalt (II) salts are the reagents for barbiturates and, more particularly, thiobarbiturates[25,63,108] and of thiacetazone.[154] Copper(II) salts are reagents for citric acid,[90] procaine amide,[183] rutin[138] and, in the form of a mixed complex with nioxime, ascorbic acid.[105] In addition, the formation of a mixed complex (manganese (II)-o-hydroxy-hydroquinonephthaleine-streptomycin) is the basis of the determination of streptomycin.[39]

Complex formation with palladium (II) ions is the basis of the determination of polythiazide,[2] pralidoxime,[79] furosemide[3] and ethionamide.[158] The method used for the determination of cephalexine and ampicillin[115] is also based on palladium (II) complexes. In this case a ligand exchange reaction takes place between a palladium(II) o-hydroxy-quinonephthaleine complex and the above antibiotics, whereupon the absorbance at the 630 nm maximum decreases in parallel with the concentration of the antibiotics. The sensitivity of the method is extraordinary ($\varepsilon = 293{,}000$). Thiourea[114] and 5-iodouracil[40] can be determined in the same way. Another antibiotic, efrotomycin, was determined on the basis of its aluminum complex.[78] The determination of flavonoid-O-glycosides[42] is also based on aluminum complexes, whereas the determination of triazolam of triazolo-benzodiazepine type can be performed in the form of its 1:1 ruthenium (III) complex, at 570 nm.[41]

From the experiences with the above methods based on complex formation and several similar methods described in the literature, the following main conclusions can be drawn. A condition for the formation of a stable complex on which the determination can be based is that the organic pharmaceutical molecule contains at least two electron pair donor groups in a steric arrangement, permitting a chelate type complex to be formed with the metal ion. The electron pair donors are, most frequently, as follows: carboxyl, hydroxyl (mainly enolic and phenolic), thiol, ketone, thioketone, amino, imino, etc., groups. Since with the majority of the above groups there is a competition for the free electron pairs of the hetero atoms between the metal ion and the hydroxon-ium ion, the complexes decompose in more or less acidic media, depending on their stability constants, or, in other words, all complexes have a pH limit below which the analysis is impossible. On the other hand, the metal ions used as reagent generally form insoluble hydroxides with increasing pH; thus, the complex starts to decompose with the formation of mixed hydroxo-complexes, and with the further increase of pH, a metal hydroxide precipitate separates from the solution. The upper limit of the pH range at which the analysis can be performed is determined by the stability constant of the complex and the solubility product of the metal hydroxide.

Thus, the rigorous development of an analytical method should include the determination of the protonation constants and of the stability constant of the complex on which the measurement is based, and at least the empirical determination of the pH range in which the spectrum to be measured does not change as a function of pH. Another important task is the determination of the composition of the complex (metal-to-ligand ratio) and, as a function of this, the optimization of the excess of metal, and a strict adherence to this excess, since there are cases when complexes of varying compositions and widely different spectroscopic properties may occur.

For a deeper insight into complex equilibria, some excellent publications are recommended.[12,20,71] The formation of the complex used for the analysis is, in most cases, instantaneous or at least very fast at room temperature, but slow complexing reactions can also be encountered. For example, in the case of the determination of efrotomycin based on the formation of an aluminum complex, discussed above,[78] the very carefully determined optimum reaction time was 15 min at 37°C.

Finally, some methods are mentioned in which, although the reagent is a metal salt or a complex, the method cannot be regarded as based on complex formation. This holds, for example, for the methods based on antimony (III) chloride color reaction, particularly significant in the analysis of Vitamin D,[11,46] since, in this reaction, color formation is due to the changes in molecular structure. Sodium-hexanitrito cobaltate is the reagent of the spectrophotometric determination of p-substituted phenol derivatives like prenalterol;[182] the basis of the measurement at 473 nm is, in this case, the formation of an o-nitrosophenol derivative, and not complex formation.

Several classical alkaloid reagents, too, are metal complexes, like Reinecke's salt (ammonium tetrathiocyanato-diamino-cobaltate)[64] or the cobalt (II)-tetrathiocyanato complex[95] itself. In these cases the reaction is not a complex formation proper, but an ion pair formation between the complex anion and the quaternary ammonium cation or the alkaloid base protonated in acidic medium. Methods based on such interactions will be discussed in the next section.

H. ION PAIR FORMATION

One of the most important methods of spectrophotometric analysis of spectrophotometrically inactive (or only UV active) organic acids, bases and quaternary ammonium salts is based on the formation of ion pairs with colored reagents and subsequent extraction. The theoretical basis[151] of this method is that the dissociation equilibrium of a BA-type electrolyte dissociating in aqueous medium according to Equation 7.23 can be shifted toward the left (association) if the associate (ion pair) is removed by extraction by means of a solvent immiscible with water.

$$BA \Leftrightarrow B^+ + A^- \qquad (7.23)$$

where B^+ is a quaternary ammonium cation or protonated primary, secondary or tertiary amine

TABLE 7.1. Extraction Constants of Ion Pairs with Tetrabutylammonium Cation in the System Chloroform/Water[151a]

Ion pair forming anion	$\log K_{ex}$
Cl^-	−0.11
Br^-	1.29
J^-	3.01
ClO_4	3.48
Acetate	−2.1
Phenylacetate	0.27
Benzoate	0.39
3-Hydroxybenzoate	−1.54
Salicylate	2.42
Picrate	5.91
p-Toluenesulphonate	2.33
Naphthalene-2-sulphonate	3.45
Anthracene-2-sulphonate	5.11
Methyl orange	5.47
Bromothymol blue	8.0

cation and A^- is an organic or inorganic acid anion. The equilibrium conditions of the extraction can be defined by the extraction constant, K_{ex}:

$$K_{ex} = \frac{BA_{org.}}{B^+_{H_2O} \cdot A^-_{H_2O}}$$ (7.24)

where the numerator contains the concentration of the ion pair transferred into the organic phase during extraction, and the denominator contains the concentrations of anion and cation after extraction in the aqueous phase.

The efficiency of extraction can be characterized by the partition ratio. If the aim is an extraction of the protonated base or the quaternary ammonium ion by means of an appropriate counter ion, A^-, the partition ratio P_B can be defined as

$$P_B = \frac{BA_{org.}}{B^+_{H_2O}} = K_{ex}\, A^-_{H_2O}$$ (7.25)

whereas if the aim is an extraction of an organic anion, the ratio is

$$P_A = \frac{BA_{org.}}{A^-_{H_2O}} = K_{ex} B^+_{H_2O}$$ (7.26)

As seen from Equation 7.25, the efficiency of the extraction of quaternary ammonium compounds or protonated bases is directly proportional to the concentration of the counter ion and the extraction constant. Of the factors affecting the latter, the structure of the counter ion should be mentioned first. The data of Table 7.1[151a] show that of the groups bearing negative charge, sulphonate has stronger ion pair formation potential than carboxylate; and, the extraction constant is increased by the size of non-polar molecular fragments, but reduced by the subsequent non-ionizing polar groups. (From a comparison of the data of the series benzoic acid–3-hydroxybenzoic acid–salicylic acid, it can be seen that the hydroxy group in the ortho position strongly increases the extraction constant, due to its ability to form an intramolecular hydrogen bridge.) Excellent ion pair forming agents are the strongly ionized phenols, too.

The size of the non-polar molecule fragment attached to the amine group affects the extraction constant in the same direction, and to the same extent, as in the case of counter ions. The extraction constant generally increases in the order primary–secondary–tertiary–quaternary. For example, the logarithms of the extraction constants of n-propylamine, di-n-propylamine and tetra-n-propylammonium picrates are -0.57, 2.33 and 4.46, respectively[151b] (solvent: dichloromethane).

Another important factor affecting extraction constant, and thus partition ratio, is the choice of water immiscible solvent of which benzene, chloroform and dichloromethane are used most frequently. The extraction constants of chloroform and dichloromethane, capable of hydrogen bonding, are greater, and this is particularly significant in the extraction of bases also containing polar groups; for the extraction of more polar bases, solvents with higher dielectric constant are required (e.g., pentanol).

The third, very important factor affecting the ion pair extraction of amines and ammonium compounds is their protolytic equilibrium state in the aqueous phase, which is determined by the pK values of the base to be extracted, the acid providing the counter ion and the pH of the solution. The case is relatively the simplest for quaternary ammonium salts, since they are practically completely dissociated, independently of pH, over the whole pH region. Thus, the pH dependence of extractibility is affected only by the counter ion. With inorganic counter ions (bromide, iodide, perchlorate) and of the analytically important inorganic ion pair forming reagents, practically any pH can be chosen with sulphonic acids. However, with the most frequently used ion pair forming reagents of weak or medium acidity, neutral or weakly basic media are usually applied, to ensure the full dissociation of the reagents.

The situation is much more complicated with primary, secondary and tertiary amines. Here, the condition of ion pair formation is that the base is present in protonated form, and thus it is recommended to choose a pH at least one unit lower than the pK_a value of the protonated base. Of course, the arbitrary reduction of pH is limited by the condition that the ion pair forming reagent should be in a dissociated state in the solution to be extracted, in order to ensure the extraction of the protonated amine and prevent the eventual extraction of the undissociated reagent. Consequently, the pH of the extraction should be optimized carefully in such cases, or, for the extraction of bases weaker than average, an ion pair forming reagent of strong acidic character should be chosen.

On the basis of Equation 7.26, similar considerations hold for the case, less frequent in pharmaceutical analysis, in which the anions of drugs of acidic character are to be extracted with basic reagents as ion pairs.

So far the requirements that the ion pair forming reagent should be water soluble at the pH of the extraction and poorly soluble in the organic solvent chosen, and that it should ensure the quantitative extraction of the analyte into the organic phase have been discussed. Since in this case our aim is the spectrophotometric determination of the extracted component,* it is also a fundamental requirement of the reagent that the ion pair formed with the analyte has an intensive ($\varepsilon =$ ca. 10,000 to 30,000) color in the organic phase, and that the reagent blank (the absorbance of the ion pair forming reagent transferred into the organic phase at the wavelength of the measurement in the absence of the analyte) is negligible with respect to the absorbance of the ion pair. The wavelengths and intensities of the absorption maxima depend, of course, on the substance to be analyzed, the reagent applied and the extracting solvent.

The application of ion pair extraction methods in pharmaceutical analysis (particularly in the analysis of alkaloids and related compounds, amines and ammonium derivatives) has been one of the most widely used photometric methods since the end of the 1940s. An excellent review of the early literature of the method has been published by Higuchi and Brochmann-Hansen.[65] In addition to the spectrophotometric measurement of the extract, some early papers contain methods according to which the ion pair of the extract is decomposed with aqueous alkali solution, and the absorbance of the indicator dye which is re-extracted into the aqueous phase is measured.[94]

Because of their excellent sensitivity, their selectivity controllable by the pH of the solution or the choice of the extracting solvent and the ion pair forming reagent, their relative simplicity and

---

* Ion pair formation is also the basis of one of the most recent high performance liquid chromatographic techniques; in this case, of course, the spectrophotometric activity of the reagent is immaterial.

unambiguous chemical foundations, the methods based on ion pair extraction have been applied widely, up to now, in pharmaceutical analysis for the determination of the types of compounds previously discussed. Papers published in the last few years indicate that the method itself hardly changes (usually a direct spectrophotometric analysis of the extract is performed), and the established ion pair forming reagents are used primarily for the solution of novel problems. Of the reagents, bromothymol blue (pK_a = 7.07), bromophenol blue (pK_a = 3.87), bromocresol green (pK_a = 4.76), bromocresol purple (pK_a = 6.12), methyl orange (pK_a = 3.40) and picric acid (pK_a = 0.38) are mentioned here, but the application of numerous other ion pair forming dyes, introduced in some cases quite recently, is also shown in Tables 10.1 and 10.3 (see p. 250 and 274).

Typical applications include, for example, the determination of bis-quaternary ammonium steroids, like pancuronium bromide[22,171] or pipecuronium bromide[46] with bromophenol blue and bromothymol blue ion pair forming reagents, respectively. The latter reagent was applied successfully for the determination of codeine, ephedrine, procaine, pyrilamine and thiamine,[27] oxprenolol,[138] choline,[9] metoclopramide,[122] butetamate[59] and tilidine.[187]

Examples of unusual techniques include the determination of benzalkonium chloride and chlorhexidine with bromothymol blue reagent. In this case extraction could be omitted: the decrease in absorbance at the absorption maximum of the free dye ion (610 nm at pH = 7.5) due to ion pair formation is directly proportional to the concentration of the quaternary ammonium derivatives.[98]

Similarly, bromothymol blue is the reagent of an interesting method[160] in which the extraction step could be omitted and the dependence of the analytical wavelength and absorptivity of the amine to be determined could be eliminated simultaneously in the determination of primary, secondary and tertiary amines. The reagent applied here is a dichloromethane solution of an equimolar mixture of butylscopolammonium bromide and bromothymol blue. Upon adding the dichloromethane solution of the amine analyte to a large excess of the reagent, the following reaction takes place:

(butylscopolamine$^+$-bromide$^-$ + bromothymol blue) + amine $\rightleftharpoons$
$\rightleftharpoons$ butylscopolamine$^+$ bromothymol blue$^-$ + amine hydrobromide ion pair

Bromothymol blue in the sulphophthaleine form is colorless in dichloromethane solution even in the presence of the quaternary ammonium salt butylscopolammonium bromide. However, an amount equivalent to the amine is converted into an open quinoidal anionic derivative.

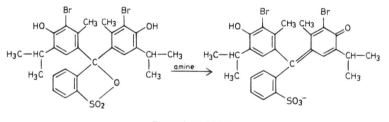

Bromothymol blue   (7.27)

Since the latter forms an ion pair with the butylscopolamine cation in large excess, the wavelength of the maximum and the molar absorptivity (λ_{max} 412 nm; ε = 18,620) are independent of the identity of the amine.

Ion pairs bromophenol blue–quinine[146] and bromocresol green–quinine[147] are reagents in the spectrophotometric determination of various quaternary ammonium derivatives. Details of this method will be given in Section 10.Q.1. Of the novel application of classical reagents, one may mention the use of picric acid in the determination of dibenzocycloheptadiene[28] and of methyl orange in the determination of chlorhexidine[134] and several alkaloids,[64] as well as bromocresol green for the measurement of dicyclomine preparations.[141]

The determination of acidic substances with basic ion pair forming agents is of much less significance in pharmaceutical analysis than that of amines as discussed so far. In this category the determination of phenylbutazone and saccharin with basic fuchsine reagent[138] can be mentioned.

Chloroform extraction is carried out at pH = 7.8 in the former, at pH = 6.0 in the latter case. Additional reagents are methylene blue,[155] methylene violet,[150] etc.

Finally, an ion pair formation between heparin, containing acidic groups, and various dyes may also form the basis of spectrophotometric methods, which were used for the determination of heparin in biological samples since the results were, to a certain extent, correlated with anticoagulant activity. Of the various dyes tested, azure A,[81] Nile blue A[164] and azocarmine G[60] are worth noting.

I. FORMATION OF CHARGE TRANSFER COMPLEXES

For the interpretation of the UV absorption spectra of some unsaturated molecules, particularly those with hetero atoms, it must be assumed that in these cases the mechanism of light absorption involves an intramolecular rearrangement due to the effect of a photon which causes electric charge flux or the formation of local dipoles within the molecule (charge transfer bands).[37,142] Particularly and analytically important cases of this type of light absorption are the absorption spectra of the so-called charge transfer complexes. During the light absorption of these complexes based on electron donor–electron acceptor interactions, charge transfer takes place between the two components associated in the complex; consequently, a new band occurs in the long wavelength region, at much longer (and in certain cases very much longer) wavelength than in the spectrum of the components. Donors include amines, sulphides, ethers, oxo-derivatives, aromatic compounds, acceptors include molecular halogens and aromatic derivatives with electron withdrawing substituents. Summaries of the theoretical foundations of charge transfer complexes have been published by Rao[142] and Foster;[37] we restrict ourselves here to the presentation of some analytical applications.

As the simplest example, the spectra of halide ions are discussed first. Chloride, bromide and iodide ions have absorption maxima at 184, 198 and 226 nm, respectively, in aqueous solution. In the formation of bands, a transfer of negative charge within the halide anion–alkali metal cation ion pair from the halide to the alkali metal can occur following excitation. If carbon tetrachloride is used as a solvent, which has low polarity and thus promotes ion pair formation, the absorption maxima shift to 246, 260 and 290 nm, respectively. Similar shifts can be observed if molecular iodine is used as acceptor in aqueous medium. The maximum of the I_2Cl^- complex is at 245 nm (ε = 23,000), whereas that of the I_2Br^- complex is at 265 nm (ε = 32,700). Although the stability constants of the complexes are very low (K = [I_2Hlg^-]/[I_2][Hlg^-] is 2.1 for the chloride complex and 11.0 for the bromide complex), under constant iodine concentration (3 $\times$ 10^{-4} M), the absorbance is a linear function of halide concentration. With chloride this occurs at 245 nm in a concentration range of 4 $\times$ 10^{-3} to 6 $\times$ 10^{-2} M, and with bromide at 265 nm in a range of 8 $\times$ 10^{-4} to 1.2 $\times$ 10^{-2} M. Since the weak absorption of iodine at the above wavelength can be compensated by reagent blank, the calibration lines pass the origin. This method was also used for pharmaceutical analysis.[66] An interesting feature of the method is that due to the low stability of the complex and the large excess of halide, only some tenths of one percent of the halide ions to be determined participate in the formation of the complex which is actually measured.

The method can be used for the determination of pharmaceuticals of quaternary ammonium salt type. In this case the quaternary ammonium ion pair is extracted with chloroform from the tablet or from an aqueous solution, and iodine dissolved in chloroform is applied as acceptor (in this case in large excess). The analysis is carried out at 263 and 368 nm for quaternary ammonium chlorides, at 280 nm for bromides and at 295 and 365 nm for iodides. The molar absorptivities are in the 10^4 order. Partly for this reason, partly for the large excess of iodine and partly for the significant stability of complexes in chloroform, the method can be used for highly sensitive determinations of a wide variety of quaternary ammonium derivatives.[1]

In contrast to the determination of quaternary ammonium derivatives by the indirect method, discussed above, on the basis of the halide ion attached to them, tertiary amines can be determined spectrophotometrically on the basis of their charge transfer complexes formed with a wide variety of donors in appropriate solvents. Thus, for example, the determination of procaine is possible in acetonitrile with iodine (λ_{max} 360 nm), with 2,4,5,7-tetranitrofluorenone (λ_{max} 535 nm) or with 7,7,8,8-tetracyanoquinodimethane (λ_{max} 745 nm).[89] Clonidine can also be determined, after the chloroform extraction of the base, as a complex with iodine (λ_{max} 295 nm).[87] These acceptors and numerous others can be used, in addition to spectrophotometric analysis, as TLC spray reagents, for example, in the selective detection of imidazoline derivatives.[88]

For the spectrophotometric determination of amines and related derivatives through the formation of charge transfer complexes, the most often applied acceptor reagent is chloranil, which has been used successfully for the determination of amines,[4,84] amino acids,[5] piperazine,[179] tranexamic acid,[180] cycloserine,[181] etc. The reactions are performed in aqueous medium at pH = 9, the absorption maxima of the complexes are in the vicinity of 350 nm. Addition of organic solvents like ethanol[69] or acetonitrile[67] to the reaction mixture shifts the maximum toward longer wavelengths and increases color intensity, which can be utilized, for example, for the determination of chlorpromazine,[69] gliclazide and tolazamide.[67]

Of the above-mentioned electron acceptor reagents, 7,7,8,8-tetracyano-quinodimethane has also been widely used for various assays in pharmaceutical analysis. Its interaction with iodide ions results in an extremely intense band (ε = 90,000) at 830 nm, and this serves as a basis for the determination of quaternary ammonium iodides and for the indirect determination of other ammonium salts which can be transformed into iodides.[169] Recent applications of the reagent include the determination of tranylcypromine sulphate, iproniazid phosphate, isocarboxazid,[70] benzocaine, procainamide, butacaine and other local anaesthetics,[112] and pholedrin.[86]

The free electron pair of the NH group adjacent to the aromatic ring of phenylhydrazones also forms complexes with iodine or chloranil acceptors. This was utilized in the indirect determination of corticosteroids after phenyl-hydrazone formation.[7]

Further acceptor reagents used in pharmaceutical analysis are chloranilic acid[102] used for the determination of quinidine, prenylamine, tolazoline, hydralazine, pindolol, etc., in acetonitrile medium on the basis of its rather weak band at 522 nm, and 2,6-dichloroquinone-4-chloroimide,[103] which also gives weak bands with various indole derivatives between 490 and 450 nm in ethanol solution.

In addition to these examples, several further examples can be found in Chapter 10 on the application of charge transfer complexes in pharmaceutical analysis.

J. KINETIC METHODS[124]

The majority of chemical reactions applied in spectrophotometric analysis proceed at a measurable rate. So far, reaction rate has been mentioned only in a context that the most important factor affecting rate is temperature (Section 6.C), or that it can also be influenced by changes in solvent, pH and reagent concentration.

Systematic kinetic studies, i.e., the investigation of the time dependence of the conversion of the reaction on which the analysis is based, may be used to develop more selective or more rapid methods of analysis. Of the differential kinetic methods aiming at selectivity, the determination of a Δ^4-3-keto impurity in $\Delta^{1,4}$-3-ketosteroids[3] is discussed. The 3-keto groups of both derivatives form thiosemicarbazones with thiosemicarbazide (λ_{max} 302 and 320 nm, respectively). As can be seen in Figure 7.3 on the example of 6α-methyl-prednisolone and 6α-methyl-hydrocortisone, the latter reacts much faster with thiosemicarbazide: this reaction is complete after 120 min, while the reaction with 6α-methyl-prednisolone proceeds only to a few percent in the same time. Under strictly defined reaction conditions, the absorbance measured at 302 nm after 120 min reaction time is a linear function of the amount of 6α-methyl-hydrocortisone impurity in 6α-methyl-prednisolone and, thus, with the simultaneous investigation of standard mixtures, even 0.5% of impurity can be determined with the desired accuracy.

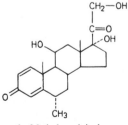

6α-Methyl-prednisolone

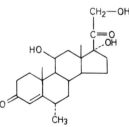

6α-Methyl-hydrocortisone

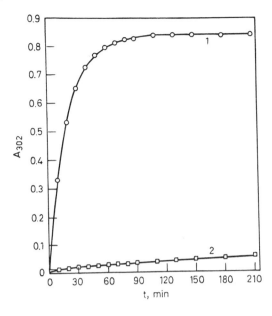

FIGURE 7.3. Absorbance vs. time curves of thiosemicarbazone formation reactions. 1: 6α-methylhydrocortisone; 2: 6α-methylprednisolone. Reagent: 0.1 M thiosemicarbazide in 0.1 M hydrochloric acid; Reaction mixture: 5 mg of steroid dissolved in 5 ml of ethanol + 1 ml of reagent. Temperature: 25°C. (From Forist, A. A. *Anal. Chem.* 1959, *31*, 913.)[36]

The epimeric contamination (containing 3α-acetoxy group) of ethynodiol diacetate can be determined on the basis of similar principles, since the acid catalyzed elimination reaction (Scheme 8.1, Section 8.C) leading to the 3,5-diene measurable at 236 nm is ca. 5 times faster for the α-epimer than for the β-epimeric main component.[2]

This and several other similar methods may now be regarded as obsolete, since their sensitivity is very limited and they are very time consuming. These analyses can be carried out much faster, with higher sensitivity and accuracy by chromatographic methods.

Another group of kinetic methods comprises the techniques utilizing the initial sections of all kinetic curves for which the concentration of the substance formed (and thus in our case, the absorbance measured at an appropriate wavelength) is a linear function of time. If sufficiently high concentrations of reagents are selected, a pseudo-first order reaction with respect to the analyte can be generated in this section of the curve, and thus the slope of the linear section of the kinetic curve is a linear function of the concentration of the analyte. A significant advantage of this method over the conventional spectrophotometric technique is that it is not necessary to wait for the completion of the reaction, which may take several hours: the evaluation of the above mentioned linear section generally takes 1 to 2 minutes. However, a disadvantage is that due to the fast reaction rates, a spectrophotometer is required (a recording spectrophotometer with stopped flow accessory) which allows the reagents to be efficiently mixed and passed into the measuring cell within a few seconds and which permits the on-line recording of time vs. absorbance curve. Because of these special demands, this technique has not become widely used.

An example of this is the determination of corticosteroids (betamethasone and its valerate, triamcinolone acetonide, fluocinolone acetonide) in ointments. The analysis is based on the tetrazolium blue reaction discussed in Sections 7.E and 10.M. The concentrations of corticosteroids can be calculated with a relative standard deviation of 0.3 to 1.9% on the basis of the slopes determined by a computer from the 30 to 70 s linear section of the kinetic curve.[85]

The second example is the determination of sulphonamides by the kinetic version of the azo-coupling color reaction[188] discussed in Section 7.D. The coupling reagent is N-(1-naphthyl)-ethylene diamine, which is added after a nitrous acid diazotization of the sulphonamides. The analysis is based on the determination of the time within which the absorbance at 545 nm rises from a predetermined value to another predetermined value (of course, both values are on the linear section of the kinetic curve). The reciprocal of this time is a linear function of sulphonamide concentration;

the slope of the curves depends on the sulphonamide. The measurement takes ca. 10 s, and thus, with automatic feeding of reagents, a full calibration curve can be taken within 30 min, even if several parallel measurements are carried out at each point.

It is noted that the techniques based on the above and similar principles are applied now primarily in automatic analyzers. As shown in Section 9.J, in both the AutoAnalyzer and the flow-injection type analyzers, reactions are often applied which do not even approach completion during the time of the analysis.

Catalytic methods can be regarded as a special area of kinetic methods. They are based on a catalytic reaction (catalyzed usually by metal ions) which can be readily followed by spectrophotometric measurement, utilizing the fact that the reaction rate under appropriate concentration conditions is proportional to the concentration of the catalyst. The micro-analysis of Vitamin B_{12} is based, for instance, on this method.[109] Cobalt(II) ions, obtained after a concentrated sulphuric acid digestion, catalyze the decolorization reaction of pyrogallol red dye with hydrogen peroxide. The decrease in absorbance at 515 nm during the 5 min reaction time is proportional to the concentration of cobalt (II) ions in the 0.2 to 13 ng/ml concentration range, therefore enabling this micro-method to be used for the analysis of very dilute Vitamin B_{12} solutions, including pharmaceutical preparations.

In contrast to these catalytic methods, the inhibitory effect of some drugs on various reactions can also serve as a basis for their quantitative determination. For example, the photochemical reduction of the dye Rose Bengal to colorless products by ethylenediamine-tetraacetic acid is strongly inhibited by epinephrine, norepinephrine, dopamine and L-dopa, and hence the time required for the photoreduction of e.g., 90% of the dye in its solution correlates with the concentration of these drugs.[106] The same principle is used for the determination of nitroprusside. In this case the dye is tetrachloro-tetrabromo-fluorescein, and its photoreduction by ethylenediamine tetraacetic acid is inhibited by the analyte.[125]

K. ENZYMATIC REACTIONS

The investigation of enzyme-catalyzed reactions is a most important field in biochemistry. Since most of these reactions are in some form connected with a change in the light absorption of the substrate or the coenzyme, spectrophotometry is one of the most important measurement methods of enzymology.[13] The determination of enzyme activity by spectrophotometric methods or the spectrophotometric investigation of enzyme reactions fall beyond the scope of this book. However, the experiences gained in connection with enzymological–biochemical studies and the techniques developed in this field can also be utilized in the analysis of a number of pharmaceuticals. These methods do not belong to the standard, widely used techniques of pharmaceutical analysis. Utilizing the outstanding specificity of enzymatic reactions, these methods are used primarily for the analysis of biological samples for which conventional methods fail in selectivity because of background problems. Of the huge biochemical literature, we restrict ourselves to short descriptions of some characteristic methods.

In some cases a spectrophotometrically inactive analyte is made active by means of an enzyme reaction. Such a reaction is, for example, the oxidation of cholesterol with atmospheric oxygen by means of cholesterol oxidase enzyme. The reaction product is spectrophotometrically active cholestenone. Although the measurement of this product at 240 nm may be the basis of cholesterol analysis, the reaction is used more often for the determination of enzyme activity,[19] and cholesterol is measured, e.g., in serum, rather on the basis of the determination of hydrogen peroxide formed according to Scheme 7.28.

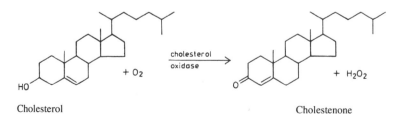

Cholesterol   Cholestenone   (7.28)

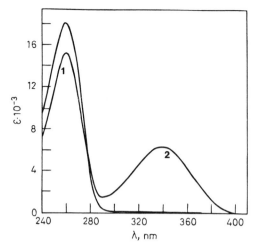

FIGURE 7.4. Spectra of NAD (nicotinamide-adenide dinucleotide, Curve 1) and of its reduced form (NADH, Curve 2). Solvent: aqueous buffer, pH = 7.6.

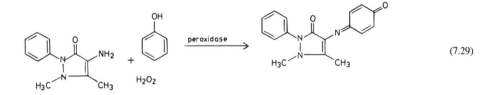

(7.29)

One widely used application of this technique is based on an oxidative condensation between 4-aminoantipyrine and phenol (Scheme 7.29), catalyzed by another enzyme (peroxidase). Because of the two selective enzyme reactions, the measurement of the resulting quinonimine-type dye at 510 nm allows cholesterol to be determined by a very selective, indirect method.[24]

Another method, based on a change in the light absorption of analyte as a result of an enzyme catalyzed reaction, is the determination of 6-mercaptopurine in urine. The oxidation catalyzed by xanthine oxidase enzyme causes a strong bathochromic shift, and this is the basis of a selective difference spectrophotometric analysis.[43] The details of the method are given in Section 8.D.

There is a method, entirely different from the above, for the determination of ascorbic acid, according to which the selectivity of a conventional spectrophotometric method is enhanced by enzymatic reaction.[148] The basis of the method is a reaction in which ascorbic acid is oxidized into dehydroascorbic acid by iron(III) salts, and the stoichiometric amount of iron(II) ions formed in this reaction can be determined in the form of their tripyridyl-s-triazine complex. The selectivity of this method is, however, insufficient for obtaining real values in plasma: other components of the plasma also convert iron(III) ions into iron(II). If, in parallel to the reactions performed in the conventional way, in a sample of the same volume taken from the plasma, ascorbic acid is oxidized selectively into dehydroascorbic acid by means of ascorbic acid oxidase enzyme, the reducing capacity of the plasma sample is split into two parts: the part due to ascorbic acid is removed, and the non-specific part is retained. Then, if an iron(III) oxidation and complex formation is carried out after this enzymatic oxidation, and the resulting solution is used as a reference in the measurement of the solutions obtained after the reactions without enzymatic oxidation, the resulting absorbance will be characteristic of the "true" ascorbic acid content.

An important group of enzymatic reactions applicable for spectrophotometric analysis comprises the processes catalyzed by dehydrogenase enzymes. The measurement is, in this case, made possible by the phenomenon that (as can be seen in Figure 7.4) the co-enzyme, NAD^+ (nicotinamide-adenine dinucleotide) has no absorption in the region above 300 nm, whereas its reduced form, NADH, has a distinct absorption band at 340 nm ($\varepsilon = 6300$). The measurement of substances capable of enzymatic dehydrogenation can be traced back to the analysis of NADH formed in equivalent amount.

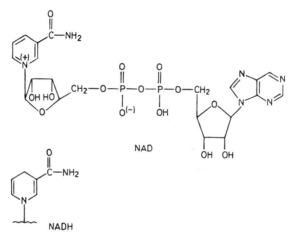

Of the numerous practical applications of this method, the determination of bile acids in serum by means of 3α-hydroxy-steroid dehydrogenase enzyme[31] is mentioned. It is noted that by means of the fluorimetric measurement of NADH, the sensitivity of the method is enhanced, whereas by means of the application of other enzymes (7α-hydroxy-steroid dehydrogenase, 12α-hydroxy-steroid dehydrogenase), the selectivity of the method can also be greatly enhanced.[100] Of the application of the method for the analysis of pharmaceutical preparations, the determination of the lactate content of infusion solutions[135] is an example. The basis of the method is the oxidation of lactic acid into pyruvic acid. The analysis can be carried out with 0.1 ml of infusion solution on the basis of the measurement of NADH formed after 20 min of reaction time.

The decrease in the concentration of NADH or of its phosphate, NAD(P)H, as measured at 340 nm, may also be the basis of spectrophotometric methods. For example, the determination of salicylic acid in serum[97,189] can be performed on the basis of the reaction given in Scheme 7.30.

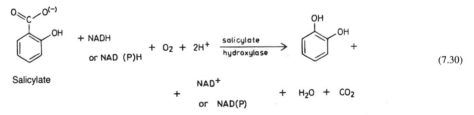

(7.30)

The wavelength of the assay can be shifted to the visible range (520 nm) if the catechol formed from the salicylate in the enzymatic reaction is subsequently converted into an indophenol-type dye with 4-aminopyrine.[116]

As a conclusion, it is noted that the spectrophotometric measurement of several enzyme-catalyzed reactions finds new analytical application with the spread of enzyme immunoassay methods, but even a summary of these methods would exceed the scope of this book.

The determination of methotrexate in serum[32] is also based on the spectrophotometric assay of NAD(P)H. The key to the method based on enzyme inhibition is the reaction in Scheme 7.31:

$$\text{Dihydrofolic acid} + 2\text{ NAD(P)H} \xrightarrow{\text{DHFR}} \text{tetrahydrofolic acid} + 2\text{ NAD(P)} \qquad (7.31)$$

The equation shows that the reduction of dihydrofolic acid is catalyzed by the enzyme dihydrofolic acid reductase. Methotrexate, by binding to the latter, inhibits the reaction, and thus the absorbance measured at 340 nm after a certain reaction time (5 min. in this case) is directly proportional to the concentration of methotrexate inhibitor.

REFERENCES

1. Abdelkader, M. A.; Taha, A. M.; Abdelfattah, S. *Pharmazie* 1980, *35*, 30.
2. Agatonovic-Kustrin, S.; Zivanovic, L. J. *J. Pharm. Biomed. Anal.* 1989, *7*, 1559.
3. Agatonovic-Kustrin, S.; Zivanovic, L. J.; Radulovic, D.; Pecanac, D. *J. Pharm. Biomed. Anal.* 1990, *8*, 983.
4. Al. Ghabsha, T. S.; Rahim, S. A.; Townshend, A. *Anal. Chim. Acta* 1976, *85*, 189.
5. Al-Sulimany, F.; Townshend, A. *Anal. Chim. Acta* 1973, *66*, 195.
6. Andermann, G.; Zimmermann, G.; Schilling, E. *Analyst* 1980, *105*, 575.
7. Ayad, M. M.; Belal, S.; El Adl, S. M.; Al Kheir, A. A. *Analyst* 1984, *109*, 1417.
8. Baggi, T. R.; Mahajan, S. N.; Rao, G. R. *J. Assoc. Off. Anal. Chem.* 1975, *58*, 875.
9. Barchuchová, A.; Gasparic, J.; Oulehlová, A. *Ceskoslov. Farm.* 1984, *33*, 343.
10. Bartos, J. *Talanta* 1961, *8*, 556.
11. Bartos, J.; Pesez, M. *Colorimetric and Fluorimetric Analysis of Steroids,* Academic Press: London, 1976, 105–115.
12. Beck, M. T.; Nagypál, I. *Complex Equilibria,* Ellis Horwood: Chichester, 1990.
13. Bergmayer, H. U., Ed. *Grundlagen der enzymatischen Analyse* Verlag Chemie: 1977.
14. Blau, K.; King, G. S., Eds. *Handbook of Derivatives for Chromatography,* Heyden: London, 1978.
15. Blau, K.; Robson, W. *Chem. Ind.* 1957, 424.
16. Bratton, A. C.; Marshall, E. K. *J. Biol. Chem.* 1939, *128*, 337.
17. Bricker, C. E.; Vail, W. A. *Ind. Eng. Chem. Anal. Ed.* 1945, *17*, 400.
18. *British Pharmacopoeia 1988,* Her Majesty's Stationery Office: London, 48.
19. Brooks, C. J. W.; Smith, A. G. *J. Chromatogr.* 1975, *112*, 499.
20. Burger, K. *Coordination Chemistry: Experimental Methods,* Butterworths: London, 1973.
21. Burger, K. *Organic Reagents in Metal Analysis,* Pergamon Press: Oxford, 1973.
22. Buzello, W. *Anaesthetist* 1974, *23*, 443.
23. Chafetz, L. *J. Pharm. Sci.* 1971, *60*, 291.
24. Cobb, S. A.; Sanders, J. L. *Clin. Chem.* 1978, *24*, 1116.
25. Connors, K. A., Derivatives of Carbamic Acid, in *Pharmaceutical Analysis,* T. Higuchi; E. Brochmann-Hanssen Eds., Interscience Publ.: New York, 1961, 236.
26. Csizér, É.; Görög, S. *Anal. Chim. Acta* 1976, *86*, 217.
27. Das Gupta, V.; Shek, O. H. *Am. J. Hosp. Pharm.* 1976, *33*, 1086.
28. Dembinski, B. *Farm. Polska* 1983, *39*, 1.
29. de Silva, J. A. F.; Strojny, N. *J. Pharm. Sci.* 1971, *60*, 1303.
30. Doulakas, J. *Pharm. Acta Helv.* 1975, *50*, 447.
31. Engart, R.; Turner, M. D. *Anal. Biochem.* 1973, *51*, 399.
32. Falk, L. C. *Clin. Chem.* 1976, *22*, 785.
33. Faeron, W. R.; Boggust, W. A. *Analyst* 1954, *79*, 101.
34. Fitzpatrick, F. A.; Siggia, S. *Anal. Chem.* 1973, *45*, 2310.
35. Folin, O.; Ciocalteu, V. *J. Biol. Chem.* 1927, *73*, 627.
36. Forist, A. A. *Anal. Chem.* 1959, *31*, 913.
37. Foster, R. *Organic Charge-Transfer Complexes,* Academic Press: London, New York, 1969.
38. Fujita, Y.; Mori, I.; Fujita, K.; Tanaka, T. *Chem. Parm. Bull.* 1987, *35*, 865.
39. Fujita, Y.; Mori, I.; Kitano, S. *Chem. Pharm. Bull.* 1983, *31*, 1289.
40. Fujita, Y.; Mori, I.; Kitano, S. *Bull. Chem. Soc. Japan* 1983, *56*, 2167.
41. Gallo, N.; Bianco, V. D.; Bianco, P.; Luisi, G. *Farmaco; Ed. Pr.* 1983, *38*, 205.
42. Glasl, H. *Z. Anal. Chem.* 1985, *321*, 325.
43. Görög, K.; Görög, S. *J. Pharm. Biomed. Anal.* 1988, *6*, 109.
44. Görög, S. *J. Pharm. Sci.* 1968, *57*, 1737.
45. Görög, S. *Anal. Chem.* 1970, *42*, 560.
46. Görög, S. *Quantitative Analysis of Steroids,* Elsevier: Amsterdam, 1983, p 430.
47. Görög, S.; Horváth, P. *Analyst* 1978, *103*, 346.
48. Görög, S.; Laukó, A.; Rényei, M.; Hegedüs, B. *J. Pharm. Biomed. Anal.* 1983, *1*, 497.
49. Görög, S.; Rényei, M. *Acta Chim. Hung.* 1983, *115*, 65.
50. Görög, S.; Rényei, M. A New Color Reaction for Bile Acids, in *Advances in Steroid Analysis '84,* S. Görög, Ed., Elsevier: Amsterdam, 1985, 555–561.
51. Görög, S.; Rényei, M.; Laukó, A. *J. Pharm. Biomed. Anal.* 1983, *1*, 39.
52. Görög, S.; Szász, Gy. *Analysis of Steroid Hormone Drugs* Elsevier: Amsterdam, 1978, pp 348–349.

53. Görög, S.; Szepesi, G. *Acta Pharm. Hung.* 1971, *41*, 25.
54. Görög, S.; Szepesi, G. *Anal. Chem.* 1972, *44*, 1079.
55. Görög, S.; Szepesi, G. *Analyst* 1972, *97*, 519.
56. Graham, R. E.; Williams, P. A.; Kenner, C. T. *J. Pharm. Sci.* 1970, *59*, 1152.
57. Grignard, R.; Kerboul, A. *Analusis* 1982, *10*, 423.
58. Gupta, R. N.; Pickersgill, R.; Stefanec, M. *Clin. Biochem.* 1983, *16*, 220.
59. Gübitz, G.; Wintersteiger, R. *Sci. Pharm.* 1976, *44*, 28.
60. Güven, K. C.; Ertan, G. *Eczacilik Bülteni* 1982, *24*, 66.
61. Hakl, J. *J. Chromatogr.* 1977, *143*, 317.
62. Hassib, S. T. *Analyst* 1980, *105*, 669.
63. Heise, E.; Himbel, K. H. *Arzneimittelforsch.* 1955, *5*, 149.
64. Higuchi, T. V.; Bodin, J. I. Alkaloids, in *Pharmaceutical Analysis,* Higuchi, T.; E. Brochmann-Hanssen, Eds., Interscience Publ.: New York, London, 1961, 410–412.
65. Higuchi, T.; Brochmann-Hanssen, E. *Pharmaceutical Analysis,* Interscience Publ.: New York, London, 1961, 413–418.
66. Hussain, A.; Bawarshi, R. *J. Pharm. Sci.* 1979, *68*, 513.
67. Hussein, S. A.; Mohamed, A. M. I.; Abdel Alim, A. A. M. *Analyst* 1989, *114*, 1129.
68. Hüttenrauch, R. *Arch. Pharm.* 1962, *295*, 721.
69. Ibrahim, E. A.; Issa, A. S.; Salam, M. A. A.; Mahrous, M. S. *Talanta* 1983, *30*, 531.
70. Ibrahim, F.; Belal, F.; Hassan, S. M.; Aly, F. A. *J. Pharm. Biomed. Anal.* 1991, *9*, 101.
71. Inczédy, J. *Analytical Applications of Complex Equilibria,* Ellis Horwood: Chichester, 1976.
72. Issopoulos, P. B. *Pharm. Acta Helv.* 1989, *64*, 280.
73. Issopoulos, P. B. *Acta Pharm. Hung.* 1991, *61*, 205.
74. Jaffery, N. F.; Ahmad, S. N.; Jailkhani, B. L. *J. Pharmacol. Methods.* 1983, *9*, 33.
75. Johnson, D. P.; Critchfield, F. E. *Anal. Chem.* 1960, *32*, 865.
76. Jordan, D. E.; Veatch, F. C. *Anal. Chem.* 1964, *36*, 120.
77. Kakac, B.; Vejdelek, Z. J. *Handbuch der photometrischen Analyse organischer Verbindungen,* Verlag Chemie: Weinheim, 1974.
78. Kaplan, L.; Fink, D. W.; Fink, H. C. *Anal. Chem.* 1984, *56*, 360.
79. Karljikovic-Rajic, K.; Stankovic, B.; Granov, A. *J. Pharm. Biomed. Anal.* 1990, *8*, 735.
80. Khattab, F. I. *Anal. Lett.* 1983, *16*(B14), 1121.
81. Klein, M. D.; Drongowski, R. A.; Linhardt, R. J.; Langer, R. S. *Anal. Biochem.* 1982, *124*, 59.
82. Knapp, D. R., *Handbook of Analytical Derivatization Reactions,* Wiley: New York, 1979.
83. Kolbezan, M. J.; Eckert, J. W.; Bretschneider, B. F. *Anal. Chem.* 1962, *34*, 583.
84. Korany, M. A.; Wahbi, A. M. *Analyst* 1979, *104*, 146.
85. Koupparis, M. A.; Walczak, K. M.; Malmstadt, H. V. *J. Pharm. Sci.* 1979, *68*, 1479.
86. Kottke, D.; Beyrich, Th.; Meincke, U. *Pharmazie* 1990, *45*, 837.
87. Kottke, D.; Beyrich, Th.; Werner, H. *Pharmazie* 1988, *43*, 99.
88. Kovar, K.-A.; Abdel-Hamid, M. *Arch. Pharm.* 1984, *317*, 246.
89. Kovar, K.-A.; Mayer, W.; Auterhoff, H. *Arch. Pharm.* 1981, *314*, 447.
90. Krzek, J. *Acta Pol. Pharm.* 1981, *38*, 589.
91. Kwong, T.; Adams, N.; Young, N. *Clin. Biochem.* 1984, *17*, 170.
92. Lange, B.; Vejdelek, Z. J., *Photometrische Analyse,* Verlag Chemie: Weinheim, 1980.
93. Lawrence, J. F.; Frei, R. W. *Chemical Derivatization in Liquid Chromatography,* Elsevier: Amsterdam, 1976.
94. Lehman, R. A.; Aitken, T. *J. Lab. Clin. Med.* 1943, *28*, 787.
95. Lemli, J.; Knockaert, I. *Pharm. Weekblad* 1958, *5*, 142.
96. Lohman, F. H. *Anal. Chem.* 1958, *30*, 972.
97. Longenecker, R. W.; Trafton, J. E.; Edwards, R. B. *Clin. Chem.* 1984, *30*, 1369.
98. Lowry, J. B. *J. Pharm. Sci.* 1979, *68*, 110.
99. Lukáts, B.; Takácsi Nagy, G. *Acta Pharm. Hung.* 1965, *35*, 116.
100. Macdonald, I. A.; Williams, C. N.; Musial, B. C. *J. Lipid Res.* 1980, *21*, 381.
101. Mader, W. J.; Buck, A. H. *Anal. Chem.* 1952, *24*, 666.
102. Mahrous, M. S.; Issa, A. S.; Abdel Salam, M. A.; Soliman, N. *Anal. Lett.* 1986, *19*, 901.
103. Manzar, A. Q. N. *Pharmazie* 1981, *36*, 685.
104. Marczenko, Z. *Separation and Spectrophotometric Determination of Elements,* Ellis Horwood: Chichester, 1986.
105. Martinez Calatayud, J.; Bosch Reig, F.; Marin Saez, R. *Pharmazie* 1984, *39*, 425.

106. Martínez-Lozano, C.; Pérez-Ruiz, T.; Tomás, V.; Val, O. *Analyst* 1991, *116*, 857.

107. Matsuda, R.; Takeda, Y.; Ishibashi, M.; Uchiyama, M.; Suzuki, M.; Takitani, S. *Bunseki Kagaku* 1986, *35*, 151.

108. Mattson, L. N., Holt, W. L. *J. Am. Pharm. Assoc.* 1949, *38*, 55.

109. Medina-Escriche, J.; Hernández-Llorens, M. L.; Llobat-Estelles, M.; Sevillano-Cabeza, A. 1987, *112*, 309.

110. Medwick, T., Organomercurials, in *Pharmaceutical Analysis,* T. Higuchi; E. Brochmann-Hanssen, Eds., Interscience Publ.: New York, London, 1961, 730–732.

111. Milun, A. J. *Anal. Chem.* 1957, *29*, 1502.

112. Mohamed, A.-M. I.; Hassan, H. Y.; Mohamed, H. A.; Hussein, S. A. *J. Pharm. Biomed. Anal.* 1991, *9*, 525.

113. Mopper, B. *J. Assoc. Off. Anal. Chem.* 1987, *70*, 42.

114. Mori, I.; Fujita, Y.; Kitano, S.; Sakaguchi, K. *Bunseki Kagaku* 1982, *31* E305.

115. Mori, I.; Fujita, Y.; Sakaguchi, K. *Chem. Pharm. Bull.* 1982, *30*, 2599.

116. Morris, H. C.; Overton, P. D.; Campbell, R. S.; Hammond, P. M.; Atkinson, T.; Price, C. P. *Clin. Chem.* 1990, *36*, 131.

117. Munson, J. W. *J. Pharm. Sci,* 1974, *63*, 252.

118. Munson, J. W.; Bilous, R. *J. Pharm. Sci.* 1977, *66*, 1403.

119. Nachtmann, F. *Z. Anal. Chem.* 1976, *282*, 209.

120. Nishina, T.; Sakai, Y.; Kimura, M. *Chem. Pharm. Bull.* 1965, *13*, 414.

121. Pankratz, R. E.; Bandelin, F. J. *J. Am. Pharm. Assoc.; Sci. Ed.* 1952, *41*, 267.

122. Patel, R. B.; Gandhi, T. P.; Patel, V. C.; Patel, S. K.; Gilbert, R. N. *Indian Drugs* 1983, *20*, 490.

123. Pawelczyk, E.; Smilowski, B. *Acta Polon. Pharm.* 1984, *41*, 351.

124. Pérez-Bendito, N.; Silva, M. *Kinetic Methods in Analytical Chemistry,* Ellis Horwood: Chichester, 1987.

125. Pérez-Ruiz, T.; Martínez-Lozano, M. C.; Tomás, V.; López-Balsera, C. *J. Pharm. Biomed. Anal.* 1991, *9*, 123.

126. Pesez, M. *J. Pharm. Pharmacol* 1959, *11*, 475.

127. Pesez, M.; Bartos, J. *Talanta* 1960, *5*, 216.

128. Pesez, M.; Bartos, J. *Talanta* 1963, *10*, 69.

129. Pesez, M.; Bartos, J. *Colorimetric and Fluorimetric Analysis of Organic Compounds and Drugs,* Marcel Dekker: New York, 1974 a. 521–531.

130. Pesez, M.; Poirier, P.; Bartos, J. *Pratique de l'analyse organique colorimetrique,* Masson: Paris, 1966.

131. Pindur, U. *Pharm. Acta Helv.* 1982, *57*, 112.

132. Pindur, U.; Schiffl, E. *Pharm. Acta. Helv.* 1983, *58*, 322.

133. Pindur, U.; Witzel, H. *Deutsche Apoth. Ztg.* 1988, *128*, 2127.

134. Pinzauti, S.; La Porta, E.; Casini, M.; Betti, C. *Pharm. Acta Helv.* 1982, *57*, 334.

135. Pokorny, E.; Menszel, J. *Acta Pharm. Hung.* 1988, *58*, 241.

136. Porter, C. C.; Silber, R. H. *J. Biol. Chem.* 1950, *185*, 201.

137. Radovic, Z.; Malesev, D. *Pharmazie* 1984, *39*, 870.

138. Radulovic, D.; Jovanovic, M.; Milosevic, R. *Acta Pharm. Jugosl.* 1984, *34*, 169.

139. Radulovic, D.; Pecanac, D.; Zivanovic, L.; Agatonovic Kustrin, S. *J. Pharm. Biomed. Anal.* 1990, *8*, 739.

140. Radulovic, D.; Pecanac, D.; Zivanovic, L.; Agatonovic Kustrin, S. *J. Pharm. Biomed. Anal.* 1991, *9*, 203.

141. Raggi, M. A.; Cavrini, V. V.; di Pietra, A. M. *J. Pharm. Biomed. Anal.* 1985, *3*, 287.

142. Rao, C. N. R. *Ultraviolet and Visible Spectroscopy,* Butterworths: London, 1975, 27–28, 162–182.

143. Rényei, M.; Görög, S. *Gyógyszerészet* 1979, *23*, 346.

144. Rosales, D.; Gómez Ariza, J. L. *Anal. Chem.* 1985, *57*, 1711.

145. Ryan, J. A. *J. Pharm. Sci.* 1984, *73*, 1301.

146. Sakai, T. *Analyst.* 1983, *108*, 608.

147. Sakai, T. *Anal. Chim. Acta* 1983, *147*, 331.

148. Samyn, W. *Clin. Chim. Acta* 1983, *133*, 111.

149. Sandell, E. B. *Colorimetric Determination of Traces of Metals,* Interscience Publ.: New York, 1959.

150. Sastry, C. S. P.; Tipirneni, A. S. R. P.; Suryanarayana, M. V. *Analyst* 1989, *114*, 513.

151. Schill, G.; Ehrsson, H.; Vessman, J.; Westerlund, D. *Separation Methods for Drugs and Related Organic Compounds,* Swedish Pharmaceutical Press: Stockholm, 1984, 1–39; a.) 10; b.) 15.

152. Schulz, E. P.; Diaz, M. A.; López, G.; Guerrero, L. M.; Barrara, H.; Pereda, A. L.; Aguilera, A. *Anal. Chem.* 1964, *36*, 1624.

153. Sethia, M. M. *Ind. J. Pharm. Sci.* 1983, *45*, 30.

154. Shah, A. K.; Agrawal, Y. K.; Banerjee, S. K. *Anal. Lett.* 1981, *14B*, 363.

155. Shih, I. K.; Teare, F. W. *Can. J. Pharm. Sci.* 1967, *1*, 35.

156. Shingbal, D. M.; Agni, R. M. *Indian Drugs* 1983, *20*, 162.

157. Siggia, S. *Instrumental Methods of Organic Functional Group Analysis,* Wiley-Interscience: New York, 1972.

158. Sikorska-Tomicka, H. *Z. Anal. Chem.* 1982, *312,* 353.

159. Snell, F. D. *Photometric and Fluorimetric Methods of Analysis, Metals, Nonmetals,* Wiley: New York, 1978–1981.

160. Stamm, H.; Hubbert, T. *Arch. Pharm.* 1983, *316,* 435.

161. Stewart, J. T.; Parks, E. H. *Int. J. Pharm.* 1983, *17,* 161.

162. Stewart, J. T.; Shaw, T. D.; Ray, A. B. *Anal. Chem.* 1969, *41,* 360.

163. Strode, C. W.; Schott, H. O.; Coleman, O. J. *Anal. Chem.* 1957, *29,* 1184.

164. Stuzka, V.; Havlová, J.; Dvorák, J.; Vávra, V. *Ceskoslov. Farm.* 1984, *33,* 412.

165. Svobodová, D.; Gasparic, J.; Fraenkl, M.; Nováková, L. *Coll. Czech. Chem. Commun.* 1976, *41,* 2176.

166. Szepesi, G. *C. Sc. Thesis,* Hungarian Academy of Sciences: Budapest, 1976.

167. Szepesi, G. *Z. Anal. Chem.* 1973, *265,* 334.

168. Szepesi, G.; Görög, S. *Acta Pharm. Hung.* 1971, *41,* 30.

169. Taha, A. M.; Abdelkader, M. A.; Abdelfattah, S. *Pharmazie* 1980, *35,* 93.

170. Taha, A. M.; El-Rabbat, N. A.; El-Kommos, M. E.; Refat, I. H. *Analyst* 1983, *108,* 1500.

171. Tanaka, K.; Hioki, M.; Shindo, H. *Chem. Pharm. Bull.* 1974, *22,* 2599.

172. The United States Pharmacopoeia XXII, USP Convention Inc.: Rockville, 1990, 111.

173. Trinder, P. *Biochem. J.* 1954, *57,* 301.

174. Umberger, E. J. *Anal. Chem.* 1955, *27,* 768.

175. Umbreit, G. R.; Houtman, R. L. *J. Pharm. Sci.* 1967, *56,* 349.

176. Vachek, J. *Ceskoslov. Farm.* 1985, *34,* 149.

177. Vejdelek, Z. J.; Kakac, B. *Farbreaktionen in der spektrophotometrischen Analyse organischer Verbindungen,* Gustav Fischer: Jena, 1980.

178. Wahbi, A. M.; Abdine, H.; Korany, M. A.; Abdel-Hay, M. H. *Analyst* 1978, *103,* 876.

179. Wahbi, A. M.; Abounassif, M.; Gad-Kariem, E. *Analyst* 1984, *109,* 1513.

180. Wahbi, A. M.; Lotfi, E. A.; Aboul-Enein, H. Y. *Talanta* 1984, *31,* 77.

181. Wahbi, A. M.; Mohamed, M. E.; Abounassif, M.; Gad-Kariem, E. *Anal. Lett.* 1985, *18B,* 261.

182. Wahbi, A. M.; Mohamed, M. E.; Kariem, R. A. G.; Aboul-Enein, H. Y. *Analyst* 1983, *108,* 886.

183. Whitaker, J. E.; Hoyt, A. M. *J. Pharm. Sci.* 1984, *73,* 1184.

184. Zimmermann, W. *Chemische Bestimmungsmethoden von Steroid-hormonen in Körperflüssigkeiten,* Springer: Berlin, 1955, 53–57, 63–75.

185. Zinner, G. *Arch. Pharm.* 1955, *288,* 129.

186. Zivanovic, L.; Agatonovic Kustrin, S.; Radulovic, D. S. *Pharmazie,* 1990, *45,* 935.

187. Zivanov Stakic, Z.; Solomun, L.; Zivanovic, L. *Acta Pharm. Hung.* 1990, *60,* 179.

188. Xenakis, A. G.; Karayannis, M. I. *Anal. Chim. Acta* 1984, *159,* 343.

189. You, K.; Bittikofer, J. A. *Clin. Chem.* 1984, *30,* 1549.

Chapter 8

DIFFERENCE SPECTROPHOTOMETRY*

A. INTRODUCTION

Because of its great importance in pharmaceutical analysis, difference spectrophotometry[23] has been taken out of the special methods of spectrophotometry discussed in Section 4.G and it is treated separately in this chapter. Since this technique depends on the application of chemical reactions, the discussion is based on the treatment of the matter in Chapters 6 and 7.

The only goal of this method is to increase the selectivity of the measurement by decreasing or eliminating the effect of foreign light absorptions on the absorbance of the analyte. As is referred to in Sections 2.G and 6.C the general way to eliminate, or at least minimize, light absorptions originating from irregularities of the instrument and the cells, from the solvent and the reagent, etc., is to prepare the reference and test solutions in exactly the same manner—the only difference being that the reference solution does not contain the analyte. In difference spectrophotometry even this restriction is lifted: the reference and test solutions contain the analyte at the same concentration being prepared by the same dilution from the same stock solution. The difference absorbance (ΔA), which is the basis of the measurement, is produced by subjecting one of the solutions (but not necessarily the test solution) to a chemical reaction resulting in the formation, elimination or, more generally, shift of an absorption band in the spectrum of the analyte. If the spectrum of the test solution prepared in this way is recorded against the reference, the difference spectrum is obtained. If the basis of the difference spectrum is the formation or elimination of a band, then it is naturally identical over a wide spectral range with the normal spectrum while in the case of difference spectrophotometric methods based on spectral shifts, entirely new spectra are formed usually possessing positive and negative maxima. This is illustrated in Figures 8.1 and 8.2. Curve "a" in Figure 8.1 is the spectrum of clioquinol (iodochloroxyquinoline) in neutral solution where the drug is present as the free base. Curve "b" is the spectrum of the solution in 1 M hydrochloric acid where even this very weak base is in the fully protonated state. Curve "a" of Figure 8.2 is the difference spectrum with the neutral solution in the reference and the acidic solution in the sample cell.[28]

Difference spectra can be prepared using both single- and double-beam spectrophotometers. In the case of recording double-beam instruments, the zero-point of the difference curve is set at a value between 0.3 and 0.5 on the absorbance scale. When a single-beam instrument is used, the positive part of the spectrum is taken by measuring the absorbances of cell No. 2 against No. 1 while for measuring the negative section, the cells are interchanged. Any kind of spectrophotometers can be used; the only problem is the diminished energy caused by the absorbance of the reference solution. If this is kept below 0.5, even instruments with less sensitive detectors can be used without significant loss in spectral purity.

The application of difference spectrophotometry is very advantageous both in qualitative and quantitative analysis. Comparison of the curves of Figure 8.1 with those of Figure 8.2 clearly demonstrates that the difference curve with its positive and negative maxima and the zero-crossing $\Delta A = 0$ point between them (corresponding to the isosbestic point of the normal spectra) is much

---

* This method is often termed ΔA spectrophotometry and in the old literature—incorrectly—differential spectrophotometry.

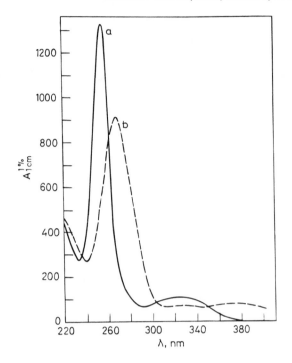

FIGURE 8.1. Spectra of iodochloroxyquinoline. Solvent: ethanol containing 10% v/v water. Curve "a": spectrum in neutral medium; Curve "b": spectrum in 1 M hydrochloric acid. (From S. Görög, *Farm. Tidende*, 1970, *80*, 321.[28])

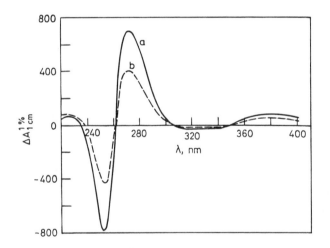

FIGURE 8.2. Difference spectra of iodochloroxyquinoline. Solvent: ethanol containing 10% v/v water. Curve "a": iodochloroxyquinoline; Curve "b": Prednisolone-J ointment. Reference solution: neutral; sample solution: 1 M HCl. (From S. Görög, *Farm. Tidende*, 1970, *80*, 321.[28])

more characteristic than the latter, thus creating better possibilities for the identification of the investigated material.

However, the real great importance of difference spectrophotometry is in quantitative analysis as a consequence of its potential to increase the selectivity of the measurement. The basis of this ability is as follows. If the spectra of accompanying substances interfering with the determination of the analyte do not change during the reaction (because they do not participate in the reaction) then their equal absorbances in the sample and reference cells are cancelled out during the difference spectrophotometric measurement.

Interfering absorbances which can be successfully cancelled out by means of difference spectro-photometry are of two types: on the one hand poorly defined, background-type absorptions and, on the other hand, spectra of well-defined accompanying substances.

As for the first type the difference spectrophotometric technique can be regarded as a highly effective alternative to the background correction methods in Section 4.C, provided that the reaction serving as the basis for the method does not affect the background absorption. In the case of background absorption of unknown origin, it is not possible to predict whether this condition will be met or not. However, in the case of well-defined components it is usually easy to predict whether they will take part in the reaction or not.

The question whether background absorptions can be cancelled out by difference spectrophotom-etry can be readily established experimentally. At first the difference spectrum of the pure analyte is taken followed by that of the analyte in the matrix in question. The two spectra are then compared. If they are identical (at least in the environment of the band selected for the measurement) then the selective determination can be carried out. If distortion is observable in the spectrum of the sample as compared to that of the pure analyte, it can be concluded that the elimination of the background was not successful. Very often further steps are possible to achieve the elimination of the background (decreasing the difference of the pH values of the reference and test solutions, applying shorter reaction times, decreasing the temperature, etc.).

What is described in the preceding paragraph is illustrated by the comparison of curves "a" and "b" in Figure 8.2. Curve "a" already mentioned before is the neutral–acidic difference spectrum of iodochloroxyquinoline while curve "b" is that of the extract of Prednisolone-J ointment taken under identical conditions. Prednisolone with its intense band at 242 nm and excipients would interfere if the assay were attempted by the normal spectrophotometric measurement of iodochloro-xyquinoline. The excellent agreement of the two spectra indicates that both interferences could be eliminated thus creating a good basis for the assay.[28] (It has to be noted that the elimination of the interference of prednisolone was easily predictable, namely it does not contain acidic or basic groups and hence its spectrum is not pH-dependent.)

Iodochloroxyquinoline (clioquinol)

The difference–absorbance measured both in the positive and negative maxima, or even their sum (the amplitude of the full wave) are suitable for the purposes of the quantitation. The sources of error are similar to those of ordinary spectrophotometry; one important point, however, merits mentioning. Especially in those instances when the difference–absorbance is measured for strongly absorbing systems, it is extremely important to check even more carefully than usual, the accuracy of the path length of the two cells, to use the same pipette during the dilution of the stock solution to prepare the sample and reference solutions, etc. It is evident that if these points are neglected, a fraction of the eliminated background is added to or subtracted from the difference–absorbance.

B. METHODS BASED ON ACID-BASE EQUILIBRIA: pH-INDUCED DIFFERENCE SPECTROPHOTOMETRY

An example of the application of acid–base equilibria (deprotonation of acids and protonation of bases) for the difference spectrophotometric determination of drugs has already been shown in the preceding section in connection with Figures 8.1 and 8.2. Some further practical applications are collected in Table 8.1.

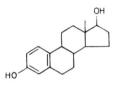

Oestradiol

As a guideline, the pH values of the test and reference solutions should be set in such a way that the analyte be in a fully protonated form in one of them and in a fully deprotonated form in the other. Especially in those cases when the elimination of the background spectrum is a delicate task, the difference between the two pH values should be kept as small as possible: it should not be more than the difference between the highest and lowest possible values, yet assuring the deprotonation and protonation, respectively, of at least 99% of the analyte. For example in the course of the determination of oestradiol esters in the presence of ketosteroids in oily injections, the selectivity of the difference spectrophotometric assay was significantly improved by using pH 9.2 and 0.1 M sodium hydroxide in the reference and sample cells[33] instead of 0.02 M hydrochloric acid and 0.6 M sodium hydroxide as prescribed by USP XIX.[53a] The basis for this is that the ionization of the phenolic hydroxyl group on which the assay is based begins at pH 9.2 only and is complete in 0.1 M sodium hydroxide.

In a few of the examples of Table 8.1, the requirement of the full protonation and deprotonation in the two solutions is not met: too short pH-intervals are applied to assure the above mentioned ionization states. This is the case e.g., with the determination of theophylline in pharmaceutical products,[48] where the pH of 9 of the test solution is not high enough to assure the full ionization of the very weak acidic analyte. This was done by the authors of the method on purpose in order to decrease the rate of the decomposition of theophylline, which is too rapid at the pH necessary for full dissociation. Thus, pH 9 can be considered as a compromise. In such cases on the one hand, the sensitivity of the measurement decreases and, on the other hand, the results will be very sensitive to even minor changes of the pH. The situation is similar in the determination of barbiturates.[55] Here pH 10.5 was selected for the reference solution (monoanionic form—see Figure 3.21); the test solution was set at pH 13.5 where the formation of the dianionic form is not yet complete but at higher pH barbiturates also decompose rapidly.

The determination of pentachlorophenol impurity in the active pentachlorophenyl esters of peptides merits attention. Here the pH of the sample solution was optimized in such a way that it assured the complete dissociation of the relatively strongly acidic pentachlorophenol (pH 6.5), avoiding at the same time the hydrolytic decomposition of the above mentioned very unstable active esters, which would take place in alkaline solution. By means of the difference spectrophoto-metric method the absorbance of the latter could be successfully eliminated.[32]

Figure 8.3 shows the spectra of vanillic acid at pH 2 and 6.5. At the latter pH the carboxyl group of the compound dissociates but the phenolic hydroxyl does not. The dissociation of the aromatic carboxyl group results in a hypsochromic shift (see Section 3.P). The difference spectrum based on this shift (Figure 8.4) with its two positive and one negative maxima is a suitable basis for the quantitative determination. Using this method, e.g., 1–2% of vanillic acid could be determined as an impurity in homovanillic acid. Although the latter does have a carboxyl group which predominantly dissociates at pH 6.5, this is isolated from the aromatic ring and hence the dissociation does not influence the difference spectrum of vanillic acid. The reason for selecting such a relatively low pH for the sample solution is that at higher values the phenolic hydroxyl groups of both vanillic and homovanillic acids would dissociate making the selective determination of vanillic acid impossible.[32]

As a curiosity it is mentioned that the difference spectrophotometric method enables extremely weak acids (e.g., 20-oximino-5,16-pregnadien-3β-ol acetate[37b]) and bases (e.g., caffeine[50]) to be determined if sufficiently high concentrations of bases and acids, respectively, are used in the sample solutions.

TABLE 8.1. Some Difference Spectrophotometric Methods Based on Acid–Base Equilibria

Compound	Acidity–basicity		Wavelength, nm		$\Delta A_{1\ cm}^{1\%}$	Sample	First author/Year	Ref.
	reference	sample	min.	max.				
Vanillic acid	pH 6.5	pH 2	241;	Acids 266,300	−160;325, 210	Impurity in homovanillic acid	Görög, 1971	32
Salicylic acid	pH 6.83	pH 13		246		Serum	Zwillenberg, 1982	57
Aspirin	pH 9	pH 13.5		300		In the presence of salicylic acid	Shane, 1967	51
Salicylamide, paracetamol	pH 6	pH 10		330	375*	Tablet (in the presence of aspirin and caffeine)	Shane, 1968	50
	pH 6	pH 10		263.5	400*			
Pentachlorophenol	pH 2	pH 7.5		256,323	335 206	Impurity in pentachlorophenyl esters of peptides	Görög, 1971	32
Eugenol	pH 3	pH 12		296	237	Pharmaceutical preparations	Demetrius, 1960	20
Morphine	pH 3	pH 12		298	82	Pharmaceutical preparations	Casinelli, 1962	4
Morphine	pH 1	pH 13		298	—	Pharmaceutical preparations	Wahbi, 1970	54
Morphine	pH 6	pH 13		299	67*	Plant extract	Rondina, 1973	47
Oestrone, oestradiol, ethinyloestradiol	pH 9.2	0.1 M NaOH		300	—	Impurities in 3-methyl ethers	Görög, 1978	37a
Oestradiol-17-valerate	0.02 M HCl	0.6 M NaOH		300	—	Oil-injectable	USP XIX, 1975	53a
Oestradiol esters	pH 9.2	0.1 M NaOH		300	—	Oil-injectables (in the presence of androgens and gestogens)	Görög, 1976	33
Diloxanide furoate	0.02 M NaOH	neutral		267	430*	Tablet (in the presence of tinidazole)	Sethi, 1988	49
Benzthiazide	acidic	basic	271;	313	−90*; 280*	—	Doyle, 1974	21
Hydrochlorothiazide	acidic	basic	305*;	339*	−90*; 125*	—	Doyle, 1974	21
Triamterene	acidic	basic	256*; 345*;	277* 385*	−420*; 720*; −890*; 860*	Tablet	Doyle, 1974	21
20-Oximino-5,16-pregnadiene-3-ol acetate	neutral	2 M NaOH		262	280	Intermediates of steroid syntheses	Görög, 1978	37b
2-methyl-5-nitroimidazol	neutral	0.01 M NaOH	290;	369	−320; 916	Impurity in metronidazol	Görög, 1970	29

TABLE 8.1. Continued.

Compound	Acidity–basicity		Wavelength, nm		$\Delta A_{1\,cm}^{1\%}$	Sample	First author/Year	Ref.
	reference	sample	min.	max.				
Amphoteric compounds								
Theophylline	pH 7	pH 9		285	108	Capsule (in the presence of theobromine, caffeine, ephedrine, etc.)	Young, 1976	56
Chlordiazepoxide, demoxepam	pH 7	0.1 M NaOH		285	126	Blood	Gupta, 1973	41
	pH 3	pH 8		269	370	Capsule, tablet (simultaneous determination)	Davidson, 1984	19
Sulphatiazole	pH 8	pH 13	251*;	263	490		Doyle, 1974	21
	basic	neutral		290*	−80*; 290*			
Nitrazepam	neutral	basic		290*		280*	Davidson, 1989	19
	acidic			263*		420*		
	0.1 M NaOH	0.1 M HCl	372;	282	632	Tablet		
Bases								
Caffeine	0.35 M HCl	pH 1.8		283	95*	Tablet (in the presence of aspirin, salicylamide, paracetamol)	Shane, 1968	50
Ampicillin	pH 9	pH 5		268	3.2	Injectables (in the presence of other penicillins)	Davidson, 1974	8
Hydroxocobalamine	pH 2	pH 8		362	62	Crude product (in the presence of Vitamin B_{12}, etc.)	Horváth, 1978	43
Metronidazole benzoate	neutral	1 M HCl		324	274	Syrup	Görög, 1976	35
	0.2 M HCl	neutral		285	999	Ear drops (in the presence of chloramphenicol)	Kannan, 1987	44
Benzocaine	0.1 M KOH	0.1 M HCl		283	500	Pharmaceutical preparations	Bonazzi, 1990	2
Minoxidil								
Tinidazol	0.1 M HCl	neutral		320	280*	Tablet (in the presence of diloxanide furoate)	Sethi, 1988	49

*Figures taken from spectra or calibration graphs.

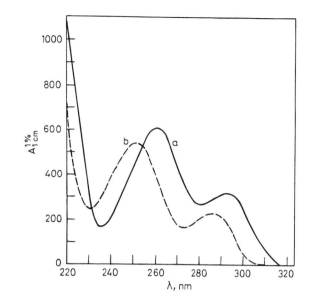

FIGURE 8.3. Spectra of vanillic acid. Solvent: water containing 2% v/v ethanol. Curve "a": 0.01 M hydrochloric acid; Curve "b": pH 6.5. (From S. Görög, *2nd Symp. Appl. Phys. Chem. Veszprém,* Akadémiai Kiadó: Budapest, 1971, p. 323.[32])

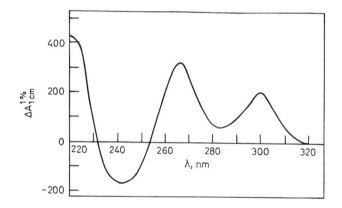

FIGURE 8.4. Difference spectrum of vanillic acid. Reference: pH 6.5; sample solution: 0.01 M HCl. Solvent and literature source: see Figure 8.3.

An interesting example taken from Table 8.1 demonstrates that the difference spectrophotometric method can be applied even in the case of ampicillin which spectrophotometrically is only a very weakly active antibiotic, where the chromophoric group is an isolated aromatic ring and a decrease of the pH results in the protonation of the benzylamine-type amino group only. The effect of this on the spectrum is restricted to a somewhat more pronounced fine structure of the benzene ring with a small increase of the intensity ($\Delta A_{1\,cm}^{1\%} = 3.8$). Even this low difference specific absorbance value is a sufficient basis for the determination of ampicillin in the presence of penicillins which do not contain the same functional group.[8]

Exploiting the differences in the dependence of the spectra of various compounds as well as the different shapes of their difference spectra with the maxima at different wavelengths enables very difficult analytical problems to be solved in a simple way. Such an example is the simultaneous determination of aspirin, paracetamol, salicylamide and caffeine in a tablet formulation[50] (see Table 8.1). The pK_a of the phenolic hydroxyl groups of paracetamol and salicylamide, respectively, are 9.0 and 8.4. This means that at the pH of the reference solution (pH 6), no dissociation takes place while in the sample solution (pH 10) the dissociation of both acids is practically completed. Thus the two compounds form a common, overlapping difference spectrum. However, as is seen in

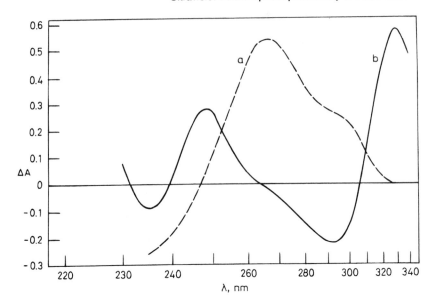

FIGURE 8.5. Difference spectra of salicylamide and paracetamol. Solvent: water. Curve "a": paracetamol (1.54 mg/ 100 ml.); Curve "b": salicylamide (1.35 mg/100 ml). Reference: pH 6; Sample solution: pH 10. (From N. Shane; M. Kowblansky, *J. Pharm. Sci.* 1968, 57, 1221.[50])

Figure 8.5, at the maximum of the difference spectrum of salicylamide, at 330 nm paracetamol does not possess difference absorption, while at 263.5 nm (which is the maximum of the difference spectrum of the latter) salicylamide only has an isosbestic point, i.e., its difference-absorbance is zero. Due to this fortunate situation the simultaneous determination of the two components at the two wavelengths can be readily carried out. Aspirin is transformed to salicylic acid by mild alkaline hydrolysis prior to the assay. In this form it does not interfere with the determinations of paracetamol and salicylamide since the pK_a of the phenolic hydroxyl of salicylic acid is 13.4 and hence its dissociation does not begin at pH 10. Consequently, aspirin does not show a difference spectrum between pH 6 and 10. The same also applies to caffeine, since it does not contain acidic functional groups. It can be determined on the basis of its very weak basicity ($pK_a = 0.61$) by measuring the difference absorbance at 283 nm between the strongly acidic reference (0.35 M hydrochloric acid) and mildly acidic sample solution (pH 1.8). The other components of the formulation do not interfere.

The method of Abounassif[1] for the assay of bromazepam tablet merits special attention. In the simple difference spectrophotometric variant of the method, the solution in 0.1 M hydrochloric acid is placed into the reference cell while that in 0.1 M sodium hydroxide is placed into the sample cell and the difference-absorbance is measured at 239 nm. In the difference-derivative spectrophotometric variant of the method (see Section 4.E.3) the assay is based on the measurement of the difference of the first derivative spectra of the above solutions, thus greatly increasing the selectivity of the assay.

C.   METHODS BASED ON ACID- AND BASE-CATALYZED IRREVERSIBLE REACTIONS

In addition to the methods based on reversible acid—base equilibria discussed in the preceding section, some difference—spectrophotometric methods depend on irreversible reactions catalyzed by acids and bases. A common feature of these methods is that spectrophotometrically inactive or only poorly active compounds are transformed by means of elimination or rearrangement-type reactions to derivatives with intense absorption bands.

At first the determination of ethynodiol diacetate based on acid-catalyzed elimination reaction is discussed.

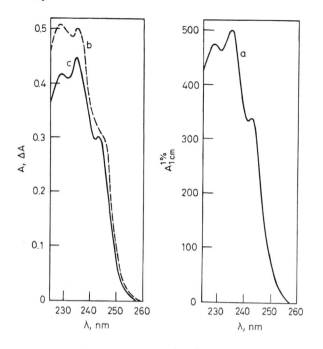

FIGURE 8.6. Spectra of ethynodiol diacetate. Solvent: Ethanol containing 5% water. Curve "a": spectrum of ethynodiol diacetate after transformation according to Equation 8.1; Curve "b": spectrum of the extract of Bisecurin tablet after transformation according to Equation 8.1; Curve "c": difference spectrum of the extract of Bisecurin tablet. Reference: neutral; sample solution: treated with HCl. (From S. Görög, É. Csizér, *Z. Anal. Chem.* 1971, *254*, 119.[31])

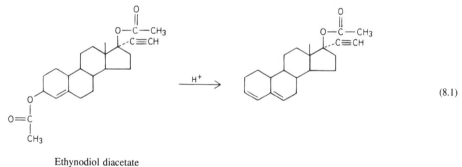

Ethynodiol diacetate

$$(8.1)$$

As can be seen in Equation 8.1, this spectrophotometrically practically inactive synthetic gestogen is transformed to the 3,5-diene derivative when treated with hydrochloric acid. This trans-diene type derivative possesses an intense absorption band at 236 nm with side maxima at 228 and 243 nm. This absorption band enables the spectrophotometric determination of the drug. As for the conditions of the elimination reaction, the pharmacopoeias prescribe 10 min boiling in a mixture of 50 ml methanol, 2 ml of water and 3 ml of concentrated hydrochloric acid.[3,52b] Görög and Csizér[31] dissolve the sample in 10 ml of ethanol and allow it to react with the freshly prepared mixture of 5 ml hydrochloric acid and 5 ml of ethanol. The latter authors described the difference spectrophotometric variant of the method for the assay[31] and content uniformity test[36] of contraceptive pills containing ethynodiol diacetate. It is shown in Figure 8.6 that due to the oestrogenic component of the pill and background absorptions in the spectrum of the tablet extract treated with hydrochloric acid the spectrum of the 3,5-diene can be recognized in distorted form only. At the same time, the difference spectrum prepared by measuring the solution treated with hydrochloric acid against the untreated solution of the same concentration is practically identical with the difference-spectrum of the pure active ingredient.

Another example of an acid-catalyzed elimination reaction successfully applied in difference spectrophotometry is the determination of 1-(2,5-dimethylphenyl)-1-phenyl-propan-1-ol in animal feedstuff used in toxicological studies. The weak spectrum of the potential drug with its isolated phenyl rings is not suitable for the solution of the problem (λ_{max} = 268 nm; ε = 795) because of the strong background absorption. This compound can be dehydrated by treatment with 1.3 M hydrochloric acid in ethanolic medium leading, within 3 hours, to 1-(2,5-dimethylphenyl)-1-phenyl-1-propene. With its styrene-like structure this derivative possesses a strong absorption band at 247 nm (ε = 13,970).[38] Because of the strong background originating from the feedstuff not even this spectrum enables the assay using simple spectrophotometry. By means of difference spectrophotometry, however, the assay can be carried out by eliminating the background in a similar manner as described with the preceding example. It is noteworthy that the elimination of the pH-dependent components of the background could be achieved in such a way that the reference and sample solutions contained not only the extract of the feedstuff at the same concentration but the hydrochloric acid, too. The only difference was that the sample solution was prepared by the 20-fold dilution of the reaction mixture after the completion of the elimination reaction, while the reference solution was first diluted before the addition of the hydrochloric acid. In the diluted solution no dehydration takes place during the measurement, thus creating a basis for the assay.[39]

The base-catalyzed elimination of the nitro group of metronidazole is the basis of the method of Sanyal[48] for the determination of its benzoate in pharmaceutical preparations. The treatment with hot alkali results in the disappearance of the band at 276 nm (see Figure 3.34). This is the wavelength of the measurement of the difference-absorbance (untreated vs. alkali-treated solutions).

Further to elimination reactions acid- and base-catalyzed rearrangement reactions can also form the basis of difference spectrophotometric methods. As an example the determination of 5(10)-ene-3-keto steroids is presented. Equation 8.2 shows that these non-conjugated ketosteroids can easily be rearranged to the conjugated 4-ene-3-keto derivatives possessing a strong absorption band (λ_{max} = 240 nm; $\varepsilon \sim$ 17,000).

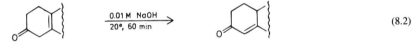

$$\text{(8.2)}$$

Both acids and bases can be used as catalysts. The base-catalyzed reaction requires much milder experimental conditions. This reaction was successfully used, among others, for the determination of norethynodrel (17α-ethinyl-17-hydroxy-5(10)-oestrene-3-one) in tablets[36] and for the determination of 5(10)-oestrene-3,17-dione impurity at the 0.2% level in 3,3-dimethoxy-5(10)-oestrene-17-one. In the latter case the composition of the reference and stock solutions was identical, the only difference being that the test solution was neutralized with acetic acid after the alkaline treatment, while acetic acid was added to the reference solution followed by sodium hydroxide, thus avoiding the interaction of the latter with the analyte.[27]

D. METHODS BASED ON REDOX REACTIONS

When an oxo group conjugated with a double bond or an aromatic ring is reduced to the corresponding alcohol, the intense conjugation band disappears from the spectrum. If sodium borohydride is used as the reducing agent, the reduction takes place smoothly and unidirectionally (usually at room temperature). Since the excess of the reagent decomposes after the reaction to a spectrophotometrically inactive product, this reaction can serve as the basis for the selective difference spectrophotometric determination of conjugated unsaturated and aromatic ketones. In this case the solution treated with sodium borohydride is placed in the reference and the untreated solution in the sample cell. In accordance with the general principles of difference spectrophotometry the composition of the two solutions can be brought closer to each other in such a way that the excess of sodium borohydride in the reference solution is decomposed by methanol or dilute hydrochloric acid; and, the same quantity of sodium borohydride is added to the sample solution, too, but it is first decomposed in the same way and the aliquot of the stock solution of the analyte is added only

TABLE 8.2. Difference Spectrophotometric Methods Based on Reduction with Alkali Borohydrides

Compound	Sample	Reducing agent	Solvent	λ_{max}, nm	$\Delta A_{1\,cm}^{1\%}$	Ref.
Hydrocortisone	Fermentation liquor	NaBH$_4$	Methanol	240	442	25,26
Nandrolone phenylpropionate	Oil-injectables	NaBH$_4$	Methanol	240	421	24,26
Progesterone	Oil-injectables	NaBH$_4$	Methanol	240	541	24,26
Prednisolone	Tablet, oil-injectables	NaBH$_4$	0.1 M NaOH/ ethanol	243	410	24,26
Norethisterone	Impurity in ethynodiol	NaBH$_4$	Methanol	240	560	26
Aromatic ketones (e.g., 4-methylpropio- phenone)	Reaction mixture of Friedel-Crafts syntheses	NaBH$_4$	Methanol	251	921	52
Norgestrel	Tablet	NaBH$_4$	Methanol	240	—	36
4-ene-3-ketosteroids,1,4- diene-3-ketosteroids	Bulk materials	LiBH$_4$	Tetrahydrofuran	240	359–510	6
4,6-diene-3-ketosteroids				284		
Fluocinolone acetonide	Ointment	LiBH$_4$	Tetrahydrofuran	238	360	7
Halcinonide, etc.	Bulk materials	NaBH$_4$	0.03 M NaOH/ methanol-pro- pylene glycol	239	—	45
Quingestanol acetate*	Oil-injectables	NaBH$_4$	Methanol	243	—	46
Norgestrel	Tablet	NaBH$_4$	Methanol	241	—	22
Benzaldehyde	Impurity in benzyl alcohol	NaBH$_4$	Methanol	246	1130	34
5-hydroxymethyl- furfuraldehyde	Degradation product in (syrups)	NaBH$_4$	Water/ethanol	283	1296	16,17

*After transformation to norethisterone acetate.

afterwards. The difference spectrum of the unsaturated or aromatic ketone is similar to the normal spectrum in the range of the conjugation band which disappears during the reduction. These ketone derivatives can be selectively determined in the presence of those materials which do not change their spectra during the interaction with sodium borohydride.

The difference spectrophotometric method based on reduction with sodium borohydride was introduced by Görög[24–26] for the determination of α,β-unsaturated 3-ketosteroids. The method was later modified by Chafetz[6] and Kirschbaum.[45] The former author suggested the use of lithium borohydride for the reduction of $\Delta^{1,4}$-3-ketosteroids which react only slowly with sodium borohy- dride while the latter added propylene glycol to the reaction mixture, thus complexing any sodium metaborate that was present as an impurity in degraded sodium borohydride lots and which would inhibit the reduction. The application of this method for the solution of various problems is summarized in Table 8.2.

Of the examples in Table 8.2 the spectra corresponding to the determination of progesterone in Limovanil oily injection are shown in Figure 8.7. As is seen, the spectrum of the methanolic extract of the injection shows strong distortion as compared with that of pure progesterone (curve "a"). The reasons for the distortion are visible in the spectrum of the reduced solution (curve "b"). The short-wavelength section originates from oestradiol benzoate being present in the injection, the α band with fine structure around 260 nm is due to the preservative benzyl alcohol, and in addition a typical background can also be seen which is the consequence of undefined components extracted from the oil. As expected, the former two compounds do not react with sodium borohydride and the excellent agreement of the difference spectrum (curve "c") with that of pure progesterone indicates that the background-type absorption does not change during the reduction either.

Another typical application is illustrated in Figure 8.8. In the normal spectrum of the extract of prednisolone ointment, the spectrum of the preservative methylparaben overlaps that of predniso- lone to such an extent that the latter is hardly recognizable (curve "a"). Since methylparaben does not react with sodium borohydride the difference-spectrum (curve "c") is suitable for the

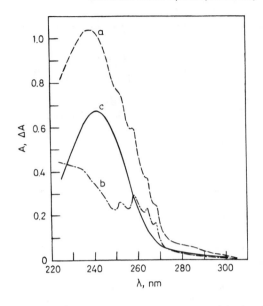

FIGURE 8.7. Spectra to the determination of progesterone in Limovanil oily injection. Solvent: methanol. Curve "a": spectrum of the methanolic extract without treatment with sodium borohydride; Curve "b": spectrum of the methanolic extract after treatment with sodium borohydride; Curve "c": difference spectrum. Reference: "b"; sample solution: "a". (From S. Görög, *J. Pharm. Sci.* 1968, *57*, 1737.[24])

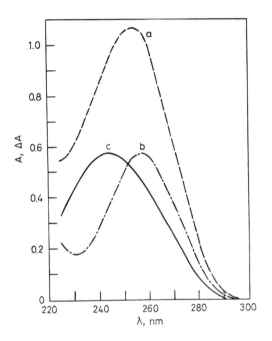

FIGURE 8.8. Spectra to the determination of prednisolone in an ointment. Solvent: ethanol. Curve "a": spectrum of the ethanolic extract without treatment with sodium borohydride; Curve "b": spectrum of the ethanolic extract after treatment with sodium borohydride; Curve "c": difference spectrum. Reference: "b"; sample solution: "a". Literature source: see Figure 8.7.

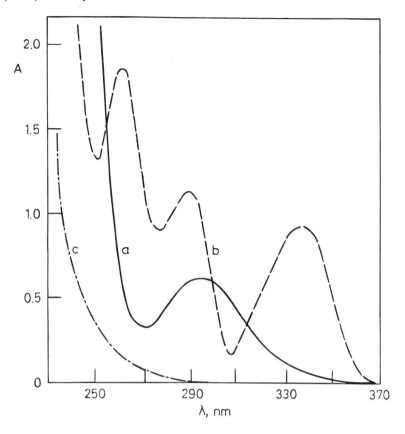

FIGURE 8.9. Spectra to the determination of promethazine after transformation to the sulphoxide. Solvent: water. Curve "a": promethazine hydrochloride (10^{-4} M/l); Curve "b": promethazine sulphoxide (10^{-4} M/l); Curve "c": diluted (1 + 19) oxidizing reagent (peroxyacetic acid). (From A. G. Davidson, *J. Pharm. Pharmacol.* 1976, *28*, 795.[9])

determination of prednisolone while the spectrum of the reduced solution (curve "b") is characteristic of methylparaben.

Oxidation reactions can also form the basis for difference spectrophotometric methods, provided that the reaction is unidirectional, the oxidizing agent does not absorb light in the range of the analytical measurement, and the spectra of the material to be determined and its oxidized form differ significantly. These requirements are entirely fulfilled in the case of the method developed by Davidson[9,10] for the determination of phenothiazine drugs. The basis of the method is the oxidation of the sulphur atom in the phenothiazine ring system to sulphoxide using peroxyacetic acid as the oxidizing agent. This reaction takes place at room temperature in glacial acetic acid and it causes the significant bathochromic shift of the spectra. As an example the spectra of untreated promethazine, its oxidized form and the oxidizing agent are shown in Figure 8.9 while the difference-spectrum (oxidized solution vs. untreated) is in Figure 8.10.

The maxima of the difference spectra of promethazine, chlorpromazine, trifluoperazine, thioridazine, promazine, etc., investigated by Davidson[9] and levomepromazine determined by Hackmann et al.[42] are between 333 and 353 nm with difference specific absorbances ranging between 56 and 166. These enable sufficiently sensitive and extremely selective assay methods to be developed for their solid and liquid dosage forms. It merits a special mention that the sulphoxides are also the main degradation products of phenothiazine drugs and the difference spectrophotometric method is specific for the undegraded drug in the presence of the degradation product. It is interesting that another type of difference spectrophotometric methods is suitable for the selective determination of the degradation product. This method (which can be considered as the reversed variant of the above described method) is based on the reduction of the sulphoxide by zinc/hydrochloric acid.[10]

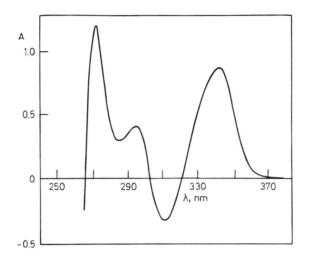

FIGURE 8.10. Difference spectrum of promethazine after transformation to the sulphoxide. Solvent: water. Reference: untreated solution; sample solution: solution treated with peroxyacetic acid (10^{-4} M/l). Literature source: see Figure 8.9.

The method of Görög and Görög[40] for the determination of 6-mercaptopurine also depends on an oxidation reaction. This drug is oxidized by atmospheric oxygen to thiouric acid in an enzymatic reaction (see Equation 8.3).

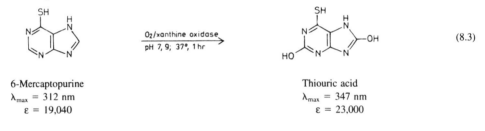

(8.3)

6-Mercaptopurine
λ_{max} = 312 nm
ε = 19,040

Thiouric acid
λ_{max} = 347 nm
ε = 23,000

Scanning the spectrum of the enzymatically oxidized solution against the reference not containing the enzyme, a difference spectrum is obtained (λ_{max} = 348 nm; ε = 22,600) which increases the selectivity of the spectrophotometric assay to such an extent that in this form, the method is suitable for the determination of 6-mercaptopurine in urine in the presence of metabolites and endogenous materials.

E. METHODS BASED ON OTHER REACTIONS

In principle any kind of spectrophotometric methods could have their difference spectrophotometric variants. On the basis of the previously described principles this can be achieved in one of the following two ways:

1. One of the reagents is omitted from the reference solution.
2. The sequence of the addition of the reagents is modified in the course of the composition of the reference solution in such a way that the possibility of the reaction be excluded, even if it contains the analyte and all reagents in the same concentration as the sample solution.

The use of difference spectrophotometry is of no importance when color reactions are applied, since in the visible range matrix-based background absorptions do not usually interfere. The real problems here are the frequent, non-specific color reactions accompanying the main reaction; this is, however, usually not overcome by means of difference spectrophotometry.

More common and successful is the application of difference spectrophotometry in combination with a variety of reactions producing reaction products which absorb in the ultraviolet range and

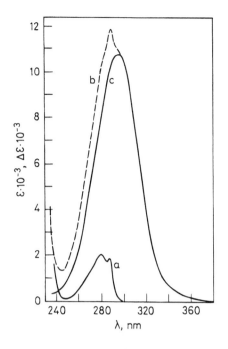

FIGURE 8.11. Spectra to the determination of oestrone-3-methyl ether impurity in its 17-ethylene ketal. Solvent: 9% v/v tert-butanol + 5% v/v cyclohexane in 0.02 M ethanolic hydrochloric acid. Curve "a": Oestrone-3-methyl ether 17-ethyleneketal; Curve "b": oestrone-3-methyl ether after treatment with diethyl oxalate; Curve "c": Difference spectrum of oestrone-3-methyl ether. Reference: solution untreated with diethyl oxalate; sample solution: solution treated with diethyl oxalate. (From S. Görög, *Analyst* 1971, *96*, 437.[30].)

are, to some extent, overlapped by the absorption band of another, isolated part of the molecule or by the spectra of other components in the sample.

As the first example the difference spectrophotometric variant of the method using diethyl oxalate as the reagent for the determination of ketones (see p. 177) is presented.[30] As is seen in reaction Equation 8.4, the active methylene group in the vicinity of the 17-keto group of oestrone-3-methyl ether selectively reacts with diethyl oxalate in the presence of the 17-ethyleneketal of the latter. The determination on the basis of the formation of a 16-glyoxalyl derivative cannot be carried out smoothly since—as seen in Figure 8.11—the absorption band of the 16-glyoxalyl-17-keto system overlaps with that of the phenol ether type A ring which is present in great excess.

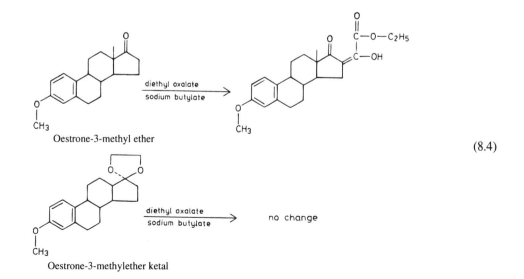

(8.4)

To avoid the effect of the latter the sequence of the addition of the reagents is varied. The test solution contains the reaction mixture of the Claisen condensation carried out in water-free medium in the presence of sodium butylate which is acidified after the completion of the reaction. The reference solution contains the sample and all reagents in the same concentration, but first, the diethyl oxalate–sodium butylate solution is acidified and the aliquot of the sample is added at the end, thus avoiding the reaction taking place. In this way the absorption of the phenol ether type absorption bands of both oestrone-3-methyl ether and of the 17-ethylene ketal, which is present in great excess when the aim of the analysis is the determination of oestrone-3-methylether as impurity in its 17-ethylene ketal, are cancelled out. Thus, the difference spectrum is characteristic of the 17-keto group only: $\Delta A_{295 \text{ nm}} = 376$. This method enables as little as 0.1% of oestrone-3-methylether to be determined in the 17-ketal.

Methyltestoterone impurity can be similarly determined in methandienone (manufactured from it) based on the same principle. In this case the ability of the active methylene group of methyltestosterone to react with diethyl oxalate is exploited, while the main component not possessing an active methylene group does not react (see Equation 8.5). The reason necessitating the use of difference spectrophotometry is that even the weak $n–\pi^*$ band of the 3-keto group of methandienone (which is present in great excess) would interfere with the simple spectrophotometric determination at 324 nm.

In section 8.B some difference spectrophotometric methods were presented for the selective determination of phenols based on their ionization in alkaline medium. In the following section an even more selective method will be described for the determination of 1,2-diphenols on the basis of the formation of cyclic complexes with boric acid[13] or germanium dioxide.[14,15] As a result of the complex formation at pH 7 and 6, respectively, an intense bathochromic shift of the spectrum of the diphenol takes place. There are two, almost equally intense bands in the difference spectra, of which the long wavelength one is used for analytical purposes. For example Davidson described $\Delta \varepsilon_{292 \text{ nm}}$ values of 2506 and 2622 for adrenaline and methyldopa, respectively, after complex formation with boric acid.

Higher sensitivity is attainable with germanium dioxide as the complex forming agent where the difference molar absorptivities of the above two compounds are 4366 and 4306, respectively, at the same wavelength.

1,2,3-Triphenols also give positive reaction, but the position of the maxima of the difference spectra—and especially their intensities—differs significantly. This is the basis of the simultaneous determination of levodopa and benserazide based on difference-absorbance measurements at 238 and 292 nm, and simultaneous equations.

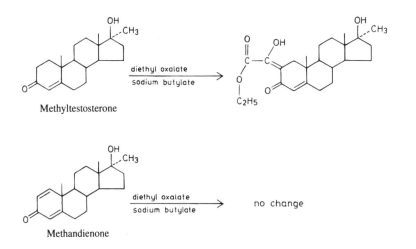

(8.5)

Although the reaction between primary aromatic amines and formaldehyde leading to Schiff's bases does not cause strong bathochromic and hyperchromic shifts, even this small shift is sufficient to serve as the basis of highly selective difference spectrophotometric method. This is demonstrated

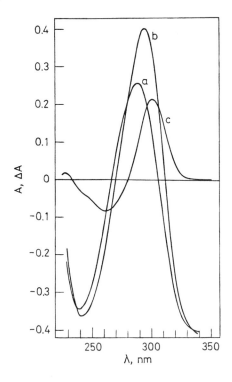

FIGURE 8.12. Spectra to the determination of procaine after reaction with formaldehyde. Solvent: water. Curve "a": spectrum of procaine (1 mg/100 ml); Curve "b": spectrum of procaine after treatment with formaldehyde (1 mg/100 ml); Curve "c": difference spectrum (1 mg/100 ml). Reference: solution untreated with formaldehyde; sample solution: solution treated with formaldehyde. (From T. D. Doyle; F. R. Fazzari, *J. Pharm. Sci.* 1974, *63*, 1921.[21])

in Figure 8.12 taken from the original paper of Doyle and Fazzari[21] where the spectra of procaine is shown before and after the reaction with formaldehyde together with the difference spectrum.

The method of Davidson[11] for the determination of traces of hydrazine in isonicotinoyl hydrazide and its preparations is based on the formation of a hydrazone with 4-dimethylaminobenzaldehyde measurable at 456 nm. The reaction between hydrazine and the reagent is avoidable by binding hydrazine with acetone prior to the addition of the reagent. The method enables traces of hydrazine on the order of 10 ppm to be determined, and hence it is suitable for monitoring the degradation of preparations of isonicotinoyl hydrazide.

The reverse principle is the basis of the method of Davidson[15] for the determination of 5-hydroxymethylfurfuraldehyde as a degradation product of carbohydrates in syrup formulations.

As the last example the method of Cavrini et al.[5] is presented for the determination of ethacrynic acid in its preparations. As seen in Equation 8.6, the terminal vinyl group of this compound easily undergoes addition reaction with N-acetylcysteine. This reaction results in hypsochromic and hyperchromic shift and these are the basis of a selective difference spectrophotometric method, suitable for stability testing (measurement at 270 nm).

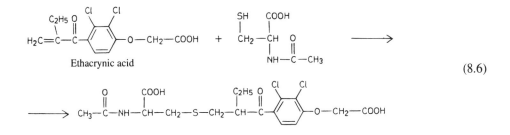

$$(8.6)$$

Finally, it has to be noted that difference spectrophotometry is a widely used and developing branch of spectrophotometric drug analysis. In addition to the few selected examples in this chapter, several other applications (mainly based on pH-changes) are presented in the sections of Chapter 10.

REFERENCES

1. Abounassif, M. A. *J. Pharm. Belg.* 1989, *44*, 329.
2. Bonazzi, D.; Di Pietra, A. M.; Gatti, R.; Cavrini, V. *Farmaco* 1990, *45*, 727.
3. *British Pharmacopoeia 1973,* The Pharmaceutical Press: London, 199.
4. Casinelli, J. L.; Sinsheimer, J. E. *J. Pharm. Sci.* 1962, *51*, 336.
5. Cavrini, V.; Bonazzi, D.; Di Pietra, A. M.; Gatti, R. *Analyst* 1989, *114*, 1307.
6. Chafetz, L.; Tsilifonis, D. C.; Riedl, J. M. *J. Pharm. Sci.* 1972, *61*, 148.
7. Coda, L.; Timallo, L. *Boll. Chim. Farm.* 1976, *115*, 515.
8. Davidson, A. G.; Stenlake, J. B. *Analyst* 1974, *99*, 476.
9. Davidson, A. G. *J. Pharm. Pharmac.* 1976, *28*, 795.
10. Davidson, A. G. *J. Pharm. Pharmac.* 1978, *30*, 410.
11. Davidson, A. G. *Analyst* 1982, *107*, 422.
12. Davidson, A. G. *J. Pharm. Sci.* 1984, *73*, 55.
13. Davidson, A. G. *J. Pharm. Biomed. Anal.* 1984, *2*, 45.
14. Davidson, A. G. *J. Pharm. Sci.* 1984, *73*, 1582.
15. Davidson, A. G. *J. Pharm. Biomed. Anal.* 1985, *3*, 235.
16. Davidson, A. G. *J. Clin. Pharm. Therap.* 1987, *12*, 11.
17. Davidson, A. G.; Dawodu. T. O. *J. Pharm. Biomed. Anal.* 1987, *5*, 213.
18. Davidson, A. G.; Dawodu. T. O. *J. Pharm. Biomed. Anal.* 1988, *6*, 61.
19. Davidson, A. G. *J. Pharm. Pharmacol.* 1989, *41*, 63.
20. Demetrius, J. C.; Sinsheimer, J. E. *J. Pharm. Sci.* 1960, *49*, 523.
21. Doyle, T. D.; Fazzari, F. R. *J. Pharm. Sci.* 1974, *63*, 1921.
22. Fell, A. F. The Quality Control of Steroids in Pharmaceutical Formulations by Higher Derivative Spectroscopy, in *Advances in Steroid Analysis,* S. Görög, Ed.; Elsevier: Amsterdam, 1982, 495–510.
23. Fell, A. F. Difference Spectroscopy, in *Practical Absorption Spectrometry,* A. Knowles; C. Burgess, Eds., Chapman and Hall: London, 1984.
24. Görög, S. *J. Pharm. Sci.* 1968, *57*, 1737.
25. Görög, S. *Steroids* 1968, *11*, 93.
26. Görög, S. *Magy. Kém. Foly.* 1968, *74*, 447.
27. Görög, S. *Acta Chim. Hung.* 1969, *61*, 341.
28. Görög, S. *Farm. Tidende* 1970, *80*, 321.
29. Görög, S.; Csizér, É. *Proc. Anal. Conf. Budapest,* Vol. 2, Akadémiai Kiadó: Budapest, 1970, 65.
30. Görög, S. *Analyst* 1971, *96*, 437.
31. Görög, S.; Csizér, É. *Z. Anal. Chem.* 1971, *254*, 119.
32. Görög, S. *Proc. 2nd Symp. Appl. Phys. Chem. Veszprém,* Akadémiai Kiadó: Budapest, 1971, 323.
33. Görög, S. *Analyst* 1976, *101*, 512.
34. Görög, S.; Sütő, J. *Acta Pharm. Hung.* 1976, *46*, 227.
35. Görög, S.; Fütő, M.; Laukó, A. *Acta Pharm. Hung.* 1976, *46*, 113.
36. Görög, S. *Zbl. Pharm.* 1977, *116*, 259.
37. Görög, S.; Szász, Gy., *Analysis of Steroid Hormone Drugs,* Elsevier: Amsterdam, 1978. a. p. 325; b. p. 355.
38. Görög, S.; Rényei, M.; Laukó, A. *J. Pharm. Biomed. Anal.* 1983, *1*, 39.
39. Görög, S.; Laukó, A.; Rényei, M.; Hegedüs, B. *J. Pharm. Biomed. Anal.* 1983, *1*, 497.
40. Görög, K.; Görög, S. *J. Pharm. Biomed. Anal.* 1988, *6*, 109.
41. Gupta, R. C.; Lundberg, G. D. *Anal. Chem.* 1973, *45*, 2403.
42. Hackmann, E. R. M.; Magalhaes, J. F.; Santoro, M. I. R. M. *Rev. Farm. Biochim. Univ. S. Paolo* 1986, *22*, 22.
43. Horváth, P.; Szepesi, G. *Acta Pharm. Hung.* 1978, *48*, 199.
44. Kannan, K.; Manavalan, R.; Kelkar, A. K. *Indian Drugs* 1987, *25*, 128.
45. Kirschbaum, J. *J. Pharm. Sci.* 1978, *67*, 275.
46. Penner, M. H.; Tsilifonis, D.; Chafetz, L. *J. Pharm. Sci.* 1971, *60*, 1388.

47. Rondina, R. V. D.; Bandoni, A. L.; Coussio, J. D. *J. Pharm. Sci.* 1973, *62,* 502.

48. Sanyal, A. K. *Analyst,* 1992, *117,* 93.

49. Sethi, P. D.; Chatterje, P. K.; Jain, C. L. *J. Pharm. Biomed. Anal.* 1988, *6,* 253.

50. Shane, N. A.; Kowblansky, M. *J. Pharm. Sci.* 1968, *57,* 1218.

51. Shane, N. A.; Routh, J. I. *Anal. Chem.* 1967, *39,* 414.

52. Szepesi, G.; Görög, S.; Szakolczay, I. *Chem. Anal.* 1971, *16,* 211.

53. *The United States Pharmacopoeia XIX,* Mack Publishing Co.: Easton, PA, 1975. a. 181; b. 191.

54. Wahbi, A. M.; Farghaly, A. M. *J. Pharm. Pharmacol.* 1970, *22,* 848.

55. Williams, L. A.; Zak, B. *Clin. Chim. Acta* 1959, *4,* 170.

56. Young Soon Chae; Shelwer, W. H. *J. Pharm. Sci.* 1976, *65,* 1178.

57. Zwillenberg, L. O.; Bösiger, G.; Zwillenberg, H. H. L. *Meth. Find. Exptl. Clin. Pharmacol.* 1982, *4,* 575.

Chapter 9

SPECTROPHOTOMETRIC DETERMINATION OF PHARMACEUTICALS IN VARIOUS MATRICES

A. INTRODUCTION

Having reviewed the possibilities for the determination of pharmaceuticals in Chapters 4 to 8, and before dealing with the problems of the spectrophotometric determination of the most important families of pharmaceuticals in Chapter 10, this chapter briefly reviews the special problems connected with the determination of pharmaceuticals in various matrices.

Thus a short discussion will be devoted to the problems of the spectrophotometric investigation of bulk materials, then to the assay of active principles in various dosage forms, in which the problem is caused in part by the interfering spectrophotometric activity of the matrix of the preparation and, in part, by the insufficient extraction of the active principle. The possibilities of spectrophotometry in the stability studies of pharmaceutical preparations are also reviewed in this chapter. Extraction and the elimination of matrix interference are the main problems in the spectrophotometric investigation of biological samples, but these problems are much more severe than with pharmaceutical preparations. With all types of applications there is a possibility of automation, and thus questions relative to this are also discussed in this chapter.

B. BULK MATERIALS

Of course, in the spectrophotometric investigation of bulk materials no matrix effect in the classical sense is encountered. In practice, however, bulk materials can be regarded as mixtures of the active principle and its impurities, and thus, in the determination of active principle the impurities, in the determination of impurities the active principle should be considered as matrix.

1. Determination of Active Principle Content

Spectrophotometric methods are often prescribed by pharmacopoeia and drug analytical standards for the analysis of active principles. If the objective is only the identification of active principle, the criterion is usually an agreement of the wavelength of the absorption maximum (minimum) with that of a standard sample. It should be noted that the value of this identification tool is generally inferior to that of infrared or NMR spectra which contain many more bands.

For quantitative analyses two approaches have become widespread. In the first, the only criterion is an agreement (generally within $\pm$ 2 to 3%) in the specific absorbance at the band maximum between the sample and a reference standard measured in parallel. Therefore, in this case the determination of the specific absorbance is regarded no more and no less than the measurement of a physical constant like melting point or specific optical rotation. The correct value of this constant is a necessary, but by no means definitive, condition for assessing the quality of the bulk material. A reason for this limitation is that, as pointed out in connection with several practical examples in Section 4.A, the impurities of an active ingredient may well have the same or similar

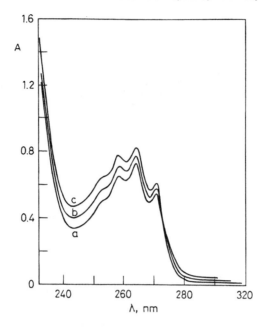

FIGURE 9.1. Spectra of pure and contaminated flumecinol. Solvent: ethanol; c = 0.237 g/100 ml. Curve a: pure flumecinol; Curve b: flumecinol containing 0.5% of 3-trifluoromethylbenzophenone; Curve c: flumecinol containing 1.0% of 3-trifluoromethylbenzophenone.

spectra as the main component, and thus the specific absorbance alone would not be sufficiently discriminating towards impure samples. If, however, as is generally prescribed by pharmacopoeia, the contents of related impurities are limited by means of a quantitative or semiquantitative chromatographic method (usually by TLC), the determination of specific absorbance provides additional information on the identification which is required for a complete specification.

The situation is more problematic if a comparison of the measured specific absorbance to that of a standard is used for the calculation of the active principle content, since on the basis of the above considerations and the examples shown in Section 4.A, the resulting active principle content is often an apparent value only, which can be regarded as true value only if all contaminants are spectrophotometrically inactive.

On the basis of the previous chapters, it is a further limitation that reliable quantitative analysis can be done only on the basis of bands above 220 nm with a molar absorptivity of at least 1000, but preferably greater than 10,000. When a spectrophotometric analysis is based on a weak absorption band, the errors due to eventual impurities may be much greater (in some cases an order greater) than in the above discussed case when the spectra of the main component and the impurity are the same. The presence of minor impurities may easily lead to apparent active principle contents significantly exceeding 100%. This is illustrated by Figure 9.1 in which the spectra of flumecinol (for its formula, see p. 45) are shown in the presence of 0.5 and 1% of 3-trifluoromethyl-benzophenone (a potential impurity). Flumecinol with isolated aromatic rings has only a weak band at 264 nm ($\varepsilon = 870$), whereas the absorption band of the latter compound at 249 nm is strong ($\varepsilon = 16,050$) and the specific absorbance at 264 nm is still 9,900. Consequently, an impurity of 0.5% increases the apparent active principle content to 106.2%! Of course, this "active principle content" exceeding 100% is an indication of the impurity of the substance, but it is easy to see that an incidental combination of the amounts of spectrophotometrically active and inactive contaminants may lead to a situation when the active principle content of a strongly contaminated sample is apparently 100%. It is, however, also evident that such a mistake can be made only if the absorbance measurement is restricted to the wavelength of absorption maximum. When a complete spectrum is recorded, the distortion of certain areas of the spectrum indicates the presence of impurities, since it can be excluded that the spectrum of a contaminant with a stronger spectrophotometric activity than that of the main component is completely identical in shape to the spectrum of the main component.

Of course, all the above considerations pertain to spectrophotometric analyses based on natural light absorption. By means of appropriate chemical reactions, substances of insufficient spectrophotometric activity may be converted into spectrophotometrically measurable substances, and the interferences due to impurities may be reduced or sometimes eliminated. However, it must be considered that the precision of methods based on chemical reactions is always poorer (with some reactions significantly poorer) than those of methods based on natural absorption. Thus, to decide on the basis of such methods whether the purity of a sample is 98, 98.5 or 99% requires very precise work (e.g., the measurement of a large number of replicates from both the sample and the standard). It can be stated in general that, although pharmacopoeias and analytical standards often prescribe such investigations, spectrophotometric measurements based on chemical reactions are not the most advantageous techniques for the determination of the active principle content of bulk materials.

2. Determination of Impurities

Although the determination of the number of impurities, their identification and then quantitation is typically a chromatographic problem in contemporary pharmaceutical analysis, in the case of favorable spectroscopic properties, a direct spectrophotometric assay of some impurities is also possible, primarily with spectrophotometrically active impurities in inactive main components. For example, in ethynodiol diacetate, the pharmacopoeia limit the absorption at 236 nm under the section of "Conjugated Dienes".[147a] The allowed maximum absorbance (0.500 in 0.05% solution) permits the presence of 1.6 and 2.0%, respectively, of the two potential impurities: 17α-ethynyl-3,5-oestradiene-17β-ol acetate and 17α-ethynyl-17β-hydroxy-4-oestrene-3-one acetate with absorption maxima at 236 and 240 nm, respectively.

In fortunate cases impurities of spectrophotometrically active compounds can also be determined, provided that their spectra are sufficiently different from those of the main component. With smaller differences, impurities in the percent concentration range can only be determined; if the differences are greater, as with the 2-amino-benzophenone derivatives present as manufacturing impurities or decomposition products in benzodiazepines (see Section 7.F), smaller amounts can also be determined.

In certain cases, for example 1,4-di-(2,5-dimethyl-phenyl)-1,4-diphenylbutane-1,4-diol impurity in 2,5-dimethyl-α-ethylbenzhydrol (RGH-3995), selective determination may also be promoted by chemical reaction (in this case an acid catalyzed water elimination, see Section 7.F),[48] but such cases are relatively rare and, as pointed out before, the determination of the impurities of bulk materials can be regarded basically as a chromatographic problem.

C. TABLETS AND CAPSULES

For the determination of the active ingredient content of tablets and capsules, UV-VIS spectrophotometry is still a frequently applied technique. This technique is of decisive importance in solving all the four types of tasks occurring in this field (determination of active ingredient content and content uniformity, i.e., the uniform distribution of active ingredient in the dosage units, determination of dissolution degree and of dissolution rate). The importance of the spectrophotometric technique here is due to the fact that the excipients and vehicles used in the manufacture of tablets and capsules are usually inactive spectrophotometrically and, on the other hand, they are generally practically insoluble in the solvents used for the extraction of active ingredients (water, dilute alkali or acid, methanol, ethanol, chloroform, and mixtures thereof). It should be stressed, however, that the absorbance of the solution obtained by the extraction of vehicles taken against pure solvent is generally not negligible at the short wavelength end of the UV region, but it decreases and then vanishes with increasing wavelength. The neglect of this absorbance, particularly in the short wavelength region, for preparations of low active ingredient content or substances with low absorptivity, may cause positive deviations. Therefore, in this field it is of particular importance that measurement below 220 nm is not recommended and that background correction, previous chromatographic separation or the application of chemical reactions may prove necessary for higher accuracy.

1. Determination of Active Agent Content

The sample is generally a finely ground powder obtained from 10 to 20 tablets or the combined and homogenized contents of the same number of capsules. An amount depending on the magnitude of active agent content and the molar absorptivity of the analyte is weighed from this sample. If solubility conditions permit, this solid is extracted with water, an appropriate buffer or dilute acid or alkali. Frequently used extracting solvents also include methanol and ethanol. Chloroform may be used only if the analytical band is above 240 nm or if solvent exchange is done after extraction. The undissolved components of the tablet should be removed from the mixture obtained after extraction. This can be carried out by means of a sintered glass filter or filter paper. Particularly with the latter, in the case of short analytical wavelengths, impurities dissolved from the filter may cause interference. Thus, it is recommended to dilute the extracted mixture to an exact volume in a volumetric flask and to filter this mixture discarding the first portion of the filtrate and to measure a sample of the rest of the filtrate directly or after appropriate dilution.

Spectrophotometric measurement can be carried out directly on the basis of natural light absorption or on a derivative formed by chemical reaction. For substances with absorption bands of appropriate wavelength and intensity, pharmacopoeia and standards generally prescribe measurement on the basis of natural absorption. (In the majority of the large number of publications dealing with the determination of the active principle of tablets on the basis of new chemical reactions, when the substance is spectrophotometrically active in its original form as well, the application of reaction hardly provides advantages, with respect to the direct measurement.) In the case of methods based on natural absorption, to eliminate the interferences due to vehicles or other factors, several techniques can be used:

1. Purification of the extract by further extraction. The following two examples are taken from USP XXII. The first is the analysis of brompheniramine maleate tablets,[147b] in which the powdered tablets are treated with an aqueous buffer at pH = 11 in a separation funnel, and then the base is extracted with hexane. The solution subjected to spectrophotometric measurement is obtained from the hexane extract by extraction with hydrochloric acid, and thus the hydrochloride of the active principle is measured at 264 nm. The other example is the analysis of chloroquinium phosphate tablet.[147c] After dissolution in water and filtering, the free base is extracted with chloroform after basification with ammonium hydroxide, then the chloroform solution is re-extracted with hydrochloric acid, and this solution is measured at 343 nm.
2. Purification by column chromatography. In the determination of the active principle of carbamazepine tablets according to USP XXI,[146a] extraction is carried out with a 95:5 mixture of ethanol and methanol, the extract is centrifuged after dilution with cyclohexane and subjected to chromatography on a 30 × 1.9 cm silica gel column. After discarding the non-polar fraction, the methanol eluate is measured spectrophotometrically at 285 nm.
3. Application of background correction methods. All methods described in Section 4.C are applicable. The simplest of these methods, the two-wavelengths technique, is even prescribed by pharmacopoeia. For example, in the analysis of amphetamine sulphate tablets, absorbance is measured not only at the maximum of the weak band of the analyte (the α-band of the isolated aromatic ring of amphetamine), at 257 nm, but also at 280 nm, where the absorption of amphetamine is negligible. When calculating the result, the difference of the two absorbances is taken into account, with both the tablet extract and the standard solution measured in parallel.[147d] Another example is the determination of chlorpromazine hydrochloride, for which the two wavelengths are 254 nm (the strong maximum of the spectrum of the phenothiazine ring) and 277 nm, where, after an extremely sloping section, the spectrum has a minimum with weak absorption (see Figure 3.26).[147e]

4. Application of derivative spectroscopy. As shown by the several examples of Section 4.F, derivative spectrophotometry offers excellent possibilities for the solution of the background problem occurring in the analysis of tablets.

Although, as mentioned above, pharmacopoeia and standards often prescribe methods based on natural light absorption for the analysis of the active principle of tablets, techniques based on chemical reactions are also very significant. These techniques are discussed in detail in the preceding and next chapters; here, the presumable reasons for applying methods based on chemical reactions are illustrated only on some examples taken from USP XXII. With compounds of low spectrophotometric activity, like benztropine mesylate[147f] or biperidene hydrochloride[147g] containing isolated phenyl groups only, the determination can be carried out with a much higher sensitivity in the visible region after derivatization than on the basis of their natural UV absorption. This requires ion pair formation of the bases with acid-type dyes (bromophenol blue, bromocresol purple) and then extraction with chloroform (see Section 7.H), resulting in derivatives absorbing at 415 and 408 nm, respectively. It is also probably the poor spectrophotometric activity of the active principle that makes a color reaction necessary for the analysis of tablets containing β-lactam-type antibiotics.[147ll] Here, the reaction is based on a hydroxamic acid formed with hydroxylamine and then the formation of its iron (III) complex (see Section 7.G). Sensitivity is not improved, but the interferences at low active ingredient content due to the relatively short wavelength of the maximum (240 nm) can be eliminated in the case of norethisterone and norgestrel by the formation of an isonicotinic acid hydrazone at the 3-keto group (see Sections 7.D and 10.M), which shifts the absorption maximum to 380 nm.[147h] The aim of the color reaction based on a sulphuric acid reagent with ethinyloestradiol and mestranol tablets (Sections 7.E and 10.M) is to improve both sensitivity and selectivity.[147i]

The simultaneous determination of the active principles of tablet combinations can be carried out either by multi-wavelength measurement and solution of a system of equations (an example is given in Section 4.D), or by applying methods based on chemical reactions selective for the individual components. However, the analysis of such preparations is performed increasingly often by chromatographic methods.

2. Content Uniformity Studies

Due to the large number of parallel measurements required here, spectrophotometric methods, which are simple, sufficiently sensitive and easy to automate, are particularly suitable for checking the variation of declared, nominal active principle content of drug preparations. In several cases, the same spectrophotometric method is prescribed by the pharmacopoeia for the determination of active ingredient content and for content uniformity studies, with differences in the manner and extent of dilution, due to the different sample weights. It often occurs, however, that active agent content is determined by highly selective chromatographic methods, whereas content uniformity is checked by the simpler and faster spectrophotometric method of lower selectivity.

3. Dissolution Tests

Modern analytical investigations of tablets and capsules also include dissolution tests, i.e., the determination whether a given part (e.g., 70–75%) of the active principle is released from the preparation during a given time, in a solvent of defined volume, using a blade or basket stirrer apparatus[147jj] at a given revolution rate. Since the solvent is usually water, dilute hydrochloric acid or mineral acid or a neutral or slightly alkaline phosphate buffer, it is possible to determine the concentration of dissolved organic substances by spectrophotometric methods in the filtrate obtained after filtering aliquot amounts. Accordingly, the analytical methods applied in dissolution tests are most often spectrophotometric measurements based on natural light absorption or, if necessary, on chemical reactions.

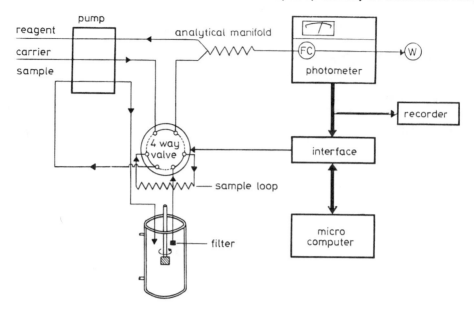

FIGURE 9.2. Instrument for the measurement of tablet dissolution profiles connected to automatic FIA instrument. (From M.A. Koupparis; P. Macheras; C. Reppas. *Int. J. Pharm.* 1984, *20*, 325.[85])

4. Determination of Dissolution Rate

In the case of controlled release preparations not only the amount of active agent released after a given time is prescribed, but also the shape and the quantitative parameters of the curve characteristic of the required time of release. This can be determined by sampling the solution at regular time intervals.[124] However, it is faster, less labor-consuming and more accurate to use an instrument designed specially for this purpose.

The Flow Injection Analysis (FIA) technique developed at the end of the 1970s and discussed in Section 9.J is excellent for these studies. The FIA instrument, cheaper and simpler than the automatic analyzers[147ee] based on different principles, can be attached directly to the system with standardized vessel and stirrer used for the determination of dissolution profiles. Controlled by a computer built into the system, aliquots can be introduced periodically, through a filter, into the solvent flow of the FIA instrument, in which the sample (directly or after passing through a reaction coil ensuring chemical reactions) passes into a detector providing a signal applicable for the determination of concentrations, which is usually an UV-VIS spectrophotometer. The speed and high flexibility of the FIA method allow samples to be taken as frequently as every minute, and thus even very fast dissolution processes can be followed. In this field the work of Koupparis and co-workers may be mentioned on the dissolution profiles of a number of tablet preparations.[7,82–87,98] Figure 9.2, illustrating the coupling of dissolution system with an FIA instrument, is taken from one of their papers; Figure 9.3 shows the dissolution profiles of paracetamol tablets manufactured by five different firms, taken by this instrument.[84]

D. SUPPOSITORIES

In the determination of the active ingredient of suppositories, UV-VIS spectrophotometry plays a significant role, similar to that in the investigation of tablets and capsules, discussed above. Of course, there are significant differences in the extraction of active ingredient and the preparation of this extract for spectrophotometric measurement.

The analytical process starts with the homogenization of several suppositories, which can be done by crushing them with a glass rod, under heating in the presence or absence of solvent. Of the solvents used, the primary role of which is to dissolve the excipients of suppositories, diethyl

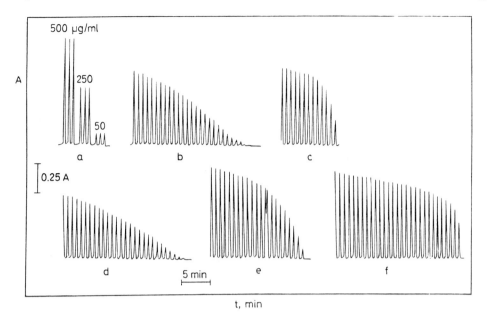

FIGURE 9.3. Dissolution profiles of paracetamol containing tablets determined by the FIA method. Curve a: calibration; Curve b: Distalgesic®; Curve c: Lonarid®; Curve d: Norgesic®; Curve e: Panatol®; Curve f: Depon®. (From M.A. Koupparis; P. Macheras; C. Tsaprounis. *Int. J. Pharm.* 1985, *27,* 349.[86])

ether is the most widely used, but chloroform is also frequently used for this purpose. Spectrophotometric measurement, also, can be done directly in chloroform if the active principle is sufficiently soluble in it and the analytical band is above 240 nm. Thus, for example, the active principle of a suppository containing acetylsalicylic acid can be determined at 280 nm after a column chromatographic purification of the chloroform solution (elution from the column with chloroform containing 1% of acetic acid).[147j] However, methods based on extraction of active ingredients of basic or salt character from the ether (or chloroform) solution with dilute aqueous acid, are more typical. Spectrophotometric determination can be carried out in the aqueous phase directly, e.g., with Bisacodyl suppositories at 263 nm,[147k] or after background correction, as with chlorpromazine[147l] or prochlorperazine,[147m] as already mentioned in connection with the spectrophotometric analysis of the tablet preparations of phenothiazine derivatives in Section 9.C (measurement at 254 and 277 nm, calculation from the absorbance differences). The efficiency of the purification required for the spectrophotometric assay can be increased by inserting further extraction steps and/or combining it with chromatographic separation. Thus, for example, with suppositories containing ergotamine tartrate and caffeine,[146c] the latter can be determined by direct spectrophotometric measurement, the former by van Urk's method, in the form of its colored derivative (see Sections 7.E and 10.F). This requires crushing the suppository under ether, extraction with aqueous boric acid, an extraction with chloroform after neutralization of the boric acid extract and then a column chromatographic separation of the chloroform extract.

E. INJECTIONS, SOLUTIONS AND OTHER LIQUID PREPARATIONS

In the determination of the active ingredient content of liquid preparations, spectrophotometry plays a role just as important as with the previous types of pharmaceuticals. The degree of difficulties in solving this task is a function of solvent and other excipients used in these units, which may occasionally strongly interfere with the direct spectrophotometric measurement.

The simplest task is, of course, the analysis of injections containing single therapeutic dosages per vial in aqueous solutions, since in this case the interferences due to solvent or the excipients discussed later are absent. With injections containing active ingredients of appropriate spectrophotometric activity, the spectrophotometric analysis can be carried out after a simple dilution, without any further treatment. Of the several methods of this kind applied in everyday practice, the analysis of Vitamin B_{12} injections according to the method of USP XXII[147n] is mentioned: the injection is diluted with water and measured at 361 nm. Colchicine with a pH-independent spectrum is determined in its injections at 350 nm after a simple aqueous dilution,[146d] whereas the assay of chloroquine is performed at 343 nm in 0.01 N hydrochloric acid.[147o]

If, for some reason, it is necessary to increase the selectivity of the analysis (separation from excipients, impurities, decomposition products), extraction purification steps may also be applied here, e.g., in the case of chlorpheniramine maleate injections, where the non-basic components are first extracted from the acidified solution with ether, then the base is extracted with ether after basification, and finally a re-extraction with dilute sulphuric acid yields solutions suitable for the spectrophotometric determination at 264 nm.[147p]

As with tablets, methods based on a wide variety of chemical reactions can be used for the determination of active ingredient content. There are several examples for such analyses described primarily in the literature but in pharmacopoeia as well. Chemical reactions are used to increase either sensitivity or selectivity. In addition to the examples given in connection with the analysis of tablets, the analysis of anileridine injection is mentioned, in which the primary aromatic amino active ingredient is determined after diazotization and coupling with N-(1-naphthyl)-ethylene-diamine,[147r] or that of the preparations containing native and hydrogenated ergot alkaloids, in which the spectrophotometric assay is carried out after performing a van Urk reaction (oxidation with iron (III) chloride and coupling with p-dimethylaminobenzaldehyde).[147s]

In the spectrophotometric analysis of solution preparations containing preservatives or antioxidants, it should be taken into account that a significant number of these agents, like p-hydroxybenzoic acid esters, benzoic acid and its salts, benzyl alcohol, benzalkonium chloride, cetylpyridinium bromide, phenol, cresol, di-tert-butyl-p-cresol, phenylethyl alcohol, thiomersal, phenylmercury nitrate, sorbic acid, ascorbic acid, ascorbyl palmitate, etc., are spectrophotometrically active. Consequently, in the majority of cases, they make the direct application of spectrophotometric measurement, based on natural absorption, impossible in the region below 300 nm. As illustrated in Section 4.F and Chapter 8 in several examples, difference and derivative spectroscopic methods often allow the active agent content of liquid pharmaceutical preparations of problematic composition to be determined on the basis of natural light absorption. Concerning the application of difference spectrophotometry to the solution of such problems, we refer to Figure 8.7 of Section 8.D, which shows that in the determination of the progesterone content of Limovanil injection the interference due, among others, to benzyl alcohol could be eliminated by the application of the sodium borohydride method.[46] The application potentials of derivative spectrophotometry can be illustrated in Figures 4.15 and 4.16. It can be seen that the direct determination of flumecinol with poor spectrophotometric activity is not feasible in Zixoryn emulsions, due primarily to the preservative, sodium benzoate. However, by means of second derivative spectrophotometry, the broad band of the latter can be eliminated and flumecinol can be determined.[47] To solve similar problems, further possibilities are offered by selective extraction already mentioned in connection with many examples, chromatographic separation prior to spectrophotometric measurement, or a shift of the spectrum of the analyte by means of an appropriately selective chemical reaction into the long-wavelength region in which there is no interference.

Similar problems occur in the analysis of oily injections or other liquid preparations containing oil or oily substances. In such cases the solvent itself interferes with the direct spectrophotometric measurement below 300 nm, or even the application of color reaction methods based on certain chemical reactions (methods based on other reactions, as shown in this example, are not affected). Therefore, the analysis of such preparations generally includes extraction. The simplest form of extraction is, e.g., in the case of injections in sunflower oil, with methanol. As an example, the active agent of oestradiol-17-valerate oily injection can be determined in phenolate form by measuring the methanol extract of the injection in alkaline medium directly at 300 nm.[146e] However, a more general method is the application of appropriate chemical reactions. In the determination of the active principle of desoxycorticosterone acetate oily injection on the basis of a stability-indicating color reaction with tetrazolium blue (see Sections 7.E and 10.M), it is necessary to extract the active principle prior to analysis by iso-octane–ethanol partition,[147t] whereas in the case of menadione injection, the oily solution, after a dilution with 1:1 ether–alcohol mixture, can be subjected to a color reaction with 2,4-dinitrophenylhydrazine and, after basification, to a spectropho-tometric measurement at 635 nm.[147u] 4-Ene-3-ketosteroids, e.g., nandrolone decanoate[147v] can be determined in a similar manner at 380 nm, after a chloroform dilution of the oily solution and a reaction with isonicotinic acid hydrazide (see Sections 7.D and 10.M).

F. OINTMENTS AND CREAMS

In the analysis of various pharmaceutical preparations, ointments and creams create the greatest number of problems. These are connected mainly with the pretreatment of the sample and do not prevent the application of spectrophotometry, which is one of the most widespread methods in this field also. Consequently, the situation in the application of spectrophotometry does not differ substantially from that encountered in the case of other pharmaceutical preparations. Accordingly, our illustrations are taken primarily from USP XXII, neglecting the very rich literature of this field. It is noted, however, that the problems of the field are discussed within two excellent reviews.[151,152]

The overwhelming majority of excipients applied in the preparation of ointments and cremes, like propylene glycol, glycerol, sorbitol, paraffins, fatty alcohols, their esters and ethers, alkyl sulphates and most of the quaternary ammonium salts can be regarded as spectrophotometrically inactive. Thus, after proper sample pretreatment, the active ingredient of suitable concentration and spectrophotometric activity can be determined on the basis of natural light absorption if the analytical wavelength is above 235 to 240 nm. However, preservatives and antioxidants widely used in these products, cause problems similar to those already discussed in Section 9.E.

The relatively frequently applied method for tablets and injections according to which the absorbance is measured directly after a dilution or simple solvent extraction of the preparation can be applied here relatively less frequently. This method was applied with three preparations, all having absorption in the long-wavelength region: the active ingredient content of hydroquinone ointment can be determined at 293 nm after trituration with methanol and filtration,[147w] nitrofurazone cream can be analyzed at 385 nm after chloroform treatment and dissolution in dimethyl formami-de,[147z] whereas gentian violet can be determined at 435 nm after dispersion in dilute hydrochloric acid and filtration.[147x]

Those cases in which the active principle should be separated from the interfering excipients prior to spectrophotometric measurement are encountered more frequently. Liquid-liquid extraction methods discussed in Sections 9.C to 9.E are applied here even more widely. Extraction with ether, a subsequent extraction of the basic active principle with hydrochloric acid, a re-extraction after basification with ether and a repeated aqueous–acidic extraction and, finally, spectrophotometric measurement form the basis of the assay of tetracaine ointment at 310 nm.[147y] In simpler cases, a water–chloroform or water–dilute hydrochloric acid extraction is sufficient for removing the non-polar excipients from the sample, as in the case of mafenide acetate[147aa] and acrisorcin[147bb] creams (measurement at 267 and 400 nm, respectively). A special partition solvent pair of corticoste-roids is acetonitrile–hexane for methyl-prednisolone acetate[147cc] and acetonitrile–i–octane for flu-methasone pivalate;[41] in both cases the acetonitrile phase is used for the assay of active principle after chemical reaction.

The column chromatographic purification of the extract is more important in the case of ointments and creams than in other pharmaceutical preparations. This method is used, for example, for the analysis of idoxuridine ointment[147ee] and nitrofurazone ointment[147ff] and for the separation of the active ingredients of benzoic acid–salicylic acid combinations from one another and from the excipients prior to measurement at 275 and 311 nm, respectively.[147gg]

Naturally, for ointments and creams there is also a wide choice of spectrophotometric methods based on chemical reactions. In the determination of dexamethasone phosphate ophthalmic ointment[146f] this reaction is simply the enzymatic hydrolysis of the phosphate ester group, enabling the resulting dexamethasone to be extracted with dichloromethane and measured at 236 nm, but in the majority of cases the reactions are used to shift the absorption maxima into the visible region. Such a reaction is, for example, that of isonicotinic hydrazide (Sections 7.D and 10.M) for the determination of clocortolone pivalate in cream preparations which is

TABLE 9.1. Spectral Data of Prazepam and its Degradation Product[23]

	λ_{max}, nm	ε methanol	λ_{max}, nm	ε 3% v/vH_2SO_4 in methanol
Prazepam	227	33,350	241	31,200
	313	2,160	285	16,300
			365	3,800
2-cyclopropylmethylamino-5-chloro-benzophenone				
	235	24,800	257	12,700
	365	2,700	365	170
	408	7,100		

applied after hot methanol extraction.[147hh] In the analysis of steroid creams, the case of estropipate is noteworthy. In this case, the extraction of the acidic, sulphuric acid ester–salt type active principle is based on a partition between chloroform and aqueous piperazine. The sulphuric acid ester extracted into the aqueous phase is hydrolyzed after acidification with hydrochloric acid, the released oestrone is shaken into chloroform, and then, after evaporating the solvent, it is treated with 90 vol% sulphuric acid as reagent (see Section 10.M) to convert it into a colored product measurable at 450 nm.[146g] Of non-steroids, sulphisoxazol-diolamine ointment is mentioned, the active principle of which can be determined at 540 nm after extraction with a heptane–aqueous alkali system, acidification of the aqueous phase, diazotization, and then coupling with N-(1-naphthyl)-ethylenediamine.[147u]

Finally, it is noted that due to their relatively poor sensitivity and selectivity, spectrophotometric methods can be used for the determination of preparations with low active agent content only with considerable difficulties (note that the concentration of certain agents, like fluorocorticosteroids, in ointments is in the concentration range of 0.01%!). Such ointments and a large number of combination ointments are analyzed in modern pharmaceutical analytical laboratories primarily by HPLC methods.

G.   STABILITY STUDIES

In contemporary pharmaceutical analysis, for the stability study of pharmaceutical preparations, either direct quantitative chromatographic methods (GC, HPLC and thin-layer densitometry) or, after a separation of the original drug and its decomposition products by TLC or column chromatography, spectrophotometric methods are almost exclusively used. A definite advantage of these methods is that the original drug and its decomposition product(s) can be determined simultaneously by them, and there is a possibility for the study of more complex (multistep or multidirectional) decomposition processes. However, for checking simpler, unambiguous decomposition processes, spectrophotometry is an often applicable simple method providing satisfactory results, provided that the decomposition process causes a change on which a quantitative spectroscopic measurement can be based or, after a suitable chemical reaction, intact and decomposed substances can be distinguished spectrophotometrically. The potentials of this technique are illustrated by some examples.

The relatively infrequent case when the spectra of the original drug and its decomposition product are different enough so that a stability study can be based on them is illustrated by the example of prazepam (the cyclopropylmethyl analog of diazepam).[23] As shown in Scheme 7.21 of Section 7.E, a hydrolytic decomposition reaction of benzodiazepines is their cleavage into a 2-aminobenzophenone and glycine. The spectral data of the undecomposed drug and its decomposition product, 2-cyclopropylmethylamino-5-chlorobenzophenone are shown in Table 9.1. It can be seen that the reaction can be followed by both the absorption maximum of the decomposition product at 408 nm and the absorption maximum of the non-decomposed active ingredient appearing at

365 nm in acidic methanol. (Due to the protonation of the amino group of the decomposition product, its long-wavelength $n-\pi^*$ band disappears, and thus the absorbance measured at 365 nm is characteristic (practically selectively) of the non-decomposed drug.)

Another interesting example is a spectrophotometric monitoring of the photochemical decomposition of sodium nitroprusside.[39] The hydration of $[Fe(CN)_5NO]^{2-}$ anion simultaneously with the reduction during decomposition into a $[Fe(CN)_5H_2O]^{3-}$ anion involves a significant change in the very low intensity but characteristic spectrum of the starting substance. In this change the intensity of the shoulder at 498 nm ($\varepsilon = 7.4$) does not change whereas the maximum at 394 nm ($\varepsilon = 20.4$) increases significantly, which is sufficient for a stability study.

In practice, such spectral changes upon drug decomposition are seldom enough for a selective analysis to be performed. Modern tools of evaluation allow, however, the selective determination of non-decomposed drugs or their decomposition products even if their spectra do not differ to an extent sufficient for the selective analysis of either of them. For example, the method of orthogonal polynomials (Section 4.C) makes the determination of chloramphenicol possible along with its decomposition product of azobenzene type, and thus it proved to be applicable for the kinetic investigation of this decomposition.[2] Derivative spectrophotometry (Section 4.F), as well, promotes the solution of several problems in this field. Thus, for example, pirenezepine[158] can be measured selectively in the presence of its acidic decomposition products on the basis of the first derivative spectrum. The determination of paracetamol and phenacetin in tablets and syrups, in the presence of their decomposition products, was achieved by the second derivative technique.[79] In acetylsalicylic acid suppositories, both the active principle and its decomposition product, salicylic acid, can be determined selectively[77] without prior separation, by means of the second derivative technique. Using a combination of second and third derivatives, four different decomposition products of phenylbutazone can be determined in the presence of one another and the original drug.[121] Even with very complex molecules like nystatin,[26] derivative spectroscopy can be applied successfully in stability studies. In the spectrum of this molecule, the long-wavelength band system between 290 and 320 nm is due to the tetraene part of the molecule which is changed during oxidative decomposition. The effect of the absorbance remaining in this region after oxidation can be eliminated by derivative spectroscopy, making the method applicable for stability studies.

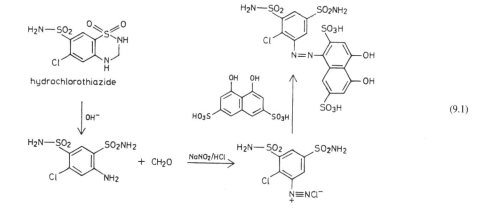

hydrochlorothiazide

(9.1)

The development of spectrophotometric methods for the selective analysis of decomposition products may often be promoted by the application of sufficiently selective reactions.[22] As a typical example for such methods, let us take the monitoring of the hydrolysis of hydrochlorothiazide (Scheme 9.1) on the basis of a coupling of the primary aromatic amino group of the resulting decomposition product (4-amino-6-chloro-1,3-disulphonamide) with chromotropic acid after diazotization,[123] or the determination of free amygdalic acid, obtained in the hydrolysis of amygdalic acid esters (e.g., homatropine, etc.), after cerium (IV) oxidation (see Scheme 9.2) in the form of benzaldehyde which has strong spectrophotometric activity.[21]

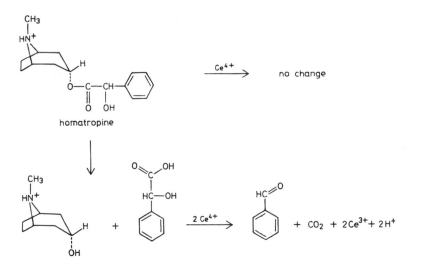

(9.2)

In addition to the methods discussed so far, specific to hydrolytic decomposition, there are several other techniques in spectrophotometric drug stability studies which are specific to oxidative decomposition. The tetrazolium method of corticosteroids (Sections 7.E and 10.M) is, for instance, specific to non-oxidized corticosteroids, since molecules oxidized in position 21 or undergoing intramolecular oxido-reduction are unable to reduce the tetrazolium reagents into colored formazans. Therefore, this method is excellent for stability studies.[25,65]

The 21-aminocorticosteroid mazipredone can be determined on the basis of similar principles, using a color reaction with o-dinitrobenzene. As the decomposition products do not reduce the reagent into a violet reduction product, this method can also be applied for investigating the stability of Depersolon preparations.[96]

H. DETERMINATION OF DRUGS AND THEIR METABOLITES IN BIOLOGICAL SAMPLES

The determination of drugs and their metabolites in biological samples (in most of the cases in plasma, serum or urine) is one of the most delicate tasks of pharmaceutical analysis. Such investigations are required in pharmacokinetic and metabolic studies, in the determinations of therapeutic doses of drugs, etc. In the early times of these investigations, the first analytical methods were colorimetric methods, which provided very significant results despite their modest sensitivities and selectivities. In the modern practice of clinical and pharmacological analysis, drugs and metabolites are determined in biological samples in the majority of cases by GC and HPLC methods. These methods enable not only the drug and its metabolites to be determined simultaneously, even in the presence of endogenous substances, but also the sensitivity of the method to be increased by 2 to 3 orders with respect to spectrophotometric techniques. The latter is due to high sensitivity detectors and to special derivatization reactions made possible by the application of such detectors. A further increase in sensitivity can be achieved by the application of various immuno-assay methods.

In spite of the dominance of chromatographic and immunoassay methods, spectrophotometry has still retained some positions in pharmacological-kinetic-toxicological analysis,[24] since the required instrument is an order cheaper than the systems required for more advanced techniques and since the methods are simple and easy to automate.

The most significant limitation of spectrophotometric analysis is its low sensitivity. It is easy to see that with a compound of molecular mass 300, if the intensity of the analytical band (either natural absorption or after chemical reaction) is $\varepsilon = 1000$, a concentration of 100 μg/ml is necessary for measuring an absorbance of 0.33 in a 1 cm cell. To provide this peak concentration in the blood of an adult, which is 5:1 in average, assuming an ideal, 100% resorption, a single dosage of 500 mg of the drug would be required. Thus, bands of such intensity (which allow several other analytical problems to be solved) cannot be applied here. In the case of $\varepsilon = 10,000$ this dosage decreases to 50 mg, which makes the method already applicable for some tasks. Since molar absorptivities very rarely exceed 20,000 to 30,000 in pharmaceutical analysis, even in an ideal case, a single dosage of 20 mg can be regarded as the approximate lower limit of dosage which ensures spectrophotometrically measurable plasma concentrations.

A direct spectrophotometric measurement of the analyte on the basis of its natural UV absorption is, of course, impossible in the great majority of cases due to the spectrophotometrically active endogenous substances always present in serum, plasma or urine. Even by selective extraction, the possibility of direct spectrophotometric determination is an exception. An example of this is the determination of amphotericin-B in serum, which is made possible by the fact that this polyene-type drug has a strong maximum at a very long wavelength, 408 nm. The majority of endogenous substances does not interfere with the analysis to a significant extent at this wavelength. Accordingly, under a bilirubin concentration of 25 mg/l, spectrophotometric assay can be carried out directly after simple acetonitrile dilution and centrifuging, but above this bilirubin concentration, a solid phase extraction step must be included.[130] Of course, with dyes absorbing in the long-wavelength visible region, nothing interferes with the direct spectrophotometric measurement: indocyanine green used as diagnostics can be determined at 800 nm simply after an aqueous dilution of the plasma.[138]

It is worth noting that in the investigation of the binding of drugs to serum proteins, the drug content of dialysates[92] and ultrafiltrates[45] proved to be measurable in the short-wavelength UV region as well.

With some drugs reaching high serum concentration and having strong pH dependence, the difference spectrophotometric method discussed in Section 8.B can be applied to advantage for solving background problems due to endogenous substances. A condition of success is that at the maximum of the spectrum of solutions with two different pH values, taken against one another (difference spectrum), the blank serum containing no drug has no significant difference absorbance. This can be achieved by combining difference spectrophotometric measurement with an efficient extraction step. This technique was used to determine salicylic acid,[137,162,166] theophylline[52] and barbiturates[17,161] in serum. The most important parameters of the methods (pH of the two solutions, wavelength) are shown in Table 8.1 of Chapter 8.

Another method for background elimination, derivative spectrophotometry discussed in Section 4.F, also plays a role in clinical analysis.[116] An example is the determination of paraquat herbicide in blood and dialyzed solution, respectively, of poisoning cases on the basis of the second and fourth derivative spectra,[34] the determination of paracetamol in serum,[28,29] the determination of nitrofurantoin in urine[120] and the determination of atracurium in plasma.[20]

Methods in which the interference of biological background is eliminated by shifting the spectrum of the analyte by means of an appropriately chosen, selective reaction into the visible region are much more significant in the practice of clinical analysis than the methods based on natural light absorption, discussed so far. These indirect methods are used most generally in the determination of the serum levels of paracetamol and salicylic acid.

Of the several methods used for the determination of the serum level of paracetamol, the most widespread are based on the nitration of the compound with nitrous acid and the assay of the resulting o-nitrophenol-type 2-nitro-4-acetaminophenol in alkaline medium, in which the phenolate derivative can be measured at 430 nm.[44,160] The interference of salicylate and other interfering components can be suppressed by selective extraction.[9] Another principle widely applied for the assay of paracetamol in biological samples is based on the reducing properties of the phenolic substance. Using phosphomolybdo-tungstic acid (Folin-Ciocalteu reagent)[53,139] or iron (III)-2,4,6-tripyridyl-s-triazine[94] as reagent, the method can be based on the colored reduced forms of the reagents. To carry out the methods discussed so far, routine "kits" for clinical analysis are commercially available. There are interesting and important papers which compare these methods from the aspects of accuracy and various interference effects (primarily the effect of salicylic acid).[67]

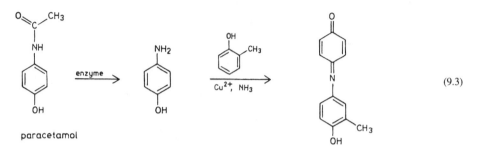

(9.3)

Further reagents for the determination of the serum levels of paracetamol are: 2-nitroso-1-naphthol-4-sulphonic acid[129] and periodic acid 2,5-dimethylphenol.[93] Excellent sensitivity and selectivity could be achieved by the method based on Scheme 9.3 and presented below. The indophenol derivative obtained after enzymatic hydrolysis and subsequent oxidative coupling with o-cresol can be measured at 615 nm.[30]

For the determination of the serum levels of salicylic acid, the most widespread method is based on a measurement of its violet complex formed with iron (III) ions. This classical method was applied by Trinder[148] to biological samples; several methods following it differ from the original mainly in the pretreatment of the sample.[33,91] The method has a number of automated versions as well, of which the ones adapted to centrifugal analyzer[142] and to flow-injection analyzer[82] are mentioned.

The other method widely used for the spectrophotometric determination of salicylic acid serum levels is based on an enzymatic oxidative decarboxylation of the acid.[95,110,118,164] The method (see Scheme 7.28 of Section 7.K) is much more selective than the one based on iron (III) complex formation, which is, of course, subject to interferences by phenol-type substances.

Spectrophotometric methods based on chemical reactions have been developed for a number of other drugs and metabolites, and the development of such methods is still in progress. Some methods are listed in Table 9.2. It is noted that the cited monograph of Chamberlain[24] contains the description of several further methods.

I. THE DETERMINATION OF DRUGS IN MEDICINAL PLANT EXTRACTS

Problems similar to the ones discussed in the previous section are faced by the analyst in the determination of the active agent content of plants or their extracts. "Background problems" are equally serious here, since the analysis is to be performed in the presence of dyestuffs of plant origin and numerous other interfering components. A particularly serious difficulty in this type of analysis is the presence of several related substances with closely similar chemical structures accompanying the drug to be analyzed. Therefore, modern methods of plant analysis are based, in the vast majority of cases, on chromatographic separation. Along with GC and HPLC, quantitative methods of TLC play an important role as well.

Apart from chromatographic separation, the potentials of direct spectrophotometric methods based on natural light absorption are generally very modest. However, compounds with favorable spectrophotometric properties, like berberine with an absorption maximum at 420 nm, can be determined from plant extracts after sufficiently selective extraction.[113] In other cases the aim of analysis may be the determination of the combined amount of substances with related structures (and thus similar spectra), like the determination of flavonoid aglycones (myricetin, quercetin, etc.) at 375 nm.[89]

In certain cases when the difference between the spectra of derivatives of related structure is large enough, a two- or multi-wavelength measurement (see Chapter 4) can be utilized for analysis. Several classical methods of this type can be found in the early textbooks of pharmaceutical analysis,

TABLE 9.2. Spectrophotometric Determination of Some Drugs and Metabolites Based on Chemical Reactions

Drug (metabolite)	Sample	Reaction	Extraction step	Reference
Isonicotinoyl hydrazide	Urine	Condensation with 9-chloro-acridine	+	136
Carbamazepine	Plasma, serum	Acid-catalyzed rearrangement to 9-methyl-acridine	+	13,35
Phenytoin metabolite (4-hydroxyphenyl)	Urine	Oxidative condensation with 4-aminoantipy-rine	+	112
Celiprolol	Plasma, urine	Hydrolysis, diazotization, coupling with naphtyl-ethylenediamine	+	143
Procainamide	Plasma	Diazotization, coupling with naphtyl-ethylenediamine	+	131
Phenylbutazone	Plasma	Oxidation to azobenzene with permanganate	+	66
Bromazepam	Urine	Hydrolysis to 2-amino-5-bromobenzoyl-pyridine	+	150
Etoposide, teniposide	Plasma	Electrochemical oxidation	+	153,154
Clavulanic acid derivatives	Serum, urine	Condensation with imidazole	−	73
Promethazine	Urine	Charge-transfer complex formation with morpholine/I_2 reagent	−	165
Erythromycin	Serum, urine	Ion-pair extraction with bromocresol purple/chloroform	+	122
Chloroquin and metabolites	Urine	Ion-pair extraction with methylorange/dichloromethane	+	111
Ephedrine, methamphetamine	Urine	Ion-pair extraction with tetrabromophenolph-talein ethyl ester/1,2-dichloroethane	+	126
Theophylline	Serum	Enzymatic oxidation with theophylline oxidase/ferricytochrom C; measured species:ferrocytochrome	−	54
6-Mercaptopurine*	Urine	Enzymatic oxidation with xanthine oxidase; measured species: 6-thiouric acid	−	50
Chloramphenicol	Serum	Enzymatic acetylation; measurement of the forming NADH by the tetrazolium method	−	57,109
Ascorbic acid*	Plasma, serum	Oxidation with Fe(III)chloride; measurement of the forming Fe (II) as the tripyridyl-s-triazine complex; enzymatic reduction of the background	−	127
Dapsone	Urine, plasma	Condensation with 4-dimethylamino-benzaldehyde	+	145
Chloroquin	Urine	Ion-pair extraction; bromothymol blue/dichloromethane, methylorange/chloro-form, tetrabromophenolphthalein/chloroform	+	12
Pyrimethamine	Urine	Formation of Schiff's base with salicylalde-hyde, complex formation with Zn (II)	+	63

*See Section 7.K.

like the book of Higuchi and Brochmann-Hanssen,[60b] but such methods are found occasionally in the recent literature, e.g., the simultaneous determination of strychnine and brucine in *Nux vomica* extracts[40,81] or the simultaneous assay of vincamine and vincine in *Vinca minor* plants,[10] etc.

Some further methods of this type are: the simultaneous determination of quinine–quinidine and cinchonine–cinchonidine in *Cinchona succirubrae* extract on the basis of measurements at 330 and 282 nm,[58] the simultaneous assay of sennosides and rhein glycosides after hydrolysis, multiple purification by extraction and measurements at 377 and 430 nm, respectively[56] or the simultaneous determination of flavonols and isoflavonols from the acidic hydrolysate of *Sophora japonica* tincture at 261 and 370 nm.[36]

The high selectivity and background elimination ability of derivative spectrophotometry can also be utilized in this field, as proved by the simultaneous determination of the quinine–quinidine and cinchonine contents of *Cinchona* extracts[31] and the simultaneous determination of strychnine and brucine in *Nux vomica* extracts.[80]

The most general method for background elimination and for the solution of selectivity problems is the insertion of a chromatographic separation step before spectrophotometric analysis. Separation is performed in several cases by column chromatography, as with the determination of the total flavonoid content of *Crataegus* after chromatographic separation on a polyamide column;[76] but, TLC is the most widespread separation method. Of the extremely large number of examples for such measurements, the ones discussed in Sections 5.D and 5.E are selected: the determination of the digoxin active principle and other components of *Digitalis lanata* convertible into digoxin by TLC separation, elution, xanthydrol reaction and subsequent spectrophotometric measurement,[60] and the determination of vinblastine in *Catharanthus roseus* extract by 2-D TLC and the direct reflection spectrophotometric measurement of the separated spot without any elution or chemical reaction.[61]

Of the further methods, the following examples are worth mentioning: the determination of condurangin in condurango after the TLC separation of the glycoside cinnamic acid ester, alkaline hydrolysis and the assay of the resulting cinnamic acid at 266 nm,[135] the measurement of juglone at 520 nm after extraction, acidic hydrolysis, TLC separation and elution,[163] moreover the analysis of sanguiritrin,[78] rutin,[6] methoxsalen,[128] silybin[140] and individual flavone glycosides.[5]

The selectivity of methods containing no chromatographic separation is enhanced usually by the inclusion of chemical reactions into the technique. Such a technique is, for example, the extremely sensitive method ($\varepsilon_{430\,nm} = 49,000$) of Baccou et al.[8] for the determination of sapogenins in *Solanum* species. This method, using a vanillin–sulphuric acid–ethyl acetate reagent, is selective enough for the determination of the combined amount of sapogenins in the presence of sterols, carbohydrates, vegetable oils, fatty acids, etc.; however, when sapogenins are to be determined individually, a TLC separation is required prior to analysis. There are analogous methods for the determination of ergot alkaloids in extracts by means of the van Urk method,[59a] or for the determination of sennosides after sodium dithionite reduction.[159]

The most important spectrophotometric method of the determination of alkaloids is the technique based on ion pair extraction. This method, discussed in detail in Sections 7.H and 10.F, is obviously applicable for the determination of the alkaloid content of plant extracts as well. Of the novel applications of the classical method, it is worth mentioning the determination of atropine and scopolamine in *Belladonna* and other plants by means of bromothymol blue,[51,55] methyl orange[19,134] and tropeolin-000,[119] the determination of thebaine by means of bromocresol green,[99] the assay of khellin and bergapten with bromothymol blue,[62] and the analysis of gangleron after sodium iodobismuthate extraction and spectrophotometric measurement.[68]

Plant analysis also utilizes a number of redox reactions. Of these, the determination of hydroxyanthracenes is an example in the form of hydroxyanthraquinones in the vicinity of 515 nm after iron (III) chloride oxidation. This method is also prescribed by the pharmacopoeia for the analysis of *Cascara sagrada*.[16,147kk] The determination of tannin in plants by means of the Folin-Ciocalteu reagent[42] is also based on the reducing ability of polyphenols. An oxidizing agent for the indirect determination of rutin is cerium (IV); the resulting cerium (III) ions are determined as a chelate formed with an arsenazo III reagent.[18] The determination of ascorbic acid in plant samples with a method using ascorbic acid oxidase, iron (III) chloride and 2,4,6-tris-pyridyl-s-triazine as reagents for measuring at 593 nm is also worth mentioning here[133] (the theory of the method was discussed in Section 7.K).

A reagent widely applied in plant analysis is 4-dimethyl-aminobenzaldehyde. In addition to its use mentioned in connection with the van Urk reaction,[59a] this reagent has been applied successfully for the determination of chamazulene in plants by means of a measurement at 642 nm[156] and in the difference spectrophotometric version of the method,[157] as well as for the determination of iridoid glycosides,[64] atropine and scopolamine.[71]

A color reaction based on the formation of an aluminum chloride complex is often used for the determination of flavonoids.[5,43,89] The determination of capsaicine in spice plants by nitration and the measurement of the nitrophenolate at 430 nm,[132] the selective determination of carvone in aroma oils with 1,3-dinitrobenzene as reagent,[72] the determination of ginsengosides in ginseng with the application of vanillin–acetic acid–perchloric acid reagent,[141] and the determination of judaicine in *Artemisia judaica* with anisaldehyde–sulphuric acid reagent are also worth mentioning.[74]

J. AUTOMATION OF SPECTROPHOTOMETRIC MEASUREMENTS

The increasing demands in the quality of pharmaceuticals in the last two decades has involved an increasing stringency in the quantitative and qualitative requirements of pharmaceutical analysis in various media. These requirements (particularly in the quantitative area) could be met only by the automation of analyses. Accordingly, serious efforts have been made in the field of the automation of pharmaceutical analysis; the basic measurement techniques had been developed already in the 1960s and have become widespread in subsequent years. The further development and extension of methods to other fields is still in progress.

UV-VIS spectrophotometric methods are excellent media for automation: the various automatic analyzers operate with spectrophotometric detectors in the practice of pharmaceutical analysis. This is one of the reasons for the fact that spectrophotometry has retained a significant role in pharmaceutical analysis, and even more so, the adaptation of various classical methods for automatic analyzers and the elaboration of new principles have started a new wave in the research of pharmaceutical analysis. Automated spectrophotometric pharmaceutical analysis has produced significant results particularly in cases where automation has allowed the analysis of a large series of samples, greatly speeding up the analysis, making it more objective and accurate, and decreasing its labor demand, as in the determination of the content uniformity of tablets, dissolution profile studies and the determination of drugs in biological samples.

The methods introduced in the first stage of the development of automatic spectrophotometric analysis were based almost exclusively on AutoAnalyzer-type instruments. In these instruments a peristaltic-proportional pump introduces and moves the solution to be analyzed and the reagents of the spectrophotometric method based on chemical reaction. The volume ratios of these reagents are determined by the cross-sections of the flexible tubes applied. The liquid portions of given volume, taken from the sample containers of the sample feeder unit, are separated from one another by air bubbles. Reagent flows are separated by air bubbles as well. The sample solution and the reagents are mixed in a mixing coil of the instrument. The more or less quantitative conversion of the reactions is ensured in the subsequent coils maintained at appropriate temperatures. Conversion can be increased: by elevating reaction temperature, increasing the length of the coil, or by decreasing flow rate. The absorbance, proportional to the concentration of the analyte, is determined in a spectrophotometer equipped with a flow cell connected to the coils. The calculations are based on the absorbances measured on standard solutions of known concentrations inserted between the sample solutions. The 100% conversion of the reactions is not critical; if exactly the same volumes and reaction conditions are ensured for the samples and the standards, the error arising from incomplete conversion can be eliminated.

It is also noted that a "Solid Prep" unit can be attached to the system, which automatically performs the disintegration and extraction of the tablets as well as the filtering and dilution of the resulting solution before feeding it into the analyzer.

In the field of sample preparation, the application of robotics provides new, significant potentials; the spread of this technique will further increase the reliability and reduce the labor and time demand of analysis.

Almost all spectrophotometric methods based on the most important chemical reactions described in the 1960s and 1970s as manual methods of pharmaceutical analysis have been adapted to automatic analyzers operating on the basis of the above principles. The wide application of the method is indicated by the fact that it is included in the most recent U.S. Pharmacopoeia. For the determination of the active principle content of β-lactam antibiotic tablets, a method based on ring opening with hydroxylamine and the measurement of the resulting hydroxamic acid in the form of its iron(III) complex is applied. The active principle content of nitroglycerine tablets is determined by a method based on a reductive hydrolysis with strontium hydroxide performed in a 12 m reaction coil immersed into a thermostat at 50°C, a diazotization reaction of the resulting nitrite ion with procaine in hydrochloric acid medium and, finally, a coupling with naphthyl ethylenediamine.[147ll] The norethisterone content of norethisterone—mestranol tablet combination is determined on the basis of its natural light absorption after the automatic splitting of the extract, whereas that of norethisterone—ethinyloestradiol tablet is determined spectrophotometrically after an isonicotinic acid hydrazide color reaction. The oestrogenic components are determined in both cases by a fluorimetric method after sulphuric acid treatment.[146h]

Of the numerous methods arising from the early AutoAnalyzer period, one may mention the determination of paracetamol in elixir and tablet preparations by nitration and the measurement of the resulting nitrophenol in alkaline medium[27] or the analysis of various steroid tablets by means of the tetrazolium, isonicotinic acid, sulphuric acid, etc. reagents.[49] Such techniques, although decreasing in number, are still being published. One may mention, for example, the method of Bongiovanni et al.[15] for the analysis of tablets with nitrate ester type active components: an alkaline hydrolysis is followed by the reduction of nitrate into nitrite and a measurement of the latter by an azo-coupling reaction; the method of Abdalla et al.[1] for the analysis of cephalosporins based on alkaline hydrolysis and the measurement of the resulting sulphide ion after cyclization with N,N-dimethyl-p-phenylenediamine–iron(III) reagent into methylene blue;[1] or the method of Kirchhoefer[75] for the selective determination of tridihexethyl chloride in tablets combined with meprobamate.

The flow diagram of the latter is shown in Figure 9.4, which is based on the extraction and measurement of the tridihexethyl base ion pair with bromocresol purple (see Section 7.G). It is worth noting in connection with this method that the AutoAnalyzer technique permits extraction steps to be inserted as well. The polystyrene cups of the sample feeder contain the aqueous extracts of the tablet and the aqueous solutions of the active principle of known concentrations. The samples are taken from these cups, mixed with phosphate buffer of pH = 5.3, segmented with air bubbles, and then mixed in another reaction coil with appropriate volumes of the alkaline dye solution and another buffer solution. Since the reaction is fast, delay coils or increase of temperature are not required. After encountering the chloroform stream, the extraction takes place in coil BM (Figure 9.4), permitting contact at a large surface area. The chloroform and the extracted aqueous phases are separated in coil SC of vertical position and glass tap B-O. The chloroform phase is passed into the flow cell of the spectrophotometer, in which the measurement is carried out at 408 nm. The method enables 30 samples to be analyzed per hour, which is generally characteristic of the throughput of such techniques.

Of the rich literature of the method, the paper of Kubin, Brehm and Ulmen[88] is worth mentioning separately. The authors divide the liquid stream of an AutoAnalyzer equipped with a "Solid Prep" unit into two flows, permitting a simultaneous photometric and HPLC measurement of the tablet extracts. This technique was used for the analysis of ergotamine tartrate (by the iron (III)-p-dimethylaminobenzaldehyde method), codeine phosphate (through the determination of phosphate content by the molybdenum blue method) and phenothiazine derivatives (in the form of their palladium (II) complexes).

The second generation of automatic analyzers applied in pharmaceutical analysis is based on the Flow Injection Analysis (FIA) technique developed at the end of the 1970s and rapidly spreading in the last decade. This technique, already covered by monographs[125,149] and reviews on its applications in pharmaceutical analysis,[97,100] is principally different from the AutoAnalyzer technique discussed so far. In FIA, a continuous, unsegmented liquid flow is produced by a pump. The sample and the reagents are injected into this flowing liquid by means of a microprocessor controlled loop injector (the sample and the reagents are moved by different pumps). The appropriate conversion of the reactions is provided for by a temperature-controlled reaction coil in this case as well. The liquid passes continuously through the cell of the detector, and the recorder tracks the spectrophotometric signal, which, due to the appropriate dimensions of the instrument and the relatively low diffusion effects at the given flow rates, is similar to a sharp chromatographic peak. Of the detectors based on the various optical and electrochemical principles, UV-VIS spectrophotometric detectors are the most common.

As a practical example, the method of Ohta et al.[117] is cited for the determination of pharmaceuticals containing N-nitroso groups. The chemical process on which the method is based is as follows.

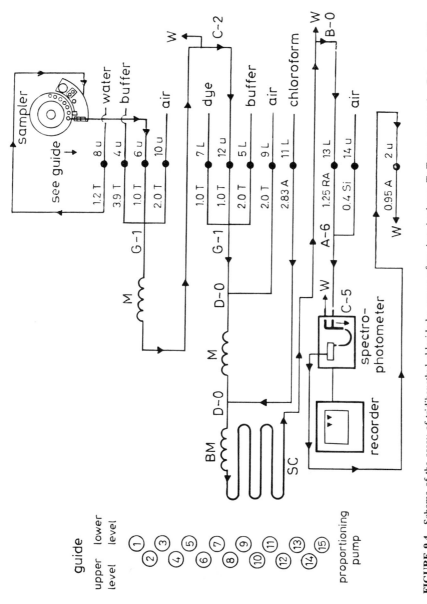

FIGURE 9.4. Scheme of the assay of tridihexethyl chloride by means of an AutoAnalyzer. T: Tygon pump tube; RA: red acidflex pump tube; Si: silicone pump tube; M: 28-turn mixer coil (2.4 mm i.d.); BM: 28-turn packed mixer coil; SC: 5-turn sedimenting coil (2.4 mm i.d.); C-2, B-0, D-0, G-1, C-5, A-6: glass taps and joints; W: waste liquid. (From R.D. Kirchhoefer. *J. Assoc. Off. Anal. Chem.* 1984, 67, 677.[75])

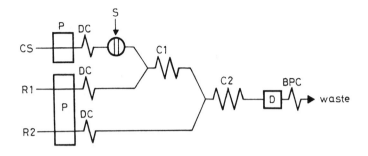

FIGURE 9.5. FIA system for the analysis of substances containing N-nitroso group. For notations see text. (From T. Ohta; N. Goto; S. Takitani. *Analyst* 1988, *113*, 1333.[117])

The N-nitroso derivative is denitrosylated with a 2.5% glacial acetic acid–hydrogen bromide solution, the resulting nitrosyl bromide converts the sulphanylamide component of the Griess reagent into a diazonium derivative, and this is coupled with naphthyl-ethylenediamine (see p. 177) to form a colored product . The scheme of the instrument is shown in Figure 9.5. Upper pump, P, feeds the solvent (ethyl acetate) at a rate of 0.7 ml/min. A 200 μl sample of the N-nitroso derivative to be analyzed is injected into this solvent flow through injector S. The lower pump feeds the denitrosylation reagent (R_1) and the Griess reagent (R_2) at rates of 0.3 ml/min each. Denitrosylation takes place in reaction coil C_1 of length 2 m, diazotization and coupling in coil C_2 of length 6 m. The volume of the flow cell of the spectrophotometric detector is 8 μl, absorbances are measured at a wavelength of 550 nm. The mixture leaves the system through a release coil (BCP). Coils denoted by DC promote the smooth, continuous flow of solutions.

FIA technique has the following advantages. The instrument is simple, easy to reassemble for a different analytical problem. The method is very sensitive (e.g., 5×10^{-8} mol/l in the above example), and the sample throughput is very high, reaching even 200 to 300 samples/h. Naturally, the quantitative conversion of the reaction cannot be ensured with this technique, either. However, this is not an absolute condition, and the tendency is, instead of insisting on this condition, to require the reproducibility of the signal, which can be done by the strict adherence to the experimental parameters. For example, in connection with the determination of the active ingredient of paracetamol tablets by the iron (III)-2,4,6-tripyridyl-s-triazine method, it has been shown by Koupparis et al.[86] that the redox reaction proceeds with a conversion of some percent only, the method is still very precise (relative standard deviation less than 0.4%).

Although concentrations are determined usually on the basis of peak height, calculation on the basis of peak area is also possible. The signal vs. concentration calibration curve is usually linear, but following an appropriate calibration procedure and a parallel measurement of a large number of standard solutions, non-linear calibration curves may also be used for analysis. Figure 9.6 shows the peaks obtained in the determination of the calibration curve prepared for the assay of promethazine.[84]

Of the several applications of the FIA method in pharmaceutical analysis, the determination of 4-amino-antipyrine upon oxidative coupling with phenol,[38] the determination of corticosteroids in drug preparations by the tetrazolium blue reaction,[90] the determination of isonicotinic acid hydrazide in tablets with the use of metavanadate reagent,[32] the determination of sulphonamides in tablets and biological samples after diazotization and coupling with N-(1-naphthyl)ethylenediamine,[83] the measurement of salicylates in tablets and biological samples on the basis of complex formation with iron (III) ions,[82] the determination of phenothiazine drugs by means of oxidation with iron (III) perchlorate reagent,[84] the determination of pharmaceuticals of chloride ion content with mercury (II) thiocyanate-iron (III) nitrate reagent,[87] the determination of bromide ions in pharmaceutical preparations and sera with chloramine T–phenol reagent,[7] the determination of pharmaceuticals containing primary aromatic amino groups by oxidative coupling with 4-N-methylaminophenol,[155] the assay of 3- and 4-substituted phenols after an oxidative coupling with 1-nitroso-2-naphthol,[41] the analysis of pharmaceuticals containing carboxy groups after a condensation with 2-nitrophenyl-hydrazine,[144] the determination of promethazine after oxidation with cerium (IV) to form a red

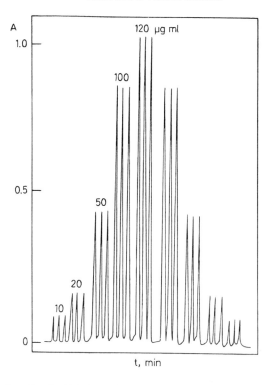

FIGURE 9.6. Curves taken for calibrating FIA: increasing and decreasing concentrations of promethazine after oxidation with iron (III) perchlorate. (From M. A. Koupparis; A. Barchuchová. *Analyst* 1986, *111*, 313.[84])

dye,[102] etc., are mentioned. An interesting feature of another version of the latter method is the use of a solid-phase reactor containing cerium (IV) arsenite as oxidant.[103]

Ion pair formation can also be utilized in the FIA of drugs. The necessity of an extraction step is eliminated by adding a surfactant to the solution, for example in the case of chlorhexidine with bromocresol green.[101] Another possibility is to allow the ion pair to precipitate and to perform turbidimetric measurement (as in the determination of amitriptyline with bromocresol purple[104] and that of diphenhydramine[106] and promethazine[105] with bromophenol blue).

In addition to the methods discussed so far, based on chemical reactions, the FIA method (like the air bubble segmented analyzers) can, of course, also be used for detection and measurement on the basis of natural absorption. In contrast to the manual method, an advantage of this technique is higher speed, and thus this method was also used for the solution of simpler problems, like the measurement of the active principle content of azintamide[4] and furosemide[3] tablets and of 9-aminoacridine preparations.[107]

When a diode array detector (see Section 2.F) is used as the UV spectrophotometric detector of FIA instruments, the high data acquisition rate permits the assay of 2-, 3- or even 4-component mixtures.[14] The quantitative evaluation of overlapping spectra can be performed by computer methods and/or by means of derivative spectrophotometry.

FIA instruments have the valuable feature that they can be connected in series with other detectors. For example, in the analysis of lovastatin tablets containing butylhydroxyanisole as antioxidant performed by Mazzo et al.,[108] the active principle was measured spectrophotometrically at 254 nm, and the antioxidant (with a negligible light absorption at this wavelength under the given concentration conditions) amperometrically by a detector connected in series after the UV detector.

The FIA method also permits a continuous liquid–liquid or membrane extraction of the spectrophotometrically determined analyte to be performed, which has been utilized, for example, in the study of drug–protein interactions[98] and in the automation of a number of analytical methods based primarily on ion pair formation.[11,37,69,70,114,115]

Finally, it is noted that the FIA technique is specially applicable, in addition to a number of other fields, for the automatic determination of the dissolution profiles of tablets. As a result of the direct sampling from the dissolution medium and the very fast analysis of the samples, it is very easy to obtain dissolution curves (see Figure 9.3 in Section 9.C), which could be acquired otherwise, by manual methods, only by a lengthy and tedious procedure. In this respect the papers of Koupparis et al.[7,82–86] are to be noted on the dissolution profiles of several tablet preparations determined by the FIA method.

REFERENCES

1. Abdalla, M. A.; Fogg, A. G.; Barber, C.; Burgess, C. *Analyst* 1983, *108*, 53.
2. Abdel Hamid, M. E.; Abuirjeie, M. A. *Analyst* 1987, *112*, 895.
3. Abdel Kader, S. A.; Abdel Moety, E. M.; Moustafa, A. A.; El Gendy, A. *Arch. Farm. Chemi. Sci. Ed.* 1986, *14*, 113.
4. Abdel Moety, E. M.; Moustafa, A. A.; Ahmad, A. K. S.; El Gendy, A. *Sci. Farm.* 1987, *55*, 259.
5. Abd El-Salam, N. A.; El-Sayed, M. A.; Khafagy, S. M. *Planta Med.* 1975, *30*, 402.
6. Abdullabekova, V. N.; Sokolov, I. V.; Sabirov, K. A.; Gunar, V. I. *Khim.-Farm. Zh.* 1989, *23*, 1367.
7. Anagnostopoulou, P. I.; Koupparis, M. A. *Anal. Chem.* 1986, *58*, 322.
8. Baccou, J. C.; Lambert, F.; Sauvaire, Y. *Analyst* 1977, *102*, 458.
9. Bailey, D. N. *Clin. Chem.* 1982, *28*, 187.
10. Bayer, J. *Acta Pharm. Hung.* 1972, *42*, 249.
11. Bergamin, F. H.; Madeiros, J. X.; Reis, B. F.; Zagatto, E. A. G. *Anal. Chim. Acta* 1978, *101*, 9.
12. Bergqvist, Y.; Eckerbom, S.; Gerhardt, W.; Rombo, L. *Therap. Drug. Monit.* 1990, *12*, 366.
13. Beyer, K. H.; Klinge, K. *Arzneim.-Forsch.* 1969, *19*, 1759.
14. Blanco, M.; Gené, J.; Iturriaga, H.; Maspoch, S. *Analyst* 1987, *112*, 619.
15. Bongiovanni, G.; Giani, C.; Innocenti, F.; Maccari, M.; Minet, E.; Pogliano, L. *Boll. Chim. Farm.* 1984, *123*, 14.
16. *British Pharmacopoeia 1988*, Her Majesty's Stationery Office: London, 1988, 671–672.
17. Broughton, P. M. G. *Biochem. J.* 1956, *63*, 207.
18. Buhl, F.; Dul Zarychta, E.; Chwistek, M. *Acta Pol. Pharm.* 1988, *45*, 369.
19. Bystrova, L. V.; Polisar, R. D.; Shatrova, V. M.; Grigoreva, V. M. *Farmatsiya* 1985, *34*, 65.
20. Carlucci, G.; Maurizi, G.; Mazzeo, P.; Varrassi, G. *Farmaco, Ed. Prat.* 1988, *43*, 297.
21. Chafetz, L. *J. Pharm. Sci.* 1964, *53*, 1162.
22. Chafetz, L. *J. Pharm. Sci.* 1971, *60*, 335.
23. Chafetz, L.; Gaglia, C. A. *J. Pharm. Sci.* 1967, *56*, 1681.
24. Chamberlain, J. *Analysis of Drugs in Biological Fluids*, CRC Press: Boca Raton, 1985.
25. Comer, J. P.; Hartsaw, P. E. *J. Pharm. Sci.* 1965, *54*, 524.
26. Coutts, A. G. *Anal. Proc.* 1985, *22*, 111.
27. Daly, R. E.; Moran, C.; Chafetz, L. *J. Pharm. Sci.* 1972, *61*, 927.
28. Dingeon, B.; Charvin, M. A.; Quenard, M. T.; Thome, H. *Clin. Chem.* 1988, *34*, 1119.
29. Edinboro, L. E.; Jackson, G. F.; Jortani, S. A.; Poklis, A. *Clin. Toxicol.* 1991, *29*, 241.
30. Edwardson, P. A. D.; Nichols, J. D.; Sudgen, K. *J. Pharm. Biomed. Anal.* 1989, *7*, 287.
31. El Sebakhi, N. A.; El Din, A. A. S.; Korany, M. A. *J. Pharm. Belg.* 1986, *41*, 222.
32. Eswara Dutt, V. V. S.; Mottola, H. A. *Anal. Chem.* 1977, *49*, 776.
33. Farid, N. A.; Born, Q. S.; Kessler, W. V.; Shaw, S. M.; Lange, W. E. *Clin. Chem.* 1975, *21*, 1167.
34. Fell, A. F.; Jarvie, D. R.; Stewart, M. J. *Clin. Chem.* 1981, *27*, 286.
35. Fellenberg, A. J.; Pollard, A. C. *Clin. Chim. Acta* 1976, *69*, 423.
36. Fekthullina, G. A.; Bullenkov, T. I. *Farmatsiya* 1984, *33*, 42.
37. Fossey, L.; Cantwell, F. F. *Anal. Chem.* 1982, *54*, 1693.
38. Fraenkl, M.; Svobodová, D.; Karlicek, R. *Ceskoslov. Farm.* 1985, *34*, 168.
39. Frank, M. J.; Johnson, J. B.; Rubin, S. H. *J. Pharm. Sci.* 1976, *65*, 44.
40. Gaudy, D.; Puech, A. *Ann. Pharm. Franc.* 1986, *44*, 49.
41. Georgiou, C. A.; Koupparis, M. A. *Analyst* 1990, *115*, 309.
42. Glasl, H. *Deutsch. Apoth. Ztg.* 1983, *123*, 1979.
43. Glasl, H. *Z. Anal. Chem.* 1985, *321*, 325.
44. Glynn, J. P.; Kendall, S. E. *Lancet* 1975, 1147.
45. Gootz, T. D.; Subashi, T. A.; Lindner, D. L. *Antimicrob. Agents Chemother.* 1988, *32*, 159.
46. Görög, K.; Görög, S. *J. Pharm. Biomed. Anal.* 1988, *6*, 109.

47. Görög, S. *J. Pharm. Sci.* 1968, *57*, 1737.
48. Görög, S.; Rényei, M.; Herényi, B. *J. Pharm. Biomed. Anal.* 1989, *7*, 1527.
49. Görög, S.; Rényei, M.; Laukó, A. *J. Pharm. Biomed. Anal.* 1983, *1*, 39.
50. Görög, S.; Szász, Gy. *Analysis of Steroid Hormone Drugs,* Elsevier: Amsterdam, 1978, 391–392.
51. Grishina, M. S.; Kovalenko, L. I.; Popov, M. D.; Samylina, I. A. *Khim.-Farm. Zh.* 1985, *19*, 1102.
52. Gupta, R. C.; Lundberg, G. D. *Anal. Chem.* 1973, *45*, 2403.
53. Gupta, R. N.; Pickersgill, R.; Stefanec, M. *Clin. Biochem.* 1983, *16*, 220.
54. Gupta, S. K.; Agarwal, A. K.; de Castro, A. F. *Clin. Chem.* 1988, *34*, 1267.
55. Guven, K. C.; Altinkurt, T.; Gulhan, S. *Aczacilik Bul.* 1977, *19*, 42.
56. Habib, A.-A. M.; El-Sebakhy, N. A. *J. Nat. Prod.* 1980, *43*, 452.
57. Hammond, P. M.; Miller, J.; Price, C. P.; Morris, H. C. *Lancet II,* 1987, 449.
58. Háznagy, A. *Acta Pharm. Hung.* 1977, *47*, 249.
59. Higuchi, T.; Brochmann-Hanssen, E., *Pharmaceutical Analysis* Interscience Publ.: New York, London, 1961. a. 427–428; b. 435–438.
60. Horváth, P. *Acta Pharm. Hung.* 1982, *52*, 133.
61. Horváth, P.; Iványi, G. *Acta Pharm. Hung.* 1982, *52*, 150.
62. Ibrahim, S. M.; Kadry, H. A.; El-Olemy, M. M. *J. Nat. Prod.* 1979, *42*, 366.
63. Idowu, O. R. *J. Pharm. Biomed. Anal.* 1989, *7*, 893.
64. Ivanov, V. D.; Georgievskii, V. P.; Grizodub, A. I.; Komissarenko, N. F.; Ldygina, E. Ya. *Farmatsiya* 1984, *33*, 30.
65. Jakovljevic, I. M.; Hartsaw, P. E.; Drummond, G. E. *J. Pharm. Sci.* 1965, *54*, 1771.
66. Jaenchen, E.; Levy, G. *Clin. Chem.* 1972, *18*, 984.
67. Jenny, R. W. *Clin. Chem.* 1985, *31*, 1158.
68. Kalugina, Z. G.; Vikhareva, E. B.; Saveleva, G. I. *Farmatsiya* 1988, *37*, 82.
69. Karlberg, B.; Johansson, P. A.; Thelander, S. *Anal. Chim. Acta* 1979, *104*, 21.
70. Karlberg, B.; Thelander, S. *Anal. Chim. Acta* 1978, *98*, 1.
71. Karawya, M. S.; Abdel-Wahab, S. M.; Hifnawy, M. S.; Ghourab, M. G. *J. Assoc. Off. Anal. Chem.* 1975, *58*, 854.
72. Karawya, M. S.; Hifnawy, M. S.; El-Hawary, S. S. *J. Assoc. Off. Anal. Chem.* 1979, *62*, 250.
73. Kenig, M. D. *Analyst* 1988, *113*, 761.
74. Khafagy, S. M.; Abdel Salam, M. A.; El-Ghazooly, M. G. *Planta Med.* 1976, *30*, 21.
75. Kirchhoefer, R. D. *J. Assoc. Off. Anal. Chem.* 1984, *67*, 677.
76. Kiseleva, T. L.; Samylina, I. A. *Farmatsiya* 1987, *36*, 30.
77. Kitamura, K.; Mabuchi, M.; Fukui, T.; Hozumi, K. *Chem. Pharm. Bull.* 1987, *35*, 3914.
78. Kolypova, I. E.; Masalova, G. A.; Perelson, M. E. *Khim.-Farm. Zh.* 1980, *14*, 58.
79. Korany, M. A.; Bedair, M.; Mahgoub, H.; Elsayed, M. A. *J. Assoc. Off. Anal. Chem.* 1986, *69*, 608.
80. Korany, M. A.; El-Din, A. A. S. *J. Assoc. Off Anal. Chem.* 1984, *67*, 138.
81. Kostennikova, Z. P. *Farmatsiya* 1986, *35*, 68.
82. Koupparis, M. A.; Anagnostopoulou, P. I. *J. Pharm. Biomed. Anal.* 1988, *6*, 35.
83. Koupparis, M. A.; Anagnostopoulou, P. I. *Anal. Chim. Acta* 1988, *204*, 271.
84. Koupparis, M. A.; Barcuchová, A. *Analyst* 1986, *111*, 311.
85. Koupparis, M. A.; Macheras, P.; Reppas, C. *Int. J. Pharm.*
86. Koupparis, M. A.; Macheras, P.; Tsaprounis, C. *Int. J. Pharm.* 1985, *27*, 349.
87. Koupparis, M. A.; Sarantonis, E. G. *J. Pharm. Sci.* 1986, *75*, 800.
88. Kubin, H.; Brehm, M.; Ulmen, J. *Pharm. Ind.* 1982, *44*, 1269.
89. Kudrin, A. N.; Maidykov, A. A.; Akimov, P. P.; Galushkina, L. R. *Farmatsiya* 1986, *5*, 55.
90. Landis, J. B. *Anal. Chim. Acta* 1980, *114*, 155.
91. Levy, G.; Procknal, J. A. *J. Pharm. Sci.* 1968, *57*, 1330.
92. Lim, C.-F.; Wynne, K. N.; Barned, J. M.; Topliss, J. D.; Stockigt, J. R. *J. Pharm. Pharmacol.* 1986, *38*, 795.
93. Liu, T. Z.; Afshari, J. T. *Clin. Chem.* 1988, *34*, 1267.
94. Liu, T. Z.; Oka, K. H. *Clin. Chem.* 1980, *26*, 69.
95. Longenecker, R. W.; Trafton, J. E.; Edwards, R. B. *Clin. Chem.* 1984, *30*, 1369.
96. Lukáts, B.; Takácsi Nagy, G. *Acta Pharm. Hung.* 1965, *35*, 116.
97. Luque de Castro, M. D.; Valcárcel, M. *J. Pharm. Biomed. Anal.* 1989, *7*, 1291.
98. Macheras, P.; Koupparis, M. A.; Tsaprounis, C. *Int. J. Pharm.* 1986, *30*, 123.
99. Maghssoudi, R. H.; Fawzi, A. B.; *J. Pharm. Sci.* 1978, *67*, 32.
100. Martinez Calatayud, J. *Pharmazie* 1986, *41*, 92.
101. Martinez Calatayud, J.; Campíns Falcó, P. *Anal. Chim. Acta* 1986, *189*, 323.

102. Martinez Calatayud, J.; Garcia Sancho, T. *J. Pharm. Biomed. Anal.* 1992, *10,* 37.

103. Martinez Calatayud, J.; Garcia Mateo, V. *Anal. Chim. Acta* 1992, *264,* 283.

104. Martinez Calatayud, J.; Martinez Pastor, C. *Anal. Lett.* 1990, *23,* 1371.

105. Martinez Calatayud, J.; Navasquillo Sarion, S.; Sanchez Sampedro, A.; Gomez Benito, C. *Microchem. J.* 1992, *45,* 129.

106. Martinez Calatayud, J.; Sanchez Sampedro, A.; Navasquillo Sarion, S. *Analyst* 1990, *115,* 855.

107. Martinez Calatayud, J.; Sanchez Sampedro, A.; Villar Civera, P.; Gomez Benito, C. *Anal. Lett.* 1990, *23,* 2315.

108. Mazzo, D. J.; Biffar, S. E.; Forbes, K. A.; Bell, C.; Brooks, M. A. *J. Pharm. Biomed. Anal.* 1988, *6,* 271.

109. Morris, H. C.; Hunter, J.; Campbell, R. S.; Hammond, P. M.; Atkinson, T. *Clin. Chem.* 1986, *32,* 1083.

110. Morris, H. C.; Overton, P. D.; Ramsay, J. R.; Campbell, R. S.; Hammond, P. M.; Atkinson, T. *Clin. Chem.* 1990, *36,* 131.

111. Mount, D. L.; Patchen, L. C.; Williams, S. B.; Churchill, F. C. *Bull. WHO* 1987, *65,* 615.

112. Moustafa, M. A. A.; Belal, F.; El Eman, A. A. *J. Pharm. Belg.* 1987, *42,* 53.

113. Naidovich, L. P.; Maslova, G. A.; Bondarenko, L. T.; Perelson, M. E.; Tolkachev, O. N. *Khim.-Farm. Zh.* 1978, *12,* 132.

114. Nord, L.; Karlberg, B. *Anal. Chim. Acta* 1980, *118,* 285.

115. Ogata, K.; Taguchi, K. *Anal. Chem.* 1982, *54,* 2127.

116. O'Haver, T. C. *Clin. Chem.* 1979, *25,* 1548.

117. Ohta, T.; Goto, N.; Takitani, S. *Analyst* 1988, *113,* 1333.

118. Opperman, E.; Johnson, G. F. *Clin. Chem.* 1988, *34,* 1269.

119. Petrishek, I. A.; Gaevsky, A. V.; Lovkova, M. Y.; Golovkin, A. B.; Grinkevich, N. I. *Khim.-Farm. Zh.* 1986, *20,* 710.

120. Poulou, M.; Macheras, P. *Int. J. Pharm.* 1986, *34,* 29.

121. Quaglia, M. G.; Carlucci, G.; Cavicchio, G.; Mazzeo, P. *J. Pharm. Biomed. Anal.* 1988, *6,* 421.

122. Regosz, A.; Dabrowska, D.; Laman, M. *Sci. Pharm.* 1986, *54,* 23.

123. Rehm, C. R.; Smith, J. B. *J. Amer. Pharm. Ass. Sci. Ed.* 1960, *49,* 386.

124. Reisner, E.; Nikolics, K. *Acta Pharm. Hung.* 1982, *52,* 179.

125. Ruzicka, J.; Hansen, E. H. *Flow Injection Analysis,* Wiley: New York, 1982.

126. Sakai, T.; Ohno, N. *Analyst* 1987, *112,* 149.

127. Samyn, W. *Clin. Chim. Acta* 1983, *133,* 111.

128. Shawl, A. S.; Vishwapaul *Analyst* 1977, *102,* 779.

129. Shibabi, Z. K.; David, R. M. *Ther. Drug. Monit.* 1984, *6,* 449.

130. Shibabi, Z. K.; Wasilauskas, B. L.; Peacock, J. E. *Ther. Drug. Monit.* 1988, *10,* 486.

131. Sitar, D. S.; Graham, D. N.; Rangno, R. E.; Dusfrense, L. R.; Ogilvie, R. I. *Clin. Chem.* 1976, *22,* 379.

132. Sivasami, K.; Rajeswari, S. *J. Ind. Chem. Soc.* 1987, *64,* 176.

133. Skaltsa, H. D.; Izakou, O. A.; Koupparis, M. A.; Philianos, S. M. *Anal. Lett.* 1987, *20,* 1679.

134. Sokolov, A. V.; Popov, D. M. *Farmatsiya* 1989, *38,* 62.

135. Steinegger, E.; Brunner, P. *Pharm. Acta Helv.* 1977, *52,* 139.

136. Stewart, J. T.; Settle, D. A. *J. Pharm. Sci.* 1975, *64,* 1403.

137. Stewenson, G. W. *Anal. Chem.* 1960, *32,* 1522.

138. Svensson, C. K.; Edwards, D. J.; Lalka, D. *J. Pharm. Sci.* 1982, *71,* 1305.

139. Swanson, M. B.; Walters, M. I. *Clin. Chem.* 1982, 1171.

140. Szilajtsis-Obieglo, R.; Kowalewska, K. *Herba Pol.* 1984, *30,* 27.

141. Tai, B.; Geng, C.; Li, J. *Yaowu Fenzi Zazhi* 1983, *3,* 234.

142. Tai Kwong; Adams, N.; Young, N. *Clin Biochem.* 1984, *17,* 170.

143. Takacs, F.; Hippmann, D. *Arzneim.-Forasch.* 1983, *33,* 5.

144. Takeuchi, T.; Kabasawa, I.; Horikawa, R.; Tanimura, T. *Analyst* 1988, *113,* 1673.

145. Tawada, J. C.; Midio, A. F. *Rev. Farm. Bioquim. Univ. Sao Paolo* 1989, *25,* 177.

146. *The United States Pharmacopoeia XXI*, USP Convention Inc.: Rockville, 1985. a. 157; b. 1146; c. 389; d. 245; e. 400; f. 293; g. 405; h. 1145–1159.

147. *The United States Pharmacopoeia XXII*, USP Convention Inc.: Rockville, 1990. Page numbers: a. 554; b. 190; c. 286; d. 85; e. 294; f. 154V g. 176; h. 958; i. 547; j. 113; k. 178; l. 292; m. 1150; n. 363; o. 286; p. 290, 1544; r. 96; s. 514; t. 391, 1532; u. 819; v. 916; w. 666; x. 605; y. 1331; z. 951; aa. 785; bb. 29; cc. 876; dd. 571; ee. 684; ff. 951; gg. 149; hh. 327; ii. 1300; jj. 1579; kk. 235; ll. 1473–1477.

148. Trinder, P. *Biochem. J.* 1954, *57,* 301.

149. Valcárcel, M.; Luque de Castro, M. D. *Flow Injection Analysis: Principles and Applications*, Ellis Horwood: Chichester, 1987.

150. Valdeon, J. L.; Escribano, M. T. S.; Hernandez, L. H. *Analyst* 1987, *112*, 1365.

151. Van de Waart, F. J.; Hulshoff, A.; Indemans, A. W. M. *Pharm. Weekblad* 1980, *115*, 1429

152. Van de Waart, F. J.; Hulshoff, A.; Indemans, A. W. M. *J. Pharm. Biomed. Anal.* 1983, *1*, 507.

153. van Opstal, M. A. J.; Blauw, J. S.; Holthuis, J. J. M.; van Bennekom, W. P.; Bult, A. *Anal. Chim. Acta* 1987, *202*, 35.

154. van Opstal, M. A. J.; Blauw, J. S.; Holthuis, J. J. M.; Van Bennekom, W. P.; Bult, A. *Pharm. Weekblad Sci. Ed.* 1987, *9*, 145.

155. Verma, K. K.; Stewart, K. K. *Anal. Chim. Acta* 1988, *214*, 207.

156. Verzár-Petri, G.; Cuong, B. N. *Acta Pharm. Hung.* 1977, *47*, 134.

157. Verzár-Petri, G.; Coung, B. N. *Sci. Pharm.* 1977, *45*, 25.

158. Wahbi, A. M.; Abounassif, M. A.; El Obeid, H. A.; Al Julani, A. M. *Arch. Pharm. Chemi. Sci. Ed.* 1986, *14*, 69.

159. Wahbi, A. M.; Abu-Shady, H.; Soliman, A. *Planta Med.* 1976, *30*, 269.

160. Walberg, C. B. *J. Anal. Toxicol.* 1977, *1*, 79.

161. Walberg, C. B. *Clin. Toxicol.* 1981, *18*, 879.

162. Williams, L. A.; Linn, R. A.; Zak, J. B. *J. Lab. Clin. Med.* 1959, *53*, 156.

163. Wojcik, E. *Farm. Pol.* 1984, *40*, 523.

164. You, K.; Bittikofer, J. A. *Clin. Chem.* 1984, *30*, 1549.

165. Youssef, A. F.; El Shabouri, S. R.; Mohamed, F. A.; Rageh, A. M. I. *J. Assoc. Off. Anal. Chem.* 1986, *69*, 513.

166. Zwillenberg, L. O.; Bösiger, G.; Zwillenberg, H. H. L. *Meth. Find. Exptl. Clin. Pharmacol.* 1982, *4*, 575.

Chapter 10

SPECTROPHOTOMETRIC ANALYSIS OF SOME IMPORTANT GROUPS OF DRUG COMPOUNDS

A. INTRODUCTION

In this chapter the practical aspects of the spectrophotometric analysis of drug compounds are briefly summarized based on the general aspects described in Chapters 1–9. Subjectivity could not be entirely excluded when the material of the chapter was arranged under 16 subtitles. In general it was our aim to bring together those compounds in one subchapter the chemistry of which can be treated in a uniform way; in some cases, however, exceptions had to be made (e.g., in the case of vitamins, antibiotics). In these instances the pharmacological effect of the drug substances was the basis of the classification. It has to be stressed that the character of this book, its scope and limitations precluded an exhaustive survey: several minor groups of drugs have been omitted and even inside a given group, no attempt has been made to discuss the spectrophotometric analysis of all drug substances belonging to the group.

The above-mentioned scope and limitation have forced us to make serious omissions in discussing the spectrophotometric analysis of a given group of drug substances, not only from the point of view of the discussed compounds but also regarding the methods used for the analysis, too.

The References of the 16 sections of Chapter 10 include about 1500 literature data. This figure represents a small portion only of the overall number of publications published on the spectrophotometric analysis of drugs from the beginning up to the present time. The modest aim in selecting the above mentioned 1500 papers for review was to give an overview to the reader about the possibilities of spectrophotometry in the analysis of the given drug compounds emphasizing those possibilities which can be regarded as more or less up-to-date even at the present time. Excellent monographs are available containing thorough summaries of the old literature of the spectrophotometric analysis of drugs (e.g., the books of Kakac and Vejdelek, Pesez and Bartos, Snell and Snell, and others which can be found in the References section at the end of Chapter 1 (11, 12, 17, 22, 28)). The spectrophotometric chapters of other, general books on drug analysis also contain useful information on the possibilities of spectrophotometry. Although the majority of the methods described there can be regarded as being obsolete, in some cases good ideas can be obtained from these sources to solve current problems. Because of the above reasons, references are only seldom made in this book to the literature covered by the above listed books: the majority of the references here are from the last 15–20 years, primarily from the last few years. This self-limitation, however, brought about various problems. Namely, in the last twenty years the publication of papers dealing with the spectrophotometric analysis of drugs has shifted from the developed countries to the least developed ones. This means that the majority of such papers are nowadays published in less accessible East European, Indian, Chinese, etc., journals. In order to facilitate the readers' orientation, the accessibility of the cited literature was considered as an important factor in the selection of the references. However, the proportion of the papers published in not widely known languages and in less widespread journals is rather high in the literature of the recent years and hence despite all efforts, many of such papers are referred to in the subsequent subchapters.

Another problem is that studying the old literature reveals that even the majority of the methods described in the past 15–20 years have their ancestors, and the differences between the "old" and "new" versions of the methods are often not too great. Since the above discussed limitations of this book precluded the presentation of the history of the development of even the most important methods, the reader can get a full picture about the possibilities and limitations of the described methods if the above listed monographs and the literature cited therein are also studied.

An important factor in the selection of the methods presented in this book i.e., the judgment whether a method can be considered to be up-to-date was that the methods which are official in the present pharmacopoeias (especially in the British and U.S. Pharmacopoeias) have been emphasized.

Even with these limitations the selection of the papers referred to in this book was made difficult as a consequence of the "confusion of abundance." For this reason further limitations have to be applied:

1. With some exceptions only methods for which the chemical background is known have been selected for presentation, i.e., which are based either on natural absorption or on well defined, stoichiometric chemical reactions.
2. Only quantitative analytical methods are discussed: color reactions serving primarily for the identification of drug substances are not discussed.
3. Although the scope of this book includes the determination of drugs in human, animal and plant samples (see 9.H and 9.I chapters), the main emphasis is still the investigation of bulk drugs and pharmaceutical formulations, and for this reason the proportion of the methods for the analysis of biological samples is less than it is in the literature itself.
4. The spectrophotometric methods associated with chromatographic separation have been discussed in Chapter 5. For this reason thin-layer densitometric methods are not mentioned in this chapter at all and spectrophotometric methods after column or thin-layer chromatographic separation are only included if the spectrophotometric method has interesting features.
5. No attempts have been made, even in the case of the most important methods, to present detailed procedures: for this, the reader is referred to the original publications cited.

The treatment of the spectrophotometric analysis of the individual groups of drugs in the subchapters of this chapter will be the following. At first, methods based on natural absorption are discussed based on and with reference to spectrum–structure correlations discussed in Chapter 3. This is followed by the description of various algebraic and derivative spectrophotometric methods, also based on natural absorption (see Chapter 4). In the second parts of the subchapters, methods based on chemical reactions are presented. Since the fundamentals of the chemical basis of the most important methods are discussed in Chapter 7, it is advisable to go back to this part of the book, too, when a solution is sought for a special problem.

B. PHENYLALKYLAMINES AND OXYPHENYLALKYLAMINES (CATECHOLAMINES)

1. Methods Based on Natural Absorption

In this subchapter the spectrophotometric determination of those compounds is discussed in which the only chromophoric group of the molecule is the phenyl group, which is separated by a carbon chain (usually two carbon atoms) from the amino group. The latter can be primary (amphetamine), secondary (ephedrine, adrenaline) or tertiary amino group (diphenhydramine) or heterocyclic tertiary amino group (clemastine). The phenyl group is usually unsubstituted but the phenol-type 3-hydroxy (phenylephrine), 4-hydroxy (bamethan), 3,4-dihydroxy (adrenaline, noradrenaline, isoprenaline), 3,5-dihydroxy (terbutaline) and 1,4-dihydroxy ether (pramoxine) derivatives are also discussed here. Derivatives containing a hydroxyl group in the α-position of the chain (ephedrine, isoprenaline) and those containing the aromatic ring(s) as the benzhydryl ether (diphenhydramine, clemastine) deserve special mention.

Of the above listed types of derivatives, mainly those containing phenolic hydroxyl groups possess natural absorption suitable for their direct spectrophotometric measurement.

For example, the following methods are official in USP XXII for the direct spectrophotometric determination of these phenol-type derivatives. The assay of isoproterenol tablet is carried out at 279 nm after extraction.[88h] Pramoxine hydrochloride is determined in jelly preparation after extraction with acid, washing the extract with ether, basification of the aqueous phase, extraction of the free base with ether, re-extraction into acidic water and absorbance measurement of the latter at 286 nm.[88r] The occasionally occurring background problems are corrected by means of the baseline method (see 4.C.1) such as e.g., in the case of phenylephrine injection,[88p] or the absorbance is measured after chromatographic separation as it is the case of isoxsuprine hydrochloride injection (measurement at 275 nm).[88l]

The possibilities for the direct determination of derivatives containing unsubstituted phenyl groups only are, of course, limited as a consequence of their low molar absorptivities ($\varepsilon < 1000$). In these cases any background problems are much more serious. In spite of this, such methods still are in use especially if the molecule contains two phenyl groups. For example, the assay of propoxyphene capsules and cyclizine tablets is carried out at 257 and 264 nm, respectively.[88e,s] The assay of ketamine injection by absorbance measurement at 269 nm after extraction[88j] and the measurement of phenmetrazine in tablets after extraction from alkaline solution and re-extraction into aqueous-acidic solution followed by absorbance measurement at 256 nm are further examples.[88o] The sensitivity of the measurement can be greatly increased if the measurement is based on the much more intense p band. Such assays, however, can only be considered reliable if the wavelength of the measurement is above 220 nm. A method of this type is the content uniformity measurement of bromdiphenhydramine tablets at 228 nm.[88c]

Of the methods in BP 1988 the determination of noradrenaline in injections after dilution and measurement at 279 nm[10a] and the assay of cyclizine tablets and injections are mentioned after ethereal extraction and re-extraction into aqueous sulphuric acid and measurement at 225 nm.[10b]

The frequently occurring background problems are overcome by using column chromatographic separation such as in the case of the assay of amphetamine sulphate bulk material and tablets.[88a] Algebraic background correction methods (in some instances combined with chromatographic separation) are also often used. For example, of the methods described in Section 4.C.2, the dual-wavelength measurement is used for the assay of amphetamine and maprotiline hydrochloride tablets: the calculation is based on ($A_{257} - A_{280}$) and ($A_{273} - A_{300}$), respectively.[88a,k]

Measurement at 5 wavelengths and the use of orthogonal polynomials enable more difficult problems to be solved. For example Korany et al.[45] used this method not only for solving background problems but for the simultaneous determination of ephedrine and diphenhydramine in formulations, too.

Derivative spectrophotometry is an eminently suitable method for the determination of benzene derivatives discussed in this chapter. Although the intensities of the zero order spectra are rather low, the fine structure of the spectra and the abundance of very steep sections between the side maxima make this technique an excellent tool for the sensitive and selective determination of these derivatives. As is described in detail in Chapter 4.E, scanning higher order derivative spectra enables even complicated background spectra to be eliminated. This means that in this case a large variety of problems can be solved in this field: the assays can be carried out without preliminary separation of the analytes. The assay of formulations with isolated phenyl chromophores is a most important field of application of derivative spectroscopy in pharmaceutical analysis.

The most thorough publications in this field are those of Davidson and Hassan[18] and Abdel-Khalek et al.[1] who described, among others, the determination of trihexyphenidyl, chlorcyclizine, cyclizine, debrisoquine, diphenhydramine, methadone, orphenadrine and tolazoline in tablets and capsules using second derivative spectroscopy. The second derivative spectra were used also by Milch and Szabó[50] for the assay of prenylamine and fendiline tablets, by Fell et al.[37] for diphenhydramine syrup and by Bedair et al.[7] for the determination of clemastine fumarate in tablets, injections and syrup formulations.

Derivative spectrophotometry is not only suitable for background problems in the field of phenylalkylamines. The "fingerprint-like" second derivative spectra are much more suitable for the characterization and identification of the individual structurally closely related derivatives than the zero-order spectra. This was sucessfully used by Lawrence and MacNeil[47] for the distinction of amphetamine, phenylethylamine, phentermine, ephedrine and meperidine. These differences in the higher order derivative spectra can be utilized also for the simultaneous determination of these

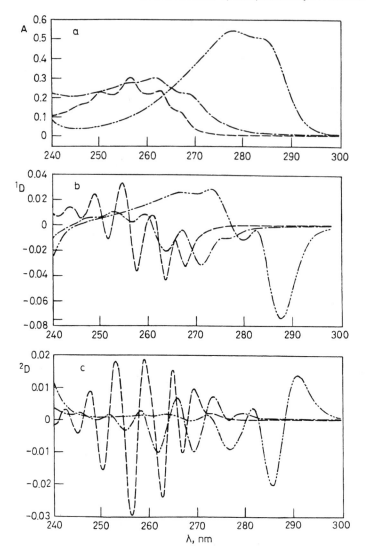

FIGURE 10.1. Spectra for the simultaneous determination of ψ-ephedrine hydrochloride (0.3 mg/ml), chlorpheniramine maleate (0.02 mg/ml) and dextromethorphan hydrobromide (0.1 mg/ml). Solvent: acetate buffer; pH 5. ψ-Ephedrine hydrochloride – –; chlorpheniramine maleate – · –; dextromethorphan hydrobromide – · · –; a: Basic spectra; b: 1st derivative spectra; c: 2nd derivative spectra. (From Murtha, J. L.; Julian, T. N.; Radebaugh, G. W. *J. Pharm. Sci.* 1988, 77, 715.[52])

derivatives in combined formulations. To illustrate this the zero-order spectra of a combined formulation containing ψ-ephedrine hydrochloride, chlorpheniramine maleate and dextromethorphan hydrobromide taken from the paper of Murtha et al.[52] are shown in Figure 10.1a.

Figure 10.1b and Figure 10.1c show the first and second order derivative spectra, respectively. As is seen, the zero-order spectra would enable only the selective determination of dextromethorphan hydrobromide. In the first and especially in the second order spectra, sufficient differences can be found which can serve as the basis for the simultaneous determination of all three components using a suitable computer program for multicomponent analysis. For the simultaneous determination of the first two of the above components, first order spectrophotometry was used.[40]

Among the reports on the successful use of second derivative spectra for the determination of the active ingredients in pharmaceutical preparations, the paper of Santoni et al.[73] on the simultaneous determination of diphenhydramine and naphazoline and that of Tan and Kolomonpunpron[84] dealing with the simultaneous determination of phenylpropanolamine and dextromethorphan are worth mentioning. The use of even higher derivative spectra enables even more complicated problems to be solved. For example, on the basis of fourth derivative spectra Davidson and Elsheikh[17] determined among others ψ-ephedrine in syrup in the presence of codeine and tripolidine. The respective spectra were shown as Figure 4.19 in Section 4.E as characteristic examples of the use of derivative spectrophotometry in the analysis of drugs. The fourth derivative spectrophotometry was also used for the determination of tranylcypromine in the presence of trifluoperazine.[46]

The amino group is isolated from the aromatic ring in the molecules of phenylalkylamines and for this reason the possibilities of difference spectrophotometry based on the acidic—basic shifts of the spectra are of limited significance here. It is, however, interesting to note that these minor changes can be greatly amplified by means of difference—derivative spectrophotometry, and this enabled Davidson and Mkoji[19] to utilize this technique for the simultaneous determination of tripolidine, ψ-ephedrine and dextromethorphan.

2. Methods Based on Chemical Reactions

Phenylalkylamines discussed in this chapter are all moderately strong bases which are almost insoluble in water as the free bases but very soluble as their salts which are used in therapy. This creates a good basis for the determination of their ion-pairs with suitable reagents and measurement of the colored derivative in the organic phase (see Section 7.H). Some of these methods are summarized in Table 10.1.

It has to be noted here that phenylalkylamines form insoluble precipitates with one of the classical reagents of alkaloids, the Reinecke-salt. This can be dissolved in acetone and this serves as the basis for the determination of e.g., diphenhydramine at 540 nm.[42]

The importance of the methods based on the formation of metal complexes is much smaller. Of these methods, the determination of adrenaline as the iron (II) complex at 530 nm[88g] or as a complex with the iron (III)–chloride–pyridine reagent at 590 nm[90] are mentioned. Phenylephrine can be determined at 415 nm after the extraction with butanol of its complex formed with the copper (II)–hydroxylamine reagent.[20] Salbutamol and phenylephrine were transformed by nitrous acid to the corresponding o-nitroso derivatives prior to complex formation with copper (II).[13,20] In addition to these methods based on the complex-forming capability of the phenolic hydroxyl groups, the determination of the ester-type meperidine after its transformation to hydroxamic acid and the formation of the iron (III) complex[38] is worth mentioning (see Section 7.G). Of the reagents for the formation of charge-transfer complexes (see Section 7.I) first of all, the use of chloranil was reported for the determination of phenylalkylamines, e.g., for the determination of ephedrine and phenylephrine,[5] adrenaline and noradrenaline[4,96] and maprotiline.[34]

Both ion-pair extraction methods and those which are based on the formation of charge-transfer complexes measure the amino group of phenylalkylamines. If this is a primary amino group, then spectrophotometrically active aldehyde or ketone-type reagents can also be used for their determination which transform the amines to the corresponding Schiff's bases as described in Section 7.D. Examples of this are the determination of amphetamine using salicylaldehyde[49] or ascorbic acid[8] reagents, the determination of noradrenaline after reaction with 4-dimethylaminocin-namaldehyde,[94] vanillin,[61] anisaldehyde or benzaldehyde.[63] In this way, since only primary amines give positive reactions, this method is suitable for the selective determination of noradrenaline in the presence of adrenaline.

The importance of these methods is that they are suitable for the determination of derivatives without phenolic hydroxyl groups, too. Other methods for their determination include the use of 2,4-dinitrofluorobenzene for the assay of phenfluramine, tranylcypromine and baclofen, where the basis of the method is the formation of 2,4-dinitrophenylamino derivative which can be measured at 395 nm,[6,80,89] the determination of phenylpropanolamine after condensation in chloroform with 7-chloro-nitrobenzofurazone (measurement at 455 nm),[81] measurement of diphenhydramine at 334 nm after treatment with a malonic acid-acetic anhydride reagent[43] and the determination of amphetamine and its derivatives at 460 nm using the Marquis reagent (sulphuric acid-formaldehyde).[3]

The reaction of β-amino derivatives containing a hydroxyl group at the α-carbon adjacent to the aromatic ring with periodate ion merits special mention. This reaction leads to benzaldehyde

TABLE 10.1. Ion-pair Extraction Methods for the Determination of Phenylalkylamines

Compound	Formulation	Reagent	pH	Extracting solvent	λ, nm	Ref.
Diphenhydramine	Capsule, injection	Bromocresol green	5	CHCl$_3$	415	48
Diphenhydramine	Capsule, injection, elixir	Dipicrylamine	5	CHCl$_3$	420	75
Diphenhydramine	Tablet, capsule, injection, cream	Methyl orange		1,2-Dichloroethane	422	22
Diphenhydramine	—	Tetrabromophenol-phthalein	8	1,2-Dichloroethane	553, 573	53
Metamphetamine, methylephedrine, ephedrine	—	Tetrabromophenol-phthalein	11	1,2-Dichloroethane	550, 570	58
Ephedrine	—	Bromothymol blue	7.5	CHCl$_3$	425	55
Ephedrine	Syrup	Bromothymol blue	6.2	CHCl$_3$	420	15,16
Ephedrine	—	Picric acid	5	CHCl$_3$	410	82
Procyclidine	Tablet	Food red 4, Acid red 14, Food yellow 3	3	CHCl$_3$	517,520,485	57
Procyclidine	Tablet	Bromocresol purple		CHCl$_3$	405	88m
Verapamil, oxyphedrine	Tablet	Food yellow 3, etc.,	3	CHCl$_3$	482	65
Oxyfedrine	—	Bromocresol green, bromophenol blue, methyl orange, eriochrom black T	dil. HCl	CHCl$_3$	420, 420, 420, 520	64
Meperidine	Tablet, injection	Bromocresol green	3	CHCl$_3$	425	35
Clemastin fumarate	Tablet	Bromocresol purple		CHCl$_3$	406	88d
Clemastin fumarate	Tablet, injection	Eriochrome black T	5	CHCl$_3$	518	83
Loperamide	Capsule	Tropeolin 00		CHCl$_3$	410	87
Tridihexethyl chloride	Tablet, injection	Bromocresol purple		CHCl$_3$	408	88t
Biperiden	Tablet, injection	Bromocresol purple		CHCl$_3$	408	88b
Methylphenidate	Tablet	Picric acid		CHCl$_3$	405	88l
Orphenadrine citrate	Injection	Picric acid		Toluene	410	88n
Pentoxiverine citrate	Tablet, syrup	Tropeolin 000	2.6	CHCl$_3$	485	95
Tilidine	Injection, suppository	Bromocresol green	3.5	CHCl$_3$	415	97

or its derivatives depending on the substituents of the aromatic ring. The spectrophotometric activity of these may serve as the basis for selective analytical methods. This classical reaction (see Equation 7.13 in Section 7.E) was investigated in detail by Chafetz.[11,12] It was found that the reaction takes place if the amino group does not have electron-attracting substituents and the phenyl group does not contain more than one hydroxyl group; (in the case of dihydroxy derivatives the formation of adrenochrome to be discussed later takes place rather than the splitting reaction). This method is eminently suitable for the determination of ephedrine and its derivatives and this method is official in USP XXII.[88f] The selectivity of the reaction enables the determination in the presence of diphenhydramine not containing an α-hydroxy group. The selectivity can be increased by using difference spectrophotometry. For example, phenylpropanolamine can be measured in the presence of guaiphenesine and dextromethorphan[85] or chlorpheniramine and pyrilamine[86] in such a way that the absorbance of the solution treated with periodate is measured against a solution not treated with this reagent. In an interesting version of this method, the oxidation is carried out in a column packed with diatomaceous earth impregnated with sodium periodate solution.

The analytical chemistry of the derivatives containing phenolic hydroxyl group(s) is much more abundant in more or less selective spectrophotometric methods than that of the simple phenyl derivatives. All the important classical colorimetric methods for phenols have been employed for their determination. As the most typical method azo-coupling can be mentioned using diazotized 4-nitroaniline,[44,67,78] sulphanilic acid,[79] benzocaine,[68] etc. as the reagents for the determination of e.g., terbutaline, salbutamol, isoxsuprine. Phosphomolybdotungstic acid (Folin-Ciocalteu reagent) was used for the determination of terbutaline.[74] Terbutaline, salbutamol and orciprenaline were transformed with nitrous acid to the o-nitrosophenol derivatives which can be measured in alkaline media as the yellow phenolates.[29,33,54] The mechanism of the reaction is probably similar when salbutamol,[92] prenalterol[93] and terbutaline[77] were determined using sodium hexanitrito-cobaltate reagent.

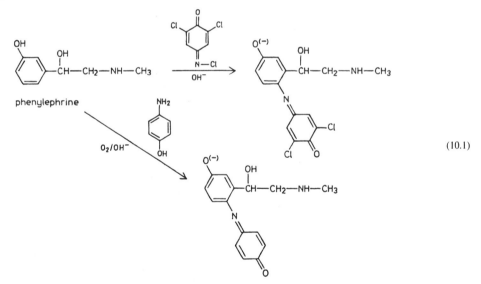

(10.1)

The reaction leading to indophenol derivatives may also serve as the basis for important spectrophotometric methods. As an example the reaction between phenylephrine and dichlorobenzo-quinonechlorimine is shown in Equation 10.1. The phenolate form of the derivative which is present in alkaline media has absorption maximum around 600 nm. The method based on the measurement of this absorption is suitable for the determination of among others phenylephrine,[9] adrenaline, noradrenaline, isosuxprine,[71] oxedrin[66] and salbutamol.[76] If the above mentioned reagent containing active chlorine is used, no oxidizing agent is necessary for the formation of the indophenol derivative. As is also seen in Equation 10.1, the reagent can also be simply 4-aminophenol. In this case, however, an oxidizing agent is also necessary, which in this case is atmospheric oxygen dissolved in the reaction mixture. This method was successfully used for the determination of phenylephrine,[69] salbutamol[70] and terbutaline.[56] As the reagent for the oxidative coupling, 4-aminoantipyrine was also used.[39,66] It is worth mentioning that the latter reaction could be used for the determination of derivatives not containing phenolic hydroxyl group, e.g., ephedrine.[39] In this case the phenolic group has to be introduced into the molecule by treating it with hydrogen peroxide and using iron (III) as the catalyst.

The oxidizability of diphenols offers excellent possibilities for the selective determination of 3,4-diphenols (adrenaline, noradrenaline, isoprenaline). These catecholamines are oxidized in neutral or slightly acidic media by various oxidizing agents to form, in the first step of the sequence of reactions, orthoquinones and these are transformed subsequently according to a complicated oxidation-ring closure-rearrangement mechanism to red adrenochrome derivatives which can easily be measured spectrophotometrically (see Equation 10.2).

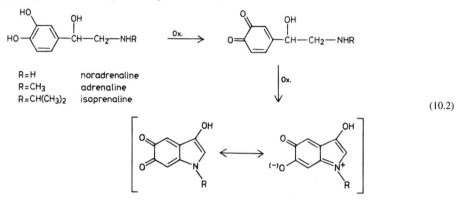

R=H   noradrenaline
R=CH₃   adrenaline
R=CH(CH₃)₂   isoprenaline

(10.2)

Derivatives with only one phenolic hydroxy group as well as 3,5-dihydroxy compounds do not give positive reaction. No positive reaction is obtained with noradrenaline either if the oxidizing agent is silver oxide.[30] With other reagents such as hypobromite,[24] bromate,[51] organic brominating agents,[2] even noradrenaline can be determined. The measurement is usually carried out in neutral media where the absorption maximum is around 485 nm and the molar absorption coefficient is in the order of some thousands. The maximum is shifted to about 400 nm in slightly alkaline medium.

A method where the oxidizing agent is triphenyltetrazolium chloride merits special mention.[28] The reaction is probably stopped in this case after the first step. Since this method is based on the measurement of the generated colored formazan (see Sections 7.E and 10.M.2), this method is more sensitive by one order of magnitude that those which are based on the measurement of the adrenochromes. The reduced form of the reagent is measured also in the indirect methods using ammonium molybdate,[32] ammonium vanadate,[62] the Folin-Ciocalteu reagent[66,74] and resazurine.[36]

The first step of reaction 10.2 leading to the 1,2-dioxo derivative occurs in alkaline media with atmospheric oxygen as the oxidizing agent. If a hydrazine derivative is added to the reaction mixture, then the corresponding hydrazone is formed rather than the adrenochrome. For example, isonicotinoyl hydrazide was used for the determination of adrenaline and isoprenaline[91] while phenylhydrazine and 4-aminophenylhydrazine for adrenaline, noradrenaline and isoprenaline.[27] The results with thiosemicarbazide were especially good: the absorption band at 490 nm enabled the sensitive determination of dobutamine ($\varepsilon = 17,000$)[25] and this reaction was used also for the determination of noradrenaline and adrenaline[60] as well as isoprenaline.[59]

The methods involving the use of 3-methylbenzothiazolin-2-one hydrazone reagent can also be mentioned here. This reagent is added to the catecholamine simultaneously with the oxidizing agent. The role of the latter is not only the oxidation of the phenol ring but also the transformation of the hydrazone reagent to an agent suitable for electrophyllic substitution. In the case of the determination of dobutamine with cerium (IV)-sulphate as the oxidizing agent, both mono- and diphenol rings in the molecule give positive reaction and this leads to a sensitive method (ε_{510nm} = 15,000).[26] For the determination of terbutaline sulphate iron (III) chloride was used as the oxidizing agent (λ_{max} = 478 nm).[31]

Using chloramine-T as the oxidizing agent and N,N-dimethyl-p-phenylenediamine as the reagent, isoxsuprin and terbutaline can be measured at 620 and 500 nm, respectively.[72]

REFERENCES

1. Abdel-Kahalek, M. M.; Abdel-Hamid, M. E.; Mahrous, M. S.; Abdel-Salam, M. A. *Anal. Lett.* 1985, *18,* 781.
2. Abou Ouf, A.; Walash, M. I.; Salem, F. B. *Analyst* 1981, *106,* 949.
3. Agarwal, S. P.; Chandrasekhara, N.; Madu, A. U. *Acta Pharm. Technolog.* 1981, *27,* 181.
4. Al-Abachi, M. Q.; Al-Ghabsha, T. S.; Shahbaz, N. A. *Microchem. J.* 1985, *31,* 272.
5. Amer, M. M.; Taha, A. M.; El-Shabouri, S. R.; Khashaba, P. Y. *J. Assoc. Off. Anal. Chem.* 1982, *65,* 894.
6. Atmaca, S.; Tosunoglu, S.; Büyüktimkin, N. *Sci. Pharm.* 1990, *58,* 281.
7. Bedair, M. M.; Korany, M. A.; Issa, A. S. *Analyst* 1988, *113,* 1137.
8. Bhatkar, R. G.; Nevrekar, V. N. S. *Indian Drugs Pharm. Ind.* 1979, *14,* 9.
9. Bhuee, G. S.; Singh, J.; Rastogi, S. N. *J. Inst. Chem. (India)* 1983, *55,* 229.
10. *British Pharmacopoeia 1988,* Her Majesty's Stationery Office: London, 1988. a. 828; b. 772, 926.
11. Chafetz, L. *J. Pharm. Sci.* 1964, *53,* 1192.
12. Chafetz, L. *J. Pharm. Sci.* 1971, *60,* 291.
13. Chatterjee, P. K.; Jain, C. L.; Sethi, P. D. *Indian Drugs* 1986, *23,* 635.
14. Clark, C. C. *J. Ass. Off. Anal. Chem.* 1980, *63,* 692.
15. Das Gupta, V.; Lara, A. J. L. *J. Pharm. Sci.* 1975, *64,* 2001.
16. Das Gupta, V.; Shek, O. H. *Am. J. Hosp. Pharm.* 1976, *33,* 1086.
17. Davidson, A. G.; Elsheikh, H. *Analyst* 1982, *107,* 879.
18. Davidson, A. G.; Hassan, S. M. *J. Pharm. Sci.* 1984, *73,* 413.
19. Davidson, A. G.; Mkoji, M. M. *J. Pharm. Biomed. Anal.* 1988, *6,* 449.
20. Deodhar, R. D.; Mehta, R. C. *Indian J. Pharm. Sci.* 1978, *40,* 167.
21. Dessouky, Y. M.; El Rub, L. N. G. *Analyst* 1976, *101,* 717.
22. Dessouky, Y. M.; Mousa, B. A.; Nour El-Din, H. M. *Pharmazie* 1974, *29,* 577.
23. Doty, J. R. *Anal. Chem.* 1948, *20,* 1166.
24. Doulakas, J. *Pharm. Acta Helv.* 1975, *50,* 66.
25. El-Kommos, M. E. *Analyst* 1983, *108,* 380.
26. El-Kommos, M. E. *Analyst* 1987, *112,* 101.
27. El-Obeid, H. A. *Anal. Lett.* 1984, *17,* 771.
28. El-Rabbat, N. A.; Omar, N. M. *J. Pharm. Sci.* 1978, *67,* 779.
29. El-Sadek, M. E.; Abdel Latef, H. E.; Aboul Kheir, A. *J. Assoc. Off. Anal. Chem.* 1987, *70,* 568.
30. El-Shabouri, S. R.; Hussein, S. A.; Abdel-Alim, A.-A. M. *J. Assoc. Off. Anal. Chem.* 1988, *71,* 764.
31. El-Yazbi, F. A.; Abdel-Hay, M. H.; Korany, M. A. *Farmaco, Ed. Prat.* 1985, *40,* 50.
32. Emmanuel, J.; Mathew, R. *East Pharm.* 1985, *28,* 129.
33. Emmanuel, J.; Shetty, A. R. *Indian Drugs* 1984, *21,* 402.
34. Ersoy, L.; Alpertunga, B. *Analyst* 1988, *113,* 1745.
35. Farsam, H.; Nadjari-Moghaddam, M. R. *J. Pharm. Biomed. Anal.* 1984, *2,* 543.
36. Fatma, B. S. *J. Pharm. Belg.* 1986, *41,* 35.
37. Fell, A. F.; Scott, H. P.; Gill, R.; Moffat, A. C. *J. Pharm. Pharmacol.* 1982, *34* Suppl., 99.
38. Giridhar, R.; Menon, S.; Agrawal, Y. K. *Indian J. Pharm. Sci.* 1987, *49,* 32.
39. Halmekoski, J.; Karki, A. *Farm. Aikak.* 1976, *85,* 125.
40. Hoover, J. M.; Soltero, R. A.; Bansal, P. C. *J. Pharm. Sci.* 1987, *76,* 242.
41. Kagan, F. E.; Mitchenko, F. A.; Kirichenko, L. A.; Koget, T. A. *Farm. Zh.* 1980, 46.
42. Kar, A.; Aniuha, G. I. *J. Pharm. Sci.* 1981, *70,* 690.
43. Kasture, A. W.; Wadodkar, S. G.; Bulbule, M. V.; Tajne, M. R. *Indian Drugs* 1984, *22,* 42.
44. Kelly, C. A.; Auerbach, M. E. *J. Pharm. Sci.* 1961, *50,* 490.
45. Korany, M. A.; Bedair, M.; El-Yazbi, F. A. *Analyst* 1986, *111,* 41.
46. Knochen, M.; Altesor, C.; Dol, I. *Analyst* 1989, *114,* 1303.
47. Lawrence, A. H.; MacNeil, J. D. *Anal. Chem.* 1982, *54,* 2385.
48. Maghssoudi, R. H.; Fawzi, A. B.; Meerkalaiee, M.; Aldeen, M. *J. Assoc. Off. Anal. Chem.* 1977, *60,* 926.
49. McCoubrey, A. *J. Pharm. Pharmacol.* 1956, *8,* 442.
50. Milch, Gy.; Szabó, É. *Analusis* 1988, *16,* 59.
51. Modamed, W. I.; Salem, F. B. *Anal. Lett.* 1984, *17,* 191.
52. Murtha, J. L.; Julian, T. N.; Radebaugh, G. W. *J. Pharm. Sci.* 1988, *77,* 715.
53. Ohno, N.; Sakai, T. *Bunseki Kagaku* 1981, *30,* 398.
54. Patel, R. B.; Patel, A. A.; Pattani, U. *Indian Drugs* 1987, *24,* 298.

55. Popov, D. M.; Raschetnova, V. I. *Farmatsiya* 1979, *28*, 33.

56. Rao, K. E.; Sastry, C. S. P. *Microchem. J.* 1985, *32*, 293.

57. Sadana, G. S.; Kunhipurayil, P. S.; Sane, R. T. *J. Indian Chem. Soc.* 1986, *63*, 238.

58. Sakai, T.; Ohno, N. *Analyst* 1987, *112*, 149.

59. Salama, R. B.; El-Obeid, H. A. *Analyst* 1976, *101*, 136.

60. Salama, R. B.; Wahba, Khalil, S. K. *J. Pharm. Sci.* 1974, *63*, 1301.

61. Salem, F. B. *Anal. Lett.* 1985, *18*, 1063.

62. Salem, F. B. *Talanta* 1987, *34*, 810.

63. Salem, F. B.; Walash, M. I. *Analyst* 1985, *110*, 1125.

64. Sane, R. T.; Joshi, S. K.; Pandit, U. R.; Doshi, V. J., Sawant, S. V.; Jukar, S.; Nayak, V. G. *Indian Drugs* 1984, *21*, 165.

65. Sane, R. T.; Kubal, M. L.; Nayak, V. G.; Malkar, V. B.; Banawalikar, V. J. *Indian Drugs* 1984, *22*, 25.

66. Sane, R. T.; Kubal, M. L.; Nayak, V. G.; Nadkarni, A. D.; Zarapkar, S. S. *Ind. J. Pharm. Sci.* 1986, *48*, 47.

67. Sane, R. T.; Malkar, V. B.; Nayak, V. G.; Sapre, D. S. *J. Assoc. Off. Anal. Chem.* 1986, *69*, 186.

68. Sane, R. T.; Nayak, V. G.; Malkar, V. B. *Talanta* 1985, *32*, 31.

69. Sane, R. T.; Narkar, V. S. *Indian Drugs* 1980, *18*, 23.

70. Sane, R. T.; Thombare, C. H.; Ambardekar, A. B.; Sathe, A. Y. *Indian Drugs* 1982, *19*, 195.

71. Sankar, D. G.; Sastry, C. S. P.; Reddy, M. N.; Prasad, S. N. R. *Indian J. Pharm. Sci.* 1987, *49*, 69.

72. Sankar, D. G.; Sastry, C. S. P.; Reddy, M. N.; Singh, N. R. P. *Indian Drugs* 1987, *24*, 410.

73. Santoni, G.; Mura, P.; Pinzauti, S.; Gratteri, P.; La Porta, E. *Int. J. Pharm.* 1989, *50*, 75.

74. Sethia, M. M. *Indian J. Pharm. Sci.* 1983, *45*, 30.

75. Shamsa, F. A.; Maghssoudi, R. H. *J. Pharm. Sci.* 1976, *65*, 761.

76. Shingbal, D. L.; Naik, S. D. *Can. J. Pharm. Sci.* 1981, *16*, 65.

77. Shingbal, D. M.; Agni, R. M. *Indian Drugs* 1983, *20*, 167.

78. Shingbal, D. M.; Joshi, S. V. *Indian Drugs* 1984, *21*, 398.

79. Shingbal, D. M.; Naik, R. R. *Indian Drugs* 1985, *22*, 273.

80. Shingbal, D. M.; Rao, V. R. *Indian Drugs* 1986, *23*, 228.

81. Street, K. W.; Abrenica, M. B. *Anal. Lett.* 1986, *19*, 597.

82. Subert, J. *Ceskosl. Farm.* 1985, *34*, 116.

83. Sulkowska, J.; Rynarzewska, G.; Malecki, F. *Pharmazie* 1989, *44*, 862.

84. Tan, H. S. I.; Kolmonpunpron, M. *Anal. Chim. Acta* 1989, *226*, 159.

85. Tan, H. S. I.; Salvador, G. C. *Anal. Chim. Acta* 1985, *176*, 71.

86. Tan, H. S. I., Salvador, G. C. *Anal. Chim. Acta* 1986, *188*, 295.

87. *The United States Pharmacopoeia XXI*, USP Convention Inc.: Rockville, 1985, 605.

88. *The United States Pharmacopoeia XXII*, USP Convention Inc.: Rockville, 1990. a. 85, 1530; b. 176; c. 188; d. 318; e. 366; f. 499; g. 504; h. 734; i. 744; j. 747; k. 804; l. 874; m. 888; n. 978; o. 1058; p. 1070; r. 1122, 1544; s. 1168; t. 1405.

89. Tosunoglu, S.; Büyüktimkin, N. *Arch. Pharm.* 1989, *322*, 445.

90. Trublin, F.; Robert, H.; Guyot-Hermann, A. M. *Bull. Soc. Pharm. Lille* 1976, *32*, 161.

91. Vanaclocha, J.; Sanchez, M.; Thomas, J. *Ars. Pharm.* 1978, *19*, 115.

92. Wahbi, A.-A. M.; Abdine, H.; Korany, M.; Abdel-Hay, M. H. *J. Assoc. Off. Anal. Chem.* 1978, *61*, 1113.

93. Wahbi, A.-A. M.; Mohamed, M. E.; Gad Kariem, E. R. A.; Aboul-Enein, H. Y. *Analyst* 1983, *108*, 886.

94. Walash, M. I.; Abou Ouf, A.; Salem, F. B. *J. Assoc. Off. Anal.* 1985, *68*, 91.

95. Weclawska, K.; Regosz, A. *Pharmazie* 1987, *42*, 483.

96. Zakhari, N. A.; Salem, F. B.; Rizk, M. S. *Farmaco, Ed. Prat.* 1987, *42*, 103.

97. Zivanov-Stakic, D.; Solomun, L.; Zivanovic, L. *Acta Pharm. Hung.* 1990, *60*, 179.

C. SALICYLIC ACID AND ITS DERIVATIVES

1. METHODS BASED ON NATURAL ABSORPTION

Salicylic acid has absorption maxima at 205, 236 and 303 nm in ethanolic solution (see Table 3.7). The latter two wavelengths are suitable for quantitative measurements. The direct spectrophotometric method is widely used for the assay of various formulations.

Clayton and Thiers[7] reported on the determination of salicylic acid, salicylamide, caffeine and phenacetin in tablets and powders. Heise and Pfeiffer[23] measured salicylic acid selectively at 308 nm in the presence of amidopyrine, phenazone, phenacetin and caffeine. Van de Vaart et al.[54] determined salicylic acid and methyl salicylate in creams in the presence of tripellenamine hydrochloride, resorcinol and clioquinol. Krzek[32] described the extraction of preparations of salicylic acid with 2-propanol and 15% sodium hydroxide prior to the measurement at 298 nm. Lactic acid, benzoic acid, lanolin, and vaseline do not interfere. Sodium salicylate, phenazone and caffeine were determined simultaneously without separation at 272, 302 and 326.8 nm by Zhou and Yang.[61]

The first derivative spectra of sodium salicylate (232 and 236 nm) and theobromine (264 and 268 nm) were used by Elsayed et al.[15] for their determination in 0.25 M sodium hydroxide solution. Tobias also used first derivative spectra for the determination of sodium salicylate and paracetamol in tablets[52] using the "zero crossing"[39] technique.

The determination of free salicylic acid in acetylsalicylic acid can be based on absorbance measurement at 300–306 nm.[21,35,41] Different sample pretreatments were used: extraction,[41] and column chromatographic separation.[21,35] Day et al.[9] determined salicylic acid at 308 nm and acetylsalicylic acid at 278 nm after extraction with chloroform.

Derivative spectrophotometry is also suitable for the determination of free salicylic acid in acetylsalicylic acid. Mazzeo et al.[36] as well as Kitamura et al.[28–30] used the second derivative spectra for this purpose without preliminary separation.

The spectrum of acetylsalicylic acid is shown in Figure 3.15. Its absorption band appearing between 272 and 280 nm in various solvents is suitable for its determination in dosage forms. For example the United States Pharmacopoeia XXII uses chloroform containing 1% acetic acid (to avoid hydrolysis) as the solvent, and the absorbance reading is carried out at 280 nm.[51b] It is interesting to note that in the course of the dissolution testing of aspirin tablets when some hydrolysis is unavoidable, its effect is cancelled out by measuring at the isoabsorption point of the spectra of salicylic and acetylsalicylic acids (265 nm).[51b]

Another possibility for the determination of acetylsalicylic acid is its hydrolysis to salicylic acid prior to the measurement. For example, Siedlanowska determined it in such a way in the presence of codeine and phenacetin.[45] Tatrai and Mühleman[49] extracted aspirin, codeine and phenacetin from suppositories using 80% ethanol. After hydrolysis the measurement was carried out at 297.5 nm. Fogg et al.[20] determined acetylsalicylic acid in the presence of paracetamol by measuring in 0.1 M sodium hydroxide at 297 nm.

Kirchhoefer[27] determined the aspirin content of intestino-solvens tablets after extraction with chloroform at 280 nm. Absorbance measurement at 278–280 nm is suitable for the determination of acetylsalicylic acid in the presence of salicylic acid, too. The minor interference of the latter can be avoided by dual-wavelength measurement[9] or by extraction[25,35,48] Second derivative spectrophotometry is also suitable for the determination of acetylsalicylic acid.[4,30]

In the course of the investigation of 4-aminosalicylic acid (PAS) according to the US Pharmacopoeia the absorbance of the two maxima at 265 and 299 nm is measured in a phosphate buffer at pH 7; the ratio of the two values should be between 1.5 and 1.56.[51a] The quantitative determination of PAS in the presence of its degradation product, aminophenol, was carried out by Moussa[37] in pH 3 borate buffer at 300 nm. Derivative spectrophotometry, too, offers excellent possibilities for the simultaneous determination of PAS and its degradation product.[58,59]

In the spectra of salicylamide in water and ethanol, maxima can be found at 235–236 and 298–300 nm. Kluczykowska and Krowczynski determined it in 0.1 M sulphuric acid medium in the presence of phenobarbital and papaverine,[31] while in the assay of formulations also containing paracetamol and codeine, the solvent was 0.01 M sodium hydroxide and the measurement of salicylamide was made at 329 nm.[16] Wahbi and Ebel[60] used the orthogonal polynomial method for the determination of salicylamide in the presence of chloroquine phosphate (wavelength range: 228–250 nm). Soliman et al.[46] used extraction with chloroform and measurement at 308 nm when determining salicylamide in analgesic formulations.

UV spectrophotometry is suitable for the determination of salicylic acid ester derivatives, too. For example Eröss-Kiss et al.[18] described the determination of ethyleneglycol monosalicylate in ointments in the presence of ethyl nicotinate and capsaicine.

The dependence of the spectra of salicylic acid derivatives on the pH of the solvent can serve as the basis for selective difference spectrophotometric methods. Some of these methods are presented in Chapter 8 in Table 8.1.

2. Methods Based on Chemical Reactions

The red color appearing when a chelate is formed in the reaction of salicylic acid and iron(III) ion was known in the last century. This fundamental reaction is widely used even today for the determination of salicylic acid and its derivatives. The main differences among the various methods are in the anion, the pH of the solution and the wavelength of the measurement. In the British Pharmacopoeia this method is prescribed for the determination of free salicylic acid in aspirin and for the assay of patches.[5] Other versions of the method are used for stability assays of aspirin preparations[55] and in vitro modeling of its hydrolytic decomposition in gastric juice,[22] the automated analysis of salicylic acid preparations,[8,33] moreover for the determination of other salicylic acid derivatives such as its 5-(2,4-difluorophenyl) derivative (diflunisal), too.[53]

Several indirect applications of the color reaction with iron(III) are based on the hydrolysis of the ester and amide derivatives and measurement of the resulting salicylic acid. This method was used for aspirin,[1] salicylamide,[19] and salicylic acid esters.[6,12] In an interesting application of this principle described by Khan,[26] alkyl salicylates were selectively determined in the presence of phenyl salicylate; the latter was transformed with secondary amines to stable salicylamide derivatives prior to the hydrolysis step.

Complex formation with cobalt(II) (400 nm) and copper(II) (520 nm) is the basis of the method of Belal et al.[3] for the determination of salicylamide. In these methods the analyte was transformed to the nitroso derivative prior to the complex formation. Another reagent for the determination of salicylic acid and PAS-sodium was molybdenum(VI).[56]

Ion-pair formation can also serve as the basis for the determination of salicylic acid. Shih and Teare[44] used methylene blue while Elsayed et al.[17] used crystal violet for this purpose.

Lacroix et al.[34] boiled PAS with trichloroacetic acid to form by decarboxylation, 3-aminophenol which can be coupled with 2,6-dichloro-p-benzoquinone-4-chloroimine. The same coupling and the formation of the indophenol dye was used by Szekeres et al.[24,48] (without decarboxylation) to salicylamide. Deodhar and Mehta[10] also used the decarboxylation reaction of PAS followed by the formation of a red product with potassium iodate.

The reactivity of the acetylated phenolic hydroxyl group in aspirin is the basis of selective methods. Verma and Jain[57] reacted it with 4-aminophenol to transform the latter to paracetamol, which can be measured after oxidation with iodyl benzoate at 430 nm. Salicylic acid, salicylamide, oxyphenbutazone and caffeine do not interfere. Deodhar et al.[11] used the iron(III) hydroxamate method (see p. 184).

Several other reactions were also used for the spectrophotometric determination of salicylic acid derivatives. The classical azo-coupling reaction was adopted by Belal et al.[2] for the determination of PAS after diazotization and coupling with ethyl acetoacetate. Coupling of PAS with 9-chloroacridine at pH 4 is the basis of the method of Przyborowski and Smajkiewich.[40] Further reagents for the determination of PAS are 5-nitrofuraldehyde[43] and N-methyl-p-benzoquinoneimine.[42]

For the determination of aspirin, the indirect method of Sultan[47] based on oxidation with dichromate in strongly acidic solution and measuring the resulting chromium(III) at 580 nm and the kinetic method of Theimer and Ciurczak[50] with the 3,3'-dimethoxybenzidine reagent are worth mentioning.

The basis of the simultaneous determination of salicylamide and paracetamol by El Kheir et al.[14] is based on the formation of nitroso derivatives and another method of the same authors[13] for salicylamide on the formation of Meisenheimer-type complexes.

The determination of salicylic acid in biological samples (9.H) is mainly based on complex formation with iron (III) or on enzymatic reactions (Section 7.K).

REFERENCES

1. Amschler, U. *Pharm. Ztg.* 1980, *125*, 555.
2. Belal, S. F.; Elsayed, M. A.; El-Nenaey, A.; Soliman, S. A. *Talanta* 1978. *25*, 290.

3. Belal, S. F.; Elsayed, M. A. H.; Elwalily, A.; Abdine, H. *Analyst* 1979, *104*, 919.

4. Bezakova, Z.; Bachrata, M.; Blesova, M.; Knazko, L. *Farm. Obz.* 1986, *55*, 257.

5. *British Pharmacopoeia 1988,* Her Majesty's Stationery Office: London, 1988, 48, 649.

6. Cates, L. A.; Brandino, T. F. *Can. J. Hosp. Pharm.* 1979, *32*, 169.

7. Clayton, A. W.; Thiers, R. E. *J. Pharm. Sci.* 1966, *55*, 404.

8. Cullen, J. F.; Peckman, D. L.; Papariello, G. J. *Ann. N.Y. Acad. Sci.* 1968, *153*, 525.

9. Day, M. D.; Harper, J. F.; Olliff, C. J. *Pharm. J.* 1967, *199*, 63.

10. Deodhar, R. D.; Mehta, R. C. *Indian J. Pharm. Sci.* 1976, *38*, 131.

11. Deodhar, R. D.; Shastri, M. R.; Mehta, R. C. *Indian J. Pharm. Sci.* 1976, *38*, 18.

12. Dutt, M. C. *J. Assoc. Off. Anal. Chem.* 1966, *49*, 854.

13. El Kheir, A. A.; Belal, S. F.; El Sadek, M; El Shanwani, A. *Analyst* 1986, *111*, 319.

14. El Kheir, A. A.; Belal, S. F.; El Shanwani, A. *J. Assoc. Off. Anal. Chem.* 1985, *68*, 1048.

15. Elsayed, M. A.-H.; Abdine, H.; Elsayed, Y. M. *Acta Pharm. Jugos.* 1977, *27*, 161.

16. Elsayed, M. A.-H.; Belal, S. F.; Elwalily, A. -F. M.; Abdine, H. *Analyst* 1979, *104*, 620.

17. Elsayed, M. A.-H.; Nwangwu, C. *Sci. Pharm.* 1981, *49*, 290.

18. Erös-Kiss, K., Hanák-Fehér, G.; Balogh, S. *Acta Chim. Hung.* 1990, *127*, 519.

19. Fink, D. W. *Anal. Chim. Acta* 1981, *131*, 281.

20. Fogg, A. G.; Sausins, P. J.; Smithson, J. R. *Anal. Chim. Acta* 1970, *49*, 342.

21. Guttman, D. E.; Salomon, G. W. *J. Pharm. Sci.* 1969, *58*, 120.

22. Guven, K. C.; Altinkurt, T.; Araman, A. *Eczacilik Bul.* 1980, *22*, 13.

23. Heise, D.; Pfeiffer, S. *Dtsch. Apot.-Ztg.* 1975, *115*, 2024.

24. Joy, J.; Szekeres, L. *Microchim. Acta* 1975, *11*, 125.

25. Juhl, W. E.; Kirchhoefer, R. D. *J. Pharm. Sci.* 1980, *69*, 544.

26. Khan, M. N. *J. Pharm. Biomed. Anal.* 1987, *5*, 515.

27. Kirchhoefer, R. D.; Jefferson, E.; Flinn, P. E. *J. Pharm. Sci.* 1982, *71*, 1049.

28. Kitamura, K.; Mabuchi, M.; Fukui, T.; Hozumi, K. *Chem. Pharm. Bull.* 1987, *35*, 3914.

29. Kitamura, K.; Majima, R. *Angew. Chem. Int. Ed. Engl.* 1978, *17*, 785.

30. Kitamura, K.; Takagi, M.; Hozumi, K. *Chem. Pharm. Bull.* 1984.

31. Kluczykowska, B.; Krowczynski, L. *Dissnes Pharm. Pharmacol.* 1972, *24*, 507.

32. Krzek, J. *Pharm. Pol.* 1980, *36*, 281.

33. Kubin, H.; Gaenshirt, H. *Pharm. Ind. Berl.* 1976, *38*, 224.

34. Lacroix, R.; Guidollet, J.; Zigliara, J. *Ann. Pharm. Fr.* 1965, *23*, 513.

35. Luber, J. R.; Visalli, A. J.; Patel, D. M. *J. Pharm. Sci.* 1979, *68*, 780.

36. Mazzeo, P.; Quaglia, M. G.; Segnalini, F. *J. Pharm. Pharmacol.* 1982, *34*, 470.

37. Moussa, A. -F. A. *Pharmazie* 1978, *33*, 460.

38. Oeydvin, K.; Sapiraa, D. E. *Meddr. Norsk. Farm. Selsk.* 1975, *37*, 115.

39. O'Haver, T. C. *Clin. Chem.* 1979, *25*, 1548.

40. Przyborowski, L.; Smajkiewicz, A. *Farm. Pol.* 1976, *32*, 43.

41. Reed, R. C.; Davis, W. W. *J. Pharm. Sci.* 1965, *54*, 1533.

42. Sastry, C. S. P.; Reddy, B. S.; Rao, B. C. *Indian J. Pharm. Sci.* 1981, *43*, 118.

43. Shah, P. R.; Raji, R. R. *Indian J. Pharm. Sci.* 1976, *38*, 52.

44. Shih, I. K.; Teare, F. W. *Can J. Pharm. Sci.* 1966, *1*, 32.

45. Sidlanowska, H. *Acta Pol. Pharm.* 1965, *22*, 93.

46. Soliman, S. A.; Salaheldin, A. *J. Pharm. Sci.* 1976, *65*, 1627.

47. Sultan, S. M. *Analyst* 1987, *112*, 1331.

48. Szekeres, L.; Harmon, R. E.; Gupta, S. K. *Microchem. J.* 1973, *18*, 101.

49. Tatrai, O.; Mühleman, H. *Pharm. Acta Helv.* 1968, *43*, 465.

50. Theimer, E. E.; Ciurczak, E. W. *J. Pharm. Sci.* 1977, *66*, 139.

51. *The United States Pharmacopoeia XXII* USP Convention Inc.: Rockville, 1990. a. 71; b. 112–113.

52. Tobias, D. Y. *J. Assoc. Off. Anal.-Chem.* 1983, *66*, 1450.

53. Vachek, J.; Svátek, E. *Ceskoslov. Farm.* 1986, *35*, 227.

54. Van de Vaart, F. J.; Hulshoff, A.; Indemans, A. W. M. *Pharm. Weekbl. Sci. Ed.* 1980, 1687.

55. Varga, P.; Farkas Mohay, J. *Gyógyszerészet* 1991, *35*, 87.

56. Velasevic, K.; Dukanovic, A.; Radovic, Z. *Arh. Farm.* 1977, *27*, 163.

57. Verma, K. K.; Jain, A. *Anal. Chem.* 1986, *58*, 821.

58. Vetuschi, C.; Ragno, G.; Mazzeo, P. *J. Pharm. Biomed. Anal.* 1988, *6*, 383.

59. Vetuschi, C.; Ragno, G. *Pharm. Acta Helv.* 1988, *63*, 290.

60. Wahbi, A. M.; Ebel, S. *J. Pharm. Pharmacol.* 1974, *26*, 317.

61. Zhou, Y.; Yang, S. *Yivao Gongye* 1986, *17*, 35.

D. SULPHONAMIDES

1. Methods Based on Natural Absorption

The spectra of disubstituted benzene derivatives were discussed in Section 3.I. In the spectrum of benzene-sulphonamide (see Table 3.5), the vibrational fine structure of both the K and the α-bands can still be observed. As a result of alkyl substitution such as in the case of 4-methyl-benzenesulphonamide, significant bathochromic and hyperchromic shifts of both bands can be observed together with the reduction of the fine structure. In the spectrum of sulphanilamide (which is the basic compound of this group of drugs) a further bathochromic shift can be found: the intense K-band appears at 261 nm in ethanolic solution greatly overlapping the α-band. In the spectra of sulphacetamide and sulphaguanidine (see Figure 3.16) the K-band is further shifted towards the longer wavelengths and its intensity increases. In the spectrum of sulphathiazole the K-band is split into two maxima. In the spectra of several other sulphonamides (sulphamethoxy-pyridazine, sulphadimethoxine, sulphadimidine, sulphasomidine) the effects of auxochromic groups are observable; in general, the effect of the sulphonamide group is pronounced only in the case of spectrophotometrically strongly active substituents. The yellow color of salazosulphapyridine is due to the presence of the salicylazo group in the molecule. As regards the spectra of sulphona-mides and their interpretation, the atlases listed in Chapter 3 and the papers of Kracmar et al.,[33,34] Napoli,[50] Garratt,[27] Bradford and Brackett,[14] Bellomonte et al.,[11] Braun and van Kerchove,[15] Wagner and Wander,[64] Mody and Naik,[46] Vignoli et al.[63] and Ramanujam[51] are recommended.

A variety of dosage forms are available for the therapeutic application of sulphonamides, very often in the form of their combinations, mainly solid dosage forms but ointments, injections and eye drops are also used.

The strong absorption band of sulphonamides in the range of 240–270 nm is suitable for their determination in various matrices. The differences between the absorption maxima of the individual sulphonamides as well as the changes of the spectra due to changes in pH of the solution create a good basis for the determination of individual sulphonamides in their combinations. The use of multi-wavelength spectrophotometry and pH-induced difference spectrophotometry and their combination are described in many publications including those of Matsuno et al.,[42] Marzys et al.,[41] Milch et al.,[43-45] Wahbi et al.,[65] Dombrowski et al.,[19] Madsen et al.,[38,39] Albasini et al.,[6] Zajac,[69] Liu,[36] Zhang et al.,[70] Chatterji et al.,[17] Wu and Luo,[66] Yao and Yi,[68] and Bezakova.[13] The combinations successfully analyzed without separation in the above listed papers include sulphadiazine-sulphamerazine-sulphamethazine, sulphadimeth-ylpyrimidine-sulphathiourea-sulphamethylthiazole, sulphadimethylpyrimidine-sulphathiourea-sul-phamethoxypyridazine, sulphamethoxazole-trimethoprim, sulphamethoxydiazine-sulphadiazine-sulphamethoxazole-trimethoprim, marphanil-sulphathiourea-sulphamerazine, etc.

Of the methods based on TLC-separation, spot elution and direct spectrophotometric determina-tion of the separated compound, the method of Vámos et al.[61] is mentioned for the simultaneous determination of sulphamethoxazole and trimethoprim in tablets.

The computer-based methods described in Section 4.D provide excellent means for facilitating the multi-wavelength analysis of the mixtures of sulphonamides and greatly improve the reliability of the method.[39]

Of the infrequent applications of the direct spectrophotometric method in pharmacopoeias, some sections of USP XXII are cited here. The dissolution studies of several sulphonamide tablets are monitored by their ultraviolet absorption such as in the case of sulphadiazine at 254 nm,[60a] sulphamerazine at 243 nm,[60b] and sulphamethazole at 267 nm.[60c] The assay of the tablet dosage form of the above mentioned sulphasalazine with its long-wavelength band is carried out in acetic acid medium after alkaline extraction at 359 nm.[60d]

Derivative spectrophotometry enables many assay problems to be solved among sulphonamides and their combinations. The first derivative spectra were used by Shen and Han[58] for the determination of sulphadimoxine, Abdellatef et al.[1] for sulphacetamide, sulphadimidine and sulphathiourea in the presence of their degradation products. The first derivative spectra were used also by Abounassif[3] for the assay of a tablet containing sulphametrole and trimethoprim.[1] D_{239} and $^1D_{315}$ are measured, the latter being suitable for the selective determination of sulphametrole while the trimethoprim content can be calculated from $^1D_{239}$ by means of the usual mathematical equation. Berzas Nevado et al.[12] applied the third and fourth derivatives for the simultaneous determination of sulphanilamide and sulphathiazole. This method is demonstrated in detail on p. 122.

2. Methods Based on Chemical Reactions

In the overwhelming majority of the spectrophotometric methods for the determination of sulphonamides based on chemical reactions, the reaction takes place on the primary amino group of the molecule. The most widely used method is based on the formation of diazonium salts followed by reacting them with suitable coupling agents to form azo-dyes, which are the basis for highly selective and sensitive determination for pharmaceutical and clinical analysis.[10,20,23,25,26,31,40,52,55–57,67,71] The details of the method are discussed in Section 7.D (see reaction Equation 7.9). Of the coupling agents N-(1-naphthyl)-ethylenediamine (Bratton-Marshall reagent) is most generally used (e.g., for the determination of sulphacetamide and sulphisoxazole in ophthalmic solution and ointment, respectively, in the United States Pharmacopoeia[60e]). In the same pharmacopoeia, the azo-coupling method is widely used for the determination of sulphonamide combinations after the TLC separation of the components.[60f]

Of the other coupling reagents used in the azo-coupling method for the determination of sulphonamides, ethyl acetoacetate and barbituric acid merit attention: Belal et al.[10] attained very good sensitivity with the latter ($\varepsilon_{380–386 \text{ nm}} = 32,200 - 44,600$). In the method of Halse and Wold[28] no coupling reagent is used: the method is based on the measurement of the intermediate of the Bratton-Marshall method, the diazonium salt itself after the decomposition of the excess nitrite.

Of the new applications of the Bratton-Marshall method its adaptation to the flow injection analysis (FIA) method (see Section 9.J) is worth mentioning. This variant of the method is most useful in serial examinations (clinical analysis, dissolution studies of solid dosage forms). Of these methods those of Koupparis et al.[32] and Esteve Romero et al.[23] are mentioned here. An interesting feature of the latter is that the rate of the reaction and the solubility of the azo-dye are greatly improved by adding sodium dodecyl sulphate to the reaction mixture.

The derivative spectrophotometric estimation of the reaction products of the Bratton-Marshall method was described by Salinas et al.[55] Measuring suitable amplitudes of the derivative spectra of various order, e.g., $^1D_{507,581}$, $^2D_{545,610}$, $^3D_{580,621}$, $^4D_{545,604}$ enables delicate problems to be solved, such as the direct determination of sulphonamides in urine.

The kinetic analysis variant of the method (see Section 7.J) has also been developed. This method described by Marquez et al.[40] was successfully used for the assay of several sulphonamide bulk materials and formulations.

The importance of the many other reagents described for the spectrophotometric determination of sulphonamides is much smaller than that of the Bratton-Marshall method; for this reason these are discussed only very briefly.

Several aldehydes form Schiff's bases with the primary amino group of sulphonamides and these are suitable for spectrophotometric determinations (vanillin and salicylaldehyde,[18] glutaconaldehyde,[35] 2-hydroxynaphthylaldehyde[16]). The mechanism of the reaction with p-benzoquinone is similar.[47]

The complex formation with metal ions such as copper (II)[7] and palladium (II)[4] can also serve as the basis for spectrophotometric methods. Several methods are based on the formation of charge transfer complexes, the reagents being tetracyanoethylene,[53] 7,7,8,8-tetracyanoquinodimethane,[29,48] o-chloranil,[65] p-chloranile[29] and p-bromanile.[65]

The indophenol reaction (see p. 179) was also used for the determination of sulphonamides.[22,62] Further reagents for this purpose include (among others) phenothiazine,[2,49] promethazine,[5] 1,2-naphthoquinone-4-sulphonic acid,[24] 9-chloroacridine,[59] o-diacetylbenzene[30] and 3-methyl-benzthiazoline-2-one hydrazone.[21]

In some interesting papers, the performance of the spectrophotometric methods for sulphonamides is compared with that of the HPLC technique.[8,31,54]

REFERENCES

1. Abdellatef, H. E.; Elbalkiny, M. N.; Aboulkheir, A. *J. Pharm. Biomed. Anal.* 1989, *7*, 571.
2. Abdine, H.; Korany, M. A.; Wahbi, A. M.; El-Yazbi, F. *Talanta* 1979, *26*, 1046.
3. Abounassif, M. A.; Hagga, M. E. M.; Gad-Kariem, E. A.; Al-Awady, M. A. *Acta Pharm. Jugosl.* 1991, *41*, 223.
4. Agatonovic Kustrin, S.; Zivanovic, L. J. *J. Pharm. Biomed. Anal.* 1989, *7*, 1559.
5. Al Abachi, M. Q.; Salih, E. S.; Salem, M. S. *Z. Anal. Chem.* 1990, *337*, 408.
6. Albasini, A.; Rastelli, A.; De Benedetti, P. G.; Mavi, G. *Boll. Chim. Farm.* 1973, *28*, 941.
7. Askal, H.; Saleh, G. A. *J. Pharm. Biomed. Anal.* 1991, *9*, 297.
8. Basci, N. E.; Bozkurt, A.; Kayaalp, S. O.; Isimer, A. *J. Chromatogr. Biomed. Appl.* 1990, *527*, 174.
9. Belal, S.; El-Sayed, M. A.; El Nanaey, A.; Soliman, S. A. *Talanta* 1978, *25*, 290.
10. Belal, S.; Soliman, A. S.; Bedair, M. *J. Drug. Res.* 1983, *14*, 195.
11. Bellomonte, G.; Calo, A.; Cardini, C. *Sci. Pharm.* 1960, *28*, 44.
12. Berzas Nevado, J. J.; Salinas, F.; De Orbe Paya, I.; Capitan-Vallvey, L. F. *J. Pharm. Biomed. Anal.* 1991, *9*, 117.
13. Bezakova, Z. *Farm. Obz.* 1981, *50*, 153.
14. Bradford, L. D.; Brackett, J. W. *Microchim. Acta 1958*, 353.
15. Braun, J.; Van Kerchove, C. *Sci. Pharm.* 1960, *28*, 53.
16. Chaban, I.; Belal, S. *Zbl. Pharm.* 1977, *116*, 919.
17. Chatterji, P. K.; Jain, C. L.; Sethi, P. D. *Indian Drugs* 1987, *24*, 351.
18. Colaizzi, J. L.; Boenigk, J. W.; Martin, A. N.; Knevel, A. M. *J. Pharm. Sci.* 1965, *54*, 564.
19. Dombrowski, L. J.; Browning, R. S.; Pratt, E. L. *J. Pharm. Sci.* 1977, *66*, 1413.
20. Dux, J. P.; Rosenblum, C. *Anal. Chem.* 1949, *21*, 1524.
21. El-Kommos, M. E.; Emara, K. M. *Analyst* 1988, *113*, 133.
22. Ellock, C. T. H.; Fogg, A. G. *Lab. Pract.* 1974, *24*, 555.
23. Esteve Romero, J. S.; Ramis Ramos, G.; Forteza Coll, R.; Cerdá Martin, V. *Anal. Chim. Acta* 1991, *242*, 143.
24. Filipeva, S. A.; Strelets, L. N.; Petrenko, V. V.; Burjak, V. P. *Farmatsiya* 1987, *36*, 39.
25. Fogg, A. G.; Fayad, N. M. *Anal. Chim. Acta* 1979, *106*, 365.
26. Funk, K. F. *Pharmazie* 1967, *22*, 341.
27. Garratt, D. C. *The Quantitative Analysis of Drugs,* Chapman and Hall: London, 1964, 608–612.
28. Halse, M.; Wold., K. *Medd. Norsk. Fram. Selskap* 1951, *13*, 103.
29. Hussein, S. A.; Mohamed, A.-A. I.; Abdel-Alim, A.-A. M. *Analyst* 1989, *114*, 1129.
30. Kagawa, M. *Jap. Analyst* 1967, *16*, 671.
31. Klimowicz, A. *Europ. J. Drug. Metab. Pharmacokin.* 1989, *14*, 181.
32. Koupparis, M. A.; Anagnostopoulou, P. I. *Anal. Chim. Acta* 1988, *204*, 271.
33. Krácmar, J.; Alvarez, M. S.; Lastovkova, M.; Krácmarová, J. *Pharmazie* 1975, *30*, 447.
34. Krácmar, J.; Lastovkova, M.; Kracmarova, J. *Ceskoslov. Farm.* 1975, *24*, 166.
35. Kunovits, G. *Microchim. Acta 1974*, 717.
36. Liu, C. *Yaowu Fenxi Zazhi* 1986, *6*, 213.
37. Madsen, B. W.; Herbison-Evans, D.; Robertson, J. S. *J. Pharm. Pharmac.* 1974, *26*, 629.
38. Madsen, B. W.; Robertson, J. S. *J. Pharm. Pharmacol.* 1974, *26*, 682.
39. Malkki, L.; Tammilehto, S. *Acta Pharm. Fenn.* 1989, *98*, 141.
40. Marquez, M.; Silva, S.; Perez Bendito, D. *Anal. Chim. Acta* 1990, *237*, 353.
41. Marzys, A. E. O. *Analyst* 1961, *86*, 460.
42. Matsuno, M.; Mishimura, M. *Yakugaku Kenkyu* 1957, *29*, 178.
43. Milch, Gy. *Acta Pharm. Hung.* 1962, *32*, 206.
44. Milch, Gy. *Acta Pharm. Hung.* 1963, *33*, 257.
45. Milch, Gy.; Borsai, M.; Mogács, I. *Acta Pharm. Hung.* 1967, *37*, 6.
46. Mody, S. M.; Naik, R. N. *J. Pharm. Sci.* 1963, *52*, 201.
47. Mohamed, A.-M. I.; Askal, H. F.; Saleh, G. A. *J. Pharm. Biomed. Anal.* 1991, *9*, 531.
48. Mohamed, A.-A. I. *J. Assoc. Off. Anal. Chem.* 1989, *72*, 885.
49. Mohamed, F. A.; Mohamed, A. I.; El-Shabouri, S. R. *J. Pharm. Biomed. Anal.* 1988, *6*, 175.
50. Napoli, I. *Boll. Chim. Farm.* 1958, *97*, 544.
51. Ramunajam, V. M. S.; Gowda, N. M. M.; Trieff, N. M.; Legator, M. S. *Microchem. J.* 1980, *25*, 295.
52. Ramis Ramos, G.; Esteve Romero, J. S.; Alvarez Coque, M. C. G. *Anal. Chim. Acta* 1989, *223*, 327.
53. Rao, G. R.; Murty, S. S. N.; Rao, P. J.; Raju, I. R. K. *Indian J. Pharm. Sci.* 1988, *50*, 138.
54. Revett, S. D.; Tenneson, M. E. *Anal. Proc.* 1984, *21*, 248.

55. Salinas, F.; Espinosa Mansila, A.; Berzas Nevado, J. J. *Anal. Chim. Acta* 1990, *233*, 289.

56. Sane, R. T.; Nayak, V. G. *Indian Drugs* 1982, *19*, 206.

57. Sanyal, A. K.; Laha, D. *J. Assoc. Off. Anal. Chem.* 1983, *66*, 1447.

58. Shen, K.; Han, Y. *Yiyao Gongye* 1986, *17*, 30.

59. Stewart, R. T.; Ray, A. B.; Fackler, A. N. B. *J. Pharm. Sci.* 1969, *58*, 1261.

60. *The United States Pharmacopoeia XXII*, USP Convention Inc.: Rockville, 1990. a. 1286; b. 1288; c. 1290; d. 1296; e. 1283; f. 1545.

61. Vámos, J.; Kóczián-Földvári, K.; Szász, Gy. *Acta Pharm. Hung.* 1986, *56*, 217.

62. Verma, K. K.; Sanghi, S. K.; Jain A. *Talanta* 1988, *35*, 409.

63. Vignoli, L.; Cristau, B.; Defretin, J.-P. *Ann. Pharm. Franc.* 1963, *21*, 477.

64. Wagner, G.; Wander, J. *Pharmazie* 1966, *21*, 105.

65. Wahbi, A.-A.; Abdine, K.; Korany, M.; El-Yazbi, F. *J. Assoc. Off. Anal. Chem.* 1979, *62*, 67.

66. Wu, Q.; Luo, G. *Yaowu Fenxi Zazhi* 1986, *6*, 10.

67. Xenakis, A. G. *Anal. Chim. Acta* 1984, *159*, 343.

68. Yao, G.; Yi, Y. *Yaowu Fenxi Zazhi* 1985, *5*, 153.

69. Zajac, M. *Farm. Polska* 1971, *27*, 11.

70. Zhang, Y.; Yang, Q.; Yu, R. *Yaoxue Xuebuo* 1984, *19*, 367.

71. Zöllner, E.; Vastagh, G. *Pharm. Zhalle* 1957, *96*, 99, 1958, *97*, 219.

E. OTHER PHENOL AND ANILINE DERIVATIVES

1. Introduction

In this section the possibilities of the spectrophotometric determination of aromatic hydroxy and alkoxy as well as amino (acylamino, alkylamino) derivatives are discussed. These functional groups are present either alone or together on aromatic rings, or they may be combined with other functional groups. Of the latter the carboxyl group is the most important (p-hydroxybenzoic acid esters, derivatives of p-aminobenzoic acid, diphenylamine carboxylic acids). Because of their great importance, o-hydroxybenzoic acid (salicylic acid) derivatives are discussed separately in Section 10.C. Some important phenol derivatives (catecholamines and related derivatives) are included in Section 10.B among the phenylalkylamines. Of the aromatic amines, sulphonamides are also discussed separately in Section 10.D.

As is described in Sections 3.G, 3.I and 3.J, the intensity of the spectra of aromatic derivatives containing auxochromic groups are sufficiently intense to serve as the basis for determinations with acceptable sensitivity. The selectivity of these methods can be increased for spectra containing free hydroxyl or amino groups on the basis of changes of the pH of the solution (Section 3.Q) especially using the difference spectrophotometric technique (8.B).

Both phenolic hydroxyl and aromatic amino groups are capable of serving as the basis of selective and sensitive spectrophotometric methods based on chemical reactions (see Chapter 7).

2. Paracetamol (Acetaminophen) and Phenacetin

Of the pharmaceutically important 4-aminophenol derivatives, paracetamol contains the phenolic hydroxyl group in free and the amino group in acetylated form. In accordance with this structure the intense band in its spectrum at 248 nm ($\varepsilon = 12,900$) undergoes a bathochromic shift in alkaline medium (λ_{max}/0.1 M NaOH = 256 nm). The α-band around 280 nm appears as a shoulder only. In the United States Pharmacopoeia the assay of the bulk substance[104a] and the monitoring of the dissolution testing of the tablet formulation[104b] are based on absorbance measurement in neutral solution, while the assay of the tablet is carried out in alkaline solution according to the method of BP-88.[13a]

The spectrum of phenacetin is similar ($\varepsilon_{250\,nm} = 16,200$). Since the phenolic hydroxyl is blocked in this molecule as the ethyl ether, the spectrum does not change upon basification.

These spectra serve as the basis for several analytical methods from the early days when e.g., the simultaneous determination of aspirin, phenacetin and caffeine was described on the basis of absorbance measurements at 226, 250 and 272 nm[64] up to the present time when by means of the above mentioned bathochromic shift, neutral/alkaline or acidic/alkaline difference spectrophotometric methods are described at 267 nm for their determination e.g., in tablets, suppositories and syrups,[35] in injections in the presence of phenazone,[117] and in combined preparations in the presence of chlorzoxazone.[16]

Derivative spectrophotometry is a useful method for the selective determination of paracetamol. Korany et al.[49] and Milch and Szabó[67] used the second derivative method for its determination in formulations in the presence of the degradation product 4-aminophenol. The same technique was used by Kir et al.[55] for the simultaneous determination of paracetamol and methocarbamol in tablets and by Edinboro et al.[23] for the determination of serum paracetamol. Its determination in the presence of degradation products is possible also by means of the orthogonal polynomial method as described by Elsayed et al.[36]

Of the methods based on chemical reactions, at first some methods which depend on the oxidation of paracetamol and phenacetin (or their hydrolysis products) by various oxidizing agents will be presented. Sultan et al.[101] and Quersi and Saeed[79] determined paracetamol in formulations using cerium(IV) sulphate–10 M sulphuric acid and cerium(IV) nitrate, respectively, as the oxidizing agents. The product of the hydrolysis and oxidation upon which the measurement is based (λ_{max} at 410 or 355 nm) is p-benzoquinone. The acetylimine derivative of the latter is the reaction product if the oxidizing agent is 2-iodylbenzoate.[107] The sensitivity of these methods is poor. The method of Morelli[70] is also based on acidic hydrolysis followed by oxidation (with ammonium molybdate). In this method the assay is based on the measurement at 670 nm of the reduced form of the reagent (molybdenum blue). The sensitivity of this indirect method is about 30 times higher than that of the methods based on the measurement of p-benzoquinone.

Many methods depend on oxidative coupling with phenol and the measurement of the blue indophenol around 620 nm (see Section 7.E). The procedures usually begin with the acidic hydrolysis of the acetamido (and in case of phenacetin, also the ethoxy) group requiring severe conditions (boiling in 10–32% hydrochloric acid) to form 4-aminophenol, which can be transformed to indophenol using e.g., potassium hexacyanoferrate[22] or atmospheric oxygen[29,75] as the oxidizing agents.

The method of Martinez Calatayud et al.[62] merits special mention. In the course of the determination of paracetamol in tablets by the FIA method (see Section 9.J) no acidic splitting was used (see Equation 10.3). Paracetamol itself was oxidized by potassium hexacyanoferrate by the formation of N-acetyl-p-benzoquinonimine which reacts with phenol to form the indophenol. In a recent variant of the method, the oxidation is carried out with the above reagent bound to an anion exchange column.[63] The basis of another FIA method described by Georgiou and Koupparis[40] is the oxidative condensation of paracetamol (and several other phenol-type drug substances) with 1-nitroso-2-naphthol (oxidizing agents: cerium(IV) or lead(IV)).

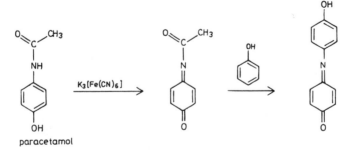

(10.3)

The methods based on the easy and (due to the substituents) unidirectional nitration or nitrosation of the aromatic ring also belong to this group of methods. Paracetamol can easily be transformed to the o-nitrosophenol derivative, which can be measured in alkaline media at 430 nm.[11] The selectivity of the method can be increased by measuring difference spectrophotometrically (alkaline vs. neutral solution[26]). The sensitivity of these methods is rather poor. The sensitivity is not much better in the case of methods based on nitration, either. Of these methods two are mentioned. Inoue et al.[43] measure paracetamol after nitration, alkaline hydrolysis and extraction with chloroform as

2-nitro-p-phenetidine. El Kheir et al.[25] prepared the dinitro derivative (under stronger reaction conditions). This reacts in alkaline solution with acetone to form a Meisenheimer complex which can be measured at 355 nm (see Equation 10.4).

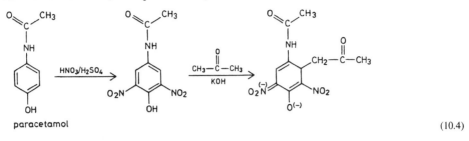

paracetamol

(10.4)

In another variant of the method[31] the nitrated derivative is reacted with a dimedone-triethylamine reagent (λ_{max} 435 nm).

The reaction of the primary aromatic amino group with aromatic aldehydes to form the Schiff's base is also the basis for several analytical methods (Section 7.D). As the reagents, among others, vanillin,[77,106] 4-nitrobenzaldehyde,[51] 2,4-dinitrobenzaldehyde,[51] 2-hydroxy-5-phenylbenzalde-hyde,[65] and 4-dimethylaminocinnamaldehyde[44] were used. These react only with the decomposition product of paracetamol, 4-aminophenol. After strong acidic hydrolysis when paracetamol is transformed to the latter, its indirect determination can be carried out. The wavelength of the measurement depends on the aldehyde reagent (385–520 nm). The sensitivity of the methods is fair.

Since both paracetamol and phenacetin contain aromatic amine and phenol functional groups, methods based on azo-coupling can also be widely used. There are two possibilities:

1. Transformation of the analyte to diazonium salts and coupling them with phenol reagents. Since the aromatic amino group is in acetylated form in the molecules of both compounds, the acetyl group has to be removed by acidic hydrolysis (similar to the methods when Schiff's bases are formed). Coupling agents used for this purpose include β-naphthol for the determination of phenacetin[7,74] and N-(1-naphthyl)-ethylenediamine for the determination of paracetamol in syrups according to the United States Pharmacopoeia.[104b]
2. Using diazotized 2-nitroaniline as the reagent and paracetamol as the phenol component in the coupling reaction, the determination of the latter can be carried out even without acidic hydrolysis. This possibility was exploited in three different ways by Belal et al.[9] The resulting azo-dye can be directly measured in alkaline media at 426 nm. The copper(II) complex of the dye can be extracted from acidic medium by chloroform for the measurement at 400 nm. Finally the copper(II) content of the above complex can be extracted as the diethyldithiocarbamate and measured in this form at 442 nm. The latter, indirect method is the most selective and sensitive of the three methods.

The acetamido group is the basis of another kind of the spectrophotometric determination. When treated with hydroxylamine in alkaline solution acethydroxamic acid is formed which can be measured as the iron(III) complex[19,45] (see Section 7.G).

Several other reagents have also been described for the determination of phenacetin and paraceta-mol, e.g., 4-nitrosoantipyrine[46] and sodium nitroprusside.[89] The determination of paracetamol can be accomplished with the aid of ion-pair extraction, too. Using Rhodamine B as the dye the ion-pair can be extracted at pH 3 and measured at 542 nm.[72] Charge-transfer complexation using chloranil as the reagent was also successfully used.[5] Several further methods can be found in two reviews on paracetamol[38] and phenacetin.[30] Section 9.H contains some more data on the determination of paracetamol in biological samples.

3. Aniline-Based Local Anesthetics and Related Derivatives

The most characteristic member of the group of local anesthetic 4-aminobenzoic acid derivatives, procaine has an intense absorption maximum at 290 nm which is suitable for its determination ($\varepsilon =$

16,100). This is utilized in USP XXII[104c] for its determination in injections. Procaine is extracted with chloroform from alkaline solution; in this solvent the maximum is shifted to 280 nm. This method is suitable for the determination even in the presence of phenylephrine.[104d] In strongly acidic media, the aromatic amino group is protonated and consequently, a benzoic acid-like spectrum is obtained ($\varepsilon_{228\ nm}$ = 10,600; ε_{279} = 1400). The decrease of the absorbance upon acidification enables selective assay methods to be developed, e.g., determination in the presence of adrenaline and phenylephrine.[54] The method of orthogonal polynomials can also serve as the basis for selective measurements, e.g., determination in the presence of 4-aminobenzoic acid.[12] For details see Section 4.C.1.

Of the methods based on chemical reactions, diazotization and coupling with various reagents is most frequently used for the determination of derivatives with primary amino groups (procaine, benzocaine and metoclopramide). In the course of the recent applications of this classical method (see Section 7.D) 2-naphthol,[78] 1-naphthylamine,[53,84] thymol,[52] N-(1-naphthyl)-ethylenediamine,[96] 3-aminophenol,[84] resorcinol,[84] 2-thiobarbituric acid,[84] rodanine[50] and ethyl acetoacetate[8] were used as the coupling agents.

A large variety of local anesthetics were determined by El-Kommos and Emara[27] using 3-methylbenzothiazolin-2-one hydrazone as the reagent after oxidation with iron(III). The structure of the blue reaction product in the case of tetracaine is as follows ($\varepsilon_{615\ nm}$ = 55,000):

(10.5)

The application of the classical method for ester groups depending on the formation of hydroxamic acid in the course of hydroxylaminolysis followed by the formation of its iron(III) complex is also worth mentioning.[71] Other classical reagents for aromatic amines used for local anesthetics include 1,2-naphthoquinone-4-sulphonic acid,[39] ammonium reineckate,[12] 4-dimethylaminobenzaldehyde,[116] 4-dimethylaminocinnamaldehyde.[28] Methods based on complex formation with copper(II) are also in use for the determination of various local anesthetics[18] and procainamide.[114]

4. Other Aniline Derivatives

Neostigmine bromide and methylsulphate containing quaternary ammonium groups can easily be determined by the ion-pair method (Section 7.H). Using bromothymol blue as the reagent, Kulikov et al.[59] extract at pH 8.3–10 with chloroform and measure the complex at 415 nm. Belal et al.[10] extract at pH 7.8 with the same dye and measure at 416 nm or alternatively carry out the re-extraction of the chloroform extract with 0.01 M sodium hydroxide and measure the absorbance of the aqueous-alkaline phase at 615 nm.

The method of Cacho Palomar et al.[14] for bromhexine is also based on ion-pair extraction. The ion-pair forming anion is tetrathiocyanatocobalt(II) and the absorbance is read after extraction with chloroform at 625 nm. In this method it is not the weakly basic aromatic amino group which takes part in the ion-pair formation but the other more basic secondary amino group in the molecule. In contrast, it is the primary aromatic amine which is the basis for the analyses depending on diazotization and coupling with N-(1-naphthyl)-ethylenediamine[88,97] as well as formation of Schiff's base with 4-dimethylaminobenzaldehyde.[15]

The intense spectrum of chlorhexidine allows its determination in various pharmaceutical preparations[103] and the presence of the strongly basic guanidino group in the molecule enables the ion-pair extraction method to be applied. After checking several sulphonephthalein-type dyes, the highest sensitivity was attained by Pinzauti et al.[76] using methyl orange at pH 6; (extraction with chloroform, λ_{max} 425). Andermann et al.[6] used bromocresol green, extracted with chloroform at pH 5.7 and measured at 410 nm. Bromocresol green was used at pH 5 by Martinez Calatayud and Campins Falco, too.[61] In this case, however, the ion-pair formation took place in the presence of the detergent Triton X-100, thus enabling the extraction step to be avoided; (ε_{630nm} = 12,500). The FIA variant of this method is also described.

Thiambutosine was determined at 540 nm after reaction with 2,3-dichloro-1,4-naphthoquinone.[21]

Diphenylamine carboxylic acids can also be regarded as aniline derivatives. Both mefenamic and flufenamic acids possess intense spectra in the ultraviolet region suitable for analytical purposes. The absorption band appearing around 350 nm in dilute acidic media is especially suitable for quantification; namely, the relatively long wavelength of this band enables selective assays to be carried out, e.g., the determination of mefenamic acid in the presence of paracetamol.[17] The selectivity of the method can be further increased using the method of orthogonal polynomials (Section 4.C.2) or pH-induced acidic–alkaline difference spectrophotometry (Section 8.B).[41] The simultaneous presence of carboxylic and amine functional groups in the molecule provides these molecules with excellent complex-forming ability. This was utilized for their determination as the iron(III) complex.[32,120] In another method their copper(II) complex is extracted followed by measurement of the copper(II) content of the latter as the diethyldithiocarbamate.[2] Ion-pair extraction is also suitable for the determination of diphenylamine carboxylic acids. Since these are stronger acids than bases, basic ion-pair forming dyes are utilized.[111]

Of the redox reactions in addition to mefenamic and flufenamic acids,[60] diclofenac can also be determined by oxidation with hexacyanoferrate.[87] Further methods for the determination of diclofenac include reaction with a 3-methyl-benzothiazolin-2-one hydrazone-cerium(IV) reagent leading to a derivative with absorption maximum at 600 nm,[91] ion-pair extraction with methylene blue at pH 6.8 with chloroform and measurement[4] at 640 nm and, finally, oxidation with iron(III) chloride and measurement of the forming iron(II) by the 2,2′-dipyridyl method.[4]

To conclude this section, four general methods are presented for the determination of aromatic amines. In the methods of Krishna[57] and Sastry et al.[90] the reagents are methol (4-methylaminophenol) oxidized with potassium dichromate or catechol oxidized with iodine. The condensation of these with the primary aromatic amino group results in derivatives with absorption maxima around 500–520 nm. The FIA method of Verma and Stewart[108] is also based on the reaction product of methol with potassium dichromate (1,4-benzoquinone-methylimine). The reagent of the method of Witzel and Pindur[115] is triethylorthoformate. On the addition of this reagent and a Lewis-acid as the catalyst three moles of the tertiary aromatic amines are condensed to form colored triarylmethyl cations.

5. β-Receptor Blocking Agents

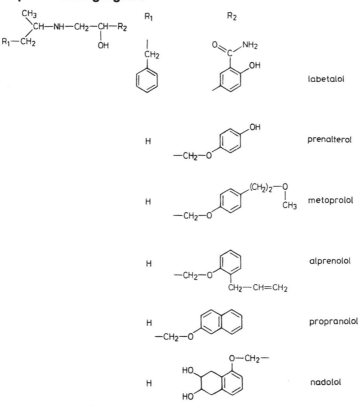

(10.6)

With their structures containing phenolic hydroxyl and phenolether groups, the moderately selective and sensitive determination of β-receptor blocking agent is possible based on their natural UV absorption. For example, pharmacopoeia methods are available for the determination of propranolol in injections[13b] at 290 nm or for the monitoring of the dissolution tests of the tablets at 289 nm.[104e] The selectivity of the method can be increased by using derivative spectrophotometry.[110]

The spectra of the derivatives with free phenolic hydroxy groups undergo in alkaline media bathochromic and hyperchromic shifts. This creates good possibilities for the application of difference spectrophotometry (Section 8.B). For example, in the case of labetalol, the maximum of the difference spectrum (0.1 M sodium hydroxide vs. 0.1 M sulphuric acid solutions) is at 332 nm ($\Delta A_{1\,cm}^{1\%} = 136.4$) and this enables (in the method of Mohamed[68]) cancelling out of the spectral background in the course of its determination in tablets. Prenalterol can be determined in a similar way at 305 nm.[113]

The secondary amine group being present in these derivatives enables them to be determined by the ion-pair extraction method (Section 7.H). For example Özden and Gümüs[73] described the determination of nadolol using bromothymol blue as the reagent. After the extraction with chloroform at pH 3.2 the absorbance is read at 412 nm. Alprenolol,[48] metoprolol[37] and propranolol[80] can be determined similarly.

The classical method for the selective determination of secondary amines, namely their transformation with carbon disulphide to dialkyldithiocarbamate, followed by the formation of copper(II)-bis-dithiocarbamate complex was adopted by Zakhari et al.[118] The molar absorptivities of a large variety of β-blockers were between 4440 and 6650 at 435 nm and this was used for their determination in tablets. Another method based on the secondary amino group is the use of 2,4-dinitrofluorobenzene as a result of which dinitroaniline derivatives measurable at 380 nm are formed, and this was used by Shingbal and Bhangle[95] for the determination of metoprolol in tablets.

The metal complex formation of β-blockers can also be utilized for analytical purposes. Radulovic et al. described the determination of oxprenolol,[81] propranolol[119] and penbutolol[82] on the basis of the formation of their mixed complexes with iron(III) and thiocyanate ions, extraction with chloroform and measurement around 477 nm.

The reducing characteristics of this group of drugs containing phenolic hydroxyls are the basis for several analytical methods. Mohamed et al.[69] determined prenalterol by reacting it with iron (III) and measuring the iron(II) thus formed by the 1,10-phenantroline method at 511 nm. The Folin-Ciocalteu reagent and 2,6-dichloroquinoneimine were also used for the determination of labetalol.[56,83]

Derivatives not containing free hydroxyl groups can also be oxidized if sufficiently strong oxidizing agents are used. In this case presumably the secondary hydroxyl group is involved in the reaction. For example, Sultan[99] determined propranolol by reacting it with potassium dichromate and measuring the reduced form (chromium(III)) of the oxidizing agent. The kinetic method[100] (see Section 7.J) where the reagent is cerium (IV) is based on the same principle. The method of Ivashkiv[47] for the very selective determination of nadolol merits special mention. Here the vicinal diol system on the cyclohexane ring is transformed to a dialdehyde by splitting with periodic acid and this intermediate is reacted with 2,4-dinitrophenylhydrazine. The derivative is extracted with chloroform and measured at 352 nm.

6. Some Other Phenol Derivatives

The natural absorption of phenol-type drugs provides good possibilities for their determination in various dosage forms. It is very advantageous to measure in alkaline media making use of the bathochromic and hyperchromic shifts.[66] The application of this method is problematic in the case of the determination of the preservative 4-hydroxybenzoic acid esters since at the high pH necessary for the complete ionization of the hydroxyl group, the ester group slowly hydrolyzes and this causes further shift of the spectrum. However, if instead of measuring at the maximum (296 nm) the isoabsorption point of the spectra of the ester and the free acid is selected for the measurement (282 nm), the determination is based on an absorbance value which does not depend on the time between preparing the test solution and the measurement.[24] On the other hand, the determination of free 4-hydroxybenzoic acid impurity in its esters can be accomplished by difference spectrophotometry: the test solution contains phosphate buffer at pH 5.9 while in the reference cell the solvent is 0.1 M hydrochloric acid. The basis of the selective determination is that at pH 5.9, the carboxyl group is completely ionized while the phenolic hydroxyl is not at all.[98]

Of the derivatives of 4-hydroxybenzoic acid, the determination of diflunisal is also described by the pH-induced difference spectrophotometric method.[1]

The selectivity of the determination of phenols can be improved by using derivative spectrophotometry. Vetuschi et al.[109] determined cloctofol (2-(2,4-dichlorobenzyl)-4-(1,1,3,3-tetramethyl-butyl)-phenol) in ointment by measuring $^2D_{298,286}$ and $^2D_{289,301}$.

Of the classical methods based on the reactions of phenols, coupling with diazonium salts (mentioned several times in previous chapters) is most important, as it is exemplified by the determination of resorcinol,[42] amyl-m-cresol[86] and chlorothymol.[85] Of the other methods the oxidative coupling with 4-aminoantipyrine (see p. 179) is used for the determination of hexachlorophene[105] and naproxene.[92] In the case of the latter the phenol ether has to be split with 5 M hydrochloric acid prior to the assay. The uses of the indophenol method (see p. 179)[86] and complexation with iron(III)[58] can also be mentioned.

The determination of clofibrate[3,93] and diloxanide furoate[94] by the ferric hydroxamate method (see p. 184) is based on the ester groups in the molecules. Of the reactions with not completely clarified mechanisms, the determination of guaiphenesine is mentioned on the basis of the red color formed with the formaldehyde-sulphuric acid reagent.[20]

Finally, it is interesting to note that even in the analytical literature of phenolphthalein which possess the well known characteristic and intense red color in alkaline media, several methods based on chemical reactions can be found, e.g., treatment with a sodium nitrite-acetic acid-cobalt(II) nitrate reagent and measurement as the nitroso derivative.[33] It is worth mentioning that the measurement on the basis of the red color in alkaline media requires careful optimization: Suzuki et al.[102] found that the best results could be obtained at pH 10.6 in carbonate buffer.

The simultaneous determination of aloin and phenolphthalein is described using a dual-wavelength (230 and 295 nm) measurement.[34]

REFERENCES

1. Abdel-Hamid, M. E.; Najib, N. M.; Suleiman, M. S.; El-Sayed, Y. M. *Analyst* 1987, *112*, 1527.
2. Aboul Kheir, A.; El-Sadek, M.; Baraka, M. *Analyst* 1987, *112*, 1399.
3. Agrawal, Y. K.; Patel, D. R. *Talanta* 1987, *34*, 365.
4. Agrawal, Y. K.; Shivramchandra, K. *J. Pharm. Biomed. Anal.* 1991, *9*, 97.
5. Al-Ghabsha, T. S.; Al-Abachi, M. Q.; Ahmad, A. K. *Microchem. J.* 1987, *35*, 66.
6. Andermann, G.; Buhler, M. O.; Erhart, M. *J. Pharm. Sci.* 1980, *69*, 215.
7. Bandelin, F. J.; Pankratz, R. E. *Anal. Chem.* 1956, *28*, 218.
8. Belal, S. F.; Elsayed, M. A. H.; El-Nenaey, A.; Soliman, S. A. *Talanta* 1978, *25*, 290.
9. Belal, S. F.; Elsayed, M. A. H.; El-Walily, A.; Abdine, H. *J. Pharm. Sci.* 1979, *68*, 750.
10. Belal, S. F.; Elsayed, M. A. H.; El-Walily, A. F. M.; Abdine, H. *Sci. Pharm.* 1981, *49*, 397.
11. Belal, S. F.; Elsayed, M. A. H.; El-Walily, A.; Abdine, H. *Analyst* 1979, *104*, 919.
12. Bhatkar, R. G.; Chodankar, S. K. *East Pharm.* 1981, *24*, 125.
13. *British Pharmacopoeia 1988*, Her Majesty's Stationery Office: London, 1988; a. 980; b. 843.
14. Cacho Palomar, J.; Nerin de la Puerta, C; Ruberte, L. *Anal. Letters* 1983, *16*, 237.
15. Chainani, M. L.; Nighojkar, A.; Naik, S. D. *Indian Drugs* 1986, *24*, 51.
16. Chatterjee, P. K.; Jain, C. L.; Sethi, P. D. *J. Pharm. Biomed. Anal.* 1989, *7*, 693.
17. Das, S.; Sharma, S. C.; Talwar, S. K.; Sethi, P. D. *Analyst* 1989, *114*, 101.
18. De Freitas, J. F. *Aust. Dent. J.* 1977, *22*, 182.
19. Deodhar, R. D.; Shastri, M. R.; Mehta, R. C. *Indian J. Pharm. Sci.* 1976, *38*, 18.
20. Desai, D. K.; Kurani, S. P.; Seshadrinathan, A. G. *Indian J. Pharm. Sci.* 1985, *47*, 210.
21. Devani, M. B.; Shishoo, C. J.; Mody, H. J. *Analyst* 1975, *100*, 178.
22. Domagalina, E.; Zareba, S. *Acta Pol. Pharm.* 1977, *34*, 627.
23. Edinboro, L. E.; Jackson, G. F.; Jortani, S. A.; Poklis, A. *Clin. Toxicol.* 1991, *29*, 241.
24. Elek, B. *Acta Pharm. Hung.* 1989, *59*, 269.

25. El Kheir, A. A.; Belal, S. F.; El Sadek, M.; El Shanwani, A. *Analyst* 1986, *111*, 319.

26. El Kheir, A. A.; Belal, S. F.; El Shanwani. A. *Pharmazie* 1985, *40*, 62.

27. El-Kommos, M. E.; Emara, K. M. *Analyst* 1987, *112*, 1253.

28. El-Kommos, M. E.; Sidhom, M. B. *Arch. Pharm. Chemi. Sci. Ed.* 1982, *10*, 57.

29. Ellcock, C. T. H.; Fogg, A. G. *Analyst* 1975, *100*, 16.

30. El-Obeid, A.; Al-Badr, A. Phenacetin, in *Analytical Profiles of Drug Substances,* (Vol. 14,) K. Florey (Ed.), Academic Press: Orlando, 1985, 551–596.

31. El Sadek, M. *Anal. Letters* 1986, *19*, 479.

32. El Sadek, M.; Baraka, M.; Aboul Kheir, A. *Indian J. Pharm. Sci.* 1987, *49*, 97.

33. El-Sayed, L.; Ellaithy, M. M.; Bebawy, L. *Farmaco, Ed. Prat.* 1985, *40*, 273.

34. El-Sayed, L.; Ellaithy, M. M.; Bebawy, L.; Amer, M. M. *Sci. Pharm.* 1986, *54*, 95.

35. Elsayed, M. A. H.; Belal, S.; El-Walily, A. F. M.; Abdine, H. *J. Assoc. Off. Anal. Chem.* 1979, *62*, 549.

36. Elsayed, M. A. H.; Korany, M. A.; Bedair, M. M.; Mahgoub, H.; Korany, E. A. *Drug Dev. Ind. Pharm.* 1990, *16*, 1011.

37. Ersoy, L.; Kocaman, S. *Arch. Pharm.* 1991, *324*, 259.

38. Fairbrother, J. E. Acetaminophen, in *Analytical Profiles of Drug Substances* (Vol. 3), K. Florey (Ed.), Academic Press: Orlando, 1972, 2–109.

39. Filipyeva, S. O.; Strilets, L. M.; Petrenko, V. V.; Buryak, V. P. *Farm. Zh.* 1988, 73.

40. Georgiou, C. A.; Koupparis, M. A. *Analyst* 1990, *115*, 309.

41. Hassan, S. M.; Shaaban, S. A. M. *Pharmazie* 1984, *39*, 691.

42. Horváth Rusvai, M.; Sperlágh, J.; Végh, M. *Gyógyszerészet* 1987, *31*, 121.

43. Inoue, T.; Tatsuzawa, M.; Hashiba, S. *Eisei Kagaku* 1974, *20*, 217.

44. Inoue, T.; Tatsuzawa, M.; Lee, S. C.; Ishii, T. *Eisei Kagaku* 1975, *21*, 313.

45. Iovchev, I.; Dryanovska, L.; Lazova, O. *Farmatsiya (Sofia),* 1976, *26*, 12.

46. Issa, A. S.; Beltagy, Y. A.; Kassem, M. G.; Daabees, H. G. *Talanta,* 1985, *32*, 209.

47. Ivashkiv, E. *J. Pharm. Sci.* 1978, *67*, 1024.

48. Jovanovic, M. S.; Radulovic, D.; Zivanovic, L. *Acta Pol. Pharm.* 1987, *44*, 322.

49. Korany, M. A.; Bedair, M.; Mahgoub, H.; Elsayed, M. A. H. *J. Assoc. Off. Anal. Chem.* 1986, *69*, 608.

50. Kvach, A. S.; Aleksandrova, V. Y. *Khim.-Farm. Zh.* 1987, *21*, 110.

51. Kalatzis, E.; Zarbi, I. *J. Pharm. Sci.* 1976, *65*, 71.

52. Kamalapurkar, O. S.; Chudasama, J. J. *Indian Drugs* 1983, *20*, 298.

53. Kamalapurkar, O. S.; Priolkar, S. R. S. *Indian Drugs,* 1982, *20*, 108.

54. Kannan, K.; Manavalan, R.; Kelkar, A. K. *Indian J. Pharm. Sci.* 1987, *49*, 120.

55. Kir, S.; Safak, C.; Tureli, A.; Temizer, A. *Z. Anal. Chem.* 1991, *339*, 264.

56. Korany, M. A.; Abdel-Hay, M. H.; Galal, S. M.; Elsayed, M. A. H. *J. Pharm. Belg.* 1984, *40*, 178.

57. Krishna, R. R.; Sastry, C. S. P. *Talanta* 1979, *26*, 861.

58. Krzek, J.; Janczenko, Z. *Chem. Anal.* 1985, *30*, 465.

59. Kulikov, S. I.; Bokovikova, T. N.; Karpova, L. K.; Chichiro, V. E. *Farmatsiya* 1987, *36*, 43.

60. Mahrous, M. S.; Badel-Khalek, M. M.; Abdel-Hamid, M. E. *Talanta* 1985, *32*, 561.

61. Martinez Calatayud, J.; Campins Falco, P. *Anal. Chim. Acta* 1986, *189*, 323.

62. Martinez Calatayud, J.; Pascual Marti, M. C.; Sagrado Vives, S. *Anal. Lett.* 1986, *19*, 2023.

63. Martinez Calatayud, J.; Sagrado Vives, S. *J. Pharm. Biomed. Anal.* 1989, *7*, 1165.

64. Mattocks, A. M.; Hernandez, H. R. *Bull. Natl. Form. Comm.* 1950, *18*, 113.

65. Mayadeo, M. S.; Banavali, R. K. *Ind. J. Chem., Sect. A.* 1986, *25*, 789.

66. Milch, Gy.; Aninger, L. *Acta Pharm. Hung.* 1976, *46*, 78, 102; 1977, *47*, 125.

67. Milch, Gy.; Szabó, É. *J. Pharm. Biomed. Anal.* 1991, *9*, 1107.

68. Mohamed, M. E. *Pharmazie* 1983, *38*, 784.

69. Mohamed, M. E.; Wahbi, A. A. M.; GadKariem, E. A. *Anal. Letters* 1983, *16*, 1545.

70. Morelli, B. *J. Pharm. Biomed. Anal.* 1989, *7*, 577.

71. Novakovic, J. *Boll. Chim. Farm.* 1989, *128*, 370.

72. Onur, F.; Acar, N. *J. Fac. Pharm. Gazi* 1990, *7*, 25.

73. Özden, S.; Gümüs, F. *Arch. Pharm.* 1988, *321*, 565.

74. Pankratz, R. E.; Bandelin, F. J. *J. Am. Pharm. Assoc. Sci. Ed.* 1956, *45*, 364.

75. Patel, H. V.; Morton, D. J. *J. Clin. Pharm. Therap.* 1988, *13*, 233.

76. Pinzauti, S.; La Porta, E.; Casini, M.; Betti, C. *Pharm. Acta Helv.* 1982, *57*, 334.

77. Plakogiannis, F. M.; Saad, A. M. *J. Pharm. Sci.* 1975, *64*, 1547.

78. Popov, D. M.; Litvin, A. A. *Khim. -Farm. Zh.* 1980, *14*, 108.

79. Qureshi, S. Z.; Saeed, A. *Ann. Chim.* 1990, *80*, 203.
80. Radulovic, D.; Jovanovic, M. S.; Zivanovic, L. *Pharmazie* 1986, *41*, 434.
81. Radulovic, D.; Pecenac, D.; Zivanovic, L.; Agatonovic-Kustrin, S. *Farmaco* 1990, *45*, 447.
82. Radulovic, D.; Pecenac, D.; Zivanovic, L.; Agatonovic-Kustrin, S. *J. Pharm. Biomed. Anal.* 1990, *8*, 739.
83. Sane, R. T.; Chandrashekhar, T. G.; Nayak, V. G. *Ind. J. Pharm. Sci.* 1986, *48*, 113.
84. Sane, R. T.; Dhamankar, A. Y. *Indian Drugs* 1981, *19*, 74.
85. Sane, R. T.; Karkhanis, P. P.; Sathe, A. Y. *Indian Drugs* 1981, *18*, 149.
86. Sane, R. T.; Kubal, M. L.; Nayak, V. G. *Indian Drugs* 1986, *23*, 466.
87. Sane, R. T.; Samant, R. S.; Nayak, V. G. *Indian Drugs* 1986, *24*, 161.
88. Santoro, M. I. R. M.; Dos Santos, M. M.; Magalhaes, J. F. *J. Assoc. Off. Anal. Chem.* 1984, *67*, 532.
89. Sastry, C. S. P.; Murty, K. V. S. S. *Indian Drugs* 1982, *19*, 158.
90. Sastry, C. S. P.; Rao, B. G.; Murthy, K. V. S. S. *J. Indian Chem. Soc.* 1982, *59*, 1107.
91. Sastry, C. S. P.; Rao, A. R. M.; Prasad, T. N. V. *Anal. Letters* 1987, *20*, 349.
92. Sastry, C. S. P.; Rao, A. R. M.; Vijaya, D. *Indian Drugs* 1986, *24*, 111.
93. Shah, G. S.; Mehta, R. C.; Ganatra, J. P. *Indian J. Pharm. Sci.* 1985, *47*, 206.
94. Shah, P. P.; Mehta, R. C. *Indian J. Pharm. Sci.* 1981, *43*, 147.
95. Shingbal, D. M.; Bhangle, S. R. *Indian Drugs* 1987, *24*, 270.
96. Shingbal, D. M.; Sawant, K. V. *Indian Drugs* 1982, *19*, 239.
97. Shingbal, D. M.; Rao, V. R. *Indian Drugs* 1985, *22*, 275.
98. Skakun, N. M.; Kazarinov, M. O. *Farm. Zh.* 1986, 47.
99. Sultan, S. M. *Analyst* 1988, *113*, 149.
100. Sultan, S. M.; Altamrah, S. A.; Alrahman, A. M. A.; Alzamil, I. Z.; Karrar, M. O. *J. Pharm. Biomed. Anal.* 1989, 7, 279.
101. Sultan, S. M.; Alzamil, I. Z.; Alrahman, A. M. A.; Altamrah, S. A.; Asha, Y. *Analyst* 1986, *111*, 919.
102. Suzuki, M.; Kanbe, K.; Asabe, Y.; Takitani, S. *Bunseki Kagaku* 1978, *27*, 525.
103. Szabolcs-Sándor, E. *Acta Pharm. Hung.* 1977, *47*, 174.
104. *The United States Pharmacopoeia XXII.* USP Convention Inc.: Rockville, 1990; a. 12; b. 16; c. 1147; d. 1148; e. 1177.
105. Van Kerchove, C.; Thielemens-Bosmans, H.; Van Geert-Verkest, I. *J. Pharm. Belg.* 1979, *34*, 13.
106. Vaughn, J. B. *J. Pharm. Sci.* 1969, *58*, 469.
107. Verma, K. K.; Gulati, A. K.; Palod, S.; Tyagi, P. *Analyst* 1984, *109*, 735.
108. Verma, K. K.; Stewart, K. K. *Anal. Chim. Acta* 1988, *214*, 207.
109. Vetuschi, C.; Ragno, G.; Losacco, V. *Boll. Chim. Farm.* 1988, *127*, 193.
110. Vetuschi, C.; Ragno, G.; Mazzeo, P.; Mazzeo-Farina, A. *Farmaco, Ed. Prat.* 1985, *40*, 215.
111. Vinnikova, A. V. *Farm. Zh.* 1979, 74.
112. Wahbi, A. -A. M.; Belal, S.; Bedair, M.; Abdine, H. *J. Assoc. Off. Anal. Chem.* 1981, *64*, 1179.
113. Wahbi, A. -A. M.; Mohamed, M. E.; Gad Kariem, E. A.; Aboul-Enein, H. Y. *Analyst* 1983, *108*, 886.
114. Whitaker, J. E.; Hoyt, A. M. *J. Pharm. Sci.* 1984, *73*, 1184.
115. Witzel, H.; Pindur, U. *Pharm. Acta Helv.* 1988, *63*, 164.
116. Xia, G.; Liu, S. *Yaoxue Tongbao* 1986, *21*, 25.
117. Yan, Z.; Liu, Y. *Yaowu Fenxi Zazhi* 1987, *7*, 120.
118. Zakhari, N. A.; Hassan, S. M.; El-Shabrawy, Y. *J. Pharm. Biomed. Anal.* 1991, *9*, 421.
119. Zivanovic, L.; Radulovic, D.; Jovanovic, M. *Pharm. Acta Helv.* 1988, *63*, 350.
120. Zommer-Urbanska, S.; Bojarowicz, H. *J. Pharm. Biomed. Anal.* 1986, *4*, 475.

F. ALKALOIDS AND RELATED DERIVATIVES

1. Methods Based on Natural Absorption

Alkaloids which are usually heterocyclic bases of plant origin are spectrophotometrically active substances, but the extent of this activity strongly depends on the characteristic skeleton of the alkaloid and its functional groups. The spectrophotometric data of the pharmaceutically and toxicologically most important alkaloids are collected in Table 10.2.

As it is seen from the data in Table 10.2, the spectrophotometric activities are really very different. For example, berberine is colored, the rubane and isoquinoline alkaloids possess very

TABLE 10.2. Spectral Data of Some Pharmaceutically or Toxicologically Important Alkaloids[14,98,104]

Alkaloid	Skeleton/ heterocycle	Chromophore*	λ_{max}, nm	ε	Solvent
Aconitine	Pyrrolidine	Benzoyloxy	234	12,800	0.5 M H_2SO_4
			275	1,150	
Apomorphine	Isoquinoline	Dihydroxybiphenyl	272	19,800	0.1 M H_2SO_4
Atropine	Tropane	Phenyl	257	230	0.5 M H_2SO_4
Benztropine	Tropane	Phenyl (2)	259	370	0.1 M H_2SO_4
Berberine	Phenyl- isoquinoline	Phenyl-dimethoxy- isoquinoline	228	20,400	0.1 M H_2SO_4
			264	19,600	
			345	17,650	
Brucine	Indole	Dimethoxy-N-acyl-aniline	267	14,100	EtOH
			301	10,000	
Caffeine	Purine	Purine	272	9,100	0.1 M HCl
Canescine	Indole	Indole + trimethoxybenzoyl	272	17,400	EtOH
Cocaine	Tropane	Benzoyloxy	233	14,200	0.1 M H_2SO_4
			275	1,150	
Codeine	Morphinane	Phenol ether	285	1,740	0.5 M H_2SO_4
Cinchonine	Rubane	Quinoline	235	33,400	0.1 M H_2SO_4
			315	6,390	
Dihydroergotamine	Ergoline	Indole	279	6,250	0.1 M HCl
Dihydrocodeinone	Morphinane	Phenol ether	280	1,110	0.1 M H_2SO_4
Emetine	Isoquinoline	Phenol ether (2)	228	12,500	1 M H_2SO_4
			282	5,390	
Ergonovine	Ergoline	Dehydro-indole	312	8,120	0.1 M H_2SO_4
Ergotamine	Ergoline	Dehydro-indole	240	25,000	0.5 M H_2SO_4
Ethylmorphine	Morphinane	Phenol ether	285	1,530	0.2 M H_2SO_4
Heroin	Morphinane	Phenol ester	278	1,440	0.1 M HCl
Hydrastine	Isoquinoline	bis-(OCH_3)-Phenyl- carboxylic ester + phenol ether	295	5,360	EtOH–H_2O 1:1
Hyoscyamine	Tropane	Phenyl	258	260	0.5 M H_2SO_4
Homatropine methylbromide	Tropane	Phenyl	258	210	0.1 M H_2SO_4
Levallorphane	Morphinane	Phenol	226	4,670	0.1 M HCl
			278	2,090	
			240	8,830	0.1 M NaOH
			299	3,030	
			310	7,100	0.1 M HCl
Lysergic acid diethylamide	Ergoline	Dehydro-indole			
Mescaline	Phenylethylamine	Trimethoxyphenyl	269	1,580	EtOH
Morphine	Morphinane	Phenol	284	1,570	0.5 M H_2SO_4
			250	7,270	0.1 M NaOH
			296	3,300	
Nicotine	Pyridine– pyrrolidine	Pyridine	260	7,940	0.1 M H_2SO_4
Noscapine	Isoquinoline	Dimethoxybenzoyl + phenol ether	290	4,000	MeOH
			310	4,750	
Papaverine	Isoquinoline	Dimethoxyisoquinoline + dimethoxyphenyl	251	62,000	1 M HCl
			284	6,540	
			310	8,600	
Pilocarpine	Imidazole	Imidazole	215	5,200	0.2 M H_2SO_4
Physostigmine	Eserine	Aminophenol ester	246	10,470	0.2 M HCl
			302	2,640	
Quinidine	Rubane	Methoxyquinoline	236	36,000	EtOH
			278	4,300	
			332	5,280	

TABLE 10.2. Continued

Alkaloid	Skeleton/ heterocycle	Chromophore*	λ_{max}·nm	ε	Solvent
Quinine	Rubane	Methoxyquinoline	236	36,000	EtOH
			278	4,300	
			332	5,280	
Rescinnamine	Yohimbane	Methoxyindole			
		+	228	49,200	
		trimethoxycinnamoyl	301	24,100	Borate/H_2O
Reserpine	Yohimbane	Methoxyindole	266	16,700	
		+ trimethoxybenzoyl	294	10,200	EtOH
Scopolamine	Tropane				0.5 M
		Phenyl	258	270	H_2SO_4
Strychnine	Indole	N-acylaniline	255	12,600	EtOH
Theobromine	Purine				0.5 M
		Purine	270	10,200	H_2SO_4
Theophylline	Purine	Purine	270	9,700	0.1 M HCl
Vinblastine	bis-Indole	Indole			
		+ methoxyaniline	267	18,300	MeOH
Vincamine	Eburnane	Indole	228	32,500	EtOH
			271	8,200	
Vinpocetine	Eburnane	Dehydro-indole	229	31,400	EtOH
			273	12,000	
			314	6,500	
Yohimbine	Yohimbane		220	37,700	
			272	7,400	0.1 M
		Indole	287	5,400	H_2SO_4

*Chromophores basically determining the spectrum. If the spectrum is the sum of two isolated chromophores, a "+" sign is placed between them.

intense and structured spectra most suitable for their spectrophotometric determination. At the same time, however, the skeleton of tropane alkaloids is inactive and the poor spectrophotometric activity of these derivatives is only due to the isolated phenyl groups of the esterifying acids. Imidazole alkaloids are also only moderately active spectrophotometrically.

The possibilities of methods based on the pH-dependence of the spectra are very limited, thus in many cases the basic center of the alkaloid skeleton is outside the chromophoric system (morphinane, tropane derivatives) or the chromophoric system does contain nitrogen, but this is not the center of the basicity: this is another nitrogen located outside of the chromophoric system (different types of indole alkaloids such as ergoline and eburnane derivatives, alkaloids with quinoline and isoquinoline skeleton). Consequently, spectral shifts which can be used for analytical purposes taking place upon mild acidification occur only in a few cases (e.g., purine derivatives, berberine). The spectra of morphinanes containing phenolic hydroxyl group are shifted in alkaline media.

The intense light absorption of alkaloids is widely used in the pharmacopoeias mainly for their determination in formulations but in some cases direct spectrophotometric determination is used for the assay of bulk drugs, too, e.g., for vinblastine[11a] and vincristine[11b] at 267 and 297 nm, respectively. One of the rare opportunities when direct spectrophotometry is used for the purity test of a bulk drug is the determination at 245 nm of apoatropine as an impurity in atropine containing only an isolated phenyl ring.[11c]

Spectrophotometric assays based on natural absorption are very often used for the assay of alkaloids in various formulations. For example vinblastine and vincristine sulphates are determined in injections,[11d,106a] morphine in injections at 284 nm,[11e] theophylline in sustained release capsules at 271 nm.[106c] The content uniformity test of quinine sulphate capsules is carried out at 345 nm[106b] and the dissolution test of the tablet of the same at 348 nm.[11f] The assay of pilocarpine in ophthalmic preparations is based on an unusually low wavelength: 215 nm.[106d] In these cases no specific sample preparation is necessary. In the case of the assay of papaverine hydrochloride injection, however, the base is extracted from alkaline media with chloroform, the solvent evaporated and papaverine measured in 0.1 M hydrochloric acid at 251 nm.[106e] Extraction enables caffeine to be determined selectively at 273 nm in a combination with aspirin.[11g]

The direct spectrophotometric method is widely applied for the more or less selective determination of alkaloids in plant extracts in the presence of related alkaloids and other components and in multi-component pharmaceutical preparations. Of the immense literature related to this field only a few examples are mentioned here.

Salama and Belal[93] described the simultaneous determination of strychnine and yohimbine in tablets and injections after extraction and dilution, respectively, and measuring at 255 and 272 nm in 0.1 M sulphuric acid medium using the method of simultaneous equations with two unknowns. Of the very frequently occurring problems of pharmaceutical analysis, the simultaneous determination of strychnine and brucine can be similarly accomplished, the two wavelengths being 262 and 300 nm. In this method described by El-Masry and Wahbi,[21] the two alkaloids are separated from other, interfering components by extraction of their ion-pair complexes with chloroform at pH 5. The extracted ion-pair is then destroyed by 0.1 M sulphuric acid and the alkaloids are measured in this phase. Further examples for the two-wavelength measurement are the simultaneous determination of theobromine and papaverine in 0.1 M hydrochloric acid at 269 and 311 nm,[112] the determination of caffeine and salicylamide in methanolic medium at 273 and 304 nm,[114] simultaneous determination of papaverine and caffeine in the presence of phenobarbital after the extraction of the two alkaloids with chloroform from a mildly alkaline medium and measurement at 328 and 276 nm.[52] The analysis of a similar papaverine-theophylline-phenobarbital tablet can be performed without a separation step by measurement at three wavelengths at two different pHs using the selective absorption of papaverine at 309 nm.[68] Milch et al.[60] described the determination of caffeine in the presence of sodium salicylate, sodium benzoate and procaine, respectively, by measuring at two wavelengths.

Other three-component mixtures can also be easily resolved if a wavelength can be found where one of the components absorbs selectively. For example at the long-wavelength maximum of the spectrum of quinine (see Table 10.2), it can be measured without separation in the presence of ascorbic acid and aspirin[77] as well as caffeine and aminophenazone.[113] Measuring at three wavelengths and using three equations with three unknowns enables the assay of three-component preparations even if their spectra more or less overlap, e.g., the simultaneous determination of caffeine, pyridoxine and meclizine.[99]

It has to be noted that the role of the additivity of absorbances of the components of mixtures is not always valid in the field of alkaloids; complex formation between the components can distort the spectra.[16] Such a phenomenon is likely to occur if one component is present in great excess. For example in the case of a formulation containing ergotamine tartrate and hundred-fold quantity of caffeine, the determination of the latter is not difficult at the maximum of its spectrum. In the case of ergotamine, however, the spectrum is distorted and hence it is advisable to measure at an isoabsorption point (312 nm) where the absorbance of ergotamine is independent of the concentration of caffeine, rather than on the maximum of the spectrum.[35]

An interesting possibility for the determination of an alkaloid in combined formulations is to use a chemical reaction for the elimination of the spectrum of the other, interfering component in the spectral range in question. For example, papaverine could be selectively determined in the presence of dipyrone after the oxidation of the latter with iodine.[69]

Multi-wavelength measurements and algebraic treatment of the data enables several analytical problems to be solved in the field of alkaloids. Using the method of orthogonal polynomials (Section 4.C.3) for the determination of quinine in the presence of dipyrone, the absorbances were read at 4 nm distances between 324 and 344.[3] For the determination of theophylline in the presence of phenobarbital,[23] the 266–286 nm range was used under similar conditions. Measuring at 0.5 nm distances and using computer-assisted workup of the results enables difficult analytical problems to be solved such as the simultaneous determination of atropine (240–250 nm) and its degradation products: tropic acid (245–266 nm), apoatropine and atropic acid (255–271 nm).[80]

Derivative spectrophotometry is especially important in the field of tropane alkaloids: their isolated phenyl group is too weak as a chromophore for quantitative analytical purposes, the fine structure of the α-band, however, presents excellent possibilities for the use of derivative spectrophotometry. For example, Konochen et al.[51] used the second derivative spectrum for the determination of homatropine methylbromide in dosage forms. The very interesting method of Hassan and Davidson[37] also uses the second derivative spectra for the determination of tropane alkaloids, in this case; however, the alkaloids themselves are not examined, but their tetraphenylborates (after the removal of the excess of the reagent by extraction with 1,2-dichloroethane). The advantage of this method is that the four phenyl groups in the tetraphenylborate part of the derivative contribute to the amplitude of the derivative spectra, thus increasing the sensitivity of the measurement.

Derivative spectrophotometry is often applicable to overcome background problems in the spectrophotometric analysis of spectrophotometrically more active alkaloids, too, and for their simultaneous determination in preparations. Some characteristic examples are as follows. Caffeine was determined in the presence of procaine[1] and sodium benzoate[110] as well as in tea, coffee and beverages.[1] The sum of quinine and quinidine was determined in cinchona extracts,[27] berberine in tablets,[111] codeine also in tablets in the presence of ten-fold quantity of paracetamol.[108] The simultaneous determination of morphine and heroin,[54] the determination of physostigmine in the presence of its oxidative degradation products[8] and the determination of theophylline in the presence of sorbic acid[63] and several other determinations were performed also by derivative spectrophotometry.

As has already been mentioned, the changes in the spectra induced by pH changes are, in most instances, insignificant in the field of alkaloids: these are greatly restricted to morphine and purine derivatives. These can be determined by pH-induced difference spectrophotometry. Some of these methods are presented in Table 8.1.

The possibilities of spectrophotometry based on natural absorption are greatly enhanced in the field of alkaloids, too, by combining the measurement with preliminary chromatographic separation. Even an outlined presentation of the vast amount of literature data of this type would be beyond the scope of this book. Instead, some characteristic examples will be presented. An example for the use of column chromatographic separation is the method of De Fabrizio.[17] The theophylline and 7-(2-hydroxyethyl)-theophylline components of a syrup were separated using an Amberlite CG-400 ion-exchange column in its hydroxyl form. The latter is first eluted with water followed by the elution of theophylline with 1 M hydrochloric acid prior to the spectrophotometric measurement.

Thin-layer chromatography is more extensively used for the separation of alkaloids for their subsequent spectrophotometric determination. The method of Agnihotri et al.[5] for the determination of thebaine in a crude mixture containing 40–55% thebaine after chromatography on silica layer and visualization with iodine vapor is a typical elution–spectrophotometric method. The main spot and that of standard thebaine on the same plate were extracted by 0.1 M hydrochloric acid and the measurement was performed at 285 nm. As the blank solution the extract of a blank zone of the same layer corresponding to the R_f value of thebaine was used. A similar method was used, among others, for the determination of the synthetic purine analog, 6-mercaptopurine in the presence of its degradation products.[75]

The application of in situ densitometric methods is also widespread in the analysis of alkaloids (see Section 5.E). A characteristic example is the simultaneous determination of reserpine and rescinnamine in pharmaceutical preparations. These alkaloids were separated on a silica layer using 24:1 mixture of carbon tetrachloride and methanol as the eluent. The wavelength of the densitometric investigation was the long-wavelength maximum of the spectrum of reserpine: 295 nm.[36]

2. Methods Based on Chemical Reactions

Of the quantitative methods for the spectrophotometric determination of alkaloids based on chemical reaction, it is undoubtedly the ion-pair extraction method which is most extensively used. As regards the theory and practice of this method, Section 7.H of this book and for the early literature of the application of this method to alkaloids the monograph of Higuchi[38] are recommended. If the table in Reference 38 is compared with Table 10.3 where some characteristic data of the methods published in the last 15 years are summarized, it can be stated that neither the reagents nor the methodology have changed significantly in the last 30–40 years, but the method is still one of the most important methods in the analysis of alkaloids. In the majority of cases, traditional

TABLE 10.3. Some Ion-Pair Extraction Methods for the Spectrophotometric Determination of Alkaloids

Alkaloid	Ion-pairing reagent	pH	Extracting solvent	λ_{max}, nm	Ref.
Atropine, scopolamine	Bromothymol blue	7.5	$CHCl_3$	430	34
Scopolamine butyl-bromide	Eriochrome black-T	5.6	$CHCl_3$	520	6
Tropane alkaloids	Methyl orange	5.4	$CHCl_3$	428–430	9
Tropane alkaloids	Bromocresol purple	3–4	$CHCl_3$	405–410	12
Scopolamine butyl-bromide	Bromocresol purple	5.3	$CHCl_3$	410	73
Tropane and quinine alkaloids, strychnine, brucine	2,6-Dichlorophenol indophenol	dil. NH_3	$CHCl_3$	580–603	2
Berberine	2,6-Dichlorophenol indophenol	8.5	Nitrobenzene	650	90
Berberine	Picric acid	11.5; 8	1,2-Dichloroethane	360	84,101
Berberine	Bromocresol purple	7.2	$CHCl_3$	590*	20
Berberine	Bromocresol purple + quinine	7	1,2-Dichloroethane	593	92
Berberine	Bromophenol blue + quinine	6.7	1,2-Dichloroethane	610	85
Berberine	Tetrabromophenol-phthalein ethyl ester	8.5	1,2-Dichloroethane	610	86,87
Morphine, codeine, thebaine	Tropeolin 00	2–4.6 (M), 2–2.8 (C, T)	$CHCl_3$	552	78
Morphine	Tropeolin 00	3.5	$CHCl_3$	543	64
Morphine (after acetylation)	Methyl orange	4	$CHCl_3$	420	13
Codeine, thebaine, noscapine, papaverine	Solochrome Green V-150	6 (c), 3.6 (T), 2 (N, P)	$CHCl_3$	520	83
Codeine	Bromocresol green	3.8	$CHCl_3$	418	29
Codeine, noscapine	Orange II	0.5–2	$CHCl_3$	485	10
Codeine	Methyl orange	3.5	$CHCl_3$	418	25
Ethylmorphine	Bromothymol blue	7.6	$CHCl_3$	597*	79
Quinine, papaverine, strychnine	Chrome Azurol S	4.1	$CHCl_3$	450–455	72
Quinine	Cresol red	5	$CHCl_3$	410	102
Quinine	Eosine/polyvinyl alcohol	2.7	No extraction	540	117
Quinine	Bromophenol blue	6.5	$CHCl_3$	590	71
	Bromocresol green	7.2	1,2-Dichloroethane	550	
Quinine	Tropeolin 00	4.35	CH_2Cl_2	543**	59
Quinine, etc.	Orange II	2	$CHCl_3$	485*	57
Quinine	Methyl orange	3.5	$CHCl_3$	464*	48
Quinine, emetine, ephedrine	Tetrabromophenol-phthalein ethyl ester	8.5	1,2-Dichloroethane	555, 570	89
Brucine, strychnine	Solochrome Green V-150	3	$CHCl_3$	520	81,82
Emetine	Methyl orange	5	$CHCl_3$	460*	94
Nicotine	Tropeolin 00	4.35	CH_2Cl_2	408, 530**	58
Papaverine	Eriochrome black T	2	CH_2Cl_2	518	103
Reserpine	Hexaiodobismuthate 1 M H_2SO_4		$CHCl_3$	475	18
Solasodine	Methyl orange	4.7	$CHCl_3$	525	15
Tubocuranine	Methyl orange	7.8	$CHCl_3$	490**	53
Veratrine	Bromophenol blue	2	$CHCl_3$	410	62

*Measurement in alkaline medium after the decomposition of the complex.
**Measurement in acidic medium after the decomposition of the complex.

reagents are used and the colored ion-pair is extracted at optimized with chloroform and the organic phase is measured spectrophotometrically. New reagents introduced from time to time occasionally afford very good sensitivity. For example a molar absorptivity of 60,800 was described for the brucine–Solochrome Green V-150 complex.[81] The early version of the method is sometimes still used where the extracted ion-pair is decomposed and re-extracted into the aqueous phase by alkaline treatment and the determination of the alkaloid is based on the measurement of the excess dye anion in the alkaline phase.[20,48,53,79] Another possibility is the decomposition of the ion-pair complex by strong acid and measurement of the undissociated dye. If the decomposition is carried out with methanolic hydrochloric acid, miscible with the organic phase then no further extraction step is necessary.[55,58,59]

The adaptation of the ion-pair extraction method to FIA analyzers (Section 9.J) merits special attention. For example Sakai[88] described the determination of berberine using tetrabromophenol-phthalein as the reagent. 1,2-Dichloromethane was used for the extraction at pH 11.

Of the other classical methods the importance of the formation of colored derivatives with various metal complex reagents has decreased considerably. The classical alkaloid reagent, Reinecke salt (diamino-tetrathiocyanato chromium(III)) forms colored precipitates in acidic media with a great variety of protonated alkaloid cations. The spectrophotometric measurement around 520 nm requires the filtration of the precipitate and its dissolution (usually in acetone) prior to the absorbance reading. This method is still used in USP XXII for the assay of homatropine methylbromide tablets.[106f]

Several other reagents of the metal complex type have also been described for the determination of alkaloids and occasional new reagents are still being introduced. In modern methods the precipitation and filtration steps are usually omitted: the reaction products are separated from the excess of the reagent by extraction. These methods can be considered to be the analogs of the ion-pair extraction methods where the dyes are replaced by the metal complex reagents. For example, cobalt(II) thiocyanate forms complexes at pH 2–2.5 with papaverine, quinine and ephedrine which can be measured at 625 nm after extraction with carbon tetrachloride-cyclohexane.[97] Aconitine can be measured with the same reagent at 320 nm after extraction with chloroform.[47] Not even an extraction step is necessary when the aminopentacyano-iron(III) and nitrosylpentacyano-iron(III) reagents are used e.g., for the determination of 6-mercaptopurine at 455 and 650 nm, respectively.[47]

Of the methods based on the formation of metal complexes the methods introduced by Japanese workers using the formation of ternary complexes excel with their extremely high sensitivities. In these methods the reagents are various metal ions and xanthene and related derivatives. The complexes are either separated by extraction from the colored reagent or, even better, the rather insoluble ternary complexes are solubilized using suitable solvents and solubilizing agents. As seen in Figure 10.2 in the example of the determination of reserpine as the complex with the uranium(III) o-hydroxy-hydroquinonephthalein reagent, the bathochromic shift taking place during the ternary complex formation is sufficient to form the basis of the measurement in the presence of the excess reagent.

The main characteristics of some of these methods are summarized in Table 10.4.

The formation of charge transfer complexes (Section 7.I) can also be utilized in the spectrophotometric analysis of alkaloids. In fact the classical methods based on interaction with halogens (mainly iodine) also belong to this category.[38] Of the new applications of the old method, the determination of emetine and lobeline on the basis of reaction with iodine in chloroform and measurement at 292 nm,[26] the determination of papaverine, ephedrine, reserpine, homatropine methylbromide, etc., in 1,2-dichloroethane at 295 nm[105] and the selective determination of reserpine in the presence of hydrochlorothiazide at 294 nm[4] are mentioned. In the course of the determination of hyoscyamine, two variants were developed.[22] The complex with iodine can be measured against the reagent (iodine in carbon tetrachloride) at 390 nm. Interchanging the test and blank solutions, the determination can be based on the absorbance measurement at 520 nm; in this case the decolorization of iodine is the basis of the method. Cytisine was determined in plant extracts by absorbance measurement at 830 nm after reaction with the potassium iodide–iodine reagent in acidic aqueous medium.[118] In the course of the determination of atropine, pilocarpine and strychnine at 530 nm the reagent was chloranilic acid.[24] Papaverine could be measured at the same wavelength after reduction with zinc/hydrochloric acid and complex formation with fluoranil.[44]

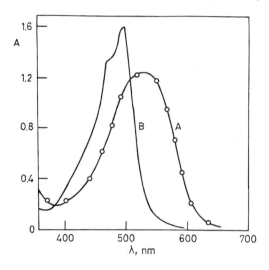

FIGURE 10.2. Spectra for the determination of reserpine as the ternary complex with uranium (VI) and o-hydroxyhydroqui-nonephthalein. Solvent: water; pH 4.6. A. Reserpine: 5.10^{-6}M; uranium (VI): $2.5 \cdot 10^{-5}$M; o-hydroxyhydroquinonephthalein: $5 \cdot 10^{-5}$M; methylcellulose: 0.075%. B. Reagent blank (without reserpine) against water. (From Fujita, Y.; Mori, I.; Kitano, S. *Bunseki Kagaku* 1984, *33*, E195.[30])

The methods discussed in the first part of this section can be considered to be more or less general methods for alkaloids. Several other methods are available and even used sometimes for the determination of certain groups of alkaloids. One example of the new applications of these classical methods is the determination of derivatives containing phenolic hydroxyl group (or phenol derivatives which can be split to form free phenols) such as morphine,[11k] thebaine[39] and capsaicine[7] using nitrous acid as the reagent which transforms them to the o-nitrosophenol derivatives which can be measured in the 430–510 nm range. If the phenol ether is not split prior to the determination, then this method is suitable for the determination of free phenolic impurities, e.g., for the determination of morphine in codeine[111] and dihydromorphine in dihydrocodeine.[11m] Phenolic alkaloids can be determined with the azo-coupling reaction, too, e.g., with diazotized sulphanilic acid at 510 nm.[28] Diazotized 4-nitroaniline was used for the determination of ergot alkaloids.[115] The determination of capsaicine at 600 nm using dichloro-p-benzoquinonechloroimine as the reagent,[74] the measurement of morphine at 450 nm after oxidation with potassium iodate[96] and the determination of ergot alkaloids and emetine after treatment with periodate and absorbance readings at 410 and 420 nm, respectively, can also be mentioned here. The latter method was adopted to the FIA analyzer (Section 9.J), too.[57]

Alkaloids containing ester or lactone groups can be determined by the iron(III)-hydroxamate method (Section 7.G), e.g., reserpine,[19,46] pilocarpine,[49] hydroxyquinuclidine esters,[61] etc. It is interesting to note that nickel (II) ions catalyze the transformation of carboxamides, too, to hydroxamic acid and this was utilized for the determination of the synthetic alkaloid analogs, pyrazinamide and morphazinamide.[76]

The most important reagent of indole alkaloids which has reserved its wide use is 4-dimethylaminobenzaldehyde (van Urk reagent). This reaction is discussed in detail in Section 7.E and the mechanism is presented in Equation 7.12. Of the applications of the method in pharmacopoeias, the assay of bulk dihydroergotamine mesylate[106g] and the injection,[106h] at 585 nm, measurement of ergonovine maleate bulk material[106i] and injection,[11h] ergotamine tartrate tablet[11i] and injection[11j,106h] are mentioned. The reaction found application in AutoAnalyzer-type automatic analyzers, too.[50] It was an early observation that in a strongly acidic medium, other alkaloids also react with 4-dimethylaminobenzaldehyde; in this case, however, red products are formed rather than the blue product with indole alkaloids. The measurement at 550 nm was used for the determination of tropane alkaloids in plant extracts.[41] Of the several other aromatic aldehydes applied for the determinations of this type, the application of 4-dimethylaminocinnamaldehyde is mentioned for

TABLE 10.4. Determination of Some Alkaloids Based on Ternary Complex Formation

Alkaloid	Reagent		pH	Extracting solvent	Solubilizing agent	λ'_{max} nm	$\varepsilon \cdot 10^{-4}$	References
	Metal ion	Complexing agent						
Reserpine	U(VI)	o-Hydroxyhydroquinonephthalein	4.6	—	Methylcellulose	535	6.8	30
Papaverine	U(VI)	o-Hydroxyhydroquinonephthalein	5.2	—	Methylcellulose	550	4.6	67
Quinine	U(VI)	o-Hydroxyhydroquinonephthalein	5.5	—	Methylcellulose	555	4.1	66
Berberine	Zr(IV)	o-Hydroxyhydroquinonephthalein + fluoride	1.8	—	Na-dodecyl sulphate	515	4.2	32
Papaverine	Pd(II)	Tetrachlorofluorescein	4.0	Membrane filter + DMSO	Polyvinyl alcohol	540	8.0	65
Adenine	Ag(I)	Eosine	5.0	—	Polyvinyl-pyrrolidone	560	11.00	33
Pilocarpine	Cu(II)	Tetrabromophenolphthalein	9.0	1,2-Dichloroethane	—	528	1.8	91
Atropine	V(IV)	Catechol	3–4	CHCl₃	—	620	1.4	100
Tubocuranine	Zr(IV)	o-hydroxyhydroquinonephthalein + fluoride	2.3	—	Na-dodecyl sulphate	515	3.0	31

the determination of brucine.[70] After a preliminary treatment with acetylacetone, caffeine can also be determined by the 4-dimethylaminobenzaldehyde reagent. The mechanism of the reaction between ergot alkaloids and ninhydrin is similar and this has also been used for analytical purposes.[116]

The classical color reaction of quinine alkaloids based on bromination (thalleiochin reaction) was made suitable for quantitative analytical purposes.[43] Chloramine T reagent was used for the determination of quinine, papaverine and codeine,[54] potassium bromate for purine alkaloids[45] and malonic acid-acetic anhydride reagent for alkaloids with tertiary amine groups (pilocarpine, emetine, noscapine).[107] A method for the determination of colchicine involves the reduction of its acetylamino group with lithium borohydride and transformation of the secondary amino group thus formed to the copper(II) dithiocarbamate.[42]

REFERENCES

1. Abdel Moety, E. M.; Mostafa, A. A. *Medd. Nor. Farm. Selsk.* 1986, *48,* 75.

2. Abdel Salam, M.; Mahrous, M.S.; Issa, A. S. *J. Pharm. Belg.* 1986, *41,* 226.

3. Abdine, H.; Elsayed, M. A. H.; Abdel-Hamed, M. *Acta Pharm. Jugosl.* 1978, *28,* 75.

4. Abdine, H.; Elsayed, M. A. H.; Elsayed, Y. M. *Analyst* 1978, *103,* 354.

5. Agnihotri, A.; Chandra Tewari, S.; Khatod, P.; Banerjee, S. *Analyst* 1984, *109,* 1413.

6. Altinkurt, T. *Eczacilik Bul.* 1982, *24,* 9.

7. Ayad, M. M.; Khayyal, S. E.; Farag, N. M. *Anal. Letters* 1985, *18,* 793.

8. Barary, M.; El Sayed, M.; Amin, E. A. *Acta Pharm. Jugosl.* 1989, *39,* 73.

9. Belikov, V. G.; Karpenko, V. A.; Stepanyuk, S. N. *Farmatsiya* 1984, *33,* 76.

10. Bosch Serrat, F.; Font Perez, Y. G. *An. Real Acad. Farm.* 1986, *52,* 279, 637.

11. *British Pharmacopoeia* 1988, Her Majesty's Stationery Office: London, 1988. a. 594; b. 595; c. 51; d. 864; e. 785; f. 1000; g. 903; h. 788; i. 940; j. 789; k. 738; l. 153; m. 196.

12. Bubon, N. T.; Senov, P. L. *Farmatsiya* 1977, *26,* 34.

13. Chichuev, Yu. A. *Farmatsiya* 1984, *33,* 70.

14. Clarke, E. G. C. *Isolation and Identification of Drugs.* The Pharmaceutical Press: London, 1969.

15. Crabbe, P. G.; Fryer, C. *J. Pharm. Sci.* 1982, *71,* 1356.

16. Davidson, A. G. *J. Pharm. Pharmacol.* 1976, *28,* 794.

17. De Fabrizio. *J. Pharm. Sci.* 1978, *67,* 572.

18. Dembinski, B. *Acta Polon. Pharm.* 1980, *37,* 435.

19. Dryanovska, L.; Iovchev, I. *Pharmazie* 1976, *31,* 130.

20. El-Masry, S.; Korany, M. A.; Abou-Donia, A. H. A. *J. Pharm. Sci.* 1980, *69,* 597.

21. El-Masry, S.; Wahbi, A. -A. M. *J. Assoc. Off. Anal. Chem.* 1978, *61,* 65.

22. Elsayed, M. A.-H. *Pharmazie* 1979, *34,* 115.

23. Elsayed, M. A.-H.; Abdine, H.; Elsayed, Y. M. *J. Pharm. Sci.* 1979, *68,* 9.

24. Elsayed, M. A.-H.; Agarwal, S. P. *Talanta* 1982, *29,* 1982.

25. Elsayed, M. A.-H.; Belal, S. F.; El-Walily, A.-F. M.; Abdine, H. *Analyst* 1979, *104,* 620.

26. Elsayed, M. A.-H.; Salam, M. A. A.; Salam, N. A. A.; Mohammed, Y. A. *Planta Med.* 1978, *34,* 430.

27. El-Sebakhy, N. A.; Seif El-Din, A. A.; Korany, M. A. *J. Pharm. Belg.* 1986, *41,* 222.

28. Emmanuel, J.; Viera, A. J. *Indian Drugs* 1986, *24,* 56.

29. Farsam, H.; Yahya-Saeb, H. H.; Fawzi, A. *Int. J. Pharm.* 1981, *7,* 343.

30. Fujita, Y.; Mori, I.; Kitano, S. *Bunseki Kagaku* 1984, *33,* E195.

31. Fujita, Y.; Mori, I.; Kitano, S.; Kamada, Y. *Bunseki Kagaku* 1983, *32,* E375.

32. Fujita, Y.; Mori, I.; Kitano, S.; Kamada, Y. *Bunseki Kagaku* 1984, *33,* E445.

33. Fujita, Y.; Mori, I.; Kitano, S.; Kawabe, H.; Kamada, Y. *Bull. Chem. Soc. Japan* 1984, *57,* 1828.

34. Grishina, M. S.; Dyukova, V. V.; Kovalenko, L. I.; Popov, D. M. *Farmatsiya* 1986, *35,* 24.

35. Halim, A. F. *Pharmazie* 1981, *36,* 157.

36. Hartmann, V.; Schnabel, G. *Pharm. Ind.* 1975, *37,* 451.

37. Hassan, S. M.; Davidson, A. G. *J. Pharm. Pharmacol.* 1984, *36,* 7.

38. Higuchi, T.; Bodin, J. I. Alkaloids and Other Basic Nitrogenous Compounds, in *Pharmaceutical Analysis;* (Higuchi, T., Brochmann-Hanssen, E., Eds.), Interscience: New York, London, 1961, 313–543.

39. Ikonomovski, K. *J. Pharm. Sci.* 1981, *70*, 102.

40. Jaksevac-Miksa, M.; Hankonyi, V.; Karas-Gasparec, V. *Acta Pharm. Jugosl.* 1979, *29*, 139.

41. Karawya, M. S.; Abdel-Wahab, S. M.; Hifnawy, M. S.; Ghourab, M. G. *J. Assoc. Off. Anal. Chem.* 1975, *58*, 884.

42. Karawya, M. S.; Diab, A. M. *J. Assoc. Off. Anal. Chem.* 1975, *58*, 1173.

43. Karawya, M. S.; Diab, A. M. *J. Pharm. Sci.* 1977, *66*, 1317.

44. Karawya, M. S.; Diab, A. M.; Swelem, N. Z. *Anal. Lett.* 1984, *17*, 69.

45. Karawya, M. S.; Diab, A. M.; Swelem, N. Z. *Anal. Letters* 1984, *17*, 77.

46. Karawya, M. S.; Sharaf A.-A. A.; Diab, A. M. *J. Assoc. Off. Anal. Chem.* 1976, *59*, 795.

47. Karawya, M. S.; Sharaf, A.-A. A.; Diab, A. M. *J. Assoc. Off. Anal. Chem.* 1976, *59*, 799.

48. Kartashov, V. A.; Knaub, V. A.; Kudrikova, L. E.; Kuzmina, E. Yu. *Farmatsiya* 1984, *33*, 37.

49. Kashkina, A.; Cakste, Z. *Latv. PSR. Zinat. Akad. Vestis, Kim. Ser.* 1986, 582.

50. Kirchhoefer, R. D.; Wells, C. E. *J. Ass. Off. Anal. Chem.* 1975, *58*, 879.

51. Knochem, M.; Bardanca, M.; Piaggio, P. *Boll. Chim. Farm.* 1987, *126*, 294.

52. Kostennikova, Z. P.; Fetkhilinna, G. A.; Grigoreva, T. V.; Raschetnova, V. I. *Farmatsiya* 1980, *29*, 34.

53. Kuzmitskaya, A. E.; Kramarenko, V. F. *Farmatsiya* 1987, *36*, 68.

54. Lawrence, A. H.; Kovar, J. *Anal. Chem.* 1984, *56*, 1731.

55. Mahmoud, Z. F.; El-Masry, S. *Sci. Pharm.* 1980, *48*, 365.

56. Manes Vinuesa, J.; Bosch Serrat, F. *An. Real Acad. Farm.* 1982, *48*, 525.

57. Martinez Calatayud, J.; Vives, S. S. *Pharmazie* 1989, *44*, 616.

58. Matsuoka, T.; Mitsui, T. *Bunseki Kagaku* 1982, *31*, 377.

59. Matsuoka, T.; Mitsui, T.; Fujimura, Y. *Eisei Kagaku* 1984, *28*, 274.

60. Milch, Gy.; Csontos, A.; Borsai, M.; Vadon, E.; Mogács, I. *Acta Pharm. Hung.* 1966, *36*, 200.

61. Minka, A. F.; Shkadova, A. I.; Kalashnikov, A. A.; Ogurtsov, V. V. *Farm. Zh.* 1988, 45.

62. Mironova, T. *Farmatsiya* 1983, *32*, 48.

63. Mitra, A. K.; Berge, S. M. *Pharm. Res.* 1989, *6*, S133.

64. Mitsui, T.; Fujimura, Y. *Eisei Kagaku* 1983, *29*, 292.

65. Mori, I.; Fujita, Y.; Kawabe, H.; Fujita, K. *Chem. Pharm. Bull.* 1986, *34*, 902.

66. Mori, I.; Fujita, Y.; Kitano, S. *Bunseki Kagaku* 1982, *31*, 475.

67. Mori, I.; Fujita, Y.; Sakaguchi, G. *Bunseki Kagaku* 1982, *31*, E77.

68. Moussa, A.-F. A. *Pharmazie*, 1978, *33*, 296.

69. Moussa, A.-F. A. *Pharmazie* 1978, *33*, 460.

70. Nabi, S. A.; Siddigi, A. R.; Ahmad, N. *Chem. Anal.* 1980, *25*, 643.

71. Ohno, N.; Sakai, T. *Bunseki Kagaku* 1985, *34*, 695.

72. Ottis, M.; Malát, M. *Cesk. Farm.* 1980, *29*, 131.

73. Pan, S. K.; Gupta, P. N.; Chib, S. K.; Thomas, K. M. *Ind. J. Pharm. Sci.* 1986, *48*, 116.

74. Pankar, D. S.; Magar, N. G. *J. Chromat.* 1977, *144*, 149.

75. Pawelczyk, E.; Majewski, W.; Wezyk, E. *Farm. Polska* 1987, *43*, 336.

76. Pawelczyk, E.; Smilowski, B. *Acta Polon. Pharm.* 1986, *43*, 340.

77. Peterdi, B.; Sperlágh, J.; Végh, M. *Acta Pharm. Hung.* 1986, *56*, 115.

78. Peterkova, M.; Matuosova, O.; Kakac, B. *Cesk. Farm.* 1981, *30*, 217.

79. Postrigan, I. G.; Petrenko, V. V.; Lutsko, P. P.; Samko, A. V.; Mikhno, V. V. *Farm. Zh.* 1987, 64.

80. Puech, A.; Kister, G.; Monleaud-Dupy, J. *Annls. Pharm. Fr.* 1975, *33*, 189.

81. Rao, N. V. R.; Tandon, S. N. *Chem. Anal.* 1977, *22*, 965.

82. Rao, N. V. R.; Tandon, S. N. *Z. Anal. Chem.* 1977, *286*, 256.

83. Rao, N. V. R.; Tandon, S. N. *Ind. J. Chem. Sect. A.* 1978, *16*, 1000.

84. Sakai, T. *Bunseki Kagaku* 1978, *27*, 444.

85. Sakai, T. *Bunseki*, 1982, 493.

86. Sakai, T. *Bunseki Kagaku* 1975, *24*, 135.

87. Sakai, T. *J. Pharm. Sci.* 1979, *68*, 857.

88. Sakai, T. *Analyst* 1991, *116*, 187.

89. Sakai, T.; Hara, I.; Tsubouchi, M. *Chem. Pharm. Bull.* 1976, *24*, 1254.

90. Sakai, T.; Hara, I.; Tsubouchi, M. *Chem. Pharm. Bull.* 1977, *25*, 2451.

91. Sakai, T.; Ohno, N.; Higashi, T.; Tanaka, M. *Anal. Sci.* 1985, *1*, 275.

92. Sakai, T.; Ojima, S.; Saeki, E; Tanaka, M. *Bunseki Kagaku* 1986, *35*, 177.

93. Salama, O.; Belal, F. *Analyst* 1986, *111*, 581.

94. Saleh, M. R. I.; El-Masry, S.; El-Shaer, N. *J. Assoc. Off. Anal. Chem.* 1979, *62*, 1113.

95. Salmeron, J. A.; Bosch Serrat, F. *An. Real Acad. Farm.* 1986, *52*, 65.

96. Sarwar, M.; Aman, T. *Microchem. J.* 1984, *30*, 304.

97. Shahine, S.; Khamis, S. *Microchem. J.* 1983, *28*, 26.

98. Sharkey, M. F.; Andres, C. N.; Snow, S. W.; Major, A.; Kram, T.; Warner, V.; Alexander, T. G. *J. Assoc. Off. Anal. Chem.* 1968, *51*, 1124.

99. Sharma, S. C.; Sharma, S. C.; Saxena, R. C.; Talwar, S. K. *J. Pharm. Biomed. Anal.* 1989, *7*, 321.

100. Shesterova, I. P.; Talipov, S. T.; Karibyan, E. E.; Yakubova, M. *Zh. Anal. Chim.* 1980, *35*, 141.

101. Shi, J.; Gu, G. *Zhongcaoyao* 1984, *15*, 105

102. Soucek, J.; Halámek, E.; Kysilka, R. *Cesk. Farm.* 1986, *35*, 388.

103. Sulkowska, J.; Szumska, B.; Staroscik, R. *Acta Polon. Pharm.* 1988, *45*, 132.

104. Sunshine, I. *Handbook of Spectrophotometric Data of Drugs*, CRC Press: Boca Raton, 1981.

105. Taha, A. M.; Gomaa, C. S. *J. Pharm. Sci.* 1976, *65*, 986.

106. *The United States Pharmacopoeia XXII*, USP Convention Inc: Rockville, 1990. a. 1448; b. 1205; c. 1349; d. 1083; e. 1013; f. 643; g. 440; h. 441; i. 512; j. 515.

107. Thomas, A. D. *J. Pharm. Pharmac.* 1976, *28*, 838.

108. Ueda, H.; Pereira-Rosario, R.; Riley, C. M.; Perrin, J. H. *J. Pharm. Biomed. Anal.* 1989, *7*, 309.

109. Vachek, J.; Kakac, B. *Cesk. Farm.* 1974, *23*, 280.

110. Vergiechik, E. N.; Saushkina, A. S.; Likhota, T. T.; Raimova, L. S. *Farmatsiya* 1986, *35*, 43.

111. Yang, Q.; Zhu, S.; Yu, R. *Nanjing Yaoxueyuan Xuebao* 1985, *16*, 50.

112. Zajac, M. *Farm. Polska* 1974, *30*, 43.

113. Zajac, M.; Baranczyk, A.; Molenda, A. *Acta Polon. Pharm.* 1985, *42*, 30.

114. Zajac, M.; Ksiezniakiewicz, B. *Farm. Polska* 1972, *28*, 959.

115. Zakhari, N. A.; Hassan, S. M.; El-Shabrawy, Y. *Anal. Lett.* 1989, *22*, 3011.

116. Zakhari, N. A.; Hassan, S. M.; El-Shabrawy, Y. *Acta Pharm. Nord.* 1991, *3*, 151.

117. Zhebentyaev, A. I.; Duksina, S. G. *Farmatsiya* 1986, *35*, 16.

118. Zschiedrich, H.; Kickuth, R. *Z. Anal. Chem.* 1984, *319*, 434.

G. BARBITURATES, THIOBARBITURATES AND HYDANTOINS

1. Methods Based on Natural Absorption

Barbiturates are 5,5-disubstituted hexahydropyrimidine-2,4,6-trione derivatives. In the case of thiobarbiturates the 2-carbonyl is replaced by a thiocarbonyl group.

The spectra of barbiturates are determined by their characteristic tautomeric forms and the two-step dissociation[29] (see Section 3.L). These equilibria are shown in the reaction Scheme (10.5).

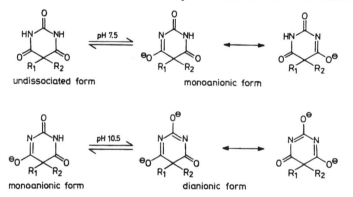

undissociated form   monoanionic form

monoanionic form   dianionic form

(10.5)

The spectra belonging to the three different ionization states are shown in Figure 10.3. The model compound selected from among the great variety of substituted barbiturates is proxibarbal where the substituents R_1 and R_2 are allyl and 2-hydroxypropyl.

As seen in Figure 10.3, the spectrum of the undisocciated form not containing conjugated bonds being present below pH 3 is not suitable for analytical measurements. The formation of the ionized

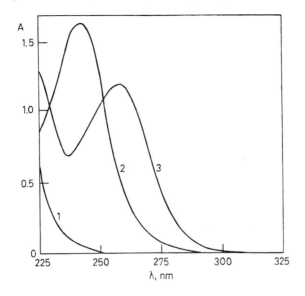

FIGURE 10.3. Spectra of proxibarbal at different pH values. Solvent: water. 1. pH 1; 2. pH 10; 3. pH 13. (From Milch, Gy., *Acta Pharm. Hung.* 1984, *54*, 230.[15])

form begins at pH 3 and is completed at 9.5. (The first dissociation constant as determined by Milch[15] simply on the basis of the change of the spectrum with pH was found to be 7.72—see Section 3.Q.4.) The absorption maximum of this form at 240 nm is sufficiently intense and selective to solve various analytical problems. The effects of the R_1 and R_2 substituents on the position of this band is negligible: it appears at 239–240 nm in the spectra of various derivatives (e.g., barbital, phenobarbital, and butobarbital). The molar absorptivities are in the range of 8,500–10,000. In the case of the 1,5,5-trisubstituted derivatives, the maximum is shifted to about 243 nm.

The second dissociation constant is lower by about 5 orders of magnitude ($pK_2 = 12.75$).[15] This means that the monoanionic form exists in a broad pH range and hence the exact setting of the pH is of no major importance: spectrophotometric measurements are usually carried out in buffers of pH 9.5–10.

As seen in Figure 10.3, the second dissociation results in a maximum at 256 nm. In the case of the 1,5,5-trisubstituted derivatives, a second dissociation is not possible and hence this bathochromic shift is not observable in their spectra.

The exploitation of the above mentioned bathochromic shift for analytical purposes is rather problematic, even in the case of the 5,5-disubstituted derivatives. Because of the very low value of the dissociation constant, the dissociation is only complete in very strongly alkaline media (above 1 M sodium hydroxide concentration). In this medium, however, the hydrolytic opening of the pyrimidine ring has to be taken into account and this results in the disappearance of the band. The rate of this reaction strongly depends on the substituents R_1 and R_2.[3,8,9,26]

On the basis of what has been described, measurement at 240 nm in moderately alkaline media is most frequently used for the determination of barbiturates based on their natural absorption. For example in the United States Pharmacopoeia, the assays of amobarbital,[31a] phenobarbital[31b] and butobarbital[31c] (bulk materials, injections, capsules and tablets, including the content uniformity test of the latter) are carried out at this wavelength in borate buffer of pH 9.6. Many papers deal with the determination of barbiturates in drug formulations, among them combined formulations based on their 240 nm band. As a consequence of the decreasing importance of barbiturates in drug therapy, these are mainly early papers, but similar papers[38] have been published occasionally even until quite recently. Several papers dealing with the assay of formulations containing barbiturates are discussed in the sections devoted to the other components of the formulation.

After appropriate separation steps, the spectrophotometric method is suitable for the determination of barbiturates in biological-toxicological samples, too.[28] The comparison of spectra at two different basicities (pH 9.9 and 0.45 M sodium hydroxide) enables the differentiation of single barbiturates, and this method used to have great importance in toxicological screening.[7,13] Problems of this type are today usually solved by chromatographic methods. It is, however, worth mentioning that the bathochromic shift occurring in mild alkaline media is widely used even now in conjunction with their identification by HPLC in formulations and biological samples. The buffer necessary for the ionization is injected into the chromatographic solution between the column and the UV detector.[4,16,22,37]

Of the modern methods for the determination of drugs in complicated matrices on the basis of their natural absorption, the method of orthogonal polynomials[33] and derivative spectrophotometry[25] found application in the analysis of barbiturates. The two-step spectral shifts in their spectra creates a suitable basis for their determination by pH-induced difference spectrophotometry (see Section 8.B). In such a method e.g., the spectral shift taking place in the course of the second dissociation was utilized (pH of the reference and test solutions 10.5 and 13.5, respectively).[36]

5,5-Disubstituted thiobarbituric acid derivatives possess a strong absorption band around 284 nm even in acidic medium and this is shifted to about 304 nm in alkaline media with about two-fold intensity relative to barbiturates.[24] This is the wavelength of the maximum in the spectrum of e.g., thiopental (5-ethyl-5-(1-methyl-butyl) derivative) in borate buffer of pH 9.4 (ε = 22,000). The USP XXII method[31d] for the assay of thiopental bulk material and injection is based on the measurement of this band. The simultaneous determination of thiobarbituric acid and barbital in blood samples was based on second derivative spectrophotometry.[25]

Of the hydantoin derivatives used in therapy, allantoin does not possess significant light absorption in acidic or neutral media. In strongly alkaline solution, however, (pH 11–13) tautomerism and ionization equilibria exist similar to those described for barbiturates, resulting in the appearance of an intense band at 226 nm[12] and this was used for its determination in formulations.[11] The situation is similar with the very important drug 5,5-diphenylhydantoin (phenytoin). Since its acidic dissociation constant is more than one order of magnitude lower than that of barbiturates, very strongly alkaline solution is necessary to use the band appearing at 235 nm for analytical purposes. Because of the presence of the two isolated phenyl groups in the molecule, their α-band appears with fine structure and low intensity at 257 nm; by measuring this, less sophisticated analytical problems can be solved, e.g., the content uniformity test of phenytoin capsules.[31e] A paper comparing the performances of spectrophotometric and HPLC methods for phenytoin merits attention.[23]

The above mentioned α-band with its fine structure is eminently suitable for applying derivative spectrophotometry. Both the first[32] and second order[19] spectra were used, significantly increasing the selectivity and sensitivity of the determination as compared to the use of zero-order spectra. The methods were used among others for the determination of phenytoin sodium in capsules.[19]

Finally, one more spectrophotometric method is mentioned from the monograph on phenytoin in USP XXII.[31f] Its benzophenone impurity content is measured at 248 nm after the selective extraction of the impurity from alkaline solution by n-hexane.

2. Methods Based on Chemical Reactions

Of the color reactions for the detection and determination of barbiturates and thiobarbiturates, those based on the formation of metal complexes are the most important. The reagent of the classical Parri[20] method was cobalt(II) nitrate in water-free methanolic medium in the presence of ammonia. In the numerous modifications of the method, mainly cobalt(II) acetate and the mixture of methanol and chloroform were employed and ammonia was replaced by various organic bases, usually by isopropylamine. As a result of these changes the stability of the color and its applicability to quantitative purposes improved considerably. The minimalization of the water content of the reaction medium also resulted in increasing color stability. The specificity of the method is moderate (other derivatives with neighboring O and N substituents also give positive reaction) and the sensitivity is weak; the reaction still found wide application in the analysis of barbiturates and thiobarbiturates in formulations and biological samples.[14,30]

The other classical method based on the formation of metal complexes is the Zwikker method[6,39] in which the reagent is copper(II) sulphate. Of the several variants of the reaction, that using aqueous copper(II) sulphate and chloroform containing pyridine as the reagent and solvent, respectively, merits attention. The selectivity of this method is superior to the Parri method although barbiturates, thiobarbiturates and hydantoins all give positive reaction. The color of the separated aqueous phase for these compounds is violet, green and blue, respectively. The modified version of the method was successfully used, especially for the determination of thiobarbiturates in biological samples.[17,18]

An indirect method for the determination of barbiturates is based on the formation of the mercury(II)-barbiturate complex. It is extracted from the aqueous solution with chloroform, and the mercury(II) content of the extract is determined using the dithizone method[1,2] (see Section 10.P).

In some of the numerous methods for the determination of phenytoin, the hydantoin ring undergoes oxidative cleavage leading to benzophenone which can be measured spectrophotometrically.[21,34,35] Using potassium permanganate as the oxidizing agent and measuring at 247 nm enables the measurement to be carried out in biological samples; moreover, using two-wavelength measurement, phenytoin and carbamazepine can be determined simultaneously. Phenobarbital, diazepam and primidone do not interfere.[10]

Methods based on the nitration of the aromatic ring are also available.[5,27] The resulting nitro group can be reduced to an amino group and the method can be based on diazotization and coupling of the diazonium salt with N-(1-naphthyl)-ethylenediamine.[5]

REFERENCES

1. Björling, C. O.; Berggren A.; Willman-Johnson, B. *J. Pharm. Pharmacol.* 1959, *11*, 297.
2. Björling, C. O.; Berggren, A.; Willman-Johnson, B.; Grönvall, A. V; Zaar, J. *Acta Chem. Scand.* 1958, *12*, 1149.
3. Brandau, R.; Neuwald, F. *Pharm. Techn.* 1964, *30*, 11.
4. Clark, C. R.; Chan, J.-L. *Anal. Chem.* 1978, *50*, 635.
5. Dill, W. A.; Kazenko, A.; Wolf, L. M.; Glazko, A. J. *J. Pharmacol Exper. Ther.* 1956, *118*, 270.
6. Flotow, L. *Pharm. Zentralhalle Dtsch.* 1949, *88*, 198.
7. Goldbaum, L. R. *Anal. Chem.* 1952, *24*, 1604.
8. Hermann, T. *Pharmazie* 1974, *29*, 453.
9. Hermann, T. *Pharmazie* 1976, *31*, 368, 618.
10. Jaffery, N. F.; Ahmad, S. N.; Jailkhani, B. L. *J. Pharmacol. Meth.* 1983, *9*, 33.
11. Lukasiak, J.; Jamrógiewicz, Z.; Sznitowska, M. *Pharmazie* 1987, *42*, 200.
12. Lutomski, J.; Jernas, B. *Pharmazie* 1976, *31*, 111.
13. Maher, J. R.; Puckett, J. *Lab. Clin. Med.* 1955, *45*, 806.
14. Matton, L. N.; Holt, W. L. *J. Am. Pharm. Assoc. Sci. Ed.* 1949, *38*, 55.
15. Milch, Gy. *Acta Pharm. Hung.* 1984, *54*, 230.
16. Minder, E. I.; Schaubhut, R.; Vonderschmitt, D. J. *J. Chromatogr. Biomed. Appl.* 1988, *428*, 369.
17. Morvay, J.; Középessy, Gy. *Acta Pharm. Hung.* 1969, *39*, 145.
18. Morvay, J.; Középessy, Gy.; Nikolasew, V. *Acta Pharm. Hung.* 1969, *39*, 54.
19. Mura, P.; Santoni, G.; Pinzauti, S. *Pharm. Acta Helv.* 1987, *62*, 226.
20. Parri, W. *Boll. Chim. Farmac.* 1924, *63*, 401.
21. Saitoh, Y.; Nishihara, K.; Nakagawa, F.; Suziki, T. *J. Pharm. Sci.* 1973, *62*, 206.
22. Scott, E. P. *J. Pharm. Sci.* 1983, *72*, 1089.
23. Shah, V. P.; Ogger, K. E. *J. Pharm. Sci.* 1986, *75*, 1113.
24. Smyth, W. F.; Jenkins, T.; Siekiera, J.; Raydar, A. *Anal. Chim. Acta* 1975, *80*, 233.
25. Soriano, J.; Jiménez, F.; Jiménez, A. I.; Arias, J. J. *Spectrosc. Lett.* 1992, *25*, 257.
26. Stainier, C. *Pharm. Acta Helv.* 1963, *38*, 587.
27. Stájer, G.; Vinkler, E.; Avar, Z. *Pharmazie* 1978, *33*, 126.
28. Stevenson, G. W. *Anal. Chem.* 1961, *33*, 1374.
29. Stimson, M. M. *J. Am. Chem. Soc.* 1979, *71*, 1470.
30. Stolman, A.; Stewart, C. P. *Analyst* 1949, *74*, 543.
31. The United States Pharmacopoeia XXII; USP Convention Inc.: Rockville, 1990. a. 77; b. 1060; c. 197; d. 1364; e. 1076; f. 1073.
32. Traveset, J.; Such, V.; Gonzalo, G.; Gelpi, E. *J. Pharm. Sci.* 1980, *69*, 629.
33. Wahbi, A.-A. M.; Belal, S.; Abdine, H.; Bedair, M. *Talanta* 1982, *29*, 931.
34. Wallace, J. E.; Briggs, J. D.; Dahl, V. E. *Anal. Chem.* 1965, *37*, 410.
35. Wallace, J. E.; Hamilton, H. E. *J. Pharm. Sci.* 1974, *63*, 1795.

36. Williams, L. A.; Zak, B. *Clin. Chim. Acta* 1959, *4*, 170.
37. Wilson, T. D.; Trompeter, W. F.; Gartelman, H. F. *J. Liqu. Chrom.* 1989, *12*, 1231.
38. Zakhari, N. A.; Walash, M. I.; Ahmed, S. M. *Farmaco* 1991, *46*, 601.
39. Zwikker, J. J. L. *Pharm. Weekblad* 1931, *68*, 975.

H. PYRAZOLE DERIVATIVES

1. Methods Based on Natural Absorption

This chapter deals with two important groups of pyrazole derivatives. The most important pyrazolone and pyrazolidine derivatives are as follows.

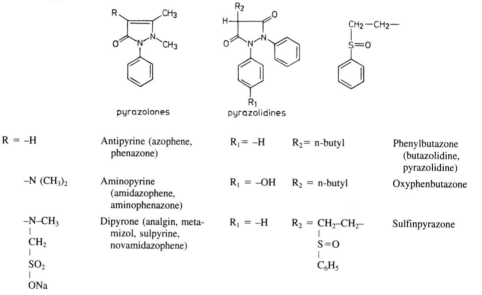

pyrazolones	pyrazolidines	

R = –H	Antipyrine (azophene, phenazone)	$R_1 = $ –H	$R_2 = $ n-butyl	Phenylbutazone (butazolidine, pyrazolidine)
–N (CH$_3$)$_2$	Aminopyrine (amidazophene, aminophenazone)	$R_1 = $ –OH	$R_2 = $ n-butyl	Oxyphenbutazone
–N–CH$_3$ | CH$_2$ | SO$_2$ | ONa	Dipyrone (analgin, meta- mizol, sulpyrine, novamidazophene)	$R_1 = $ –H	$R_2 = $ CH$_2$–CH$_2$– | S=O | C$_6$H$_5$	Sulfinpyrazone

The spectra of pyrazolones are discussed in Section 3.K and the spectrum of the main representative (phenazone, antipyrine) is shown in Figure 3.18. In methanolic or ethanolic media they have
two intense maxima around 234–236 and 265–275 nm, respectively ($\varepsilon \sim 10.000$). The spectra are
pH-dependent.

Quantitative measurements based on the above discussed spectra do not play important roles
in the various pharmacopeias. For example the USP XXII[58a] prescribes the determination of the
absorbance of antipyrine at 266 nm in methanolic solution. Many more publications can be
found in the literature describing the determination of pyrazolones in simple dosage forms and in
combinations based on their natural absorption. Of these methods the determination of antipyrine
and aminopyrine in the presence of salicylates and caffeine,[27] the determination of aminopyrine
in various combinations after extraction with chloroform,[50] the determination of the latter in the
presence of its degradation products[46] and the assay of the dosage form of dipyrone are mentioned.[65]
Spectrophotometric methods combined with TLC separation suitable for studying photochemical
degradation (which is of great importance in this family of drugs) merit special attention. Papers
of this type include the determination of antipyrine,[35–37] propiphenazone,[38] aminopyrine[51] and
isopropylaminophenazone.[52]

The application of algebraic or computer-assisted methods make the methods more reliable and
rapid. A method of this type is described in the paper of Santoni et al.[54] for the determination of
antipyrine and procaine in ear drops on the basis of absorbances at 242 and 290 nm and the use
of the absorbance ratio method.

The selectivity of the measurement can be greatly improved by using difference spectrophotometry based on the above mentioned pH-induced shifts in their spectra. For example aminopyrine

was determined in the presence of mepyramine, nialamide and chloroquine; the pH of the test solution was set at 8.5 while the reference solution was in 0.05 M sulphuric acid, thus cancelling out the spectra of the above listed other components.[71] A similar method was described for the assay of dipyrone formulations (pH 10 solution vs. pH 1 reference).[20]

In addition to the pH-induced difference spectrophotometric procedure other methods of this type have also been described. For example, dipyrone was determined in such a way that its spectrum in the reference solution was destroyed by the addition of bromine.[20]

The spectrum of phenylbutazone which is the main representative of pyrazolidines is discussed in Section 3.L and its bathochromic shift in alkaline media as the consequence of the dissociation of its enolic form is the characteristic example for this type of spectral changes in Section 3.Q.2. The tendency to enolization is the basis for the various spectrophotometric methods for the determination of derivatives in the pyrazolidine group.

Since the enol form of phenylbutazone is an acid of approximately similar strength as acetic acid (pK = 4.8) the diketo form is present only in organic solvents, e.g., in ethanol or in aqueous solution below pH 2. Although the spectrum of the diketo form ($\varepsilon_{240\ nm}$ = 15,000) is also suitable for the spectrophotometric determination of phenylbutazone, the enolic form being present in mildly alkaline media is more frequently used for the selective and sensitive determination ($\varepsilon_{264\ nm}$ = 20,400). The selectivity can be further increased by using the difference spectrophotometric variant of the measurement: the absorption of accompanying substances can be cancelled out if the absorbance of the mildly alkaline solution is determined against an acidic solution of the same concentration presuming that the accompanying substances do not change their spectra in this pH range. Similar possibilities exist for the determination of the substituted derivatives of phenylbutazone.[33]

Of the applications in pharmacopoeias of the spectrophotometric methods based on natural absorption, the United States Pharmacopoeia prescribes the determination of the absorbance of bulk phenylbutazone at 264 nm and uses the same wavelength for the dissolution testing of tablets.[58b] The wavelength of the dissolution testing of sulfinpyrazone capsules is 259 nm.[58c] The assay of oxyphenbutazone ointment according to the British Pharmacopoeia,[15] involves extraction with petroleum ether, re-extraction of the analyte into aqueous sodium carbonate solution, purification of the extract with ethyl ether and, finally, measurement at 254 nm. Reviews on phenylbutazone[34] and oxyphenbutazone[6] contain several other methods for their spectrophotometric determination.

In addition to assays of simple formulations in ethanolic medium,[56] measurement at two wavelengths enables two-component formulations to be analyzed such as phenylbutazone-aminopyrine,[3] oxyphenbutazone-dipyrone[10,17] and the determination of phenylbutazone in the presence of its degradation products.[41]

The above mentioned difference spectrophotometric methods are also widely used for the assay of various formulations, e.g., selective measurement of phenylbutazone in the presence of aminopyrine and prednisolone,[21] simultaneous determination of oxyphenbutazone and paracetamol,[40] determination of phenylbutazone, oxyphenbutazone, sulfinpyrazone, kebuzone and tribuzone in their formulations.[1,9,13]

First derivative spectrophotometry was also used e.g., for the determination of phenylbutazone and oxyphenbutazone in dosage forms[31,32] while third and fourth derivative spectra were used for the determination of the degradation products of phenylbutazone.[48]

2. Methods Based on Chemical Reactions

Many methods are available for the determination of pyrazolone derivatives based on chemical reactions. Antipyrine affords red color with 4-dimethylaminobenzaldehyde[14,24,42,53] and blue color with 4-dimethylaminocinnamaldehyde.[60] The prerequisite of a positive reaction is that the carbon atom at position 4 in the pyrazolone ring should be free, and hence no positive reaction is given by aminopyrine. Moreover, the reaction is suitable for the determination of antipyrine impurity in aminopyrine.[22] This classical color reaction has been recently re-evaluated and it was stated that quinoidal polymethyne derivatives are formed rather than the originally believed simple condensation products. The absorption maximum of the red product is at 510 nm and that of the blue is at 610 nm.[45]

One of the classical color reactions of antipyrine used for identification purposes is the formation of its 4-nitroso derivative with absorption maximum at 600 nm; this reaction was used for quantitative analytical purposes, too.[30,39] A further modification of this method is the reaction of the nitroso derivative with 8-hydroxy-1-naphtylamine-4-sulphonic acid and the measurement of the forming condensation product at 565 nm.[26]

Another old identification reaction of antipyrine is the formation of red color when it is reacted with iron(III) ions in acidic medium. This reaction was later applied to quantitative analytical purposes.[16,28] It is worth mentioning as a curiosity that the iron(III) antipyrine complex was used as a reagent for the indirect determination of fluoride ions: the latter inhibit the formation of the complex and, hence, the decrease of the intensity of its red color is proportional to the concentration of fluoride.

As weak or moderately strong bases, antipyrine and aminopyrine can be determined after their extraction with acidic dyes. In the case of antipyrine e.g., methyl orange was used as the ion-pairing reagent at pH 5.5[45] while aminopyrine was extracted with chloroform at pH 2.8 using tropeolin 00 as the reagent.[4]

Some redox reactions are also available for the determination of aminopyrine. Szász et al.[57] used potassium permanganate as the oxidizing agent; it opens the ring in alkaline medium. The reaction product can be extracted from the reaction mixture by chloroform and it can be measured at 248 nm; see Equation 10.6.

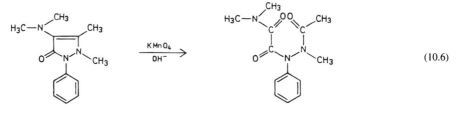

(10.6)

In the course of an indirect procedure, aminopyrine is oxidized with iron(III) and the resulting iron(II) is measured as the 2,2'-dipyridyl[61] or phenanthroline[59] complexes.

Of the other methods described for the determination of aminopyrine, the use of azo-coupling is mentioned, the reagent being diazotized 4-nitroaniline.[25] This reaction was recently applied to the determination of 4-aminoantipyrine[5] (intermediate in the synthesis of aminopyrine and important phenol reagent—see p. 179). The FIA variant of the latter method is also described.[7]

Similar to the methods for the determination of antipyrine and aminopyrine, the spectrophotometric methods in the analytical chemistry of dipyrone are based mainly on redox and complex formation reactions. The reaction leading to the 4-nitroso derivative has been used in this case, too.[2,18] In an indirect method, neotetrazolium chloride is reduced to a colored formazane which can be measured at 435 nm.[63] At first it was believed that the reducing agent is dipyrone itself. Recently, however, it was found that in the alkaline reaction mixture dipyrone first decomposes, and the reducing agent is sulphite which is one of the decomposition products.[62] In another indirect method, dipyrone is reacted with iodate and the resulting iodine is extracted with chloroform and measured spectrophotometrically.[49] The use of 4-dimethylaminobenzaldehyde is also worth mentioning (λ_{max} = 420 nm).[64]

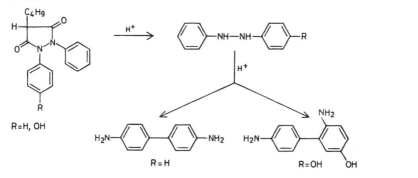

(10.7)

The indirect method of Fujita et al.[23] excels with its extremely high sensitivity. The basis of the reaction is the ligand exchange reaction between cerium(IV) o-hydroxyhydroquinonephthalein complex and dipyrone. The absorbance at 570 nm is inversely proportional to the concentration of dipyrone. The apparent molar absorptivity is 120,000. In the majority of the spectrophotometric methods based on chemical reactions in the analysis of phenylbutazone and its derivatives, the procedure begins with the hydrolytic splitting of the pyrazolidindione ring. In the course of this reaction (boiling with concentrated hydrochloric acid in various organic solvents) hydrazobenzene is forming which (under the given experimental conditions) undergoes a benzidine rearrangement (see Equation 10.7). The diaminobiphenyl derivatives thus obtained can serve as the basis for various methods. The primary aromatic amino groups form Schiff's bases with p-dimethylaminobenzaldehyde; the absorption maximum is at 460 nm for phenylbutazone and at 436 nm for oxyphenbutazone.[67] Further reagents of this kind are naphthoquinone[29] and p-dimethylaminocinnamaldehyde (λ_{max} = 525–540 nm; ε = 24,000–65,000).[19] The indophenol reaction was successfully applied for the determination of oxyphenbutazone: the 4-aminophenol type hydrolytic decomposition product (see Equation 10.7) forms intensively colored indophenol with phenol in alkaline medium in the presence of atmospheric oxygen ($\varepsilon_{635\ nm}$ = 14,900).[66,68] Ortho-cresol can also be used as the reagent.[69] Another possibility is based on diazotization of the hydrolysis product, followed by coupling with 1-dimethylaminonaphthalene.[47] It has to be noted that in the case of oxyphenbutazone containing a phenolic hydroxyl group, the azo-coupling can be applied even without preliminary hydrolytic splitting. Belal et al.[11] used diazotized 2-nitroaniline as the reagent ($\varepsilon_{435\ nm}$ = 5000). The sensitivity was similar but the selectivity was improved when the copper(II) complex of the resulting azo derivative is extracted with chloroform. The sensitivity can be greatly increased if, in the latter version, the copper(II) content of the extract is transformed to the diethyldithiocarbamate.[11]

Some of the numerous other methods for the determination of oxyphenbutazone are as follows. It can be directly transformed to the 2-nitrosophenol derivative using nitrous acid as the reagent. This derivative can be measured in alkaline medium, either directly at 358 nm[8] or 380 nm[3], or its copper(II) or cobalt(II) complexes are extracted with methylisobutylketone and measured at 357 and 330 nm, respectively.[12] The reducing properties of the phenol ring have also been exploited for analytical purposes. As examples the molybdenum blue[43] reaction, oxidation with iron(III) ions and measurement of the resulting iron(II) with potassium hexacyanoferrate,[43] oxidative coupling with 4-aminoantipyrine[3] and oxidation to azobenzene derivative with alkaline permanganate reagent and measurement at 314 nm[70] are mentioned.

As acidic materials, phenylbutazone and oxyphenbutazone form ion-pairs with suitable cations. Their ion-pairs with ferroin (iron(II) complex of o-phenanthroline) can be extracted with nitrobenzene and measured at 500 nm.[55]

REFERENCES

1. Abdine, H.; Elsayed, M. A. H.; Abdel-Hamid, M. E. *Indian J. Pharm. Sci.* 1979, *41*, 118.
2. Abdine, H.; Soliman, S. A.; Magdi, G. M. *J. Pharm. Sci.* 1973, *62*, 1834.
3. Abou-Ouf, A.; Walash, M. I.; Hassan, S. M.; El-Sayed, S. M. *Analyst* 1980, *105*, 169.
4. Adamski, R.; Pawelczyk, E. *Farm. Pol.* 1967, *23*, 295.
5. Al-Abachi, M. Q.; Al-Delami, A. M. S.; Al-Najafi, S. *Analyst* 1988, *113*, 1661.
6. Al-Badr, A. A.; El-Obeid, H. A.: Oxyphenbutazone, in *Analytical Profiles of Drug Substances*, Vol. 13, K. Florey, (Ed.) 1984, 333.
7. Alwehaid, A.-A. M. *Analyst* 1990, *115*, 1419.
8. Amer, S. M.; Ellaithy, M. M.; El-Tarasse, M. F. *Pharmazie* 1982, *37*, 182.
9. Amer, S. M.; Hassan, S. M.; El-Tarrasse, M. F. *Anal. Letters* 1980, *13*, 1625.
10. Bagavant, G.; Rowhani, V. *Indian Drugs* 1986, *23*, 373.
11. Belal, S.; El Kheir, A. A. A.; El Shanwani, A. M. *Analyst* 1985, *110*, 205.
12. Belal, S.; El Kheir, A. A. A.; El Shanwani, A. M. *Anal. Letters* 1985, *18*, 617.
13. Bezakova, Z.; Bachrata, M.; Blesova, M.; Knazko, L. *Farm. Obz.* 1980, *49*, 157.

14. Bontemps, R.; Parmentier, J.; Diessé, M. *J. Pharm. Belg.* 1968, *23*, 222.
15. *British Pharmacopoeia 1988,* Her Majesty's Stationery Office: London, 1988, 689.
16. Celechovsky, J.; Svobodova, D. *Cesk. Farm.* 1959, *8*, 380.
17. Chatterjee, P. K.; Jain, C. L.; Sethi, P. D. *Ind. J. Pharm. Sci.* 1987, *49*, 111.
18. El-Kheir, A. A.; Belal, S.; Shanwani, A. *Pharmazie* 1985, *40*, 62.
19. El-Kommos, M. E. *Analyst* 1983, *108*, 1144.
20. Elsayed, M. A.-H.; Abdine, H.; Abdel-Hamid, M. E. *Analyst* 1979, *104*, 568.
21. Elsayed, M. A.-H.; Abdine, H.; Abdel-Hamid, M. E. *Indian J. Pharm. Sci.* 1978, *40*, 52.
22. *Europäisches Arzneibuch. 1. Ausgabe. Band II.* Deutscher Apotheker-Verlag: Stuttgart, 1975, 144.
23. Fujita, Y.; Mori, I.; Kitano, S. *Bunseki Kagaku* 1984, *33*, E383.
24. Hahn, M.; Kolsek, J.; Perpar, M. *Z. Anal. Chem.* 1956, *151*, 104.
25. Haslinger, R.; Struntz, W. *Arzneimittelforsch.* 1951, *5*, 61.
26. Hentrich, K.; Pfeiffer, S. *Pharm. Zhalle* 1967, *106*, 735.
27. Heise, D.; Pfeiffer, S. *Dtsch. Apoth. Ztg.* 1975, *115*, 2024.
28. Horn, D. *Pharmazie* 1951, *6*, 330.
29. Issa, A. S.; Beltagy, Y. A.; Kassem, M. G.; Daabees, H. S. *Talanta* 1985, *32*, 657.
30. Juhl, L. *Arch. Pharm. Chemi* 1969, *76*, 819.
31. Korany, M. A.; Wahbi, A. M.; Elsayed, M. A.; Mandour, S. *Anal. Letters* 1984, *17*, 1373.
32. Korany, M. A.; Wahbi, A. M.; Elsayed, M. A.; Mandour, S. *Farmaco Ed. Prat.* 1984, *39*, 243.
33. Krácmar, J.; Krácmarová, J.; Remsová, M.; Kovarová, A. *Pharmazie* 1988, *43*, 681.
34. Laik Ali, S. Phenylbutazone, in *Analytical Profiles of Drug Substances,* Vol. 11. (K. Florey (Ed.)) 1982, 483.
35. Marciniec, B. *Pharmazie* 1984, *39*, 103.
36. Marciniec, B. *Acta Polon. Pharm.* 1984, *40*, 185.
37. Marciniec, B. *Pharmazie* 1985, *40*, 180.
38. Marciniec, B.; Wojciechowska, B. *Acta Polon. Pharm.* 1989, *46*, 248.
39. Nieth, H.; Thiele, P. *Klin. Wochensch.* 1958, *36*, 832.
40. Parimoo, P. *Drug Dev. Ind. Pharm.* 1987, *13*, 127.
41. Pawelczyk, E.; Wachowiak, R. *Chem. Anal.* 1969, *14*, 925.
42. Pesez, M. *Ann. Chim. Anal.* 1942, *24*, 11.
43. Peterkova, M.; Kakac, B.; Matousova, O. *Cesk. Farm.* 1977, *26*, 255.
44. Pietura, A. *Farm. Pol.* 1973, *29*, 425.
45. Pindur, U. *Arch. Pharm.* 1980, *313*, 301.
46. Pohloudek-Fabini, R.; Gundermann, P. *Pharmazie* 1986, *35*, 685.
47. Pulver, R. *Schweiz Med. Wschr.* 1950, *80*, 308.
48. Quaglia, M. G.; Carlucci, G.; Cavicchio, G.; Mazzeo, P. *J. Pharm. Biomed. Anal.* 1988, *6*, 421.
49. Qureshi, S. Z.; Sajed, A.; Hasan, T. *Talanta* 1989, *36*, 869.
50. Réffy, M.; Küttel, D.; Szigetváryné Takácsy Nagy, M. *Gyógyszerészet* 1970, *14*, 256.
51. Reisch, J.; Abdel-Khalek, M. *Pharmazie* 1979, *34*, 408.
52. Reisch, J.; Ekiz, N.; Güneri, T. *Arch. Pharm.* 1986, *319*, 973.
53. Rózsa, P. *Acta Pharm. Hung.* 1957, *27*, 246.
54. Santoni, G.; Mura, P.; Pinzauti, S.; Lombado, E.; Gratteri, P. *Int. J. Pharm.* 1990, *64*, 235.
55. Sarma, B. S. R.; Rao, K. B.; Raju, D. J. K. *Indian J. Pharm. Sci.* 1990, *52*, 188.
56. Stuzka, V. *Ceskoslov. Farmac.* 1989, *38*, 446.
57. Szász, Gy.; Kovács, Á.; Ladányi, L. *Acta Pharm. Hung.* 1978, *48*, 221.
58. *The United States Pharmacopoeia XXII,* USP Convention Inc: Rockville, 1990, a. 105; b. 1067.
59. Tonooka, K.; Sato, M.; Mizukami, H.; Yarimizu, S. *Sankyo Kenkyusho Nempo* 1966, *18*, 160.
60. Tulus, R.; Aydogan, Y. *Istanbul Univ. Eczacilik Fak. Mecm.* 1967, *3*, 168.
61. Vasileva-Alexandrova, P. *Mikrochim. Acta* 1971, 582.
62. Vasileva-Alexandrova, P.; Alexandrov, A. *Pharmazie* 1990, *45*, 348.
63. Vasileva-Alexandrova, P.; Shishmanov, P. S. *Anal. Chem.* 1975, *47*, 1432.
64. Végh, A.; Szász, Gy.; Kertész, P. *Acta Pharm. Hung.* 1961, *31*, 49.
65. Végh, A.; Szász, Gy.; Kertész, P. *Pharmazie* 1962, *17*, 512.
66. Verma, K. K.; Jain, A. *Analyst* 1985, *110*, 997.
67. Verma, K. K.; Jain, A.; Patel, N.; Sanghi, S. K. *Farmaco, Ed. Prat.* 1987, *42*, 185.
68. Verma, K. K.; Sanghi, S. K. *Farmaco, Ed. Prat.* 1988, *43*, 13.
69. Viswanath, K. K.; Rao, A. S.; Sivaramakrishnan, M. V. *Indian Drugs* 1986, *24*, 170.

70. Wallace, J. E. *J. Pharm. Sci.* 1968, *57*, 2053.
71. Wahbi, A. M.; Abdine, H.; Blaih, S. M. *Pharmazie* 1978, *33*, 96.

I. BENZODIAZEPINES

1. Methods Based on Natural Absorption

The formulae of 1,4-benzodiazepines with a common structure and most frequently used in therapy are summarized in Table 10.5.

Of the derivatives not represented in Table 10.5, chlordiazepoxide has a double bond between C_1 and C_2, a methylamino group and chlorine at positions 2 and 7, respectively; C_3 is unsubstituted while the nitrogen at position 4 is in the N-oxide form. The characteristic features of bromazepam are the 5-pyridyl and 7-bromo substituents. Medazepam is the 2-deoxo analog of diazepam. There is a cyclohexenyl group at position 5 in the molecule of tetrazepam while estazolam and triazolam contain a condensed triazole ring attached to positions 2 and 3. The 2-thione analogs are also worth mentioning.

As it is discussed in detail in Section 3.K, 1,4-benzodiazepines are spectrophotometrically strongly active materials (see the spectra of diazepam and nitrazepam in Figure 3.25). Since the imino group of the derivatives not bearing functional groups at position 1 is of acidic character and the nitrogen at position 4 is basic, the spectra show strong pH-dependence.[24,44] The bathochromic shift in acidic media is especially suitable for their identification and quantification. Many methods for the assay of formulations containing benzodiazepines are based on absorbance measurements in acidic solutions. For example, the assay of diazepam tablet and capsule was carried out in USP XXI[48] by absorbance measurement at 285 nm after extraction with chloroform, purification of the extract and changing the solvent to dilute sulphuric acid. The dissolution test is still based on the same absorption band (284 nm) even in USP XXII.[49a] The British Pharmacopoeia[7a] uses the same band for the determination of diazepam in solid dosage forms after extraction with the mixture of dilute acid and methanol. The diazepam content of injections and elixirs is also measured in dilute acidic medium after extraction with chloroform and change of the solvent. It is interesting that the wavelength of the measurement in these cases is 368 nm, which is the maximum of the least

TABLE 10.5. Some Important 1,4-benzodiazepine Derivatives

	R_1	R_2	R_3	R_4
Diazepam	CH_3	H	H	Cl
Nitrazepam	H	H	H	NO_2
Oxazepam	H	OH	H	Cl
Clonazepam	H	H	Cl	NO_2
Flunitrazepam	CH_3	H	F	NO_2
Flurazepam	$(CH_2)_2-N(C_2H_5)_2$	H	F	Cl
Lorazepam	H	OH	Cl	Cl
Prazepam	$CH_2-CH\begin{smallmatrix}CH_2\\|\\CH_2\end{smallmatrix}$	H	H	Cl
Clorazepate	H	COOH	H	Cl
Temazepam	CH_3	OH	H	Cl
Lormetazepam	CH_3	OH	Cl	Cl

intense long-wavelength band which, in turn, enables a more selective measurement and its intensity is still sufficient for these purposes. The assay of chlordiazepoxide tablet and capsule is carried out in acidic medium either at 308 nm[7c] or at 245 nm[49b] where higher sensitivity is attainable. For the determination of flurazepam and nitrazepam in tablets, acidic-methanolic solvent and measurement at 284 and 280 nm, respectively, are prescribed.[7d,e] In the case of nitrazepam the deprotonation of the free imino group causes a bathochromic shift and this enables its very selective and sensitive determination at 366 nm.[43] Several more spectrophotometric methods are available for the determination of the above and some other benzodiazepines (tetrazepam, medazepam, clorazepate, 2-thione analogs) with optimized solvent, pH and wavelength.[19,30,45,50] The spectrophotometric method enabled even such a difficult analytical problem to be solved at an acceptable level as the determination of chlordiazepoxide in plasma. In the method of Jatlow[25] quantitative results were obtained after extraction with chloroform, purification of the extract by alkaline extraction, re-extraction into dilute hydrochloric acid, measurement in acidic and alkaline media and the application of corrections.

As described previously in this chapter, the majority of analytical problems occurring in the course of the analysis of benzodiazepines in their formulations can be solved by simple absorbance measurements at the wavelengths of their absorption maxima. In some cases, more advanced measurement or calculation methods are necessary, such as, e.g., in the case of the determination in the presence of their degradation products. The hydrolytic decomposition of benzodiazepines leading to 2-aminobenzophenone derivatives is described in Sections 7.F and 9.G. In the presence of the latter the determination of the undecomposed drug by means of a single-wavelength measurement is not possible. Measurement at several wavelengths and the use of orthogonal polynomials is a suitable method for the solution of problems of this type in the field of benzodiazepines (see Section 4.C.3). These methods were developed by Korany et al.[5,16,27] The measurements were carried out in 0.1 M hydrochloric acid where the spectra of intact benzodiazepines undergo bathochromic (while those of the degradation products-hypsochromic) shifts. The wavelength range was 272–360 for diazepam and 230–318 nm for chlordiazepoxide and the absorbances were determined at 8 nm intervals for the selective determination. Nitrazepam, prazepam and clorazepate were determined similarly.[16] The method of orthogonal polynomials is suitable for the assay of certain combined formulations, too, e.g., for the determination of oxazepam in the presence of dipyridamole.[27]

The other main method to overcome the difficulties caused by spectral background-problems, namely derivative spectrophotometry, is also suitable for the determination of benzodiazepines in the presence of their degradation products. Abdel Hamid and Abuirjeie[1] used the fourth derivative spectra for the selective determination of diazepam and oxazepam. Richter[42] measured $^2D_{237,247}$ for the assay of bromazepam tablet. Corti et al.[10] used first and second order derivatives for the determination of several benzodiazepines and their binary mixtures. Using higher derivative spectra enables even more delicate problems to be solved. Randez-Gil et al.[40] determined nitrazepam and clonazepam in urine in alkaline medium using $^5D_{388}$ and $^6D_{384}$, respectively. For the determination of the same in blood plasma $^4D_{402}$ and $^4D_{384}$ were measured.

As a consequence of the strong and characteristic spectral shifts due to changes of the pH of the solvents (outlined in the first part of this section) the possibilities of difference spectrophotometry are also good. For example, Davidson[11] determined demoxepam and chlordiazepoxide in the presence of each other on the basis of the difference spectra between pH 8 and 13 as well as 3 and 8, respectively. When nitrazepam was determined in tablets by the same author, the pH of test and reference solutions was 1 and 13, respectively, and the wavelength was 282 nm.[12]

Difference-derivative spectrophotometry (see Section 4.E.3) was used by El Yazbi et al.[17] for the determination of oxazepam and phenobarbital in the presence of dipyridamole and by Abounassif[2] for the assay of bromazepam tablet.

Another, more generally applicable solution for background problems is the chromatographic separation of benzodiazepines prior to the spectrophotometric determination. For example, Peeran et al.[38] used column chromatographic separation in the determination of diazepam in a pharmaceutical formulation containing oxyphenbutazone and paracetamol. Pawelczyk et al.[37] determined clonazepam in the presence of its impurities and degradation products by TLC separation, spot elution and spectrophotometric measurement. By means of the combination of column chromatographic separation and second derivative spectrophotometry, Martinez and Gimenez[33] measured diazepam, nitrazepam, chlordiazepoxide and clorazepate in blood and urine.

Clonazepam was determined by Lattore et al.[29] using flow injection analysis with UV monitoring. The FIA technique was used by Fehér et al.,[18] too, in the dissolution studies of nitrazepam tablets.

2. Methods Based on Chemical Reactions

Although a great variety of problems in the analysis of benzodiazepines can be solved by the various methods based on their natural absorption, methods based on chemical reactions, leading mainly to colored derivatives, had some importance especially in the early stage of the development and this importance to some extent remains up to the present time.

The basis for the most selective determination is their acid catalyzed hydrolytic cleavage leading to substituted 2-aminobenzophenones and glycine (see Sections 7.F and 9.G). The completion of the reaction requires prolonged boiling with rather concentrated (3–6 M) hydrochloric acid. The reaction time can be reduced by using a microwave oven.[13] The simplest way for the subsequent measurement is an absorbance reading of the reaction mixture itself at the maximum wavelength of the resulting 2-aminobenzophenone, e.g., in the case of nitrazepam at 360 nm.[46] This kind of measurement does not have great advantages either from the point of view of the wavelength of the measurement or the sensitivity. More advantageous from both points of view is to measure after extraction with chloroform[3]; when diazepam was determined in such a way the wavelength of the determination was 410 nm. Of the application of this method, the work of de Silva and Strojny[15] merits special mention. In the course of the simultaneous determination of flurazepam and its metabolites in urine, the analytes were hydrolyzed by hydrochloric acid as described above, separated by TLC, eluted and measured at the individual maxima of the benzophenone derivatives.

In the interesting method of de la Guardia et al.,[14] the detergent Nemol Y® 1030 was added to the reaction mixture after the hydrolytic cleavage. The spectral shifts thus obtained could be utilized in the simultaneous determination of binary mixtures of benzdiazepines.

In the majority of methods based on acid-catalyzed hydrolysis, the resulting 2-aminobenzophe-nones are diazotized and coupled with N-(1-naphthyl)-ethylenediamine,[26] β-naphthol,[4] thymol,[35] 8-hydroxyquinoline,[36] etc. The sensitivity and selectivity of these methods are good. It is to be noted that this version of the method is applicable only for those derivatives which afford primary amines upon hydrolysis; N_1-substituted derivatives, e.g., diazepam, do not give positive reaction. In such cases, however, the measurement can be traced back to the determination of the other reaction product, glycine, by using the color reaction with ninhydrin.[41]

Another important group of photometric methods for the determination of benzodiazepines is based on their weakly basic character and their tendency to form ion-pairs with acidic dyes (see Section 7.H). Manes et al.[31] determined several benzodiazepines by means of extraction from strongly acidic solution with chloroform of their ion-pairs with Orange II and other dyes. The dye applied for the determination of medazepam by Mitsui et al.[34] was Erio Green B (extraction with chloroform at pH 4.1). The same solvent and the dye alizarin violet and extraction at pH 1.2 were used by Mangala et al.[32] for the determination of diazepam, nitrazepam and chlordiazepoxide. The same benzodiazepines were determined by Popovici et al.[39] by ion-pairing with picric acid in chloroform or benzene without extraction. Several acidic dyes such as methyl orange, tropeolin 00 and 000, eriochrome black, etc., were used for the determination of benzodiazepines by Blazsek-Bodó et al.[6]

The sensitivity and—if the pH is properly optimized—the selectivity of the methods based on ion-pair formation are good. The sensitivity is much lower in the case of analytical methods based on the formation of metal complexes; (the molar absorptivities are below 1000). Of these methods the determination of triazolam[20] and estazolam[21] is based on the formation of ruthenium(III)-complexes; bromazepam is determined as the iron(II)-complex,[47] while the interaction of chlordizep-oxide and diazepam with diamino-tetrathiocyanato-chromium(III) was utilized for their determination.[22]

Of other methods, that of Büyüktimkin[8,9] for the determination of oxazepam and lorazepam excels with its interesting mechanism. As seen in Equation 10.8, if these benzdiazepines are boiled with glacial acetic acid, ring contraction takes place rather than ring cleavage leading to quinazoline-

2-carboxaldehyde derivatives. The latter form hydrazones with 2,4-dinitrophenylhydrazine and the measurement of this in alkaline media at 520 nm is the basis of a very selective and sensitive determination.

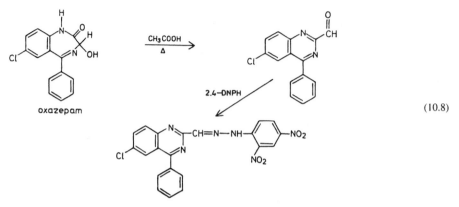

(10.8)

To conclude this section, three of the less important methods are also mentioned. Wassel and Diab[51] adopted the iron(III)-hydroxamate method (see p. 184). Larini et al.[28] developed an indirect method for the determination of several benzodiazepines. This is based on their precipitation with reduced molybdophosphoric acid and the spectrophotometric determination of the excess of the latter. The nitro group of nitrazepam can be reduced with zinc/hydrochloric acid to an aromatic amino group which can be determined by azo-coupling.[23]

REFERENCES

1. Abdel Hamid, M. E.; Abuirjeie, M. A. *Analyst* 1988, *113*, 1433.
2. Abounassif, M. A. *J. Pharm. Belg.* 1989, *44*, 329.
3. Baggi, T. R.; Mahajan, S. N.; Rao, G. R. *J. Assoc. Off. Anal. Chem.* 1975, *58*, 875.
4. Baumler, J.; Rippstein, S. *Helv. Chim. Acta* 1961, *44*, 2208.
5. Bedair, M.; Korany, M. A.; Abdel Hamid, M. E. *Analyst* 1984, *109*, 423.
6. Blazsek-Bodó, Á.; Kiss, I.; Józsa, J. *Farmacia* 1985, *33*, 15.
7. *British Pharmacopoeia 1988*, Her Majesty's Stationery Office: London, 1988. a. 633; b. 730, 783; c. 628, 914; d. 637; e. 642.
8. Büyüktimkin, N.; Büyüktimkin, S. *Sci. Pharm.* 1985, *53*, 147.
9. Büyüktimkin, N.; Büyüktimkin, S. *Acta Pharm. Turc.* 1986, *28*, 87.
10. Corti, P.; Aprea, C.; Corbini, G.; Dreassi, E.; Celesti, L. *Pharm. Acta Helv.* 1991, *66*, 50.
11. Davidson, A. G. *J. Pharm. Sci.* 1984, *73*, 55.
12. Davidson, A. G. *J. Pharm. Pharmacol.* 1989, *41*, 63.
13. de la Guardia, M.; Salvador, A.; Gomez, M. J.; Banzo, Z. A. *Anal. Chim. Acta* 1989, *224*, 123.
14. de la Guardia, M.; Galdú, M. V.; Monzó, J.; Salvador, A. *Analyst* 1989, *114*, 509.
15. de Silva, J. A. F.; Strojny, N. *J. Pharm. Sci.* 1971, *60*, 1303.
16. El Yazbi, F. A.; Abdel Hay, M. H.; Korany, M. A. *Pharmazie* 1986, *41*, 639.
17. El Yazbi, F. A.; Korany, M. A.; Bedair, M. *J. Pharm. Belg.* 1985, *40*, 244.
18. Fehér, Zs.; Kolbe, I.; Pungor, E. *Analyst* 1991, *116*, 483.
19. Fernández-Arciniega, M. A.; Rosas, J.; Hernández, L.; Alonso, R. M. *Boll. Chim. Farm.* 1985, *124*, 19.
20. Gallo, N.; Bianco, V. D.; Bianco, P.; Luisi, G. *Farmaco, Ed. Prat.* 1983, *38*, 205.
21. Gallo, N.; Bianco, V. D.; Doronzo, S. *Farmaco, Ed. Prat.* 1985, *40*, 77.
22. Ganescu, I.; Várhelyi, Cs.; Brinzan, G.; Boboc, L. *Pharmazie* 1976, *31*, 259; 1977, *32*, 52.
23. Hassan, S. M.; Belal, F.; Sharaf El Din, M.; Sultan, M. *Analyst* 1988, *113*, 1087.
24. Japp, M.; Garthwaite, K.; Geeson, A. V.; Osselton, M. D. *J. Chromatogr.* 1988, *439*, 317.
25. Jatlow, P. *Clin. Chem.* 1972, *18*, 516.
26. Kamm, G.; Baier, R. *Arzneim-Forsch.* 1969, *19*, 213.

27. Korany, M. A.; Haller, M. *J. Assoc. Off. Anal. Chem.* 1982, *65*, 144.

28. Larini, L.; Salgado, P. E. D.; De Carvalho, D. *Trib. Farm.* 1979, *47*, 75.

29. Lattore, C.; Blanco, M. H.; Abad, E. L.; Vicente, J.; Hernandez, L. *Analyst* 1988, *113*, 317.

30. le Petit, G. *Z. Anal. Chem.* 1977, *283*, 199.

31. Manes, J.; Civera, J.; Font, G.; Bosch Serrat, F. *Cien. Ind. Farm.* 1987, *6*, 333.

32. Mangala, D. S.; Reddy, B. S.; Sastry, C. S. P. *Indian Drugs* 1984, *21*, 526.

33. Martinez, D.; Gimenez, P. M. *J. Anal. Toxicol.* 1981, *5*, 10.

34. Mitsui, T.; Matsuoka, T.; Fujimura, Y. *Bunseki Kagaku* 1985, *34*, 72.

35. Ossman, A. R. E. N.; El Hassani, A. H. *Pharmazie* 1975, *30*, 257.

36. Ossman, A. R. E. N.; El Hassani, A. H. *Pharmazie* 1976, *31*, 744.

37. Pawelczyk, E.; Plotkowiak, Z.; Mikolajczak, P. *Farm. Polska* 1987, *43*, 261.

38. Peeran, M. T.; Baravani, G. S.; Bhalla, V. S. *Indian Drugs* 1985, 1985, *23*, 52.

39. Popovici, I.; Dorneanu, V.; Cuciureanu, R.; Stefanescu, E. *Rev. Chim.* 1983, *34*, 554, 653.

40. Randez-Gil, F.; Daros, J. A.; Salvador, A.; De la Guardia, M. *J. Pharm. Biomed. Anal.* 1991, *9*, 539.

41. Rao, G. R.; Kanjilal, G.; Srivastava, C. M. R. *Indian J. Pharm. Sci.* 1980, *42*, 63.

42. Richter, P. *Int. J. Pharm.* 1991, *72*, 207.

43. Sanghavi, N. M.; Jivani, N. G. *Talanta* 1979, *26*, 63.

44. Schütz, H., *Benzodiazepines,* Springer Verlag: Berlin, Heidelberg, New York, 1982.

45. Seitzinger, R. W. T. *Pharm. Weekblad.* 1975, *110*, 1073.

46. Shingbal, D. M.; Agni, R. M. *Indian Drugs* 1983, *20*, 162.

47. Smyth, W. F.; Scannell, R.; Goggin, T. K.; Lucas-Hernandez, D. *Anal. Chim. Acta* 1982, *141*, 321.

48. *The United States Pharmacopoeia XXI*, USP Convention Inc.: Rockville, 1985, 225.

49. *The United States Pharmacopoeia XXII*, USP Convention Inc.: Rockville, 1990, a. 414; b. 282.

50. Volke, J.; Ellaithy, M. M.; Manousek, O. *Talanta* 1978, *25*, 209.

51. Wassel, G. M.; Diab, A. M. *Pharmazie* 1973, *28*, 790.

J. PHENOTHIAZINES

1. Methods Based on Natural Absorption

Of the many phenothiazine derivatives used in therapy some are summarized in Table 10.6.

The ring system of the phenothiazines possesses an intense and characteristic spectrum (see Figure 3.26 in Section 3.M). Since the majority of substituents, R_1 in Table 10.6 are spectrophotometrically inactive or only slightly active (and this applies essentially to the R_2 substituents, too), the spectra of the majority of phenothiazines are very similar and can be characterized by an intense band at 250–255 nm and a group of less intense bands around 300–310 nm.[13,27] For example, in the case of chlorpromazine, the molar absorptivities of the bands appearing at 254 and 306 nm in slightly acidic medium are 33,600 and 3,900, respectively. The intensity ratio of the two bands is characteristic of the purity of phenothiazines. For example, the USP XXII prescribes the ratio A_{252}/A_{301} for promazine to be between 7.1 and 7.9 in 0.1 M hydrochloric acid.[56a]

Since the ring system of phenothiazines does not possess acidic or basic properties, and the amine groups bearing the basicity of the derivatives are positioned in the side chain and outside of the chromophoric system, the pH-dependence of the spectra are generally low.

The intense spectra of phenothiazine-type pharmaceuticals present wide possibilities for their quantitative determination in various formulations. In the majority of cases, the pharmacopoeias prescribe spectrophotometric measurement in dilute acidic media, usually using the above mentioned high-intensity band for the assay of the dosage forms. Among others the assays of promazine tablet[6a] and injection,[6b] promethazine tablet[6c] and injection,[6d] perphenazine tablet[6e] and injection,[6f] chlorpromazine tablet,[6g] injection,[6h] solution[6i] and suppository,[6j] fluphenazine tablet,[6k] trifluoperazine tablet[6l] and trimeprazine tablet[6l] are based on absorbance measurements in the range of 249–258 nm. In the case of thioridazine tablet[6m] and suspension,[56b] the main band undergoes a slight bathochromic shift (as a consequence of the methylthio substitution). In these assays the analytical wavelengths are 264 and 266 nm, respectively.

From among the background correction methods described in Section 4.C.2, the simplest one (namely the two-wavelength measurement) is frequently used. For example the measurement of

$(A_{254}-A_{277})$ is prescribed for the assay of chlorpromazine tablet, injection and suppository.[56c] For the assay of the formulations of trifluoperazine, $(A_{255}-A_{278})$ is measured[56d] while for trimeprazine syrup and tablet: $(A_{251}-A_{276})$.[56e]

When measuring at the long-wavelength, low intensity band the sensitivity is naturally lower but the selectivity is advantageous. This possibility is used for the assay of bulk promethazine and its injection formulation (measurement at 301 nm)[56a] and also for the syrup at 298 nm.[56f]

Of the highly efficient background correction methods, measurement at 6 wavelengths is suggested for the assay of chlorpromazine and promethazine tablets in the presence of their degradation products.[57]

One of the degradation products which is the sulphoxide (product of the autoxidation of phenothiazines) accompanying the drug substances both in the bulk drug state and also in their formulation merits special discussion. As seen in Figure 10.4 in the example of chlorpromazine, the difference between the spectra of this impurity and those of the parent drugs is significant. These spectra, however, overlap to a great extent and hence the selective determination of one of the components or their simultaneous determination on the basis of simple absorbance measurements is a difficult task. Fasanmade and Fell[17] solved this problem by using derivative spectrophotometry: from the third derivative spectrum of the mixture of unoxidized chlorpromazine and its sulphoxide, both components could be simultaneously determined. As seen in Figure 10.5 the amplitude $^3D_{259,267}$ is suitable for the determination of the intact drug, namely the ^3D values of the sulphoxide at these wavelengths are identical. On the basis of similar considerations, $^3D_{350,361}$ was selected for the selective determination of the sulphoxide. Derivative spectrophotometry was successfully used for the analysis of other phenothiazine formulations, too.[48]

TABLE 10.6. Some Important Phenothiazine[a] Derivatives

	R_1	R_2
Promazine	$-(CH_2)_3-N\big\langle{}^{CH_3}_{CH_3}$	H
Chlorpromazine	$-(CH_2)_3-N\big\langle{}^{CH_3}_{CH_3}$	Cl
Perphenazine	$-(CH_2)_3-N\big\langle\;\big\rangle N-(CH_2)_2-OH$	Cl
Prochlorperazine	$-(CH_2)_3-N\big\langle\;\big\rangle N-CH_3$	Cl
Promethazine	$-CH_2-CH-N\big\langle{}^{CH_3}_{CH_3}$ (CH_3)	H
Fluphenazine	$-(CH_2)_3-N\big\langle\;\big\rangle N-(CH_2)_2-O-\overset{O}{\overset{\|}{C}}-(CH_2)_5-CH_3$	CF_3
Trifluoperazine	$-(CH_2)_3-N\big\langle\;\big\rangle N-CH_3$	CF_3
Trimeprazine	$-CH_2-CH-CH_2-N\big\langle{}^{CH_3}_{CH_3}$ (CH_3)	H
Thioridazine	$-CH_2-CH_2-\big\langle\;\big\rangle$ N (CH_3)	SCH_3

[a]See formula in Equation 10.9, page 296.

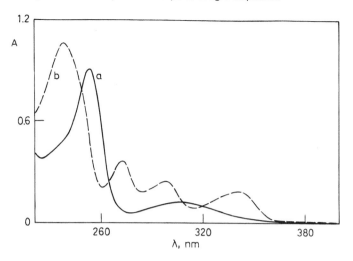

FIGURE 10.4.   Spectra of chlorpromazine (curve "a") and chlorpromazine sulphoxide (curve "b"). Solvent: water. (From Fasanmade, A. A.; Fell, A. F. *Analyst* 1985, *110*, 1117.[17])

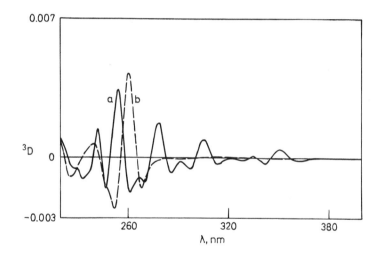

FIGURE 10.5.   Third derivative spectra of chlorpromazine (curve "a") and chlorpromazine sulphoxide (curve "b"). Solvent: 0.1 M hydrochloric acid. Source: see Figure 10.4.

2.   Methods Based on Chemical Reactions

The most important methods for the determination of phenothiazines based on chemical reactions are ion-pair extraction methods using the basicity of the side chains, complexation methods based on the tendency of the donor groups in the side chain and in the heterocycle to form metal complexes and methods based on the oxidation of the heterocycle.[5,9]

The application of the ion-pair extraction method (Section 7.H) to phenothiazines does not present special problems. The great many papers of this type include those describing the use of picric acid[37] and flavianic acid[55] with extraction using benzene and the use of bromocresol green with chloroform.[16]

Chloroform was used for the extraction when the dyes alizarin red S[54] and nitroso-R[26] as well as the ion-pair forming complex anion iodobismuthate[31] were used as the reagents. The common characteristics of these methods are that the extraction is carried out from more or less acidic solution and the absorbance of the organic phase is directly measured in the visible range. The sensitivity of the measurement is medium to good. Amaranth and tropaeolin 000 were also used associated with extraction with chloroform or dichloromethane.[3,4] In these cases, however, the complex in the organic phase was decomposed by treatment with neutral or alkaline aqueous buffers and the dye transferred to the aqueous phase is measured.

From among the complex forming metal ions, the use of palladium(II) is most widespread in the circle of phenothiazine derivatives. This classical method[33,46] which selectively measures the unoxidized phenothiazines is still widely used in pharmacopoeias for the assay of phenothiazine formulations, e.g., for the assay of perphenazine injection, syrup and tablet,[56g] fluphenazine tablet and elixir,[56h] promethazine injection[56i] and trimeprazine elixir.[6n] The analytical wavelength is in the range of 450–500 nm. The automated version of the method is also described for several phenothiazines.[51] The interesting methods of Mori and Fujita can be regarded as the combination of the methods based on metal complex and ion-pair formation. For example in the course of the determination of chlorpromazine where the reagents were palladium(II)-chloride and eosin in slightly acidic medium, the measured species is the ion-pair formed between the Pd (chlorpromazine)$_2^{2+}$ cation and the anion of eosin.[18] Advantages of this method are its high sensitivity, ($\varepsilon_{545\ nm} = 57,000$) and that no extraction is necessary if the measurement is carried out in the presence of non-ionic surfactants e.g., methylcellulose. Chlorpromazine can also be determined on the basis of the same principles using palladium(II) chloride and o-hydroxyhydroquionephthalein as the reagents ($\varepsilon_{525\ nm} = 22,000$).[29] The use of tin(IV),[36] germanium(IV),[30] copper(II)[38] as the metal ions with various complex-forming anions is also the basis for the determination of phenothiazine derivatives based on ternary complex formation. These methods which require the extraction of the colored complexes are highly selective but less sensitive than the above discussed other methods.

Charge-transfer complex formation (see 7.I) can also form the basis of the spectrophotometric determination of phenothiazines: the sulphur and nitrogen atoms in the ring system possess excellent donor properties for this. The reagents adopted include chloranil,[2] 2,3-dichloro-5,6-dicyano-p-benzoquinone,[1] iodine,[10,25] iodine monochloride,[35] morpholine-iodine,[58] and morpholine-N-bromosuccinimide.[15]

The electron donor properties of phenothiazines lead to possibilities for very selective determinations: the use of various oxidizing agents which transform phenothiazines via one-electron oxidation to relatively stable, colored radical cations which can be measured spectrophotometrically. This reaction (see Equation 10.9) was thoroughly investigated kinetically and because of its possible role in the mechanism of pharmacological activity of phenothiazines.[20] The reaction has found wide analytical applications, too.

$$\text{(10.9)}$$

Using a wide range of oxidizing agents the reaction is completed within a few minutes. The absorption maxima of the radical cations are in the range of 480–550 nm depending on the substituents of the phenothiazine ring system, the solvent, the pH and the counterion of the radical cation. The molar absorptivities range between 5,000 and 20,000. The most important task in the analytical applications is the stabilization of the radical cation, i.e., its protection against further oxidation, disproportionation and hydrolysis. One of the possibilities to achieve this is the application of extremely high concentration of acids. For example, Taha et al.[53] carried out the oxidation with N-bromosuccinimide in 30–70 v/v% sulphuric acid media.

Heteropolyacids are widely used as the oxidizing agents: arsenomolybdic acid,[41] phosphomolybdic acid,[21,52] and vanadoboric acid.[44] In these cases the anion of the excess reagent stabilizes the radical cation. Further oxidizing agents leading to spectrophotometrically measurable radical cations in acidic media are sodium vanadate,[45] sodium nitrite,[49] chloramine-T,[24,34] potassium iodate,[39,40] sodium hexanitritocobaltate[28] and iron(III)-chloride.[19,32] The latter reagent was combined with 2,2′-dipyridyl[8] or 1,10-phenanthroline[7] which form complexes with the resulting Fe(II) ions. The maxima of the spectra of the latter are in the same spectral range as that of the radical cation (510–520 nm). The method of El Kommos and Emara[14] merits special mentioning. Here very high sensitivity at very high wavelength (700–740 nm) was attained in such a way that in addition to the oxidizing agent (iron(III) chloride), 3-methylbenzothiazolin-2-one hydrazone was also added to the reaction mixture ($\varepsilon = 19,700$–$40,000$). The reaction product with these good spectrophotometric properties (and also high stability) is the product of the oxidative coupling of the radical cation with the hydrazone reagent. The same reagent was used by Sastry et al.[50] for the determination of various phenothiazine derivatives. Using sodium hypochlorite or potassium persulphate as the oxidizing agents, similar sensitivity was attained as described above. In the method of Rychlovsky and Nemcová[47] where the oxidizing agent is copper(II)-chloride, two molecules of phenothiazines are coupled (λ_{max} = 635 nm). This method gives positive results only with N-unsubstituted phenothiazines.

In general it can be stated that the methods based on charge-transfer complexes (in the classical sense of the word) cannot be sharply separated from those based on the oxidation to the radical cation. Moreover, in many cases the radical cation undergoes further, irreversible transformations. The discussion of the often divergent views about the mechanism of these reactions would be beyond the scope of this book.

The chemical basis of the methods in the preceding paragraphs was that under special circumstances, using certain oxidizing agents, the oxidation is stopped at the radical cation stage which involves the attraction of one electron. Under different conditions, using other oxidizing agents or in the case of phenothiazines and their formulations exposed to atmospheric oxygen, or in the course of their metabolism, two-electron oxidation takes place leading to sulphoxides. For this reason the simultaneous determination of the intact phenothiazines and their sulphoxides is an important task. The difference spectrophotometric method of Davidson presents an elegant procedure to solve this problem.[11,12] The determination of the unoxidized drug is based just on the spectral shift taking place as the consequence of the oxidation. As was shown in Figure 8.9 the spectrum of the sulphoxide differs considerably from that of the unoxidized phenothiazine. Accordingly, if the spectrum of a solution in which the phenothiazine is quantitatively oxidized to the sulphoxide with peroxyacetic acid is recorded against an untreated solution of the same concentration, a very characteristic difference spectrum can be obtained. As is seen in Figure 8.10 with the example of promethazine, measuring the difference absorbance at 337 nm presents a good possibility for quantitative determinations. In the case of other phenothiazines this wavelength is in the range 342–354 nm. In this way the absorbance of sulphoxides, originally present, is cancelled out and hence the method selectively measures the intact phenothiazines. This method was successfully applied to almost all of the important phenothiazines[22,23] and its automated version has also been described.[22]

For the selective determination of the sulphoxide it is reduced quantitatively to the parent phenothiazine using zinc/hydrochloric acid as the reducing agent. In the course of the difference spectrophotometric determination the unreduced solution is measured against the reduced. In this way the absorbance of the phenothiazine is cancelled out and the measured difference absorbance corresponds to the sulphoxide content.[12]

REFERENCES

1. Abdel-Salam, M.; Issa, A. S.; Mahrous, M.; Abdel-Hamid, M. E. *Anal. Letters* 1985, *18*, 1391.
2. Al-Ghabsa, T. S.; Ibrahim, S. K.; Al Abachi, M. Q. *Microchem. J.* 1983, *28*, 501.
3. Ayad, M. M.; Moussa, A. F. -A. *J. Drug. Res.* 1984, *15*, 95.
4. Beltagy, Y. A.; Issa, A.; Rida, S. M. *Pharmazie* 1976, *31*, 484.
5. Blazek, J.; Dymes, A.; Stejskal, Z. *Pharmazie* 1976, *31*, 681.
6. *British Pharmacopoeia 1988*, Her Majesty's Stationery Office: London, 1988. a. 994; b. 842; c. 995; d. 843; e. 834; f. 982; g. 918; h. 727; i. 771; j. 888; k. 948; m. 1011; n. 754.
7. Buhl, F.; Chwistek, M. *Chem. Anal.* 1984, *29*, 581.
8. Buhl, F.; Mazur, U.; Chwistek, M. *Chem. Anal.* 1976, *21*, 121.
9. Chagonda, L. F. S.; Millership, J. S. *J. Pharm. Biomed. Anal.* 1989, *7*, 271.
10. Comby, F.; Jambut-Absil, A. -C.; Buxeraud, J.; Raby, C. *Chem. Pharm. Bull.* 1989, *37*, 151.
11. Davidson, A. G. *J. Pharm. Pharmacol* 1976, *28*, 795.
12. Davidson, A. G. *J. Pharm. Pharmacol.* 1978, *30*, 410.
13. De Lenheer, A. *J. Assoc. Off. Anal. Chem.* 1973, *56*, 105.
14. El Kommos, M.; Emara, K. M. *Analyst* 1988, *113*, 1267.

15. El-Shabouri, S. R.; Youssef, A. F.; Mohamed, F. A.; Rageh, A. M. I. *J. Assoc. Off. Anal. Chem.* 1986, *69,* 821.

16. Enami Khoi, A.-A. M. *J. Pharm. Sci.* 1983, *72,* 704.

17. Fasanmade, A. A.; Fell, A. F. *Analyst,* 1985, *110,* 1117.

18. Fujita, Y.; Mori, I.; Fujita, K.; Nakahashi, Y.; Tanaka, T. *Chem. Pharm. Bull.* 1987, *35,* 5004.

19. Gandhi, V. N.; Shah, H. *Indian Drugs* 1984, *21,* 354.

20. Gasco, M. R.; Carlotti, M. E. *Pharm. Acta Helv.* 1977, *52,* 296.

21. Gowda, H. S.; Ramappa, P. G.; Nayak, A. N. *Anal. Chim. Acta* 1979, *108,* 277.

22. Gurka, D. F.; Kolinski, R. E.; Myrick, J. W.; Wells, C. E. *J. Pharm. Sci.* 1980, *69,* 1069.

23. Hackmann, E. R. M.; Magalhaes, J. F.; Santoro, M. I. R. M. *Rev. Farm. Bioquim. Univ. S. Paolo* 1986, *22,* 22.

24. Issa, A. S.; Beltagy, Y. A.; Mahrous, M. S. *Talanta* 1978, *25,* 710.

25. Jambut-Absil, A. C.; Buxeraud, J.; Lagorce, J. F.; Raby, C. *Int. J. Pharm.* 1987, *35,* 129.

26. Jayarama, M.; D'Souza, M. V.; Yathirajan, H. S. *Talanta* 1986, *33,* 352.

27. Krácmar, J.; Blazek, J. *Pharmazie* 1968 *23,* 651.

28. Mahrous, M. S.; Abdel-Khalek, M. M. *Talanta* 1984, *31,* 289.

29. Mori, I.; Fujita, Y.; Kitano, S. *Bunseki Kagaku* 1983, *32,* E1.

30. Myszczynska, B.; Puzanowska-Tarasiewicz, H. *Farm. Pol.* 1989, *45,* 87.

31. Nayak, A. N.; Ramappa, P. G.; Yathirajan, H. S. *An. Quim. Ser. C.* 1982, *78,* 86.

32. Pasich, J.; Wojdak, H.; Jurczyk, T. *Pharm. Pol.* 1978, *34,* 359.

33. Pesez, M.; Bartos, J., *Colorimetric and Fluorimetric Analysis of Organic Compounds and Drugs,* Marcel Dekker: New York, 1974, 388.

34. Puzanowska-Tarasiewicz, H.; Kojlo, A. *Farm. Pol.* 1983, *39,* 141.

35. Puzanowska-Tarasiewicz, H., Karpinska, J.; Golebiewski, Z. *Pharmazie* 1989, *44,* 350.

36. Puzanowska-Tarasiewicz, H.; Myszczynska, B. *Acta Pol. Pharm.* 1988, *45,* 311.

37. Puzanowska-Tarasiewicz, H.; Staniszewska, E.; Tarasiewicz, M. *Farm. Pol.* 1981, *37,* 495.

38. Puzanowska-Tarasiewicz, H.; Woliniec, E.; Poltorak, J. *Acta Pol. Pharm.* 1989, *46,* 55.

39. Ramappa, P. G.; Basavaiah, K. *J. Inst. Chem.* 1984, *56,* 107.

40. Ramappa, P. G.; Basavaiah, K. *Indian J. Pharm. Sci.* 1985, *47,* 125.

41. Ramappa, P. G.; Gowda, H. S.; Nayak, A. N. *Analyst* 1980, *105,* 663.

42. Ramappa, P. G.; Gowda, H. S.; Nayak, A. N. *Microchem J.* 1983, *28,* 586.

43. Ramappa, P. G.; Gowda, H. S.; Nayak, A. N. *Z. Anal. Chem.* 1979, *298,* 160.

44. Ramappa, P. G.; Nayak, A. N.; Basavaiah, K. *Indian Drugs* 1984, *21,* 448.

45. Ramappa, P. G.; Rao, N. R. R. *Indian Drugs* 1987, *24,* 441.

46. Ryan, J. A. *J. Am. Pharm. Assoc. Sci. Ed.* 1959, *48,* 240.

47. Rychlovsky, P.; Nemcová, I. *Ceskosl. Farm.* 1988, *37,* 104.

48. Rychlovsky, P.; Nemcová, I. *Ceskosl. Farm.* 1989, *38,* 241.

49. Sane, R. T.; Kamat, S. S.; Narkar, V. S.; Sathe, A. Y.; Mhalas, J. G. *Indian Drugs* 1980, *18,* 19.

50. Sastry, C. S. P.; Prasad Tipirneni, A. S. R.; Suryanarayana, M. V. *J. Pharm. Biomed. Anal.* 1990, *8,* 287.

51. Sawada, M.; Motoyama, I.; Yamada, S.; Kanaya, Y. *Bunseki Kagaku* 1986, *35,* T16.

52. Stan, M.; Dorneanu, V.; Ghimicescu, G. *Talanta* 1977, *24,* 140.

53. Taha, A. M.; El-Rabbat, N. A.; El- Kommos, M. E.; Refat, I. H. *Analyst* 1983, *108,* 1500.

54. Tarasiewicz, M.; Staniszewska, E.; Puzanowska-Tarasiewicz, H. *Chem. Anal.* 1980, *25,* 591.

55. Tarasiewicz, M.; Staniszewska, E.; Puzanowska-Tarasiewicz, H. *Farm. Pol.* 1980, *37,* 427.

56. *The United States Pharmacopoeia XXII,* USP Convention Inc.: Rockville, 1990. a. 1156; b. 1365; c. 292–294; d. 1409; e. 1416; f. 1160; g. 1048–1050; h. 586–587; i. 1159.

57. Xiang, B.; Xu, J.; Lu, M.; An, D. *Nanjing Yaoxueyuan* 1985, *16,* 33.

58. Youssef, A. F.; El-Shabouri, S. R.; Mohamed, F. A.; Rageh, A. M. I. *J. Assoc. Off. Anal. Chem.* 1986, *69,* 513.

K. OTHER HETEROCYCLIC DERIVATIVES

1. Furan Derivatives

In this section antimicrobial nitrofuran derivatives are mainly discussed but the spectrophotometric analysis of the diuretic drug furosemide and the H_2-receptor antagonist ranitidine is also summarized here.

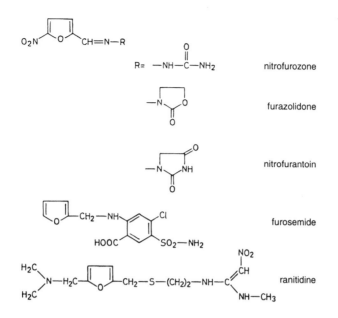

The furan ring itself is spectrophotometrically only slightly active (see Section 3.K). The characteristic and intense spectrum of nitrofuran derivatives (causing their yellow color) is the consequence of the conjugation of the nitro and aldehyde-hydrazone groups with the furan ring. Characteristic features of these spectra are two bands around 263 and 370 nm, of which the more intense latter is used for quantitative analytical purposes. The spectral data of the most important nitrofurans are collected in Table 10.7.

In the spectrum of furosemide the contribution of the furan ring is almost negligible: it is determined by the tetrasubstituted aromatic ring (see Figure 3.17). The band at 314 nm in the spectrum of ranitidine which is suitable for quantitative analytical purposes is due to the ethylene moiety substituted by nitro and amino groups.

On the basis of the data in Table 10.7, it is evident that using the intense long-wavelength band, a large variety of analytical problems can be solved. For example in the United States Pharmacopoeia the dissolution study of nitrofurantoin tablets is monitored at 375 nm,[228a] the assay of bulk nitrofurazone and its tablet dosage form is carried out in aqueous medium at the same wavelength, while the cream is determined in dimethylformamide at 285 nm.[228a] In the case of furazolidone bulk material, tablet and suspension the assay is carried out in aqueous medium at 367 nm.[228b] The assay of the above mentioned three derivatives[36f] and that of nitrofurantoin tablets[36g] are carried out spectrophotometrically in the British Pharmacopoeia, too. Due to the favorably long analytical wavelength, it is often the case that in combined formulations the other component does not interfere with the assay of nitrofuran derivatives, e.g., nitrofurantoin can be determined at 375 nm in the presence of ephedrine.[126] Measurement at two wavelengths and the use of two equations

TABLE 10.7. UV-VIS Spectral Data of Some Nitrofuran Derivatives

	λ_{max}, nm	ε
2-Nitrofuran	315	18,500
Furazolidone	259	13,000
	366	17,000
Nitrofurantoin	266	14,100
	367	18,460
Nitrofurazone	261	13,510
	375	16,300

with two unknowns enables more complicated problems to be solved: e.g., at 340 and 365 nm, nifuroxime and furazolidone can be simultaneously determined in suppositories.[70] The wavelengths for the simultaneous determination of procaine and nitrofurazone were 290 and 373 nm.[66]

The intense spectrum of furosemide is also suitable for its determination in pharmaceutical formulations. It is determined in injections in 0.1 M sodium hydroxide ($\varepsilon = 19,180$).[36d] pH-induced difference spectrophotometry[4] and derivative spectrophotometry were also used; with the latter method the assay of a furosemide-triamterene combination was successfully carried out.[149]

The absorption maximum of ranitidine at 314 nm was utilized among others for monitoring the dissolution test of its tablet formulation.[228c]

Since nitrofuran derivatives are yellow themselves, color reactions have only limited importance in this field. The basis of the relatively most frequently used method is the rearrangement of these derivatives in strongly alkaline media to orange-red products. This color produced with sodium hydroxide[54] or tetraethylammonium hydroxide[164] can be stabilized with phenol or resorcinol.[54] After hydrolysis dinitrophenylhydrazone[52] or isonicotinoyl hydrazone,[206] or after hydroxylaminolysis, iron(III) hydroxamate[12] can be formed; these are futher possibilities for their determination.

Several methods based on chemical reactions are available for the assay of the formulations of furosemide. It forms an ion-pair with amethyst violet which can be extracted with chloroform and measured at 550 nm.[154] Further possibilities are the simultaneous use of 3-methyl-2-benzthiazolinone hydrazone and oxidizing agents,[193] complex formation with copper[131] and iron,[132] the indirect molybdenum blue method based on the reducing properties of furosemide[102] and the use of phloroglucinol in hydrochloric acid-acetic acid medium.[42]

Of the colorimetric methods available for the determination of ranitidine, ion-pair extraction procedures are the most important. Ranitidine is a strongly basic drug which can be extracted as the ion-pair formed with acidic dyes such as bromothymol blue, bromophenol blue, bromocresol green, bromocresol purple[171] and Rose Bengal.[167] The extraction is carried out from neutral or slightly acidic medium using chloroform[171] or dichloromethane.[167] Other possibilities include the use of 4-dimethylaminobenzaldehyde reagent.[89]

2. Imidazoline and Imidazole Derivatives

Both imidazole and imidazoline are spectrophotometrically only slightly active (see Section 3.K). Consequently, ultraviolet spectrophotometry is not suitable for the determination of those derivatives which do not contain other chromophoric groups increasing the spectrophotometric activity (e.g., cimetidine). Due to the fact that chromophoric or auxochromic substituents are present in the majority of cases, spectrophotometry based on natural absorption has found wide application among the pharmaceutically-active imidazole and imidazoline derivatives. Substituents of this type are the nitro group in the molecule of metronidazole (in 0.1 M hydrochloric acid, $\varepsilon_{277\,nm} = 6,500$) or the thiol group in methimazole (in 0.05 M sulphuric acid, $\varepsilon_{251\,nm} = 17,400$). Examples for chromophoric systems isolated from the imidazoline ring are the naphthalene ring in naphazoline ($\varepsilon_{281\,nm} = 6,530$; for the spectrum with its naphthalene-like fine structure see p. 31 and the N,N-dialkylaniline-like moiety in antazoline (in 0.1 M hydrochloric acid, $\varepsilon_{241\,nm} = 17,200$), etc. In the case of tolazoline where the isolated chromophore is a simple phenyl group, the spectrophotometric acitivity remains poor.

In the case of the above listed derivatives and related compounds, spectrophotometric determinations based on natural absorption provide an approach for many analyses, e.g., the determination of naphazoline in solution dosage form,[228d] metronidazole in tablets[228e] in methanol and 0.1 M hydrochloric acid, respectively, at the above mentioned wavelengths.

Further to these simple problems which can be solved by single-wavelength measurement, dual-wavelength measurements are suitable for the assay of combined formulations, e.g., simultaneous determination of diloxanide furoate and metronidazole in 0.01 M aqueous sodium hydroxide at 247 and 320 nm[227] or in ethanol at 258 and 310 nm,[46] furazolidone and metronidazole in aqueous solution at 367 and 320 nm,[217] metronidazole and diiodohydroxyquinoline in 0.02 M aqueous sodium hydroxide at 320 and 267 nm[48] as well as antazoline and naphazoline in aqueous solution at 271 and 281 nm.[143] For the latter purpose, a computer-based algebraic method was also applied, using the absorbance measurements at 2 nm intervals in the spectral range around the maxima.[234]

Another example for the application of computer-based methods is the simultaneous determination of antazoline and its main degradation product (N-(2-aminoethyl)-2-(N-benzylanilino)-acetamide) where the difference between the absorption maxima of the intact drug and the degradation product is only 5.7 nm.[187]

Of course, chromatographic separation prior to the spectrophotometric measurement affords practically unlimited possibilities for the simultaneous determination of the active ingredients of the above mentioned and similar combinations. In Section 5.B dealing with this subject, it is the example of the spectrophotometric assay of a naphazoline-antazoline combination after column chromatographic separation which demonstrates this possibility.[127]

Derivative spectrophotometry has also been applied to imidazole and imidazoline derivatives. Examples are the determination of miconazole and econazole in various formulations (see Table 4.6),[44] simultaneous determination of naphazoline and diphenhydramine in eye drops in the presence of methylene blue based on the measurement of $^2D_{273}$ and $^2D_{255}$,[118] the simultaneous determination of naphazoline and thonzylamine by measuring $^3D_{280}$ and $^3D_{258}$[232] and the measurement of cimetidine on the basis of $^2D_{216}$.[23]

In the case of the assay of those derivatives where the spectrophotometric activity is associated with the substituted imidazole ring, the selectivity of the measurement can be increased by using various versions of difference spectrophotometry. The simplest possibility is to base the measurement on the spectral shifts due to changes of the pH of the test solution. The dependence of the spectrum of metronidazole on the pH is shown in Figure 3.32. On this basis, metronidazole can be determined in the presence of furazolidone by measuring the difference absorbance at 322 nm in such a way that the neutral solution is measured against that in 0.01 M hydrochloric acid.[47] The same hypsochromic shift due to acidification was used also for the determination of metronidazole benzoate in a suspension in the presence of spectrophotometrically active excipients.[86]

A byproduct in the synthesis of metronidazole (2-methyl-1-hydroxyethyl-5-nitroimidazole) is the 4-nitro isomer. The spectra of the two isomers, including their change upon decreasing the pH are also very similar. In spite of this, the difference spectrophotometric technique is suitable for the determination of the 4-nitro isomer in metronidazole. The basis for this is that metronidazole is a much stronger base than the isomer, the difference being about 2 pK units. This means that in 0.2 M hydrochloric acid, protonation of metronidazole is quantitative while it is only partial for the 4-nitro isomer. In 2 M hydrochloric acid both isomers are protonated. If the latter solution is placed into the reference cell and the solution in 0.2 M hydrochloric acid is measured against it, the quantity of 4-nitro isomer impurity in metronidazole can be determined at 325 nm.[87]

Another version of the difference spectrophotometric determination of metronidazole is the reduction of its nitro group by zinc/hydrochloric acid and measurement of the difference absorbance of the unreduced solution against the reduced, poorly absorbing solution. The characteristic and rather intense difference spectrum thus obtained enables various analytical problems to be solved.[93]

Several different color reactions have also been used for the determination of imidazoline derivatives. Ion-pair extraction methods (see Section 7.H) make use of their basic properties. The acidic dyes used for the formation of extractable ion-pairs include bromocresol green,[45,182,245] bromophenol blue,[182] methyl orange[182] and Rose Bengal.[16] With these dyes as the reagents, oxymetazoline,[16] xylometazoline,[16,182] cimetidine,[72] antazoline[182] and naphazoline[182] have been determined. In the method of Guven et al.[90] for the determination of cimetidine, the reagent is copper(II) acetate; the resulting complex is extracted, its copper content is transformed to the diethyldithiocarbamate and measured at 433 nm. Precipitation with the Reinecke salt and the dissolution of the precipitate in acetone followed by measurement at 525 nm can also be used e.g., for the determination of oxymetazoline.[65]

The most characteristic reagents of this group of compounds are various nitrite or nitroso derivatives, first of all sodium nitroprusside, which is utilized in USP XXII[228f] for the assay of tolazoline injection. The application of the same reagent for the assay of cimetidine[17] and tolazoline[211] tablets and for the determination of phentolamine and naphazoline[101] in various solid and liquid dosage forms as well as methimazole in tablets and injections[109] has also been described. Sodium nitrite was used for the determination of antazoline[15] and sodium cobaltinitrite for the assay of various formulations of oxymetazoline.[205]

The classical azo-coupling method can also be adopted. Diazotized amines (sulphanilamide, sulphophenazol, sulphacetamide, benzocain or 4-aminoacetophenone)[181] and phenylhydrazine-4-

sulphonic acid[121] have been used for the determination of cimetidine in tablets. Just the opposite of this reaction sequence can be used for the determination of metronidazole: here the nitro group is reduced with zinc/hydrochloric acid and the amino group thus obtained is diazotized, coupled with 2-naphtol and measured at 480 nm.[79] In a further version the nitro group is eliminated by alkaline treatment and the nitrite ions thus formed are used for the diazotization of 4-aminobenzoic acid[75] or sulphanilamide,[88] and the resulting diazonium salts are coupled with N-(1-naphtyl)-ethylenediamine.

The reduction of the nitro group (e.g., in tinidazole) to the primary aromatic amine by using the above mentioned zinc/hydrochloric acid reagent can be used also for the condensation of the amino group with aromatic aldehydes such as vanillin,[176] salicylaldehyde[112] or 4-dimethylamino-benzaldehyde.[111] The colored Schiff's bases can easily be measured.

The formation of charge-transfer complexes is also the basis of analytical methods for the determination of imidazole and imidazoline derivatives. 2,3-Dichloro-5,6-dicyano-1,4-benzoqui-none reagent was used for the determination of clonidine,[71] iodine and chloranil for the assay of antazoline and naphazoline formulations (even their combined formulation).[28]

The use of several other reagents was also reported. These include 1,2-naphthoquinone-4-sulphonic acid for the determination of cimetidine[165] and naphazoline,[220] 3,5-dichloro-p-benzoqui-nonechloroimine for the determination of cimetidine[209] and methimazole,[83] vanillin-phosphoric acid[137] and cerium(IV) perchloric acid[141] reagents for antazoline, the Folin-Ciocalteu reagent[129] and ninhydrin for cimetidine, 2,6-dichlorophenolindophenol reagent for antazoline, tolazoline, xylomethazine and naphazoline.[3,23] Finally, the extremely sensitive method for the determination of methimazole is based on the catalysis of the iodide-azide reaction.[51]

3. Antituberculotic Pyridine Derivatives

The natural absorption of the isonicotinic acid derivatives discussed in this section is suitable for their determination in various matrices. As is characteristic of pyridine derivatives in general, the position of the main band in their spectra is practically not affected by the protonation of these drug molecules; at the same time, however, the intensity of the band markedly increases. For this reason their spectrophotometric determination is usually carried out in 0.01 M or 0.1 M hydrochloric acid medium where the maximum of the spectrum of isoniazid (isonicotinic acid hydrazide, INH) is at 265 nm ($\varepsilon = 5,750$). The strong bathochromic shift in the spectrum of INH observable in alkaline media (above pH 12) is worth mentioning ($\varepsilon_{300 \text{ nm}} = 4,000$). Spectrophotometric measurement in acidic medium is used in the dissolution testing of INH tablets,[228g] its selective determination in a combined preparation in the presence of cycloserine,[242] etc. Measurement at two wavelengths was applied for the assay of other combinations such as INH-rifampycin at 264 and 335 nm.[201]

The above-mentioned bathochromic shift in alkaline media enables more selective measurements to be carried out. The spectral shift takes place at different pH values in the case of different derivatives, and this enables the simultaneous determination of INH and its glucosehydrazone at pH 11 on the basis of the measurement of A_{302}/A_{240}. This method could be used for the investigation of the degradation of INH-glucosehydrazone.[197]

As in many other cases, here too, derivative spectrophotometry enhances the possibilities of the spectrophotometric determination. For example, in the method of Kitamura et al.[117] the determination of INH in the presence of its lactosylhydrazone was carried out by TLC separation, spot elution and spectrophotometric measurement. The interference caused by parasitic absorption due to impurities coeluted from the plate was eliminated by second derivative spectrophotometry.

The majority of spectrophotometric methods based on chemical reactions use reactions of the substituents of the pyridine ring rather than those of the ring itself. The classical method for the determination of INH is based on the reaction of the free –NH$_2$ group of the carboxylic acid hydrazide moiety with aromatic aldehydes and the spectrophotometric determination of the resulting hydrazone. As the aldehyde reagents, vanillin,[9,33] 4-nitrobenzaldehyde,[200] pyridoxal[200] and 4-dimethylaminobenzaldehyde[151] were used. In another variant of the latter method, hydrazine is hydrolyzed by vigorous treatment with hydrochloric acid, and the aldehyde reagent reacts then at both ends of the hydrazine molecule. The use of the same reagent for the difference spectrophotomet-ric determination of traces of free hydrazine in INH and its formulations in the method of Davidson[59] was described on p. 215.

It is worth mentioning that in another field of drug analysis, INH plays the role of the spectrophotometric reagent: as described in Section 10.M.2, INH is the most generally used reagent for the determination of unsaturated ketosteroids.

Various aromatic reagents containing active halogens also attack INH at the free $-NH_2$ group. The derivative formed with the 1-fluoro-2,4-dinitrobenzene[198,233] has an absorption maximum at 530 while the absorption maxima of the derivatives with 9-chloroacridine,[219] 2,3-dichloro-1,4-naphthoquinone[62] and 6,7-dichloroquinoline-5,8-dione[69] are at 500, 640 and 645 nm (in the case of the latter two derivatives in alkaline media). The molar absorptivities after these and the previously discussed reactions range between 5 and 15,000. The use of 1,2-naphthoquinone-4-sulphonic acid[155] is also worth mentioning ($\varepsilon_{480\ nm} = 10,100$).

Another group of methods exploits the tendency of the analytes to form metal complexes. Ethionamide can be determined as the palladium(II) complex at 410 nm[212] or as the reaction product with osmium tetroxide at 380 nm.[213] The sensitivity of these methods is medium and this applies to the method for ethionamide based on hydroxylaminolysis and spectrophotometric measurement of the iron(III) hydroxamate complex.[199] On the other hand, the method of Fujita et al.[77] for the indirect determination of INH excels with its extraordinary high sensitivity. In this method INH undergoes a ligand-exchange reaction with the ternary complex palladium(II)-o-hydroquinonephthalein-cetylpyridinium chloride, and this leads to the decrease of the absorbance at 635 nm, the extent of which is proportional to the concentration of INH. The apparent molar absorptivity is as high as 620,000. INH can be determined also by the sodium nitroprusside reagent.[153]

Several methods are based on the ready oxidation of INH with the formation of isonicotinic acid and nitrogen. These are indirect methods: the reduced form of the oxidizing agent is measured. The reagents include ammonium vanadate,[166,215] iron(III) chloride,[215] potassium ferricyanide,[215] various tetrazolium reagents,[105–107,119] ammonium paramolybdate[104] and iron(III) chloride-o-phenanthroline.[108]

4. Some Other Pyridine Derivatives

The spectrum of chlorpheniramine is characteristic of the isolated pyridine ring in its molecule: $\varepsilon_{265\ nm} = 6,600$ in 0.1 M sulphuric acid. This enables its determination in various formulations.[36e,228i] The selectivity of the measurement can be improved by using derivative spectrophotometry ($^2D_{265,262}$).[125] Chlorpheniramine maleate was determined in tablets in the presence of 15-fold quantity of ψ-ephedrine hydrochloride on the basis of the differences in their derivative spectra. This study was conducted using a computerized diode-array spectrophotometer with multi-component analysis software.[97]

Several methods based on chemical reactions are also available for the determination of chlorpheniramine. Its pyridine ring can be opened by cyanogen bromide and the resulting aldehyde condensed with aniline to form a yellow derivative with absorption maximum at 482 nm. Although the sensitivity of this method is poor, its selectivity allows the determination of chlorpheniramine in the presence of codeine and ephedrine.[30] The dimethylamino group in the molecule is suitable for the formation of ion-pairs with suitable reagents such as bromophenol blue,[140] bromocresol green,[54] methyl orange[21] and zincon.[95] The method of Sakai[173] for the simultaneous determination of chlorpheniramine and dibucaine merits special attention. In this method the ion-pair reagent was tetrabromophenolphthalein ethyl ester and the measurement at 555 and 575 nm was based on the shift of the spectra between 25 and 50°C (for the detailed description of thermochromism see Section 2.K.4).

Similar to other amines, chlorpheniramine also forms a charge-transfer complex with iodine in chloroform. Measurement of the complex at 293 nm is the basis of its quantitative determination.[91] Of the spectrophotometric methods for the determination of chlorpheniramine, the procedure of Fujita et al.[78] excels from the points of view of both selectivity and sensitivity ($\varepsilon_{580\ nm} = 65,000$). In this method the measured species is the ternary complex formed with titanium(IV), o-hydroxyhydroquinonephthalein and fluoride.

The pyridine aldoxime derivatives of the cholinesterase reactivator type (pralidoxime, obidoxime, etc.) possess intense bands at 242 and 292 nm which are due to the conjugation of the pyridine ring and the aldoxime group. In strongly alkaline media the ionization of the latter results in a bathochromic shift (λ_{max} = 332 nm). These spectra are suitable for the solution of various analytical problems.[130]

Of the methods based on chemical reactions for the determination of pyridine aldoximes, those in which their ability to form metal complexes is exploited are the most important. Pralidoxime forms a 1:1 complex with palladium(II). Its measurement at 327 nm (pH 6.45) enables the assay of the tablet[115] and injection[116] formulations (ε = 10,500). Obidoxime can be determined similarly.[114] Their reactions with nickel(II)[246] and amino- or nitroso-pentacyanoferrate reagents[39,40] produce colored products, but the sensitivity of these methods is rather poor.

Ester derivatives of nicotinic acid can easily be determined on the basis of their natural absorption. For example, methyl nicotinate was determined in an ointment in the presence of diethylammonium salicylate and parabens by TLC separation, spot elution and measurement at 264 nm.[1] Inositol nicotinate and bamethan sulphate were simultaneously determined in a tablet formulation by using first derivative spectrophotometry.[56] Further methods for their determination involve the use of the cyanogen bromide–sulphanilic acid,[29] hydroxylamine-iron(III)[55] and copper(II)[122] reagents.

Of the applications of derivative spectrophotometry to other pyridine derivatives, the simultaneous determination of pirbuterol hydrochloride and butorphanol tartrate by the first derivative method[134] and the selective measurement of pirenzepine hydrochloride in the presence of its degradation products also by first derivative spectrophotometry[235] are mentioned. pH-induced difference spectrophotometry can also be used for the determination of the latter. The maximum at 280 nm in 0.1 M hydrochloric acid is shifted to 295 nm in 0.1 M sodium hydroxide; at the maximum of the difference spectrum (300 nm) pirenzepine can be selectively measured.[236]

5. Derivatives of Piperidine, Piperazine and Other Saturated Heterocycles

Spectrophotometric methods for the determination of piperidine derivatives containing spectrophotometrically active groups were treated in some of the previous sections (mainly in Section 10.B) and the generally applicable ion-pair extraction method, where their determination was also discussed. Some further methods will be presented here for the determination of loperamide, haloperidol and some other derivatives.

The two isolated phenyl and one 4-chlorophenyl groups in the molecule of loperamide are eminently suitable to serve as the basis for derivative spectrophotometric measurement. Dol et al.[67] used $^2D_{259}$ or $^2D_{265.5}$ for its determination in tablets and capsules. In addition to tropeolin 00 (see Table 10.1) other acidic dyes and other ion-pair forming reagents were also used for its determination such as methyl orange,[168] bromocresol purple[18] and tetrathiocyanato-cobalt(III).[207] Quantitative determination of loperamide via charge-transfer complex formation with iodine was also reported.[35]

Tolperisone and haloperidol are spectrophotometrically active compounds due to the presence of aromatic ketone moieties in their molecules ($\varepsilon_{257\,nm}$ = 15,750 and ε_{243} = 13,000—see Figure 3.13). This band is utilized in USP XXII for the determination of haloperidol in tablets and injection.[228i] This absorption band can be eliminated by reduction with sodium borohydride, and this creates a good possibility for the application of the difference spectrophotometric method described in Section 8.D, thus greatly increasing the selectivity of the measurement.[224] Haloperidol can be extracted with chloroform at pH 2 as the ion-pair with e.g., bromophenol blue and measured at 400 nm.[120] Other color reactions for its determination are the condensation of the ketone group with phenylhydrazine (λ_{max} = 335 nm),[133] reaction with 1, 2-naphthoquinone-4-sulphonic acid and measurement at 410 nm[163] and the formation of charge transfer complex with chloranil (λ_{max} = 540 nm).[203]

Of the other piperidine derivatives, meperidine can be determined via its ester group using the iron(III)-hydroxamate method,[84] thenalidine maleate is determined by ion-pair extraction of the base using tropeolin 00 as the reagent at pH 4 (solvent: 1, 2-dichloroethane; λ_{max} = 410 nm)[61] and ion-pair extraction method was described for the determination of heptacaine and its N-oxide, too, the reagent being picric acid.[222]

Analogs with larger rings can also be extracted using suitable ion-pairing reagents. For example, guanethidine forms ion-pair with bromophenol blue at pH 3.5 which can be extracted with chloroform.[110] If the same compound is determined using diacetyl and 1-naphthol as the reagents then the reaction takes place at the guanidine moiety.[161] In the case of pentazocine the phenol part of the molecule is more suitable for the determination than the piperidine ring partly because of the possibility to measure it at 278 nm and, on the other hand, the classical phenol reagents such as sodium nitroprusside[196] or the Folin-Ciocalteu reagent[204] can also be used. Another possibility for the phenol part of the molecule is its treatment with nitrous acid and absorbance reading in alkaline solution of the resulting nitroso derivative at 434 nm.[50] The selectivity of this method can be increased if the o-nitrosophenol derivative is measured as the cobalt(II) or copper(II) chelate.[49]

The classical method for the determination of piperazine is its precipitation as the reineckate, filtration and measurement at 525 nm after the dissolution of the precipitate in acetone.[210] It can be transformed to the N-nitroso derivative by treatment with nitrous acid and this derivative can be measured at 239 nm.[7] The reagents of the methods based on the formation of charge transfer complexes[169,170] are iodine, chloranil, chloranilic acid with absorbance maxima at 264, 545 and 343 nm. Using p-benzoquinone as the reagent, the maximum is at 516 nm.[237]

The most generally used method for the determination of piperazine derivatives such as diethyl-carbamazine citrate,[192] cyclizine[186] and centpropazine[231] is measurement in the visible spectrum range after ion-pair extraction with suitable reagents: bromophenol blue,[186,231] bromocresol green,[186] Orange II[192] and Fast Green FCF.[192] Diethyl carbamazine citrate was determined also in the form of its charge transfer complex with iodine directly[239] or after hydrolysis with sulphuric acid, using chloranilic acid as the reagent.[170] In the latter case the reacting species is N-methylpiperazine formed during the hydrolysis. The reagents for the determination of cinnarizine via charge transfer complexation are iodine and 2,3-dichloro-5,6-dicyanobenzoquinone.[174]

Hexamethylenetetramine can be cleaved to formaldehyde and ammonia by acidic hydrolysis and hence their spectrophotometric determination can be based on the spectrophotometric measurement of these products. For example, in the United States Pharmacopoeia, the resulting formaldehyde is measured using the classical color reaction with chromotropic acid (see Section 7.E).[228j] The direct determination of hexamethylenetetramine is also possible by means of charge transfer complex formation: using iodine as the reagent in 1,2-dichloroethane as the solvent, an intense band appears at 273 nm ($\varepsilon = 29,000$), which is suitable for quantitation. The reaction with alloxane leading to murexide which can be measured at 532 nm is also worth mentioning.[148]

The determination of flucytosine (4-amino-5-fluoro-2(1H)-pyrimidinone) in the presence of its synthetic precursor and degradation product 5-fluorouracil was performed by derivative spectrophotometry. The measurement of $^2D_{310.8}$ enables the intact drug to be measured while from $^2D_{292.4,267.2}$ the quantity of 5-fluorouracil could be calculated.[43]

6. Tricyclic Antidepressants and Related Derivatives

Of the derivatives treated in this section imipramine and trimipramine can be regarded as diphenyl-amine derivatives from the point of view of their UV spectra. The conjugated system of these derivatives is extended by an ethylene bridge linking the two benzene rings and an N-carbamoyl group. The 1,1-diphenylethylene type is represented by amitriptyline and nortriptyline. The insertion of oxygen or sulphur atoms into the second ring introduces phenol or thiophenol character to these derivatives with the spectroscopic consequences of this. All the above listed derivatives possess intense—although very different—UV spectra. This is not the case with maprotiline where the second ring does not contain either hetero atom or double bond, and hence the UV activity is restricted to the weak absorption of the two isolated benzene rings. It is to be noted here that phenothiazines closely related to this group are discussed in a separate section: 10.J.

The wide possibilities for the exploitation of the above discussed spectrophotometric activity to the quantitative analysis of tricyclic antidepressants are illustrated with examples taken from the United States and British Pharmacopoeias. The assay of imipramine tablet and injection is carried out at 250 nm ($\varepsilon = 6,700$)[231k] while the wavelength for the assay of bulk carbamazepine[36a] and the tablet[36b] and for monitoring the dissolution test of the latter[228l] is 285 nm ($\varepsilon = 11,600$). The dissolution test of amitriptyline and nortriptyline tablets[228m,n] as well as the assay of amitriptyline suspension[36c] is carried out at 239 nm ($\varepsilon = 13,200$ and 12,300). Doxepin tablet and solution are measured at 292 nm[228o] where the molar absorptivity is 3,350 while the wavelength for the investigation of chlorprothixene tablet and injection is 324 nm ($\varepsilon = 2,800$).[228p]

First- and second-derivative spectra enable difficult analytical problems to be solved such as the simultaneous determination of amitriptyline and perphenazine in blood plasma[74] and, further, the simultaneous determination of imipramine and amitriptyline,[81] moreover, amitriptyline, imipramine and perphenazine in formulations and blood plasma.[80]

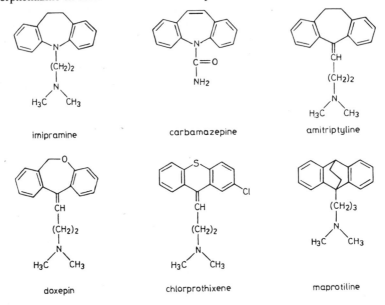

imipramine   carbamazepine   amitriptyline

doxepin   chlorprothixene   maprotiline

The overwhelming majority of the derivatives discussed in this section (including the tricyclic antidepressants) bear a side chain containing tertiary or secondary amino groups. This amino group attached to the bulky, apolar ring system makes these derivatives eminently suitable for being determined by the ion-pair extraction method (see Section 7.H). A large variety of acidic dyes have been used for this purpose including picric acid,[230] bromophenol blue,[180] bromothymol blue,[76,180] bromocresol green,[31,180] methyl orange,[63] bromocresol purple,[31] Chrome Azurol S,[152] Orange II, croceine orange, lithol red, Fast Red A,[150] etc.

Inorganic ion-pairing reagents were also used, e.g., iodobismuthate for the determination of imipramine[60] and cobalt(III)thiocyanate for amitriptyline.[185] After extraction from acidic media, the associates were measured at 490 and 625 nm, respectively. Another inorganic reagent, rarely used in quantitative drug analysis is tetrabromoaurate. Although the absorption maximum of the adduct formed with this reagent is in the ultraviolet range, the sensitivity is very good ($\varepsilon_{253\ nm}$ = 33,000).[32] An interesting feature of this method is the high selectivity attainable: at pH values below 4, a wide range of adducts can be extracted with chloroform; between pH 5 and 6, however, only desipramine and nortriptyline having secondary amino groups are extracted if cyclohexane is used for the extraction.

Maprotiline[73] and chlorprothixene[241] were determined as the charge-transfer complexes formed with chloranil and tetracyanoethylene reagents, respectively. Less important methods include the determination of doxepin[156] and carbamazepine[159] after oxidation with potassium permanganate and periodic acid, respectively. The nitro group of nitroxazepin can be reduced by zinc, and the resulting amino group gives color reaction with the sodium pentacyano-aminoferrate reagent.[121] The carbamoyl group of carbamazepine can be determined by the iron(III) hydroxamate method.[10]

7. Antimicrobial Quinoline and Acridine Derivatives

In this section the spectrophotometric methods for the determination of the halogen derivatives of 8-hydroxyquinoline (clioquinol and related derivatives), antimalarial aminoquinoline derivatives such as chloroquine and amodiaquine as well as aminoacridine derivatives (aminacrine, ethacridine) are summarized.

The spectra of quinoline derivatives which can be derived from the spectrum of naphthalene are discussed in Section 3.M. The very intense and strongly pH-dependent spectra (see Figure 8.1) enable various analytical problems to be solved. The clioquinol content of its powder formulation is determined in 3 M hydrochloride at 267 nm[228u] where the molar absorptivity is 30,400. The intensity of this band is even higher in neutral medium ($\varepsilon_{255\ nm}$ = 40,000). This is the basis of the acidic-neutral difference spectrophotometric method of Görög[85] for the determination of clioquinol in an ointment in the presence of prednisolone (see Figure 8.2). In the case of another combination (diiodohydroxyquinoline and metronidazole) measurement in alkaline medium at 267 and 320 nm and the use of equations with two unknowns were the basis for the simultaneous determination.[48]

The active ingredient content of chloroquine injection is determined in dilute hydrochloric acid medium at 343 nm (ε = 16,000).[228r] The spectral shifts in the derivatives due to the different substituents allow the simultaneous determination of the active ingredients of combined preparations with the aid of equations with two unknowns. A typical example is the assay of primaquine-amodiaquine tablet; the absorbances at 282 nm and 342 are measured in 0.1 M hydrochloric acid.[92] The latter wavelength has been selected for the assay of bulk primaquine, too.[228t] 8-Hydroxy-1-methylquinolinium methylsulphate is determined at pH 3 at 365 nm (ε = 27,000).[233]

It is of interest that the strong dependence of the spectra on the pH in this group of drug substances allows, not only the optimization of the selectivity and sensitivity of the measurement and the application of pH-induced difference spectrophotometry, but—as shown by Takla and Dakas[226] in the example of chloroquine—the investigation of the impact of excipients on the protonation of the drug and in such a way, the investigation of the influence of this factor on its absorption from the gastrointestinal tract.

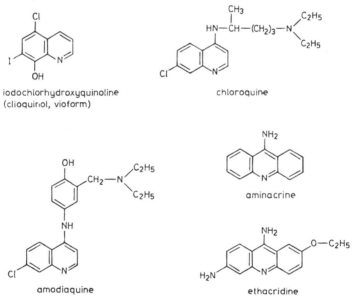

iodochlorhydroxyquinoline
(clioquinol, vioform)

chloroquine

amodiaquine

aminacrine

ethacridine

The intensity of the main band in the spectra of the yellow aminoacridine derivatives is extremely high: for aminacrine $\varepsilon_{260\ nm}$ = 64,000. The intensity of the long-wavelength band is smaller but the selectivity at this wavelength is very advantageous. For example in the case of ethacridine lactate $\varepsilon_{360\ nm}$ = 15,400 and at this wavelength even the stability assay of liquid dosage forms could be performed.[34,96] Similarly intense long-wavelength band with fine structure can be found in the spectrum of aminacrine (see Figure 10.6). This band is the basis for the determination of aminacrine in acrisorcine cream after extraction with chloroform, further extraction of the chloroform phase with 0.3 M hydrochloric acid and measurement of the latter at 400 nm.[228s] Spectral background can be corrected using the Allen correction method described in detail in Section 4.C.2. In the method of Bunch[37,38] the following formula is used to calculate the corrected absorbance: A_{402} − 0.5 (A_{412} − A_{389}).

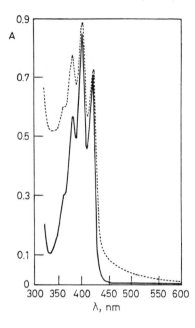

FIGURE 10.6. Spectra for the determination of aminacrine in formulations. Solvent: 1 + 99 mixture of concentrated hydrochloric acid and 95% ethanol. Concentration: 1.6 mg/100 ml. Aminacrine —; Aminacrine cream – –. (From Bunch, E. A.; *J. Assoc. Off. Anal. Chem.* 1987, 70, 560.[38])

Derivatives with a basic side chain containing diethylamino group can be determined by ion-pair extraction method using acidic dyes as the reagents. Dequalinium chloride was determined in tablets at pH 7 using picric acid as the ion-pairing reagent. After extraction with 1,2-dichloroethane, the absorbance is measured at 342 nm.[124] Other derivatives (chloroquine, amodiaquine, primaquine) were extracted with chloroform using Food Green at pH 5 or Acid Orange 7 at pH 1 as the reagents. The absorbance maxima were at 630 nm and 495 nm, respectively.[190] Using bromocresol green, the optimum of the pH is at 1.3 and the maximum of the extracted ion-pair in chloroform is at 415 nm.[123] Mount et al.[136] determined chloroquine in urine using methyl orange as the reagent. Of the inorganic ion-pairing reagents, cobalt(II)thiocyanate is mentioned which forms ion-pair with chloroquine, extractable with methylisobutylketone and measurable at 625 nm.[160] Reinecke salt is also suitable for the determination of chloroquine and related derivatives. The precipitates are dissolved in acetone prior to the spectrophotometric measurement.[58] While the methods applying organic reagents are sufficiently sensitive, the sensitivity of the latter two methods is poor.

Similarly poor is the sensitivity of the methods for the determination of clioquinol based on the formation of metal chelates with iron,[142] vanadium[13] and uranium.[172]

The reagents for the determination of aminoquinolines by charge-transfer complexation are iodine,[2] chloranil[223] and chloranilic acid.[128] Reagents for further color reactions include 2,6-dichloro-quinonechloroimine,[6] 1,2-naphthoquinone-4-sulphonic acid,[6] 3-methylbenzthiazolin-2-on hydra-zone cerium(IV),[191] ammonium molybdate,[189] chloramine T[157] and periodic acid.[177]

The majority of the photometric methods for the determination of 8-hydroxyquinoline derivatives is based on the phenol character of their molecules. The reagents are 2,6-dichloroquinonechloroi-mine ($\varepsilon_{650\,nm} = 15,300$),[25] 4-aminophenol,[178] diazotized benzocaine[179] and potassium ferricyanide 4-aminoantipyrine.[26,100]

8. Miscellaneous Derivatives

The intense spectra of *hydralazine* and *dihydralazine* enable their determination even in multicompo-nent dosage forms often administered in the course of their therapeutic use. Beyrich et al.[22] used computer-assisted method for the simultaneous determination of dihydralazine sulphate (λ_{max} 219, 310 nm), hydrochlorothiazide (226, 271 and 316 nm) and reserpine (218, 267 and 294 nm). Derivative spectrophotometry offers further possibilities. Bedair et al.[24] and Barary et al.[20] used first and second derivative spectra for analyzing formulations containing hydralazine with hydrochlorothiazide and propranolol, respectively. The basis of the selective determinations is the fact that $^2D_{256}$ is zero for hydralazine, while this wavelength is very close to the maximum of the second derivative spectrum of hydrochlorothiazide. On the basis of the same principle (zero-crossing spectra) from $^2D_{317}$ the hydralazine content can be selectively determined. The selective wavelengths for the determination of the active ingredients of the above mentioned other combination are 265 and 284 nm.

The methods involving the use of chemical reactions include the reactions between the hydrazine group(s) and aromatic aldehydes, which lead to colored hydrazones. After trying several aldehyde reagents,[138,243] 4-dimethylaminocinnamaldehyde proved to be the most suitable for the analysis of hydralazine preparations ($\varepsilon_{450\ nm}$ (chloroform) = 20,000).[138] The reaction product was also of the hydrazone type when ninhydrin was used as the reagent.[68]

Equation 10.10 shows the reaction of dihydralazine with 1,2-naphthoquinone-4-sulphonic acid leading to a product which can be measured at 455 nm.[175]

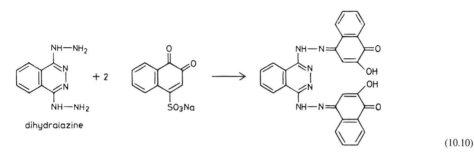

(10.10)

The color reaction with 9-chloroacridine[218] is also based on the substitution of the —NH$_2$ group of the hydrazine moiety. Several methods are available in which the ready oxidation of the hydrazine group is exploited for analytical purposes. The reagents for these indirect methods are Blue tetrazolium,[229] iron (III)-o-phenantroline,[103] and paramolybdate.[103] The interesting ring closure reaction which takes place when nitrous acid is the oxidizing agent is shown on p. 177.[135] Finally, the use of tetracyanoethylene reagent is mentioned which forms colored charge-transfer complexes with several hydrazine and pyrazolone derivatives (among them with hydralazine, too) which are the basis for their determination.[99]

In the course of the discussion of the spectrophotometric assay methods for dosage forms containing hydralazine and dihydralazine, the determination of *hydrochlorothiazide* was also mentioned as a component of these formulations.[22,24] Using second derivative spectra this can be determined in the presence of amiloride, too.[145] A formulation containing polythiazide and prazosin was also successfully analyzed: the former was calculated from $^2D_{236}$ and the latter from $^2D_{346,356}$.[144]

Of the colorimetric methods for the determination of hydrochlorothiazide and related benzothiadiazine-type diuretics, in the most selective methods, the hetero-ring is opened by acid hydrolysis, the resulting aniline derivative is diazotized and coupled with suitable reagents. In addition to the classical reagents generally used for such purposes, ethyl acetoacetate was also successfully applied.[27] The reaction scheme is shown in Equation 10.11. Hydroflumethiazide was determined at 725 nm after reaction with potassium ferricyanide reagent.[11]

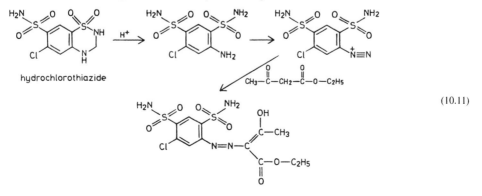

(10.11)

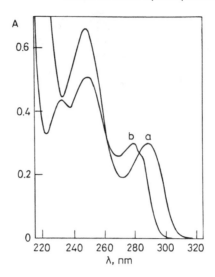

FIGURE 10.7. Spectra of benperidol for the difference spectrophotometric assay. Solvent: a. 0.1 M sodium hydroxide; b. 0.05 M sulphuric acid. Concentration: 1.5 mg/100 ml. (From Gassim, A. E. H.; Takla, P. G.; James, K. C. *Analyst* 1986, *111*, 923.[82])

Indomethacin has an intensive absorption maximum in neutral or weakly basic media at 267 nm ($\varepsilon = 16,500$). The spectra of its hydrolytic decomposition products (2-methyl-5-methoxyindol-3-yl acetic acid and 4-chlorobenzoic acid) sufficiently differ from each other and from that of the parent drug, and hence the spectrophotometric method was found to be suitable for the quantitative evaluation of the degradation of the drug.[146] For the same purpose a derivative spectrophotometric method is also available which enables the estimation of the degradation products at the 0.1% level.[41] Many color reactions are also available for the determination of indomethacin involving the use of vanillin-sulphuric acid,[195] 4-dimethylaminobenzaldehyde–4-toluenesulphonic acid,[19] chloramin T–3-aminophenol[194] and resorcinol–hypochlorite[194] reagents.

Of the *benzimidazole* derivatives, mebendazol can easily be determined in its tablet formulation by means of absorbance reading at 288 nm in 0.01 M perchloric acid solution; $\varepsilon_{288 \, nm} = 16,700$.[240] In the case of its 4-fluoro derivative (flubendazole), the short-wavelength but more intense band was recommended for the assay of the pharmaceutical preparations.[88] The classical alkaloid method (precipitation of the tetraiodobismuthate, filtration, dissolution of the precipitate in acetone and measurement at 430 nm) could be applied to mebendazole.[113] The iron(III) hydroxamate method measures the carbamate moiety of the molecule (absorbance reading at 520 nm).[158]

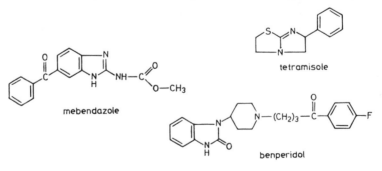

Another benzimidazole derivative, benperidol can be determined in pharmaceutical formulations by the difference spectrophotometric method of Gassim et al.[82] As seen in Figure 10.7 the spectrum is the sum of the contributions of the two isolated chromophoric systems. Of these, that of the fluoroacetophenone part is not changed when going from acidic to the alkaline media while the band at 279 nm, characteristic of the benzimidazolone moiety is shifted to 288 nm. The maximum of the difference spectrum at 292 nm ($\Delta \varepsilon$ = 5,700) can serve as the basis for a sufficiently sensitive and highly selective determination.

Tetramisole and levamisole can be determined as colored ion-pairs after extraction with chloroform from weakly acidic media. As the reagents bromophenol blue,[139,183] bromothymol blue, bromocresol purple, bromocresol green, Solochrome blue and black[183] were used. The use of sodium nitroprusside reagent has also been reported.[184,202]

The simplest possibility for the spectrophotometric determination of nomifensine is the difference spectrophotometric method of Wahbi et al.[238] The intense band at 290 nm (characteristic of the aniline moiety of the molecule) disappears upon strong acidification (see Section 3.Q.3) and hence the difference spectrum (0.1 M sodium hydroxide solution vs. 0.1 M hydrochloric acid) is intense and characteristic enough for the selective and sensitive assay method of its formulations ($\Delta \varepsilon_{292\ nm}$ = 1,120). The same authors also described the analytical application of the formation of charge-transfer complex with iodine. The sensitivity of the method is higher than with the Δ A method: $\varepsilon_{293\ nm}$ = 34,700 and $\varepsilon_{366\ nm}$ = 22,900. As an aromatic amine, nomifensine can also be determined by diazotization and coupling with N-(1-naphthyl)-ethylenediamine or 4-aminosalicylic acid (λ_{max} = 470 and 435 nm, respectively).[5]

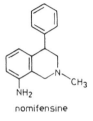

nomifensine

The benzofuran-type anti-arrhythmic drug *amiodarone* was determined in pharmaceutical preparations by derivative spectrophotometry ($^1D_{250,232}$ or $^2D_{254,243}$).[64] The benzopyrane-type *khellin* and 3-dimethylaminomethylkhellin were treated with perchloric acid in glacial acetic acid and then measured at 410 and 430 nm, respectively.[8] The latter could be determined by ion-pair extraction, too (methyl orange, pH 5, chloroform, 420 nm).[8]

Methaqualone was determined in blood by derivative spectrophotometry.[214] The assay of *famotidine* tablets was based on reaction with sodium nitroprusside. The absorption maximum is at 498 nm; the sensitivity of the method is poor (ε = 590).[14] For the determination of *diltiazem* the iron (III)-hydroxamate method was adopted.[244] Preliminary TLC separation was necessary when the method was applied to formulations.

REFERENCES

1. Abdel Moety, E. M.; Moustafa, A. A.; Ismaiel, S. A.; Bebers, M. S. *Zbl. Pharm.* 1988, *127*, 583.
2. Abdel Salam, M.; Issa, A. S.; Lymona H. *J. Pharm. Belg.* 1986, *41*, 314.
3. Abdel Salam, M.; Issa, A. S.; Mahrous, M. S.; *Anal. Letters* 1986, *19*, 2207.
4. Abdine, H.; Elsayed, M. A.-H.; Elsayed, Y. M. *J. Assoc. Off. Anal. Chem.* 1978, *61*, 695.
5. Abounassif, M. A.; Mohamed, M. E.; Aboul-Enein, H. -Y. *J. Pharm. Biomed. Anal.* 1989, *7*, 23.
6. Abou Ouf, A. A.; Hassan, S. M.; Metwally, M. E. -S. *Analyst* 1980, *105*, 1113.
7. Abu Ouf, A. M.; Walash, M. I.; Rizk, M. S.; Ibrahim, F. A. *J. Assoc. Off. Anal. Chem.* 1979, *62*, 1138.
8. Abu-Shady, H.; Girgis, E. H. *J. Pharm. Sci.* 1978, *67*, 618.
9. Adil, A. S.; Sarwar, M. *J. Chem. Soc. Pak.* 1987, *9*, 313.
10. Agrawal, Y. K.; Giridhar, R.; Menon, S. K. *Indian J. Pharm. Sci.* 1989, *51*, 75.
11. Agrawal, Y. K.; Giridhar, R.; Menon, S. K. *J. Pharm. Biomed. Anal.* 1989, *7*, 759.
12. Agrawal, Y. K.; Patel, D. R. *Anal. Letters* 1986, *19*, 1289.
13. Agrawal, Y. K.; Patel, D. R. *Indian J. Pharm. Sci.* 1985, *47*, 207.
14. Agrawal, Y. K.; Shivramchandra, K.; Singh, G. N.; Rao, B. E. *J. Pharm. Biomed. Anal.* 1992, *10*, 521.
15. Amer, M. M.; Tawakkol, M. S. *Analyst* 1974, *99*, 487.
16. Anjaneyulu, Y.; Chandra Sekhar, K.; Anjaneyulu, V.; Sarma, R. N. *Indian Drugs* 1985, *22*, 655.

17. Aromdee, C.; Raksrivong, K.; Vathanasanti, A. *Analyst* 1987, *112*, 1523.
18. Arora, S. P.; Pandey, P. N. *Indian J. Pharm. Sci.* 1987, *49*, 119.
19. Baggi, T. R.; Mahajan, S. N.; Rao, G. R. *Indian J. Pharm.* 1976, *38*, 101.
20. Barary, M. H.; Elsayed, M. A.; Mohamed, S. M. *Drug. Dev. Ind. Pharm.* 1990, *16*, 1639.
21. Barary, M. H.; Wahbi, A. M. *Drug. Dev. Ind. Pharm.* 1991, *17*, 457.
22. Beyrich, Th.; Feldmeier, H. G.; Herberg, H. *Pharmazie* 1989, *44*, 36.
23. Bedair, M. M.; Elsayed, M. A.; Korany, M. A.; Fahmy, O. T. *J. Pharm. Biomed. Anal.* 1991, *9*, 291.
24. Bedair, M.; Korany, M. A.; El-Yazbi, F. A. *Sci. Pharm.* 1986, *54*, 31.
25. Belal, F. *Analyst* 1984, *109*, 615.
26. Belal, F. *Talanta* 1984, *31*, 648.
27. Belal, F.; Rizk, M.; Ibrahiem, F.; Sharaf El-Din, M. *Talanta* 1986, *33*, 170.
28. Belal, S.; Elsayed, M. A. -H.; Abdel-Hamid, M. E.; Abdine, H. *Analyst* 1980, *105*, 774.
29. Bhuee, G. S. *J. Inst. Chem.* 1985, *57*, 95.
30. Biswas, A. *Analyst* 1980, *105*, 353.
31. Blazsek-Bodó, A.; Józsa, J.; Kis, I. *Farmacia* 1982, *30*, 193.
32. Bosch Serrat, F.; Lagarda, M. J.; Barberá, R.; Sanmartin, J. *Cienc. Ind. Farm.* 1988, *7*, 144.
33. Boxenbaum, H. G.; Riegelman, S. *J. Pharm. Sci.* 1974, *63*, 1191.
34. Brüggmann, J.; Risse, G. *Krankenhauspharm.* 1989, *10*, 364.
35. Bruno, L.; Suresh, K. *Indian J. Pharm. Sci.* 1989, *51*, 14.
36. *British Pharmacopoeia 1988,* Her Majesty's Stationery Office: London, 1988. a. 95; b. 912; c. 718; d. 794; e. 770, 917; f. 263, 392; g. 976.
37. Bunch, E. A. *J. Ass. Off. Anal. Chem.* 1983, *66*, 140.
38. Bunch, E. A. *J. Ass. Off. Anal. Chem.* 1987, *70*, 560.
39. Burger, N.; Karas-Gasparec, V. *Talanta* 1977, *24*, 704.
40. Burger, N.; Karas-Gasparec, V. *Talanta* 1984, *31*, 169.
41. Carlucci, G.; Mazzeo, P.; Quaglia, M. G.; Vetuschi, C. *Int. J. Pharm.* 1989, *49*, 79.
42. Cassassas, E.; Fabregas, J. L. *Anal. Chim. Acta* 1979, *106*, 151.
43. Cavrini, V.; Bonazzi, D.; Di Pietra, A. M. *J. Pharm. Biomed. Anal.* 1991, *9*, 401.
44. Cavrini, V.; Di Pietra, A. M.; Gatti, R. *J. Pharm. Biomed. Anal.* 1989, *7*, 1535.
45. Cavrini, V.; Di Pietra, A. M.; Raggi, M. A. *Pharm. Acta Helv.* 1981, *56*, 163.
46. Chatterjee, P. K.; Jain, C. L.; Sethi, P. D. *Indian. J. Pharm. Sci.* 1986, *48*, 25.
47. Chatterjee, P. K.; Jain, C. L.; Sethi, P. D. *Indian Drugs* 1987, *24*, 264.
48. Chatterjee, P. K.; Jain, C. L.; Sethi, P. D. *Indian J. Pharm. Sci.* 1986, *48*, 195.
49. Chatterjee, P. K.; Jain, C. L.; Sethi, P. D. *Indian J. Pharm. Sci.* 1987, *49*, 34.
50. Chatterjee, P. K.; Sethi, P. D. *Indian Drugs* 1985, *22*, 391.
51. Ciesielski, W. *Acta Pol. Pharm.* 1987, *44*, 202.
52. Chowdary, K. P. R.; Murty, K. V. R.; Sriramamurthy, A. *Indian Drugs* 1985, *22*, 482.
53. Chowdary, K. P. R.; Rao, G. N.; Rao, V. B. *Indian Drugs* 1987, *24*, 361.
54. Chowdary, K. P. R.; Sundari, P. V. *Indian Drugs* 1986, *23*, 577.
55. Corti, P. *Boll. Chim. Farm.* 1981, *120*, 484.
56. Corti, P.; Sciarra, G. *Boll. Chim. Farm.* 1981, *120*, 657.
57. Cross, A. H. J.; Hendey, R. A.; Stevens, S. G. E. *Analyst* 1960, *85*, 355.
58. Dalal, R. R.; Bulbule, M. V.; Wadodkar, S. G.; Kasture, A. V. *Indian Drugs* 1982, *19*, 361.
59. Davidson, A. G. *Analyst* 1982, *107*, 422.
60. Dembinski, B. *Acta Pol. Pharm.* 1977, *34*, 509.
61. Dessouky, Y. M.; Mousa, B. A.; Nour El-Din, H. M. *Pharmazie* 1974, *29*, 579.
62. Devani, M. B.; Shishoo, C. J.; Patel, M. A.; Bhalara, D. D. *J. Pharm. Sci.* 1978, *67*, 661.
63. Devriendt, A.; Weemaes, I.; Jansen, F. H. *Arzneim. Forsch.* 1973, *23*, 863.
64. Di Pietra, A. M., Cavrini, V.; Gatti, R.; Raggi, M. A. *Pharm. Res.* 1988, *5*, 709.
65. Dixit, R. K.; Misra, S. K.; Awasthi, B. B. *Indian Drugs* 1984, *22*, 31.
66. Dmitrenko, T. S.; Suranova, A. V.; Luttseva, A. I.; Chichiro, V. E.; Tereschchenkova, I. A.; Istranov, L. P.; Aboyants, R. K. *Farmatsiya* 1986, *36*, 41.
67. Dol, I.; Knochen, M.; Altesor, C. *Boll. Chim. Farm.* 1989, *128*, 18.
68. Dutt, M. C.; Ng, T.-L.; Long, L.-T. *J. Assoc. Off. Anal. Chem.* 1983, *66*, 1455.
69. El-Kommos, M. E.; Yanni, A. S. *Analyst* 1988, *113*, 1091.
70. Elsayed, L.; Hassan, S. M.; Kelani, K. M.; El-Fatatry, H. M. *J. Assoc. Off. Anal. Chem.* 1980, *63*, 992.

71. El-Yazbi, F. A.; Bedair, M.; Korany, M. A. *Analyst* 1986, *111*, 477.
72. Emmanuel, J.; Naik, P. N. *Indian Drugs* 1982, *20*, 33.
73. Ersoy, L.; Alpertunga, B. *Analyst* 1988, *113*, 1745.
74. Fernández, P.; Bermejo, A. M.; López-Rivadulla, M. *Anal. Letters* 1988, *21*, 1045.
75. Fink, D. W; Fox, A. *Anal. Chim. Acta* 1979, *106*, 389.
76. French, W. N.; Matsui, F.; Truelova, L. F. *Can. J. Pharm. Sci.* 1968, *3*, 33.
77. Fujita, Y.; Mori, I.; Kitano, S. *Bunseki Kagaku* 1983, *32*, E199.
78. Fujita, Y.; Mori, I.; Kitano, S. *Bunseki Kagaku* 1983, *32*, 327.
79. Gandhi, T. P.; Patel, P. R.; Patel, V. C.; Patel, S. K.; Gilbert, R. N. *J. Inst. Chem.* 1984, *56*, 127.
80. Garcia-Fraga, J. M.; Jiménez Abizanda, A. I.; Jiménez Moreno, F.; Arias León, J. J. *Anal. Chim. Acta* 1991, *252*, 107.
81. Garcia-Fraga, J. M.; Jiménez Abizanda, A. I.; Jiménez Moreno, F.; Arias León, J. J. *J. Pharm. Biomed. Anal.* 1991, *9*, 109.
82. Gassim, A. E. H.; Takla, P. G.; James, K. C. *Analyst* 1986, *111*, 923.
83. Gatti, R.; Cavrini, V.; Balboni, B.; Roveri, P. *Farmaco, Ed. Prat.* 1985, *40*, 71.
84. Giridhar, R.; Menon, S.; Agrawal, Y. K. *Indian J. Pharm. Sci.* 1987, *49*, 32.
85. Görög, S. *Farm. Tidende* 1970, *80*, 321.
86. Görög, S.; Fütō, M.; Laukó, A. *Acta Pharm. Hung.* 1976, *46*, 113.
87. Görög, S.; Csizér, É. *Proc. III. Anal. Chem. Conf., Budapest, 1970*, Vol. 2. Akadémiai Kiadó: Budapest, 65.
88. Gratteri, P.; Pinzauti, S.; La Porta, E.; Mura, P.; Papeschi, G. *Farmaco* 1990, *45*, 707.
89. Guvener, B. *Acta Pharm. Turc.* 1986, *28*, 35.
90. Guven, K. C.; Guvener, B.; Sunam, G.; Ozdemir, O. *Eczacilik Bul.* 1982, *24*, 30.
91. Hady, M. A.; Salam, M. A. *Acta Pharm. Jugosl.* 1978, *28*, 27.
92. Hassan, S. M.; Metwally, M. E. E.; Ouf, A. M. A. *J. Assoc. Off. Anal. Chem.* 1983, *66*, 1433.
93. Hassan, S. M.; Sharef El Din, M.; Belal, F.; Sultan, M. *J. Pharm. Pharmacol* 1988, *40*, 798.
94. Hassib, S. T.; Safwat, H. M.; El-Bagry, R. I. *Analyst* 1986, *111*, 45.
95. Hattori, T.; Washio, Y.; Inoue, M. *Bunseki Kagaku* 1978, *27*, 707.
96. Hecker-Niediek, A. *Krankenhauspharm.* 1988, *9*, 497.
97. Hoover, J. M.; Soltero, R. A.; Bansal, P. C. *J. Pharm. Sci.* 1987, *76*, 242.
98. Horváth, G.; Milch, Gy. *Pharm. Zentralh.* 1965, *104*, 647.
99. Ibrahim, F. A.; Rizk, M. S.; Belal, F. *Analyst* 1986, *111*, 1285.
100. Ismaiel, S. A.; Sharaby, N. A.; Soliman, R. F.; Abdel-Moety, E. M. *Zentralbl. Pharm.* 1983, *122*, 815.
101. Ismaiel, S. A.; Tawakkol, M. S. *Pharmazie* 1974, *29*, 544.
102. Issopoulos, P. B. *Z. Anal. Chem.* 1989, *334*, 554.
103. Issopoulos, P. B. *Int. J. Pharm.* 1990, *61*, 261.
104. Issopoulos, P. B. *Pharm. Acta Helv.* 1989, *64*, 280.
105. Issopoulos, P. B. *Int. J. Pharm.* 1991, *70*, 201.
106. Issopoulos, P. B. *Acta Pharm. Jugosl.* 1991, *41*, 123.
107. Issopoulos, P. B.; Economou, P. T. *Analusis* 1992, *20*, 31.
108. Issopoulos, P. B.; Economou, P. T. *Internat. J. Pharm.* 1989, *57*, 235.
109. Jaksevac-Miksa, M.; Hankonyi, V.; Karas-Gasparec, V. *Acta Pharm. Jugosl.* 1979, *29*, 91.
110. Kadyrova, R. D.; Ikramov, L. T.; Tegisbaev, E. T. *Farmatsiya* 1988, *37*, 44.
111. Kamalapurkar, O. S.; Chudasama, J. J. *East. Pharm.* 1983, *26*, 207.
112. Kamalapurkar, O. S.; Menezes, C. *Indian Drugs* 1984, *22*, 164.
113. Kar, A. *Analyst* 1985, *110*, 1031.
114. Karljikovic-Rajic, K.; Stankovic, B.; Binenfeld, Z. *J. Pharm. Biomed. Anal.* 1987, *5*, 141.
115. Karljikovic-Rajic, K.; Stankovic, B.; Granov, A. Binenfeld, Z. *J. Pharm. Biomed. Anal.* 1988, *6*, 773.
116. Karljikovic-Rajic, K.; Stankovic, B.; Granov, A. *Pharmazie*, 1989, *44*, 497.
117. Kitamura, K.; Hatta, M.; Fukuyama, S.; Hozumi, K. *Anal. Chim. Acta* 1987, *201*, 357.
118. Korany, M. A.; Bedair, M. M.; El Gindy, A. *Drug. Dev. Ind. Pharm.* 1990, *16*, 1555.
119. Kovatscheva, E.; Vasileva-Aleksandrova, P.; Ivanova, V. *Microchim. Acta* 1983, 483.
120. Kramarenko, V. P.; Turkevich, O. D. *Farm. Zh.* 1981, 67.
121. Krishnan, M. V. S.; Rao, A. S. *Indian Drugs* 1986, *23*, 469.
122. Krzek, J. *Chem. Anal.* 1979, *24*, 643.
123. Kulkanjanatorn, P.; Shaipanich, C. *Varasarn Paesachasarthara* 1978, *5*, 107.
124. Leung, C.-P.; Kwan, S.-Y. *Analyst* 1979, *104*, 143.
125. Leung, C.-P.; Law, C.-K. *Analyst* 1989, *114*, 241.

126. Li, X. *Yaoxue Tongbao* 1984, *19*, 26.

127. Lopez Silva, F.; Bocic Vildosola, R.; Vallejos Ramos, C.; Alvarez Lueje, A. *An. Real Acad. Farm.* 1990, *56*, 19.

128. Mahrous, M. S.; Abdel Salam, M.; Issa, A. S.; Abdel-Hamid, M. *Talanta* 1986, *33*, 185.

129. Mazza, P.; Gianelli, P.; Rosai, A.; Pecori Vettori, L. *G. Med. Mil.* 1980, *130*, 121.

130. May, J.; Zvirblis, P.; Kondritzer, A. *J. Pharm. Sci.* 1965, *54*, 1508.

131. Mishra, P.; Katrolia, D.; Agrawal, R. K. *Curr. Sci.* 1989, *58*, 503.

132. Mishra, P.; Katrolia, D.; Agrawal, R. K. *Indian. J. Pharm. Sci.* 1991, *52*, 155.

133. Misztal, G. *Farm. Pol.* 1988, *44*, 206.

134. Mohamed, M. E. *Anal. Lett.* 1986, *19*, 1323.

135. Mopper, B. *J. Ass. Off. Anal. Chem.* 1987, *70*, 42; 1988, *71*, 1121.

136. Mount, D. L.; Patchen, L. C.; Williams, S. B.; Churchill, F. C. *Bull. World Health Org.* 1987, *65*, 615.

137. Moussa, A.-F. A. *Pharmazie* 1977, *32*, 801.

138. Nakashima, K.; Shimada, K.; Akiyama, S. *Chem. Pharm. Bull.* 1985, *33*, 1515.

139. Nechaeva, L. Y.; Galatenko, N. A.; Bufius, N. N. *Farmatsiya* 1989, *38*, 24.

140. Nino, N.; Kapitanova, D. *Farmatsiya* (Sofia), 1974, *24*, 1.

141. Omar, N. M. *J. Pharm. Sci.* 1978, *67*, 1610.

142. Ossmann, A. R. E.-N. *Sci. Pharm.* 1981, *40*, 193.

143. Othman, S. O. *Drug. Dev. Ind. Pharm.* 1987, *13*, 1257.

144. Panzova, B.; Ilievska, M.; Trendovska, G.; Bogdanov, B. *Int. J. Pharm.* 1991, *70*, 187.

145. Parissi-Poulou, M.; Reizopoulou, V.; Koupparis, M.; Macheras, P. *Internat. J. Pharm.* 1989, *51*, 169.

146. Pawelcyzk, E.; Knitter, B. *Farm. Pol.* 1986, *32*, 357.

147. Peterkova, M.; Kakac, B.; Matousova, O. *Cesk. Farm.* 1980, *29*, 73.

148. Petrenko, V. V. *Ukr. Khim. Zh.* 1980, *46*, 1225.

149. Piippo, H. *Acta Farm. Fenn.* 1987, *96*, 129.

150. Pitarch, B.; Manes, J.; Bosch Serrat, F. *An. Real Acad. Farm.* 1986, *52*, 279.

151. Poole, N. F.; Meyer, A. E. *Proc. Soc. Exp. Biol. Med.* 1958, *98*, 375.

152. Popelkova-Mala, Z.; Malat, M. *Cesk. Farm.* 1985, *34*, 422.

153. Prakash, A.; Murthy, A. K.; Pathak, V. N.; Shukla, I. C. *Acta Chim. Hung.* 1983, *112*, 21.

154. Prasad, T. N. V.; Sastry, B. S.; Venkata Rao, E.; Sastry, C. S. P. *Pharmazie* 1987, *42*, 135.

155. Pratt, E. L. *Anal. Chem.* 1953, *25*, 814.

156. Przyborowski, L.; Misztal, G. *Farm. Pol.* 1987, *43*, 404.

157. Ramana Rao, G.; Pulla Rao, Y.; Raju, I. R. K. *Analyst*, 1982, *107*, 776.

158. Rana, N. G.; Dave, R. V.; Patel, M. R. *Indian Drugs* 1981, *18*, 333.

159. Rao, G. R.; Murty, S. S. N. *East. Pharm.* 1982, *25*, 111.

160. Rao, G. R.; Pullarao, Y.; Raju, I. R. K. *Indian Drugs* 1982, *19*, 162.

161. Rao, G. R.; Raghuveer, S. *Indian J. Pharm. Sci.* 1980, *42*, 141.

162. Rao, G. R.; Raghuveer, S. *Indian Drugs* 1981, *18*, 408.

163. Rao, G. R.; Raghuveer, S. *Indian Drugs* 1982, *19*, 408.

164. Rao, G. R.; Raghuveer, S.; Murty, S. S. N.; Bajrangrao, B. *Indian Drugs* 1979, *17*, 50.

165. Rao, G. R.; Raghuveer, S.; Rao, Y. P. *J. Inst. Chem.* 1982, *54*, 146.

166. Rao, P. V. K.; Rao, G. B. B.; Murti, P. S. *Michrochim. Acta* 1974, 979.

167. Raut, K. N.; Sabnis, S. D. *Indian J. Pharm. Sci.* 1987, *49*, 65.

168. Rivaya, M.; Batlle, R. *Cien. Ind. Farm.* 1984, *3*, 178.

169. Rizk, M. S.; Walash, M. I.; Ibrahim, F. A. *Analyst* 1981, *106*, 1163.

170. Rizk, M.; Walash, M. I.; Ibrahim, F. *Spectrosc. Letters* 1984, *17*, 423.

171. Sadana, G. S.; Sane, R. T.; Ozarkar, S. G.; Sapre, D. S.; Nayak, V. G. *Indian Drugs* 1986, *23*, 573.

172. Saha, U.; Sen, A. K.; Das, T. S. *Analyst* 1988, *113*, 1653.

173. Sakai, T. *Analyst* 1982, *107*, 640.

174. Saleh, G. A.; Askal, H. F. *Pharmazie* 1990, *45*, 220.

175. Salman, A. *Sci. Pharm.* 1987, *55*, 255.

176. Sanghavi, N. M.; Joshi, N. G.; Saoji, D. G. *Indian J. Pharm. Sci.* 1979, *41*, 226.

177. Sanghi, S. K.; Verma, A.; Verma, K. K. *Analyst* 1990, *115*, 333.

178. Sane, R. T.; Ambardekar, A. B.; Shastry, V. K. *Indian Drugs* 1981, *18*, 290.

179. Sane, R. T.; Malkar, V. B.; Nayak, V. G. *Indian Drugs* 1982, *19*, 487.

180. Sane, R. T.; Malkar, V. B.; Nayak, V. G.; Sapre, D. S.; Banawalikar, V. J. *Indian Drugs* 1984, *22*, 16.

181. Sane, R. T.; Nadkarni, A. D.; Joshi, S. K. *Indian Drugs* 1983, *20*, 333.

182. Sane, R. T.; Sane, S. *Indian Drugs* 1979, *16,* 239.

183. Sane, R. T.; Sapre, D. S.; Nayak, V. G. *Talanta* 1985, *32,* 148.

184. Sane, R. T.; Sawant, S. V.; Jashi, V. J.; Mhalas, J. G. *Indian Drugs* 1983, *20,* 460.

185. Sane, R. T.; Shinde, B. R.; Parikh, A. K.; Tikekar, S. P. *Indian Drugs* 1984, *21,* 257.

186. Sane, R. T.; Vaidya, U. M. *Indian J. Pharm. Sci.* 1979, *41,* 73.

187. Santagati, N. A.; Villari, A.; Spadaro, A.; Puglisi, G. *Int. J. Pharm.* 1990, *65,* 137.

188. Sanyal, A. K. *J. Ass. Off. Anal. Chem.* 1988, *71,* 849.

189. Sastry, B. S.; Rao, E. V.; Sastry, C. S. P. *Indian J. Pharm. Sci.* 1986, *48,* 71.

190. Sastry, B. S.; Rao, E. V.; Tummuru, M. K.; Sastry, C. S. P. *Indian Drugs* 1986, *24,* 105.

191. Sastry, C. S. P.; Aruna, M. *Pharmazie* 1988, *43,* 361.

192. Sastry, C. S. P.; Aruna, M.; Reddy, N.; Sankar, D. G. *Indian J. Pharm. Sci.* 1988, *50,* 140.

193. Sastry, C. S. P.; Prasad, T. N.; Sastry, B. S.; Rao, E. V. *Analyst* 1988, *113,* 255.

194. Sastry, C. S. P.; Mangala, D. S.; Rao, K. E. *Analyst* 1986, *111,* 323.

195. Sastry, C. S. P.; Mangala, D. S.; Rao, K. E. *Acta Cienc. Indica Ser. Chem.* 1986, *12c,* 17.

196. Sastry, C. S. P.; Rao, A. R. M. *Indian J. Pharm. Sci.* 1987, *49,* 95.

197. Savitskaya, A. V.; Paschhenko, L. A.; Dobrotvorsky, A. E. *Farmatsiya* 1989, *38,* 39.

198. Scott, P. G. W. *J. Pharm. Pharmacol.* 1952, *4,* 681.

199. Shah, A. K.; Agrawal, Y. K.; Banerjee, S. K. *Anal. Letters* 1981, *14,* 1449.

200. Shah, P. R.; Raje, R. R. *J. Pharm. Sci.* 1977, *66,* 291.

201. Sharma, S. C.; Das, S.; Talwar, S. K. *J. Assoc. Off. Anal. Chem.* 1987, *70,* 679.

202. Shingbal, D. M.; Joshi, S. V. *Indian Drugs* 1984, *21,* 396.

203. Shingbal, D. M.; Joshi, S. V. *Indian Drugs* 1985, *22,* 326.

204. Shingbal, D. M.; Naik, R. R. *Indian J. Pharm. Sci.* 1985, *23,* 124.

205. Shingbal, D. M.; Naik, S. D. *East. Pharm.* 1983, *26,* 201.

206. Shingbal, D. M.; Natekar, G. B. *East. Pharm.* 1980, *23,* 117.

207. Shingbal, D. M.; Rao, V. R. *Indian Drugs* 1986, *23,* 472.

208. Shingbal, D. M.; Rao, V. R. *Indian Drugs* 1986, *23,* 638.

209. Shingbal, D. M.; Sawant, K. V. *Indian Drugs* 1982, *20,* 104.

210. Shirsat, P. D. *Indian J. Pharm. Sci.* 1975, *37,* 101.

211. Siegel, K.; Vogel, E. *Pharm. Prax.* 1986, *41,* 116.

212. Sikorska-Tomicka, H. *Z. Anal. Chem.* 1982, *132,* 353.

213. Sikorska-Tomicka, H. *Microchim. Acta* 1985, 151.

214. Singh, V.; Shukla, S. K.; Mahanwal, J. S.; Ram. J. *Pharmazie* 1989, *44,* 229.

215. Siraj, P.; Krishna, R. R.; Murty, S. S. N.; Reddy, B. S.; Sastry, C. S. P. *Talanta* 1981, *28,* 477.

216. Slack, S. C.; Mader, W. J. *J. Amer. Pharm. Assoc. Sci. Ed.* 1957, *46,* 742.

217. Srinath, V.; Bagavant, G. *Indian Drugs* 1986, *24,* 173.

218. Stewart, J. T.; Chang, Y. -C. *J. Assoc. Off. Anal. Chem.* 1979, *62,* 1107.

219. Stewart, J. T.; Settle, D. A. *J. Pharm. Sci.* 1975, *64,* 1403.

220. Strelets, L. M.; Filipeva, S. O.; Petrenko, V. V.; Buryak, V. P. *Farm. Zh.* 1986, 64.

221. Strelets, L. M.; Filipeva, S. O.; Petrenko, V. V.; Burgak, V. P. *Farm. Zh.* 1987, 65.

222. Subert, J.; Cizmarik, J. *Zentralbl. Pharm.* 1984, *123,* 397.

223. Sulaiman, S. T.; Amin, D. *Int. J. Environ. Anal. Chem.* 1985, *20,* 313.

224. Szepesi, G.; Görög, S.; Szakolczay, I. *Chem. Anal.* 1971, *16,* 211.

225. Taha, A. M.; El-Rabbat, N. A.; Abdel Fattah, F. *Analyst* 1980, *105,* 568.

226. Takla, P. G.; Dakas, C. L. *Int. J. Pharm.* 1988, *43,* 225.

227. Talwar, S. K.; Sharma, S. C.; Das, S. *J. Pharm. Biomed. Anal.* 1986, *4,* 511.

228. *The United States Pharmacopoeia XXII,* USP Convention Inc.: Rockville, 1990. a. 950; b. 596; c. 1210; d. 917; e. 892; f. 1385; g. 729; h. 290; i. 628; j. 847; k. 686; l. 223; m. 74; n. 966; o. 476; p. 296; r. 286; s. 29; t. 79; u. 326;

229. Urbányi, T.; O'Connel, A. *Anal. Chem.* 1972, *44,* 565.

230. Vachek, J. *Cesk. Farm.* 1987, *36,* 168.

231. Varma, V.; Tewari, N.; Sircar, K. P.; Rao, K. M.; Ghatak, S. *Indian J. Pharm. Sci.* 1986, *48,* 78.

232. Vetuschi, C.; Ragno, G. *Farmaco* 1990, *45,* 771.

233. Vicente, J.; Hernandez, P.; Hernandez, L. *An. Real Acad. Farm.* 1985, *51,* 301.

234. Wahbi, A. M.; Abdine, H.; Korany, M. A.; El-Yazbi, F. A. *J. Pharm. Sci.* 1978, *67,* 140.

235. Wahbi, A. M.; Abounassif, M. A.; El-Obeid, H. A.; Al-Juliani, A. M. *Arch. Pharm. Chem. Sci. Ed.* 1986, *14,* 69.

236. Wahbi, A. M.; Abounassif, M. A.; El-Obeid, H. A.; Gad-Kariem, E. A. *Farmaco, Ed. Prat.* 1985, *40*, 334.

237. Wahbi, A.-A. M.; Abounassif, M. A.; Gad-Kariem, E. A. *Talanta* 1986, *33*, 179.

238. Wahbi, A.-A. M.; Abounassif, M. A.; Gad-Kariem, E. A.; Aboul-Enein, H. Y. *Talanta* 1987, *34*, 287.

239. Wahbi, A. M.; El-Obeid, H. A.; Gad-Kariem, E. A. *Farmaco, Ed. Prat.,* 1986, *41*, 210.

240. Wahbi, A.-A. M.; Onsy, S. *Talanta* 1978, *25*, 716.

241. Walash, M. I.; Rizk, M.; El-Brasy, A. *Pharm. Weekbl. Sci. Ed.* 1986, *8*, 234.

242. Woodside, J. M.; Piper, I.; Leary, J. B. *J. Am. Pharm. Assoc. Sci. Ed.* 1957, *46*, 729.

243. Zak, S. B.; Bartlett, M. F.; Wagner, W. E.; Gilleran, T. G.; Lukas, G. *J. Pharm. Sci.* 1972, *63*, 225.

244. Zivanov-Stakic, D.; Agbaba, D.; Vladimirov, S.; Ciric, Lj. *Farmaco* 1992, *47*, 393.

245. Zivanov-Stakic, D.; Panic, Lj.; Agbaba, G. *Farmaco* 1990, *45*, 381.

246. Zommer-Urbanska, S.; Lesz, K. *Acta Pol. Pharm.* 1980, *37*, 225.

L. AMINO ACIDS AND RELATED DERIVATIVES, PEPTIDES AND PROTEINS

1. Methods Based on Natural Absorption

The subject of this chapter is the spectrophotometric analysis of protein-forming α-amino acids, their simple derivatives (levodopa, methyldopa, EDTA, captopril, etc.), peptides and related compounds (enalapril, etc.) and some pharmaceutically important aspects of the very wide-ranging literature of the spectrophotometric analysis of proteins are also outlined. Alkaloids and antibiotics containing peptide units are discussed in the respective chapters.

Of the protein-forming amino acids only phenylalanine, the phenolic tyrosine and the indole derivative tryptophan possess ultraviolet spectra which are suitable for their quantitative determination. As seen in Figure 10.8 the determination of phenylalanine containing an isolated phenyl group only is not possible in the presence of the other two. As a consequence of their overlapping spectra the simultaneous determination of tyrosine and tryptophan is not an easy task either. The solution of the problem is based on the fact that in alkaline media only the spectrum of tyrosine

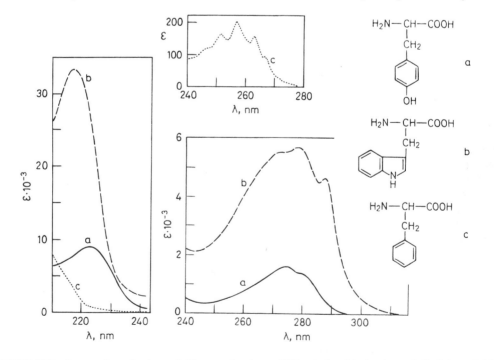

FIGURE 10.8.   Spectra of tyrosine (curve "a"), tryptophan (curve "b") and phenylalanine (curve "c"). Solvent: 0.1 M hydrochloric acid.

containing the phenolic hydroxyl group undergoes bathochromic and hyperchromic shift ($\varepsilon_{293\ nm}$ = 2,480 in 0.01 M sodium hydroxide). (Similar shifts can be observed in the spectra of the drug derivatives of tyrosine such as levodopa, methyldopa etc.)

The above mentioned shift in the spectrum of tyrosine offers two possibilities for the analysis of mixtures with tryptophan. In the classical method of Goodwin and Morton[17] the measurement is performed in 0.1 M sodium hydroxide at two wavelengths: at 280 nm which is the absorption maximum of tryptophan and at 294.4 nm which is the isosbestic point of the two spectra. From the latter value the sum of the molar concentrations of tyrosine and tryptophan can be calculated using the molar absorptivity value of 2465 for the calculation. From the two absorbances at the two wavelengths and the three molar absorptivity values, the molar ratio of tyrosine and tryptophan can also be calculated. Although in principle this method is suitable only for the determination of the two amino acids (even in the presence of spectrophotometrically inactive amino acids), in practice it is widely used for the quantitative determination of peptides containing tyrosine and/ or tryptophan and for the characterization of their purity by the ratio of the two amino acids. Similar measurements are made with proteins, too. As a consequence of the slight but not negligible distortion of the spectra incorporated into the peptide or protein chain, the values thus obtained are of limited value.

In the method of Demeester et al.[7] the only analytical wavelength is 294.4 nm but two measurements are carried out. The solution in 0.1 M sodium hydroxide is measured against a reagent blank and then against another solution of the same concentration in 0.1 M hydrochloric acid. From the absorbance and the difference absorbance, the two concentrations can be calculated using two equations with two unknowns.

The disadvantage of these and several other related methods[8] is that they are restricted to two amino acids only and because of the medium intensity of the long-wavelength bands, their sensitivity is not very high. In spite of this simple spectrophotometric measurements around 280 nm are widely used in those instances when spectrophotometrically active amino acids, their drug derivatives, peptides, proteins, enzymes or protein hydrolyzates have to be determined in samples where they are present at measurable concentration and no interference from other substances absorbing in the same spectral range has to be taken into account.

Of the application of the method in pharmaceutical analysis (moreover in pharmacopoeias) a few are mentioned. Levodopa is determined in capsules by measurement in 0.1 M hydrochloric acid at 280 nm.[54a] Methyldopa ethyl ester injection is investigated at 283 nm after purification with solvent extraction.[54b] An example of the non-specific assay methods of enzymes is the determination of hyaluronidase enzyme in suppositories: after aqueous extraction the absorbance is measured at 280 nm.[32] The pentapeptide pentagastrine containing tryptophan is determined at the same wavelength.[3a]

Measurement at 280 nm has its importance even in the case of specific enzyme assay methods. For example the protease activity of pancreatin enzyme in its preparations is determined in such a way that the solution of the protein casein is digested at 40°C for 60 min with the enzyme preparation to be determined. After the precipitation of the undigested protein with trichloroacetic acid, the absorbance at 280 nm is proportional to the concentration of the degraded protein, i.e., to the activity of the enzyme preparation.[54c]

At shorter wavelengths (around 210 nm) where the aromatic amino acids possess more intense absorption bands than that around 280 nm and the –NH–CO– groups of the peptide chain greatly contribute to the overall absorptivity, much higher sensitivity can be attained. Several methods of this type are described in the monograph of Demchenko.[8] The value of these results is, however, limited, partly because of the general problems of absorbance measurements at short wavelengths and also as a consequence of the interference from other constituents (mainly nucleic acids). It has to be noted however, that measurement around 210 nm is of immense importance in monitoring the high-performance liquid chromatograms of amino acids and derivatives, peptides and proteins by the UV spectrophotometric detector.

Another excellent possibility of improving the selectivity of the simultaneous determination of aromatic amino acids (which in this case includes phenylalanine) and to cancel out the interfering background absorptions is the use of derivative spectrophotometry (see Section 4.E). This technique was introduced by Fell for the analysis of proteins[13] and enzymes.[12] The simultaneous determination of arginine and the antibiotic drug aztreonam by means of second-derivative spectrophotometry is an application of this technique.[37] Higher order (2,4,6,8) derivative spectroscopy was found to be a useful tool for the discrimination of bovine and porcine insulin.[58]

As for the specific determination of the active ingredient content of enzyme preparations, it can be stated that spectrophotometry is the most generally used method for this purpose. The general sequence of steps in these procedures is as follows: first, a substrate is to be selected the absorbance of which significantly and characteristically changes at a given wavelength as a result of the enzymatic reaction. The solution containing this substrate at a suitable concentration in a buffered medium at optimized pH is added into the spectrophotometer cell thermostatted to an accuracy of 0.1°C. After the addition of the enzyme to be investigated to the solution, the absorbance at a suitable wavelength is monitored as a function of the reaction time. The rate of change of the absorbance is a measure of the enzyme activity. Even an outlined treatment of this matter would be beyond the scope of this book. As a characteristic example, the analysis of preparations containing trypsine is presented. In this assay the substrate is N-benzoyl-L-arginyl methyl ester, the pH is 7.6 and the temperature is 25°C. The enzymatic reaction is the hydrolytic splitting of benzoic acid from the molecule and this is monitored at 253 nm.[54d]

2. Methods for the Determination of Amino Acids Based on Chemical Reactions

Since the majority of amino acids are spectrophotometrically inactive, the application of chemical reactions leading to colored products are of special importance in their spectrophotometric analysis. Of the numerous methods available for this purpose the classical method based on the ninhydrin reaction is still the most important.[48] The mechanism of the reaction (see Equation 10.12) involves the oxidative deamination of the α-amino acids by the ninhydrin reagent followed by the condensation of the liberated ammonia with one molecule of reduced and another molecule of unreduced ninhydrin to form the colored reaction product (Ruhemann purple).

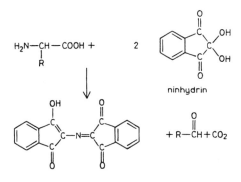

ninhydrin   (10.12)

The reaction is usually carried out by heating for a short time in a mixture of water and an organic solvent (2-propanol, butanol, methylcellosolve). The reaction product is measured between 550 and 580 nm depending on the reaction conditions. α-Iminoacids (proline, hydroxyproline) are exceptions; because of the different reaction mechanism the reaction products of these amino acids have absorption maxima at 440 nm. The sensitivity of the method is sufficient: molar absorptivities of the reaction products are about 5000 and the sensitivity is more or less the same for all α-amino acids. When judging the selectivity of the method it has to be taken into consideration that under the given conditions amines, peptides having free amino groups and free ammonia also produce colored products.

Regarding the immense quantity of early literature dealing with mechanism and optimization of the reaction and the innumerable practical applications, an important monograph of Kakac and Vejdelek is in Reference 27a. As has already been mentioned, this method is the most important in the photometric analysis of amino acids and, in addition, it has found wide fields of application in related areas such as in thin-layer chromatography where it is widely used for the visualization of the spots and for their densitometric evaluation, e.g., in the paper of Fabre and Blanchin.[11] The method is easily automated and this enables the quantitative measurement of the individual amino acids, e.g., in protein hydrolyzates after column chromatographic separation. The ninhydrin method is the basis of the quantitative evaluation in the majority of automated amino acid analyzers.

In addition to the ninhydrin method many others are also suitable for the transformation of amino acids to spectrophotometrically measurable derivatives. Amino acids react with 1,2-naphthoquinone-4-sulphonic acid to produce their 1,2-naphthoquinone derivatives according to the general scheme presented in Chapter 7 as Equation 7.7.[14,27b] The intense blue color forming in the course of complex formation of charge-transfer complex with chloranil,[1] reaction with 1-fluoro-2,4-dinitrobenzene which transforms the amino acids to the N-2,4-dinitrophenyl derivatives,[50] transformation to phenylthiohydantion (PTH) derivatives using phenylisothiocyanate reagent,[9] condensation with o-phthalaldehyde,[47] N-dansylation[18] and dabsylation[33] using dansyl and dabsyl chlorides (1-dimethylaminonaphthalene-5-sulphonyl chloride and 4-dimethylaminoazobenzene-4'-sulphonyl chloride) as the reagents have found wide application in the analysis of amino acids, not only in the spectrophotometric analysis but also in high-performance liquid chromatography where these reactions are the standard pre-column (and in some cases post-column) derivatization reactions in the analysis of amino acids by this technique.[30]

Of the other methods, at first the USP method for the selective determination of arginine in injection formulation is mentioned.[54e] This is based on the Sakaguchi reaction where a red product is formed in the reaction of the guanidine group of arginine and the 2,4-dichloro-α-naphthol reagent. Tyrosine reacts with nitrous acid to form an aromatic nitro derivative which can be measured in alkaline medium at 400 nm ($\varepsilon = 9,440$).[56]

3. Methods for the Determination of Amino Acid Derivatives, Peptides and Proteins Based on Chemical Reactions

Amino acid derivatives, peptides and proteins usually contain a variety of functional groups which may serve as the basis for spectrophotometric methods based on chemical reactions and for this reason their spectrophotometric analysis is abundant in these types of methods.

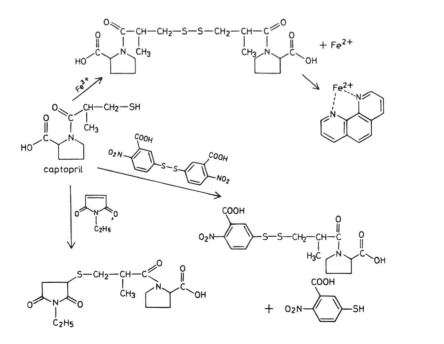

(10.13)

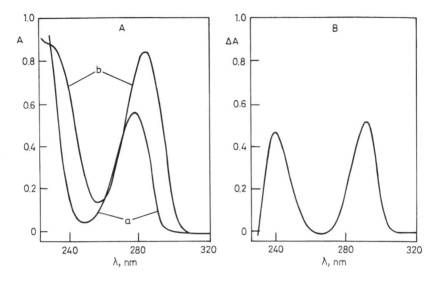

FIGURE 10.9. Spectra for the difference spectrophotometric determination of levodopa based on complex formation with boric acid. Concentration: 4 mg/100 ml. A. Basic spectra. a: Solvent: phosphate buffer; pH 7; b: Solvent: phosphate-borate buffer; pH 7. B. Difference spectrum (solution "b" against "a"). (From Davidson, A.G. *J. Pharm. Biomed. Anal.* 1984, 2, 45.[4])

Of the simple tyrosine derivatives (substituted at the α-position or at the aromatic nucleus) several methods are available for the determination of levodopa, methyldopa and carbidopa. The methods of Davidson[4,5,6] are based on their main characteristic, the 1,2-diphenol group. As seen in Figure 10.9 a strong bathochromic shift occurs if the vicinal phenolic hydroxyl groups are transformed to a cyclic boric acid ester. The difference spectrum recorded in such a way that the solution treated with boric acid is placed into the sample cell and a solution of the same concentration and pH but not containing boric acid into the reference cell has a sufficiently intense maximum at 292 nm to serve as the basis for an extremely selective method for the determination of the above listed derivatives. ($\Delta\varepsilon_{levodopa} = 2,544$, $\Delta\varepsilon_{methyldopa} = 2,622$ and $\Delta\varepsilon_{carbidopa} = 1,863$). Monophenol derivatives naturally do not give positive reaction. The method was used for the assay of the formulations of these derivatives.[4,58] Based on the same principle but using germanium oxide to replace boric acid, the maximum of the difference spectrum is at 292.5 nm but a two-fold increase in the sensitivity of the method is attainable.[5] Benserazide with its 1,2,3-triphenol moiety also gives a positive reaction, but the shape of the difference curve is different and hence on the basis of measurement at two wavelengths (238 and 292.5 nm) the simultaneous determination could be carried out in pharmaceutical formulations.[6] Using tungstic acid as the reagent, the maximum is shifted to 300 nm.[44]

In addition to these UV methods many others based on colored products are also available. Some of the reagents are iron (II) to form complexes,[3b] chloranil to form charge transfer complexes,[31] iron (III)-phenylfluorone to form a ternary complex with carbidopa ($\varepsilon_{636\ nm} = 109,800$),[25] thiosemicarbazide after autoxidation in alkaline medium,[29] ninhydrin to react in the similar way as with ordinary amino acids[53] and p-dimethylaminobenzaldehyde to react selectively with the hydrazine group of carbidopa.[53] The reducing properties have also been exploited for the determination of levodopa and related derivatives. The reagents of the mainly indirect methods are ammonium vanadate,[49,52] molybdophosphate,[21] phosphotungstate,[51] iron(III) 1,10-phenantroline,[23] tetrazolium blue,[22] tetrazolium violet,[24] cerium(IV) sulphate 3-methylbenzthiazolin-2-one hydrazone[45] and metaperiodate.[10]

Of the other, pharmaceutically important amino acid derivatives, the methods for the determination of ethylenediaminetetraacetic acid are usually based on its ability to form very stable complexes. Since its complexes are usually colorless, the methods are indirect and are based on the decrease of the color of suitable complexes as a result of ligand exchange reactions. As the colored complexes copper(II) 1-pyridylazo-2-naphthol[39] and the mixed complex scandium(III) Chrome Azurol S-1-ethoxycarbonylpentadecyltrimethylammonium bromide[26] were used.

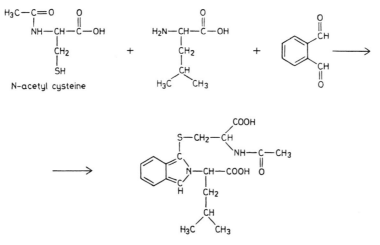

(10.14)

Of the other amino acid derivatives, drugs containing a mercapto group merit special attention. Raggi et al.[42] described three methods for the determination of captopril based on the reactions of its mercapto group. As seen in Equation 10.13, one of the methods is based on the oxidation of the mercapto group to disulphide with iron (III) chloride followed by the measurement of the resulting iron(II) as the 1,10-phenantroline complex. This reaction was also utilized by Mohamed et al.[36]

The reagent of the second method is 5,5′-dithio-bis-2-nitrobenzoic acid. In the course of a redox exchange reaction between captopril and the reagent, 2-nitro-5-mercaptobenzoic acid is formed which can be measured in a solution of pH 8 at 412 nm.

In the third, indirect and much less sensitive method addition takes place at the double bond of N-ethylmaleimide. The decrease of the absorbance of the reagent at 300 nm is proportional to the concentration of captopril.

Of the above described three methods, the iron(III) 1,10-phenantroline method was adopted for the assay of acetylcysteine and mercaptopropionylglycine preparations.[40] The latter was determined using complexation reaction with palladium(II)[43] and the above mentioned 5,5′-bis-2-nitrobenzoic acid reagent,[41] too. The reaction product of amino acids containing mercapto groups with the o-phthalaldehyde isoleucine reagent absorbs at 355 nm (see Equation 10.14).[2] The FIA variant of this method is also described for the analysis of pharmaceutical formulations.[19]

In the course of analyzing pharmaceutically important peptides, it is often the case that the well known reactions of functional groups of the individual amino acids in the peptide chain can be utilized for the assay of the peptide. For example the octapeptide hypertensine contains tyrosine and the phenolic ring of the latter can be coupled with diazotized 4-chloroaniline to form the azo derivative intensively absorbing at 420 nm ($\varepsilon = 17,000$). On the basis of this reaction Vogt et al.[57] developed a sensitive method for the determination of hypertensine in its formulations. The method of Ozol[38] for the determination of aspartam (L-aspartyl-L-phenylalanine methyl ester) is based on the condensation of the terminal primary amino group of the aspartyl unit with 4-dimethylamino-benzaldehyde. The resulting Schiff's base can be measured at 407 nm. The same functional group in the same peptide is the basis of the method of Vachek.[55] The charge-transfer complex with the 1,4-benzoquinone reagent has an absorption maximum at 480 nm.

The methodology of functional group analysis enables difficult problems to be solved such as the determination of ACTH-sulphoxide in ACTH. In the method of Rényei and Görög[46] this impurity where the –S–CH₃ group in methionine is oxidized to sulphoxide is reduced to ACTH with hydrogen iodide in strongly acidic medium and the determination of the impurity is traced back to the spectrophotometric determination of the iodine forming during the reaction. It is interesting to note that the bioassay of this peptide drug is also made by spectrophotometry: the ascorbic acid liberated by it in animal test is measured at 520 nm by using the indophenol reaction (see Section 10.O.2).[54f]

Of the *peptide derivatives* enalapril is an example. The basis for its determination is the ability of the secondary amino group in its molecule to form ion-pairs with acidic dyes. Kato[28] used bromothymol blue as the reagent at pH 3.2, extracted the complex with dichloromethane and measured the absorbance at 405 nm. Ikenishi et al.[20] used 3,5-dibromosalicylaldehyde reagent for the determination of 1-(2-o-chlorobenzoyl-4-chlorophenyl)-5-glycylaminomethyl-3-dimethylami-nocarbonyl-1H-1,2,4-triazole and 4-chloro-2-(o-chlorobenzoyl)-N-methyl-N α-glycylglycinanilide. The investigation of the mechanism of the reaction leading to a product with an absorption maximum at 550 nm revealed that the formation of the Schiff's base at the terminal amino group was followed by the attack of a second molecule of the reagent and oxidation by atmospheric oxygen to form substituted 1-imidazoline-4-ones.

A detailed review of the color reactions used for the quantitative determination of proteins would be beyond the scope of this book: the majority of the literature deals with this question from the point of view of biochemical and clinical, rather than pharmaceutical, applications. A good summary of the early literature is in the book of Kakac and Vejdelek.[27d]

Of the classical methods the most important and widespread is the Folin-Ciocalteu method[15] modified by Lowry et al.[34] which is widely used up to the present time. The reagent of this method is phosphomolybdotungstic acid originally developed for the indirect determination of phenols based on their reducing properties. The blue color which is the basis of the measurement is the reduced form of the reagent. It was found that using copper (II) as the catalyst, not only the phenol-type tyrosine residues of proteins react with the reagent but other groups, too, thus greatly improving the sensitivity of the method.

The complexes of proteins with various metals and dyes also form the basis for several methods. An interesting feature of a sensitive method based on their interaction with dyes is that the proteins are adsorbed on glass fiber filters and stained with suitable dyes (e.g., Commassie blue). The excess dye is removed by washing, the protein-dye complex is then destroyed by treatment with alkaline methanol and the liberated dye is measured after acidification at 590 nm. This method, introduced by McKnight[35] is ten-fold more sensitive than the Lowry procedure[34] and is suitable for the determination of as little as 0.1 μg/spot. Even more sensitive is the method of Fujita et al.[16] which is based on the formation and measurement at 680 nm of the ternary complex molyb-denum(VI) pyrocatechol violet-protein in the solution state.

REFERENCES

1. Al-Sulimany, F.; Townshend, A. *Anal. Chim. Acta* 1973, *66*, 195.
2. Alvarez Cocque, M. C.; Hernandez, M. J. M.; Villanueva Camanas, R. M.; Fernandez, C. M. *Analyst* 1989, *114*, 975.
3. *British Pharmacopoeia 1988*, Her Majesty's Stationery Office: London, 1988, a. 418; b. 966.
4. Davidson, A. G. *J. Pharm. Biomed. Anal.* 1984, *2*, 45.
5. Davidson, A. G. *J. Pharm. Sci.* 1984, *73*, 1582.
6. Davidson, A. G. *J. Pharm. Biomed. Anal.* 1985, *3*, 235.
7. Demeester, J.; Bracke, M.; Vochten, R.; Lauwers, A. *J. Pharm. Sci.* 1978, *67*, 729.
8. Demchenko, A. P. *Ultraviolet Spectroscopy of Proteins*, Springer Verlag: Berlin, Heidelberg, 1986.
9. Edman, P. *Acta Chem. Scand.* 1950, *4*, 283.
10. El Kommos, M. E.; Mohamed, F. A.; Khedr, A. S. *J. Assoc. Off. Anal. Chem.* 1990, *73*, 516.
11. Fabre, H.; Blanchin, M. D. *Pharm. Acta Helv.* 1983, *58*, 52.
12. Fell, A. F. *UV Spectrometry Group Bull.* 1979, *7*, 5.
13. Fell, A. F. Derivative spectroscopy in the analysis of amino acids: in *Amino Acid Analysis* (J. M. Rattenbury, Ed.) Ellis Horwood—John Wiley: Chichester—New York, 1981, 86–118.
14. Folin, O. *J. Biol. Chem.* 1922, *51*, 377, 393.
15. Folin, O.; Ciocalteu, V. *J. Biol. Chem.* 1927, *73*, 627.
16. Fujita, J.; Mori, I.; Kitano, S. *Chem. Pharm. Bull.* 1984, *32*, 4161.
17. Goodwin, T. W.; Morton, R. A. *Biochem. J.* 1946, *40*, 628.

18. Gray, W. R.; Hartley, B. S. *Biochem. J.* 1963, *89*, 59P.
19. Hernandez, M. J. M.; Alvarez Cocque, M. C.; Bonet Domingo, E.; Villanueva Cammanas, R. M. *Pharmazie* 1990, *45*, 745.
20. Ikenishi, R.; Kitagawa, T.; Hirai, E. *Chem. Pharm. Bull.* 1984, *32*, 609, 748.
21. Issopoulos, P. B. *Pharm. Acta Helv.* 1989, *64*, 82.
22. Issopoulos, P. B. *Pharm. Weekblad. Sci. Ed.* 1989, *11*, 213.
23. Issopoulos, P. B. *J. Anal. Chem.* 1990, *366*, 124.
24. Issopoulos, P. B.; Economou, P. T. *Indian Drugs* 1992, *29*, 1.
25. Issopoulos, P. B.; Economou, P. T. *J. Anal. Chem.* 1992, *343*, 318.
26. Jurkeviciute, J.; Malat, M. *Cesk. Farm.* 1980, *29*, 78.
27. Kakac, B.; Vejdelek, Z. J. *Handbuch der Kolorimetrie*, VEB Gustav Fischer Verlag: Jena, 1966. Bank III. a. 7–16; b. 18–20; c. 20–23; d. 233–272.
28. Kato, T. *Anal. Chim. Acta* 1985, *175*, 339.
29. Khalil, S. K.; Salama, R. B. *J. Pharm. Pharmacol.* 1974, *26*, 972.
30. Knapp, D. R. *Handbook of Analytical Derivatization Reactions*, Wiley-Interscience: New York, 1979.
31. Korany, M. A.; Wahbi, A.-A. M.; *Analyst* 1979, *104*, 146.
32. Kovalenko, L. I.; Stekolnikov, L. I.; Litvinova, T. P.; Ignateva. N. S. *Khim.-Farm. Zh.* 1981, *15*, 114.
33. Lin, J.-K.; Chang, J.-Y. *Anal. Chem.* 1975, *47*, 1634.
34. Lowry, O. H.; Rosenbrough, N. J.; Farr, A. L.; Randall, P. J. *J. Biol. Chem.* 1951, *193*, 265.
35. McKnight, G. S. *Anal. Biochem.* 1977, *78*, 86.
36. Mohamed, M. E.; Tawakkol, M. S.; Aboul-Enein, H. Y. *Zbl. Pharm.* 1983, *122*, 1163.
37. Morelli, B. *J. Pharm. Sci.* 1990, *79*, 261.
38. Ozol, T. *Acta Pharm. Turc.* 1984, *26*, 59.
39. Paris, F.; Pradeau, D.; Hamon, M. *Ann. Pharm. France* 1985, *43*, 133.
40. Raggi, M. A.; Cavrini, V.; Di Pietra, A. M. *J. Pharm. Sci.* 1982, *71*, 1384.
41. Raggi, M. A.; Cesaroni, M. R.; Di Pietra, A. M. *Farmaco; Ed. Prat.* 1983, *38*, 312.
42. Raggi, M. A.; Cavrini, V.; Di Pietra A. M.; Lacché D. *Pharm. Acta Helv.* 1988, *63*, 19.
43. Raggi, M. A.; Nobile, L.; Cavrini, V.; Di Pietra, A. M. *Boll. Chim. Farm.* 1986, *125*, 295.
44. Rao, G. R.; Avadhanulu, A. B. *Indian Drugs* 1986, *23*, 699.
45. Rao, G. R.; Raghuveer, S.; Khadgapathi, P. *Indian Drugs* 1985, *22*, 579.
46. Rényei, M.; Görög, S. *Gyógyszerészet* 1979, *23*, 346.
47. Roth, M. *Anal. Chem.* 1971, *43*, 880.
48. Ruhemann, S. *J. Chem. Soc.* 1910, *97*, 1438, 2025; 1911, *99*, 792, 1111, 1486; *101*, 780.
49. Sane, R. T.; Bhounsule, G. J.; Sawant, S. V. *Indian Drugs* 1987, *24*, 207.
50. Sanger, F. *Biochem. J.* 1945, *39*, 507.
51. Scott, S.; Behrman, E. J. *Anal. Lett.* 1988, *21*, 183.
52. Shukla, S.; Pathak, V. N.; Shukla, I. C. *J. Inst. Chem.* 1985, *57*, 115.
53. Steup, A.; Metzner, J.; Voll, A. *Pharmazie* 1986, *41*, 739.
54. *The United States Pharmacopoeia XXII.*, USP Convention Inc.: Rockville, 1985. a. 757; b. 869; c. 1007; d. 1430; e. 109; f. 354.
55. Vachek, J. *Cesk. Farm.* 1984, *33*, 217.
56. Verma, K. K.; Jain, A.; Gasparic, J. *Talanta* 1988. *35*, 35.
57. Vogt, W.; Koczorek, Kh. R.; Vogt, H.-H.; Knedel, M.; Reindfleisch, G. *Z. Anal. Chem.* 1974, *269*, 187.
58. Yücesoy, C. *J. Fac. Pharm. Gazi.* 1990, *7*, 43.

M. STEROID HORMONES[28–30,35,36]

1. Methods Based on Natural Absorption

The majority of native and synthetic steroid hormones are spectrophotometrically active compounds. Table 10.8 gives an overview of the functional groups bearing the spectrophotometric activity. For more detailed data some important spectral collections are referred to in References 22, 23, 57, 61, and 69.

The spectrophotometric activities reflected in the data of Table 10.8 enable various quantitative analytical problems to be solved mainly in the investigation of bulk steroids and their pharmaceutical

TABLE 10.8. Some Important Chromophoric Groups of Steroid Hormones (Solvent: Ethanol)

Group	λ_{max}, nm	ε	Occurrence	Typical derivatives
4-ene-3-keto	240	17,000	Androgens, anabolics, gestogens, corticosteroids	Testosterone, 19-nortestosterone esters, progesterone, hydrocortisone
1,4-diene-3-keto	243	16,000	Anabolics, corticosteroids	Methandienone, prednisolone
4,6-diene-3-keto	285	25,000	Gestogens	Megestrol acetate
3-acyloxy or alkyloxy 3,5-diene	238	20,000	Intermediates	
1,3,5,(10)-triene-3-ol (phenolic ring A)				
neutral	280	2,000	Oestrogens	Oestrone, oestradiol, ethinyl oestradiol
alkaline	300	2,800		
1,3,5(10)-triene-3-alkyloxy	278	1,900	Oestrogens	Mestranol
	287	1,750		
1,3,5(10)-triene-3-acyloxy	268	800	Oestrogens	Oestradiol dipropionate

formulations. These methods, however, are not suitable for the analysis of biological samples: the selectivity and sensitivity are not sufficient for such purposes.

The majority of pharmacopoeias employ the measurement of the absorbance of bulk steroid hormones at the maximum of the characteristic band of the α,β-unsaturated 3-keto group, which is the most common functional group in androgens, gestogens and corticosteroids. As a consequence of the poor selectivity of this measurement, the $A_{1\ cm}^{1\%}$ value around 240 nm can be regarded as a physico-chemical characteristic rather than a suitable value for the calculation of the active ingredient content of the bulk drug.* In some cases, however, this assay method is prescribed in pharmacopoeias, e.g., in the case of methyltestosterone.[12a] To increase the selectivity, TLC separation is often carried out prior to the absorbance determination. The assay of several 4-ene-3-keto steroids is based in the United States Pharmacopoeia on the absorbance at around 240 nm after spot elution from the TLC plates,[76a] e.g., prednisolone hemisuccinate,[76b] nandrolone decanoate[76c] and testosterone.[76d]

A specific application of the measurement of the absorbance around 240 nm is the determination of free corticosteroid impurities in their 21-phosphates. The free corticosteroids are extracted from the aqueous solution with dichloromethane. This solvent is then evaporated and the absorbance measured after dissolution in methanol. This is the official method in USP XXII, for example, for the determination of free betamethasone in betamethasone sodium phosphate.[76e]

The possibilities for using this method are greater for the analysis of pharmaceutical dosage forms, because the relatively high molar absorptivities allow the determinations of the active ingredient content of the majority of even the low-dosed formulations. Of the numerous applications in various pharmacopoeias two are mentioned: the assay of methyltestosterone tablet and capsule[76f] and testosterone suspension.[76d] TLC separation and spot elution enables this method to be used in those cases, too, when background problems interfere with the simple application of the spectrophotometric method, e.g., in the case of the assay of prednisolone ointment.[48]

TLC separation prior to the absorbance measurement is a useful tool for the simultaneous determination of combined preparations, too. A typical method of this type is described in detail as Reference 43 in Chapter 5.

In advantageous cases the active ingredients of combined formulations can be determined simultaneously by measuring at two wavelengths and using simultaneous equations with two unknowns such as in the case of hydrocortisone (241 nm) and chloramphenicol (273 nm)[65] as well

---

* The detailed discussion of this problem (through the example of prednisolone) can be found on p. 95.

as spironolactone (239 nm) and hydrochlorothiazide (318 nm)[63] formulations. Computer-based methods offer wider possibilities for solving problems of this type. Of these, the use of the method of orthogonal polynomials (Section 4.C.3.) has been reported for the determination of prednisolone and hydrocortisone, respectively, in solutions in the presence of neomycin or phenylephrine.[2]

The difference spectrophotometric method based on the reduction of the 3-keto group with sodium borohydride can be successfully used to eliminate interferences caused by parasitic background absorptions and by the spectrum of other components of the formulation. This method is described in detail in Chapter 8 on p. 208. The applications of these methods, collected in Table 8.2 contain, among others, the determination of nandrolone phanylpropionate oily injection, determination of progesterone in a similar formulation in the presence of benzyl alcohol and oestradiol benzoate, the determination of prednisolone in an ointment in the presence of methylparaben and in a tablet formulation in the presence of phenylbutazone,[31] the determination of norgestrel in contraceptive pills,[34] determination of very low concentrations of fluorocorticosteroids in ointments,[19] etc.

As is shown in Section 4.E.2 derivative spectrophotometry is also a useful tool to overcome the difficulties caused by selectivity and background problems in the analysis of 4-ene-3-ketosteroids. In that section the possibilities of the technique is demonstrated in Figure 4.14 in the example of the assay of spironolactone tablet. In addition, the determination of various gestogens in the presence of oestrogens,[21] the determination of testosterone and progesterone in oily injections[25] and the assay of corticosteroid combined preparations[26] using derivative spectra have also been described.

As a consequence of the approximately ten-fold lower intensity of their band around 280 nm, than the intensity of the 240 nm band of unsaturated ketosteroids, the use of this band for quantitative determination of oestrogens is a difficult task. The situation in this case is made even more difficult by the fact that oestrogens are administered in very low doses both in their tablet and injection formulations. The simple absorbance measurement can be used in the case of bulk oestrogens but the value of these results is even more questionable than in the above discussed case of unsaturated ketosteroids; thus, this possibility is only seldom exploited for the investigation of bulk oestrogens; (e.g., the assay of oestriol at 281 nm[76g]).

When investigating formulations of oestrogens, background problems are always encountered. Since the other component of the formulations is usually ketosteroid in great excess (gestogens, androgens) the selectivity and precision of the measurement can be greatly increased if the keto groups of the latter are reduced with sodium borohydride, thus making the interfering ketosteroids spectrophotometrically inactive.[6,37,60]

For the determination of oestrogens with free hydroxyl groups the difference spectrophotometric method based on the ionization of the phenolic hydroxyl at pH 13 is also applicable. Because of the easy hydrolyzibility of the phenol esters this method can be used for the determination of 3-acyloxy derivatives, too, but, of course, not for the 3-ether derivatives.[33]

Due to the fine structure of their spectra the possibilities of using derivative spectrophotometry for the determination of oestrogens are better than with the 4-ene-3-keto derivatives. This possibility has been exploited for determinations which included oestradiol esters,[25] ethinyl oestradiol and mestranol[21] in pharmaceutical preparations in the presence of large quantities of ketosteroids, for the assay of conjugated oestrogens[62] and for the determination of 0.1% mestranol impurity in norethisterone.[39] A difference–derivative spectrophotometric method is also described for the determination of ethinyloestradiol in the presence of a great excess of norethisterone.[55]

Of the spectrophotometric investigations of steroids not containing any of the two characteristic chromophoric systems (4-ene-3-keto group and phenolic ring A) a few are also mentioned here. Bulk danazol and its capsule formulation are determined at 285 and 287 nm, respectively; in the case of the capsule in chloroform medium.[76h] Oxymetholone (bulk material and tablets) are measured at 315 nm[76i] while the wavelength for the assay of bulk megestrol acetate is 287 nm.[12b] Stanozolol was determined in tablets by a derivative spectrophotometric method: measurement of $^1D_{239}$ or $^2D_{247,231}$.[17]

The solvent for the assays discussed so far has been ethanol or methanol in the majority of cases. It is worth mentioning that in the classical period of steroid analysis in the 1950s, the spectra of steroids in concentrated sulphuric acid were thoroughly investigated and even a spectral atlas was published containing a large number of such spectra.[70] These spectra were of great importance both in the qualitative and quantitative analysis of steroids since large differences could be found even in the case of closely related steroids (e.g., epimeric pairs) the alcoholic spectra of which are practically identical. However, during the dissolution of steroids in concentrated sulphuric acid, chemical reactions take place and hence the shape and intensity of the spectrum depends to some extent on the time between the dissolution and the measurement and on other factors which are difficult to control. Because of the problematic reproducibility of the spectra, this technique does not seem to be in general use today.

2. Methods Based on Chemical Reactions

Because of the limited sensitivity and selectivity of the methods based on natural absorption, in the case of many practical analytical problems it is mandatory to combine the analytical measurement of steroid hormones with preliminary chemical reactions, which increase their spectrophotometric activity and/or shift their spectra to the more favorable visible spectral range. The importance of these methods can be characterized by the fact that Bartos and Pesez[5] published a book on this topic. Of the other general books, that of Higuchi and Brochmann-Hanssen contains a comprehensive chapter by Forist and Johnson[27] which is an important source of the early literature of the pharmaceutical aspects of steroid analysis. The standard book of the classical period of clinical steroid analysis is the monograph by Jayle.[52] Both books reflect the great importance of spectrophotometric methods at the time of their publication. Of the newer books, Görög's monographs are mentioned dealing with the pharmaceutical,[28] general[29] and industrial[30] aspects of steroid analysis, all containing comprehensive chapters on UV-VIS methods based on chemical reactions.

From among the functional groups in steroid hormone drugs the following are mainly involved in the reactions leading to derivatives with favorable spectrophotometric properties:

1. keto group, (mainly 4-ene-3-keto and 1,4-diene-3-keto) and the adjacent active methylene groups (in the case of 4-ene-3-keto group the 2- and in the case of 17-keto group the 16-methylene groups)
2. phenolic ring A of oestrogens
3. the reducing α-ketol side chain at position 17 of corticosteroids.

The color reactions of steroids with various reagents containing high concentrations of mineral acids cannot usually be attributed to well-defined functional groups.

Table 10.9 summarizes some of the methods which are of importance in pharmaceutical steroid analysis. The classical reactions of the spectrophotometric analysis of steroids for determining endogenic steroids in biological samples such as the Zimmermann method[91] for the determination of 17-ketosteroids with the 1,3-dinitrobenzene reagent or the innumerable variants of the Kober-Haenni-Ittrich method[46,49,50,54] using the iron(II) phenol (quinol) sulphuric acid reagent are not discussed because they are beyond the scope of this book (and more or less obsolete).

Of the derivatization reactions of keto groups, the Umberger method[81] using isonicotinic acid hydrazide (INH) is by far the most important. The reaction in Equation 10.15 takes place in methanol containing dilute hydrochloric acid. The latter plays a double role: it catalyzes the condensation and protonates the hydrazone to produce the yellow color which is suitable for analytical purposes. Saturated ketones cannot be determined under the conditions of 4-ene-3-ketones (20 min reaction time at room temperature). The absorption maximum of the latter is at 380 nm with molar absorptivities around 11,500. 1,4-Diene-3-ketones and 4,6-diene-3-ketones absorb at longer wavelengths (around 405 to 415 nm) and the intensities of the bands are also higher ($\varepsilon \sim 18,000$). 1,4-Diene-3-ketones react slowly. To achieve complete reaction either the temperature has to be raised to 50°C[67] or the concentrations of the reagent and hydrochloric acid have to be increased.[16]

TABLE 10.9. Some Photometric Methods for the Determination of Steroid Hormones

Author/Year	Sample	Reagent	λ_{max}, nm	ε
		Reactions of the Keto Group		
Umberger, 1955	Pharmaceutical preparations	Isonicotinic acid hydrazide–HCl	380–405	11,000–18,000
Walash, 1987	Tablets, ointment	Semicarbazide	269	25,000–30,000
Talbot, 1955	Biol. samples	Thiosemicarbazide	299–325	34,000–39,000
Walash, 1984	Tablets, ointment	Oxalyldihydrazide	300	14,500–21,500
Woodson, 1971	Tablets	2,4-Dinitrophenylhydrazine	370–410	22,000–50,000
Zivanov-Stakic, 1989	Mesterolone tablet	3-Acetylaminobenzaldehyde thiosemicarbazone	544	7,900
Görög, 1970	Pharmaceutical preparations	Diethyl oxalate + Na-tert. butylate	289–324	5,000–15,000
Belal, 1986	Tablets	Hydroxylamine + iodine	300	5,000–5,500
		hydroxylamine + chloranil	430	6,200–6,300
		Reactions of the 17-Ethinyl Group		
Szepesi, Görög, 1974	Bulk steroids	Na-tert. butylate + diethyl oxalate	294	10,600–10,800
Walash, 1985	Tablets	Na-tert. butylate + 1,3-dinitrobenzene	512	11,300–17,700
		Reactions of the Phenol-Type Ring A		
Urbányi Mollica, 1968	Pharmaceutical preparations	Diazotized 4-NH_2-6-Cl-m-benzenedisulphonamide	500	8,050
Eldawy, 1975	Tablets	Diazotized 5-Cl-2,4-di-nitroaniline	450	11,300
Tsilifonis, Chafetz, 1967	Pharmaceutical preparations	H_2SO_4–methanol 7:3 v/v	540	36,000 (ethinyl oestradiol)
Szabó, 1977	Tablets	H_2SO_4–ethanol 65:35 v/v	535	53,000 (ethinyl oestradiol)
		Reactions of the Corticosteroid Side Chain		
Callahan, 1962	Tablets	Triphenyltetrazolium chloride	485	16,500–17,150
Kunze, Davis, 1964	Tablets	Blue tetrazolium	520	22,000–24,000
Chong Kwan, Schott, 1984	Prednisolone tablet	Phenylhydrazine–H_2SO_4	410	14,600
Ayad, 1984	Tablets	Phenylhydrazine + iodine	310	7,800
		phenylhydrazine + chloranil	450–460	1,900–3,800
Görög, Szepesi, 1972	21-OH-corticosteroids	Cu(II)acetate + 4,5-dimethyl-o-phenylenediamine	350	11,800–12,600
Görög, Szepesi, 1972	21-Amino-corticosteroids	Hg(II)chloride + 4,5-dimethyl-o-phenylenediamine	351	12,500
Tokunaga, 1976–1977	Pharmaceutical preparations	Cu(II)acetate + pyrrole	570–600	26,600–42,600
Bundgaard, 1978	Tablets	Cu(II)acetate + 3-methylbenzthiazol-2-one hydrazone	394–400	22,900–23,700
Görög, Tuba, 1972	21-Hlg and active ester derivatives	Pyridine + tetramethylammonium hydroxide	413–417	20,240–21,900

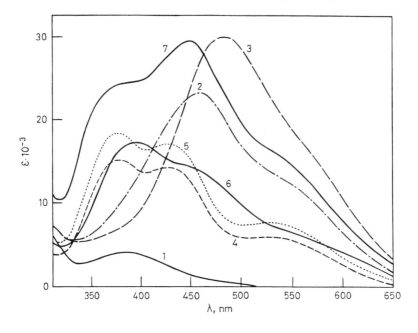

FIGURE 10.10. Spectra of ketosteroid 2,4-dinitrophenylhydrazones. Solvent: 0.1 M sodium hydroxide in the 4:1 mixture of methanol and water. 1: 2,4-dinitrophenylhydrazine; 2: norethisterone (4-ene-3-keto); 3: methandienone (1,4-diene-3-keto); 4: dehydroepiandrosterone (17-keto); 5: 3β,17-dihydroxy-5-pregnen-20-one 3-acetate (20-keto); 6: 3β-hydroxy-5,16-pregnadiene-20-on acetate (16-ene-20-keto); 7: progesterone (4-ene-3,20-dione). (From Görög, S.; Szász, Gy. *Analysis of Steroid Hormone Drugs,* Akadémiai Kiadó, Budapest; Elsevier: Amsterdam, 1978, 292.[28])

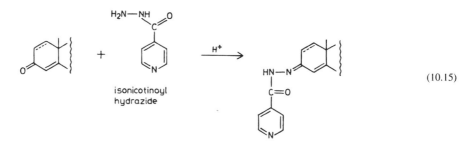

$$\text{(10.15)}$$

An advantageous feature of the reaction is that it can be conducted in other solvents, too, (ethanol, chloroform); moreover, small quantities of oil used for the preparation of injections do not interfere. Hence in addition to tablets[1,88] and ointments[1] the method is suitable for the direct assay of oil-injectables, too.[81,86] The preliminary separation of the oil is necessary only in the case of very low-dose injections.[81]

The INH method is in use in pharmacopoeias up to the present time for the investigation of bulk ketosteroids and their formulations.[76j,k,l] Several automated versions of the methods have also been described.[3,15,68,77]

In contrast to the INH method the other four derivatization reagents in Table 10.9 (semicarbazide, thiosemicarbazide, 2,4-dinitrophenylhydrazine, 3-acetylaminobenzaldehyde-thiosemicarbazone) are suitable for the determination of saturated ketosteroids, too. The 2,4-dinitrophenylhydrazones are rearranged to the quinoidal form in alkaline media and this enables selective determinations to be carried out (see Figure 10.10).

The mechanism of the method where the reagent is diethyl oxalate is quite different. Here the reagent attacks the steroid molecule at the active methylene groups in the vicinity of the keto groups. From the spectra shown in Figure 10.11 it is evident that selective measurements can be carried out. The selectivity is further increased by the fact that the Claisen condensation takes place in the case of 4-ene-3-ketosteroids only if there is no double bond at the 1 position or in

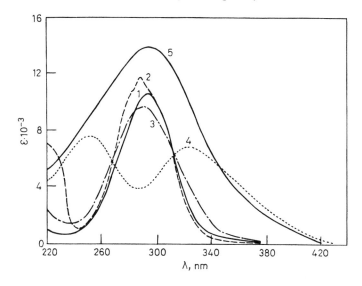

FIGURE 10.11. Spectra of ketosteroids after their reaction with diethyl oxalate. 1: dehydroepiandrosterone (17-keto); 2: oestrone (17-keto + aromatic ring A); 3: pregnenolone (20-keto); 4: methyltestosterone (4-ene-3-keto); 5: progesterone (4-ene-3,20-dione). (From Görög, S. *Anal. Chem.* 1970, *42*, 560.[32])

the case of 20-ketosteroids if there is no substituent at the 21 position. On this basis, 4-ene-3-keto impurities can be determined in 1,4-diene-3-ketones:[32] this is the basis of the difference spectrophotometric method described in Chapter 8.

17α-Ethinyl-17-hydroxy steroids can be de-ethinylated by treatment with sodium tert-butylate to form the 17-keto derivative. Since sodium tert-butylate is the condensing agent of the above discussed Claisen condensation with diethyl oxalate, too, the determination of the 17α-ethinyl group can be traced back to the 17-keto group in the form of the 16-glyoxalyl derivative.[73] As seen in reaction scheme 10.16 the de-ethinylation reaction also enables the use of the Zimmermann reaction for the indirect determination of 17-ethinyl steroids.[85]

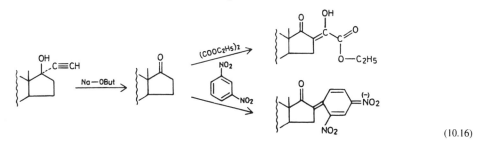

(10.16)

The most widely used reagents for the determination of oestrogens containing a phenolic hydroxyl group in the pharmaceutical steroid analysis are various dilutions of sulphuric acid with alcohols. The 7:3 v/v mixture of sulphuric acid and methanol, introduced by Tsilifonis and Chafetz[80] is an especially selective and sensitive reagent for ethinyl-oestradiol and mestranol. The reaction has a surprising selectivity: oestrone and oestradiol not containing the 17-ethinyl group do not give positive reactions. This method is in use even in pharmacopoeias, e.g., for the assay of bulk mestranol[76n] and ethinyl oestradiol tablet.[76m] In the course of the application of the method to contraceptive pills, the oestrogenic component is usually separated by chromatography from the gestogen which is in large excess.[43,89]

The classical Kober-Haenni reagent (iron(II) phenol sulphuric acid)[46,54] is prescribed even in the latest revision of the United States Pharmacopoeia for the determintion of oestradiol in suspension formulation;[76o] the ethinyl-oestradiol content of its combination with norgestrel is determined in the same revision using 80% aqueous sulphuric acid as the reagent[76p] (λ_{max} = 520 and 536 nm, respectively).

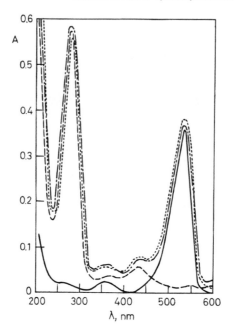

FIGURE 10.12. Spectra for the assay of Ovidon® contraceptive pill. Active ingredients: 0.25 mg norgestrel + 0.05 mg ethinyloestradiol/tablet. Solvent: 65:35 (v/v) mixture of sulphuric acid and ethanol; norgestrel 1 mg/100 ml – –; ethinyloestradiol 0.2 mg/100 ml —; mixture of the active ingredients – · –; tablet extract ···· . (From Szabó, A. *Acta Pharm. Hung.* 1977, *47*, 24.[72])

Ethanol-sulphuric acid reagents are also widely used.[47,71,72] Using the 65:35 v/v mixture of sulphuric acid and ethanol, Szabó[72] determined norgestrel and ethinyl-oestradiol in contraceptive pills. As seen in Figure 10.12, norgestrel almost selectively absorbs at 290 nm while the three-fold more intense band of ethinyloestradiol appears at 535 nm enabling the simultaneous assay to be carried out. The automated version of the determination of ethinyl oestradiol in tablets using the 9:1 mixture of sulphuric acid and ethanol is also described.[9]

Another widely used method for the determination of oestrogens making use of their phenolic character is their coupling with diazotized aromatic amines. Although—as seen in Table 10.9—the sensitivity of this method is inferior to that of the methods using sulphuric acid reagents, the selectivity is good and these methods could be successfully used in pharmaceutical steroid analysis.

The photometric determination of corticosteroids is usually based on their side chain at position 17. Although they contain 4-ene-3-keto or 1,4-diene-3-keto groups, too, the methods based on the side chain are preferred. The reason for this is that the α-ketol type side chain is sensitive to oxidation and hydrolysis, and hence the methods based on the measurement of the intact side chain are suitable for stability assays, while the same cannot be said about their measurement via the INH assay of the unsaturated 3-keto group.

The most widespread methods for the measurement of the side chain are based on its reducing properties. Depending on the oxidizing power of the reagent, the oxidation of the side chain can afford various reaction products. In the practice of spectrophotometric analysis, mild oxidizing agents are preferred leading to 21-aldehydes (see Equation 10.17) and not 17-ketones or 17-carboxylic acids.

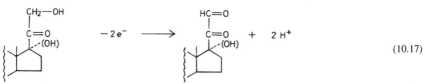

$$(10.17)$$

This is the reaction product of the most generally used tetrazolium methods.[38] As described in Section 7.E, the basis of the measurement in these methods are the colored formazans, the reduced forms of the colorless tetrazolium reagents, which are used as the mild oxidizing agents. In pharmaceutical analysis triphenyl tetrazolium chloride or the dimeric derivative Blue tetrazolium are usually used. As seen in the data in Table 10.9, the indirect determination of corticosteroids can be performed with both reagents. The disadvantage of the triphenyl tetrazolium chloride reagent is that because of its sensitivity to atmospheric oxygen the reaction has to be carried out under nitrogen.[53] This is not the case with the Blue tetrazolium reagent; in this case, however, the reagent is sensitive to light and the reaction has to be carried out under light protection.[58] The advantages and disadvantages of the two reagents are in equilibrium and hence both are widespread for the analysis of bulk corticosteroids and their formulations[11,15,58,84] and for stability assays.[20,45,51] It is interesting to note that the British Pharmacopoeias traditionally apply triphenyl tetrazolium chloride[12c,d,etc.] while in the United States Pharmacopoeias the use of Blue tetrazolium is prescribed.[76r,s,etc.]

The color which is the basis of the measurement develops at room temperature within 60–90 min. Since the reaction is first order with respect to the corticosteroid,[44] the reaction rate at the initial period of the reaction, i.e., the slope of the reaction curve is a linear function of its concentration. This was exploited for the quantitative analysis in the kinetic version of the method: running the reaction in the cell and recording the absorbance as a function of the reaction time, the measurement can be carried out within 30–90 seconds. This version of the method was utilized for the assay of corticosteroid formulations.[56,64]

The rate of the reaction can be increased by raising the temperature: this is done in the automated[8,10,13] and FIA[59] versions of the method.

The method is not specific for the 21-hydroxyl derivatives: 21-acyloxy derivatives also give positive reaction, but with considerably lower reaction rate. This is because the reaction mixture of the tetrazolium reaction is strongly alkaline and in this medium the hydrolysis of the corticosteroid esters takes place to produce the free corticosteroids. This does not apply to the important 21-phosphates: these have to be hydrolyzed enzymatically prior to the use of the tetrazolium reagents.[8,76,83] The free phosphate impurity of the latter can be determined spectrophotometrically using the molybdenum blue method.[76e]

As regards the role of the 17-hydroxy group, it can be stated that its presence is not a prerequisite of the reaction (see Equation 10.18) and hence the tetrazolium methods are equally suitable for the determination of glucocorticoids containing and mineralocorticoids not containing the 17-hydroxy group.

The selective method for the determination of glucocorticoids is the Porter-Silber method which is mainly used in biological-clinical analysis.[66] In this method the reagent is the strongly acidic solution of phenylhydrazine. The first step of the reaction is the intramolecular oxidoreduction catalyzed by the sulphuric acid in the reaction mixture. The resulting 21-aldehyde undergoes condensation with the reagent to form the 21-phenylhydrazone which rearranges in the strongly acidic reaction mixture to the enol-phenylhydrazonium cation strongly absorbing at 410 nm (see Equation 10.18). New versions of the method have also found applications in pharmaceutical analysis, e.g., in monitoring the dissolution studies of prednisolone tablet[18] (see Table 10.9) and for the investigation of hydrocortisone sodium phosphate.[76t] Under less drastic reaction conditions (without sulphuric acid) the 20-keto (and eventually the 11-keto) groups react with phenylhydrazine. The formation of charge transfer complexes between the hydrazones thus formed and iodine or chloranil is also the basis of selective spectrophotometric methods.[4]

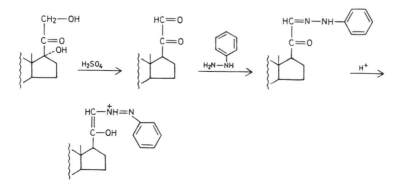

(10.18)

The 21-hydroxy group can be oxidized to aldehyde also by copper (II). As seen in Table 10.9, the 20-keto-21-aldehyde system thus formed undergoes condensation reaction with various amine reagents creating the basis for selective spectrophotometric determinations.[14,40,41,74,78] As seen in Equation 10.19, 21-hydroxy derivatives can be determined, for example, with copper(II) acetate 4,5-dimethyl-o-phenylenediamine reagent in the form of quinoxaline derivatives.[40,74] It is interesting to note that replacing copper(II) acetate by mercury(II) chloride as the oxidizing agent enables the selective measurement of 21-amino corticosteroids after the same condensation reaction, namely the 21-hydroxy derivatives are not oxidized by mercury(II)-chloride.[41,74]

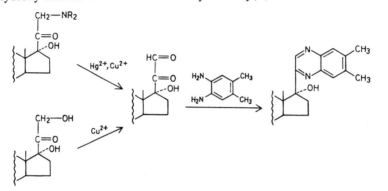

(10.19)

REFERENCES

1. Agbaba, D.; Zivanov-Stakic, D.; Vladimirov, S.; Zubac, K. *Acta Pol. Pharm. Drug. Res.* 1990, *47*, 15.

2. Amer, M. M.; Hassan, S. M.; Mostafa, A. A. *Pharmazie* 1978, *33*, 344.

3. Angermann, B.; Sonderdruck, Nr. 952, *Technicon Symposium, 1971.*

4. Ayad, M. M.; Belal, S.; El Adl, S. M.; Al Kheir, A. A. *Analyst,* 1984, *108*, 1417.

5. Bartos, J.; Pesez, M. *Colorimetric and Fluorimetric Analysis of Steroids,* Academic Press: London, 1976.

6. Bastow, R. A. *J. Pharm. Pharmac.* 1967, *19*, 41.

7. Belal, S.; El Kheir, A. A.; Ayad, M. M.; Al Adl, S. A. *Analyst* 1986, *111*, 1039.

8. Beyer, W. F. *J. Pharm. Sci.* 1966, *55*, 200.

9. Beyer, W. F. *J. Pharm. Sci.* 1968, *57*, 1415.

10. Beyer, W. F.; Smith, E. W. *J. Pharm. Sci.* 1970, *59*, 248.

11. Bracey, A.; Garrett, L.; Weiss, P. J. *J. Pharm. Sci.* 1966, *55*, 113.

12. *British Pharmacopoeia 1988,* Her Majesty's Stationery Office: London, 1988. a. 372; b. 355; c. 73; d. 290.

13. Brower, J. F. *J. Ass. Off. Anal. Chem.* 1977, *60*, 27.

14. Bundgaard, H. *Arch. Pharm. Chem. Sci. Ed.* 1978, *6*, 127.

15. Callahan, J. J.; Litterio, F.; Britt, E.; Rosen, B. D.; Owens, J. *J. Pharm. Sci.* 1962, *51*, 333.

16. Cavina, G.; Cingolani, E.; Amormino, V.; Giraldez, H. *Farmaco, Ed. Pr.* 1962, *17*, 393.

17. Cavrini, V.; Di Pietra, M.; Raggi, M. A.; Maioli, M. G. *Analyst* 1987, *112*, 1671.

18. Chong Kwan, L.; Schott, H. *J. Pharm. Sci.* 1984, *73*, 157.

19. Coda, L.; Timallo, L. *Boll. Chim. Pharm.* 1976, *115*, 515.

20. Comer, J. P.; Hartsaw, P. E. *J. Pharm. Sci.* 1965, *54*, 524.

21. Corti, P.; Lencioni, E.; Sciarra, G. F. *Boll. Chim. Farm.* 1983, *122*, 281.

22. Dorfman, L. *Chem. Rev.* 1953, *53*, 47.

23. Dusza, J. P.; Heller, M.; Bernstein, S., Ultraviolet Absorption, in *Physical Properties of the Steroid Hormones,* Engel, L. L. (Ed.); Pergamon: Oxford, 1963, 69–287.

24. Eldawy, M. A.; Tawfik, A. S.; Elshabouri, S. R. *J. Pharm. Sci.* 1975, *64*, 1221.

25. El-Yazbi, F. A. *Anal. Letters* 1985, *18*, 2127.

26. El-Yazbi, F. A.; Korany, M. A.; Abdel Razak, O.; Elsayed, M. A. *J. Ass. Off. Anal. Chem.* 1986, *69*, 614.

27. Forist, A. A.; Johnson, J. L., Steroids, in *Pharmaceutical Analysis,* Higuchi, T.; Brochmann-Hanssen, E. (Ed.). Interscience Publ.: New York, 1961, 39–136.

28. Görög, S.; Szász, Gy. *Analysis of Steroid Hormone Drugs,* Akadémiai Kiadó, Elsevier: Budapest-Amsterdam, 1978.

29. Görög, S. *Quantitative Analysis of Steroids,* Akadémiai Kiadó, Elsevier: Budapest-Amsterdam, 1983.

30. Görög, S. (Ed.), *Steroid Analysis in the Pharmaceutical Industry*, Ellis Horwood: Chichester, 1989, 5–16.
31. Görög, S. *J. Pharm. Sci.* 1968, *57*, 1737.
32. Görög, S. *Anal. Chem.* 1970, *42*, 560.
33. Görög, S. *Analyst* 1976, *101*, 512.
34. Görög, S. *Zbl. Pharm.* 1977, *116*, 259.
35. Görög, S. *CRC Crit. Rev. Anal. Chem.* 1980, *9*, 333.
36. Görög, S. *Z. Anal. Chem.* 1981, *309*, 97.
37. Görög, S.; Csizéer, É. *Acta Chim. Hung.* 1970, *65*, 41.
38. Görög, S.; Horváth, P. *Analyst* 1978, *103*, 346.
39. Görög, S.; Rényei, M.; Herér.yi, B. *J. Pharm. Biomed. Anal.* 1989, *7*, 1527.
40. Görög, S.; Szepesi, G. *Anal. Chem.* 1972, *44*, 1079.
41. Görög, S.; Szepesi, G. *Analyst* 1972, *97*, 519.
42. Görög, S.; Tuba, Z. *Analyst* 1972, *97*, 523.
43. Graham, J. H. *J. Ass. Off. Anal. Chem.* 1975, *58*, 202.
44. Graham, R. E.; Biehl, E. R.; Kenner, C. T. *J. Pharm. Sci.* 1976, *65*, 1048.
45. Graham, R. E.; Williams, P. E.; Kenner, C. T. *J. Pharm. Sci.* 1970, *59*, 1472, 1552.
46. Haenni, E. O. *J. Am. Pharm. Ass. Sci. Ed.* 1950, *39*, 544.
47. Hassan, S. S. M.; Abdel Fattah, M. M.; Zaki, M. T. M. *Z. Anal. Chem.* 1976, *281*, 371.
48. Heyde, R.; Schwalbe, R.; Illig, G. *Pharmazie* 1990, *45*, 860.
49. Ittrich, G. *Z. Physiol. Chem.* 1958, *312*, 1.
50. Ittrich, G. *Acta Endocrinol.* 1960, *35*, 34.
51. Jakovljevic, I. M.; Hartsaw, P. E.; Drummond, G. E. *J. Pharm. Sci.* 1965, *54*, 1771.
52. Jayle, M. F. (Ed.) *Analyse des Steroides Hormonaux*, Masson: Paris, 1962.
53. Johnson, C. A.; King, R.; Vickers, C. *Analyst* 1960, *85*, 714.
54. Kober, S. *Biochem. Z.* 1931, *239*, 209.
55. Korany, M. A.; El-Yazbi, F. A.; Abdel-Razak, O.; Elsayed, M. *Pharm. Weekblad. Sci. Ed.* 1985, *7*, 163.
56. Koupparis, M. A.; Walczak, K. M.; Malmstadt, H. V. *J. Pharm. Sci.* 1979, *68*, 1479.
57. Krácmar, J. Krácmarová, J. Bokovikova, T. N.; Ciciro, V. E.; Nesterova, G. A.; Suranova, A. V.; Trius, N. V. *Pharmazie* 1991, *46*, 253.
58. Kunze, F. M.; Davis, S. J. *J. Pharm. Sci.* 1964, *53*, 1170, 1259.
59. Landis, J. B. *Anal. Chim. Acta* 1980, *114*, 155.
60. Legrand, M.; Delaroff, V.; Smolik, R. *J. Pharm. Pharmac.* 1958, *10*, 683.
61. Neudert, W.; Röpke, H. *Atlas of Steroid Spectra*, Springer: Berlin, 1965.
62. Novakovic, J.; Nemcová, I. *Pharmazie* 1990, *45*, 439.
63. Nowakowska, Z. *Farm. Pol.* 1989, *45*, 454.
64. Oteiza, R. M.; Krottinger, D. L.; McCracken, M. S.; Malmstadt, H. V. *Anal. Chem.* 1977, *49*, 1586.
65. Parmentier, A.; Barthélémy, J. -F.; Andermann, G. *Labo-Pharma—Probl. Techn.* 1984, *32*, 536.
66. Porter, C. C.; Silber, R. H. *J. Biol. Chem.* 1950, *185*, 201.
67. Reed, S. R.; Walters, S. M. *Anal. Lett.* 1970, *3*, 585.
68. Russo-Alesi, F. M. *Ann. N. Y. Acad. Sci.* 1968, *153*, 511.
69. Scott, A. I. *Interpretation of the Spectra of Natural Products*, Pergamon: Oxford, 1964, 363–431.
70. Smith, L. L.; Bernstein, S. Absorption Spectra in Concentrated Sulphuric Acid, in *Physical Properties of the Steroid Hormones*, Engel, L. L., (Ed.), Pergamon: Oxford, 1963, 69–287.
71. Swendsen, R.; Nedergaard, M. *Farm. Tidende* 1964, *42*, 749.
72. Szabó, A. *Acta Pharm. Hung.* 1977, *42*, 24.
73. Szepesi, G.; Görög, S. *Analyst* 1974, *99*, 218.
74. Szepesi, G.; Görög, S. *Boll. Chim. Farm.* 1975, *114*, 98.
75. Talbot, N. B.; Ulick, S.; Koupreianow, A.; Zygmuntowich, A. *J. Clin. Endocrinol. Metab.* 1955, *15*, 301.
76. *The United States Pharmacopoeia XXII*, USP Convention Inc.: Rockville, 1990. a. 1545; b. 1132; c. 915; d. 1326; e. 164; f. 880; g. 534; h. 379; i. 993; j. 487; k. 964; l. 1328; m. 547; n. 834; o. 531; p. 965; r. 1532; s. 1545; t. 658–659.
77. Thornton, L. K. *FDA Drug Autoanalysis Manual, Method Nr.* 1972, 22.
78. Tokunaga, H.; Kimura, T.; Kawamura, J. *Bunseki Kagaku* 1976, *25*, 392; 1977, *26*, 154.
79. Tokunaga, H.; Tanno, M.; Kimura, T. *Chem. Pharm. Bull.* 1987, *35*, 1118.
80. Tsilifonis, D. C.; Chafetz, L. *J. Pharm. Sci.* 1967, *56*, 625.
81. Umberger, E. J. *Anal. Chem.* 1955, *27*, 768.
82. Urbányi, T.; Mollica, J. A. *J. Pharm. Sci.* 1968, *57*, 1257.

83. Van Dame, H. C. *J. Ass. Off. Anal. Chem.* 1974, *57*, 731.

84. Veeman, G. E. *Pharm. Weekblad.* 1981, *116*, 109.

85. Walash, M. I.; Rizk, M.; Zakhari, N. A.; Toubar, S. Colorimetric Determination of 17α-Ethinyl Steroids, in *Advances in Steroid Analysis '84*, Görög, S. (Ed.), Akadémiai Kiadó, Budapest; Elsevier: Amsterdam, 1985, 563.

86. Walash, M. I.; Rizk, M.; Zakhari, N. A.; Toubar, S. *Anal. Lett.* 1984, *17*, 817.

87. Walash, M. I.; Zakhari, N. A.; Rizk, M.; Toubar, S. *Farmaco, Ed. Prat.* 1987, *42*, 81.

88. Woodson, A. L. *J. Pharm. Sci.* 1971, *60*, 1538.

89. Wu, J. Y. P. *J. Ass. Off. Anal. Chem.* 1970, *53*, 831; 1971, *54*, 617; 1977, *60*, 922.

90. Wu, J. Y. P. *J. Ass. Off. Anal. Chem.* 1974, *57*, 747; 1975, *58*, 75.

91. Zimmermann, W. *Chemische Bestimmungsmethoden von Steroid-hormonen in Körperflüssigkeiten*, Springer:Berlin, 1955, 53–57, 63–75.

92. Zivanov-Stakic, D.; Milosevic, Lj.; Agbaba, D. *J. Pharm. Biomed. Anal.* 1989, *7*, 1983.

N. ANTIBIOTICS

In this section the spectrophotometric analysis of the pharmaceutically most important antibiotics is summarized following the classification of Bérdy.[23] This branch of their analytical chemistry is discussed in various books.[15,268,269]

1. Carbohydrate Antibiotics

Aminoglycosides.   The most important representatives of this group are the streptomycins, neomycins, kanamycins, tobramycin and the gentamicins. These antibiotics do not have characteristic absorption above 200 nm: their spectrophotometric analysis is based on color reactions.

Wahbi et al.[260] transformed dihydrostreptomycin and streptomycin to spectrophotometrically active derivatives by boiling with 0.1 M sulphuric acid and 0.5 M sodium hydroxide, respectively followed by multi-wavelength measurement and the use of the orthogonal polynomials for the calculation. The classical "maltol reaction" for the determination of streptomycin[69] also begins with alkaline hydrolysis. As seen in the reaction scheme 10.20 it is the streptose part of the molecule which takes part in the reaction; this is transformed to maltol which affords, after acidification, a violet-colored complex with iron(III) ions measurable at 525 nm. This reaction is used by the British Pharmacopoeia[36a] for the determination of streptomycin.

The guanidine group of streptomycin can be determined by the modified Sakaguchi reaction: this method is based on the measurement of the red product of the oxidative condensation with α-naphthol.[163]

The methods of Duda[67] and Divakar et al.[64] are also based on oxidation followed by reaction with thiobarbituric acid and brucine reagents, respectively. When iodine was used by the latter workers for the oxidation of streptomycin (and also penicillins, cefalosporins and grizeofulvin) its excess was measured spectrophotometrically using methol and sulphanilamide.[222]

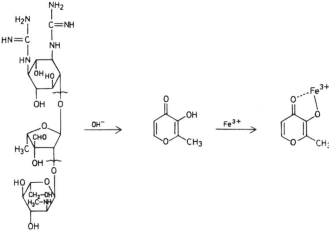

(10.20)

Streptomycin A

The formation of ion pairs and metal complexes (mainly ternary complexes) also play an important role in the spectrophotometric analysis of streptomycins. With their strongly basic groups they easily form extractable ion pairs with acidic dyes e.g., with tropaeolin 000.[22] Reagents for the ternary complex formation are praseodymium(III) phthalexone S[8] and aluminum(III) catechol violet.[9] The interesting methods of Mori et al.[80,82,158] are also based on ternary complex formation. In these very sensitive methods the reagents are the complexes of o-hydroxyhydroquinonephthalein with various metals. The spectra of these complexes are shifted towards the longer wavelengths on reacting with the antibiotic derivatives, and the absorbance at a suitable wavelength is proportional to their concentration. Manganese(II) was used for the determination of streptomycin,[80] and uranium(VI) for neomycin, tobramycin and gentamicin.[82,158]

The spectrophotometric analysis of *neomycins* was reviewed by Heyes.[90] Aminoglycosides—among them neomycins—give positive ninhydrin reactions at pH 5–8 (see Section 10.L.2).[14,41,189] Ribose formed during the acid-catalyzed hydrolysis can be determined using the orcine iron(III)[115] and phloroglucinol[92] reagents. Agrawal et al.[5] used the copper(II) tartrate reagent to form 1:1 complexes with neomycins.

Doulakas[66] oxidized neomycins with sodium hypobromite in alkaline medium and the resulting aldehyde was reacted with the phloroglucinol reagent to form the colored derivative. Csiba[53] and Das Gupta[57] adopted the Hantzsch reaction (formation of dihydrolutidine derivatives—see p. 365) for the spectrophotometric assay. Chloranil forms charge-transfer complexes with neomycin B, kanamycin, tobramycin and amikacin, and this was utilized by Rizk and Younis[202] for their determination. The widely used amine reagent 1-fluoro-2,4-dinitrobenzene also found application: Marques et al.[136] determined neomycin while Ryan[207] determined kanamycin, tobramycin, amikacin and gentamicin with this reagent.

Kanamycin, tobramycin, apramycin were determined as their ion pairs with bromothymol blue; the complex can be extracted with chloroform at pH 6–6.5.[30,266] Das Gupta[56] measured tobramycin in its formulations using the complex formation reaction with copper (II) in alkaline medium. The primary amino groups of tobramycin react with the common reagents of amino acids such as o-phthalaldehyde, dansyl chloride and fluorescamine and this was utilized by Sampath and Robinson, to form spectrophotometrically measurable derivatives.[216] These antibiotic derivatives can also be determined with ninhydrin,[12] orcine[31] and some of the already mentioned other reagents.[9,53,57]

Gentamicin was determined as the copper(II) complex[246] and as the dihydrolutidine derivative.[172] The three components of gentamicin C were determined densitometrically by Török and Paál[248] after the reaction of the separated components with ninhydrin.

Hygromycin B was determined by Bashkovich et al. at 282 nm after boiling with sulphuric acid, and by using the anthrone sulphuric acid reagent[19] or ion pairing with bromothymol blue.[20]

Other Glycosides.

Vancomycin was determined by Fooks et al.[77] by the Folin-Ciocalteu reagent. The reaction could be made selective to vancomycin even in the presence of phenolic derivatives.

Miscellaneous.

Lincomycin was determined after hydrolysis with sulphuric acid using the 5,5'-dithio-bis(2-nitrobenzoic acid) reagent (see p. 319).[188] Other methods for the determination of lincomycin and clindamycin include acid-catalyzed hydrolysis,[21] reaction with carbazole, anthrone or 1-naphthol after the hydrolysis[32] or the use of complex formation in acidic medium with palladium (II) ions.[68] The latter method was applied to the determination of lincomycin in fermentation broth.

2. Macrocyclic Lactam Antibiotics

Macrolides. The most important representatives of this group are the *erythromycins*. These are spectrophotometrically inactive: their spectra are restricted to the weak n–π* ketone band at 289 nm.

The simplest color reactions for the determination of erythromycins are based on their hydrolysis. In the classical method of Ford et al.[79] the reagent producing yellow color is 13.5 M sulphuric acid, while alkaline hydrolysis affords a derivative absorbing at 236 nm. The latter method has found application in pharmacopoeias for monitoring the dissolution test of erythromycin tablets.[247a] In a modified version cyclic erythromycin carbonate is hydrolyzed in a buffer of pH 6.8.[112]

As basic compounds erythromycins form colored ion pairs with acidic dyes which can be extracted with chloroform and measured at different wavelengths in the visible range. The most frequently used reagents are tropaeolin 00,[96,238] bromothymol blue,[233,234,238] bromophenol blue,[200,233,238] bromocresol green,[233] bromocresol purple[55,247a] and eriochrome black.[218,242]

The formation of Schiff's bases with various aromatic aldehydes,[10,219,220] and reaction with the Nelson reagent[182] are also basis for spectrophotometric methods.

The chemical and spectral characteristics of *oleandomycin* are similar to those of erythromycin. Oleandomycin and troleandomycin also form extractable ion pairs with methyl orange[96,111] and bromothymol blue.[236] Acidic treatment followed by alkaline hydrolysis leads to a derivative with absorption maximum at 236 nm. Erythromycin does not interfere with the method based on this change.[45] The color reaction with xanthydrol in glacial acetic acid can also be mentioned.[169]

Tylosin-type antibiotics possess a conjugated dienone chromophore and hence can be determined at the absorption maximum at 279–283 nm.[205]

Ansa macrolides contain a 17–18-membered ring. *Rifamycins* belonging to this group have characteristic UV-VIS spectra originating from their substituted naphthalene or naphthoquinone moieties. As a consequence of the phenolic hydroxyls, the spectra are pH-dependent. The chromophoric groups of the various rifamycins and their absorption maxima are summarized in the paper of Rizk et al.[203] Several direct and difference spectrophotometric methods are available making use of the natural absorption of rifamycins.[43,177,178,206] Measurement at their long-wavelength maxima is prescribed by pharmacopoeias for the assay of capsules[36b,247b] and suspension.[36c]

Of the color reactions for the determination of rifamycins, ion pair formation with methylene blue,[24] complex formation with zirconium(IV), lanthanum(III), nickel(II), cerium(III), thorium(IV) and uranium(VI),[221] formation of hydrazones[179] and the use of 3,5-dichlorobenzoquinonechloroimine[197] can be mentioned.

Polyene Macrolides.   In this group of antibiotics, the macrocyclic lactam ring contains 4–7 conjugated double bonds (tetra-, penta-, hexa- and heptaenes) accompanied among others with hydroxyl, carbonyl and carboxyl groups. Their UV spectra usually contain three bands whose intensities can be calculated using an empirical formula.[165]

Nystatin which is the most important tetraene has absorption maxima at 291, 305 and 319 nm, all of them being suitable for quantitative measurements.[25,27,247c] For the determination of nystatin in the presence of its degradation products the method of orthogonal polynomials[11,15,128] and first derivative spectrophotomery[50] were used successfully. Amphotericin B and griseofulvin can be investigated similarly.[128]

Further reagents for the determination of nystatin include aluminum (III) chloride in dimethylformamide,[168] 4-aminoacetophenone applied after hydrolysis,[10] Basic Violet 3 dye to form in ion pair in the presence of chromate[228] and iron(III) chloride sulphuric acid.[47,48]

An important problem in the analytical investigation of the heptaene-type amphotericin B is the determination of the tetraene impurities (amphotericin A, nystatin). For this purpose the pharmacopoeias prescribe measurement at 282 and 304 and calculation by means of equations with two unknowns.[36d,247d] In the method of Wahbi et al.[258] the ratio of the coefficients of the Fourier and orthogonal polynomials is measured in the spectral range between 264 and 294 nm in a mixture of dimethylsulphoxide and methanol.

Natural absorption of amphotericin B at 408 is suitable for its determination even in biological samples.[229] For the determination in tissue samples the fourth derivative spectrum was used.[42] For the assay of amphotericin B tablets, extraction with dimethyl sulphoxide-methanol followed by treatment with trichloroacetic acid and measurement at 430 nm[198] is described.

3. Quinone Antibiotics

Two important groups belonging to this family are tetracyclines with four, and anthracyclines with three, linearly condensed rings. *Tetracyclines* contain two isolated chromophoric units. In acidic solution (pH 2) two intense maxima appear in their spectra which are widely used for their determination in various matrices. In the case of oxytetracycline the two maxima are at 269 and 352 nm with molar absorptivities of 19,900 and 13,500. Measurement at the latter wavelength is prescribed in the USP XXII for monitoring the dissolution tests of oxytetracyclin tablets.[247e] The wavelength of the same test for doxycycline capsules is 345 nm.[247f]

Of the main degradation pathways of tetracyclines, epimerization affects the first of the above mentioned two bands while both bands are shifted in the course of dehydration to form anhydrotetracycline.[142,240] This degradation product can be measured directly at 430 nm[63] or after extraction with chloroform.[94]

Regosz and Zuk[201] determined tetracycline in the presence of its degradation products on the basis of difference absorbances. The classical absorbance ratio method of Pernarowski (see p. 126) was also used for the determination of tetracycline[183] and oxytetracyclines[98] in the presence of degradation products.

Derivative spectrophotometry naturally greatly enhances the possibilities of using natural absorption of tetracyclines for their quantitative determination. This method can be successfully applied even in their qualitative analysis: remarkable differences can be found in the second derivative spectra of those tetracyclines, the zero-order spectra of which can hardly be distinguished from each other.[91] Of the several quantitative analytical applications the method of Wahbi et al.[259] for the determination of tetracycline and oxytetracycline is based on the effect of pH changes on their first derivative spectra. The same authors measured anhydrotetracycline and 4-epianhydrotetracycline in bulk tetracycline and its formulations using the first derivative spectrum[261] and by the combination of the first derivative spectrophotometry with the compensation technique (see p. 102) in 0.01 M hydrochloric acid at 460 nm.[257] For the same purpose Vetuschi and Ragno[255] utilized the third and fourth derivative spectra taken in ethanol or dioxane. Salinas et al.[215] determined oxytetracycline and doxycycline in various pharmaceutical preparations and also in urine on the basis of the first derivative spectra taken at pH 3.8.

As has already been mentioned, tetracyclines easily undergo dehydration reactions to form anhydrotetracyclines. The equation of this acid-catalyzed reaction is depicted in Equation 10.21. The dehydration product has an absorption maximum at 430 nm. If the reaction is completed, the measurement at this wavelength is suitable for the determination of tetracyclines. The application of this reaction in the quantitative analysis of tetracyclines was described by Levine et al.[124] in the early period of the analysis of antibiotics and it is still used today.[71]

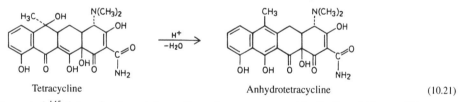

Tetracycline   Anhydrotetracycline   (10.21)

Monastero et al.[145] determined oxytetracycline after treatment with alkali at 440 nm. Chiccarelli et al.[46] described the determination of 6-demethyltetracycline in chlortetracycline by treatment with nitric acid followed by extraction with butanol.

Of the several color reactions described for the quantitative determination of tetracyclines, procedures based on metal complex formation play an important role. Sakaguchi et al.[97,212–214] investigated the complexes of tetracyclines formed with magnesium(II), cobalt(III), copper(II), uranyl, zirconium(IV) and thorium(IV) ions. These reactions have been utilized for the determination of tetracyclines in pharmaceutical preparations.[129,187] After having investigated various complexes of chlortetracycline, Oxford[171] found those formed with calcium(II) and cobalt(III) ions to be most suitable for quantitative analytical purposes. In order to find the optimum conditions for the analytical application, the iron(III) and copper(II) complexes of oxytetracycline were investigated by Hassan et al.[89] The complex formed with iron(III) ions in acidic media with its absorption maximum at 435 nm was found to be suitable by Alwarthan et al.[7] for the determination of oxytetracycline using the FIA method (see Section 9.J). Saha[209] determined the tetracycline content of capsules, tablets, injections, ointments and syrups using the complex formation in alkaline medium with copper (II) ions. Other metal ions used for the analysis of tetracyclines include titanium(III) for the determination of chlortetracycline and oxytetracycline[265] and cerium(III) in alkaline media to determine oxytetracycline[105] and chlortetracycline.[120,121] Ternary complex formation with the uranium(VI) Orange Red reagent[6] was also used for the assay of various formulations of tetracycline.

Indirect methods based on the oxidation of the phenolic tetracyclines and measurement of the reduced species of the oxidizing agent are also used. As the reagents molybdate,[110,157,243] phosphomolybdate[199] and arsenomolybdate[181,182] were used. Under different experimental conditions, complex formation rather than redox reaction takes place with the same or similar reagents such as molybdate[108] and tungstate.[107,194,254]

The wealth of functional groups on the ring system of tetracyclines form the basis for various other types of chemical reactions for their determination. The basic dimethylamino groups are suitable for ion pair formation. The method of Ottis and Malat[170] uses dioctylsulphosuccinic acid as the reagent at pH 2.1, while various chromium(III) thiocyanato and mixed thiocyanato-aniline, morpholine complex anions were used as the reagents by Ganescu et al.[83] The basis of an indirect method is the decrease of the absorbance of the crystal violet $SbCl_6^-$ complex on the effect of tetracyclines.[228] Of the classical condensation reactions of phenols coupling with diazotized sulphanilamide[87] and the oxidative condensation with the ferricyanide–4-aminoantipyrine reagent[17] were used. The iron (III)-hydroxamate method (see p. 184) measures the carboxamide moiety.[118] A kinetic micro-method for the determination of tetracycline is based on its inhibitory effect on the oxidation of Azorubine S with hydrogene peroxide catalyzed by molybdenum(VI).[224] Mainly phenol reagents have been described for the determination of the semi-synthetic tetracycline derivative, doxycycline: cobaltinitrite,[74] molybdenate,[106] modified Folin-Ciocalteu reagent,[196] ferricyanide 4-aminoantipyrine[75] and benzocaine, sulphanylamide, 4-aminoacetophenone in an azo-coupling method.[217]

Minocycline was determined as a mixed complex with the zirconium(IV) o-hydroxyhydroquinonephthalein-fluoride reagent[81] or the gallium(III)-eosin reagent.[159]

Of the anthracycline glycosides, the anticarcinogenic agents daunomycin (daunorubicin) and adriamycin (doxorubicin) are mentioned. Due to their anthraquinone structure, the spectra have several bands in the UV-VIS range (λ_{max} = 236, 255, 293, 482 and 498 nm)[269] and the compounds are red. The analytical methods for the determination of adriamycin were reviewed by Bosanquet.[35] In the method of Plumier-Vandenbussche and Dryon[184] the determination of daunorubicin is based on complex formation with iron(III) or copper(II) ions in alkaline media.

4. Amino Acid- and Peptide-Antibiotics

β-Lactam Antibiotics.   In this section the spectrophotometric methods for the determination of natural and semi-synthetic penicillins and cephalosporins are summarized.

Penicillins usually do not possess characteristic and intense UV spectra. In the spectra of derivatives containing an isolated phenyl group in the acylamido side chain, the weak benzenoid band appears around 255 nm. The intensity of the band around 270 nm is somewhat higher in the case of derivatives with a phenol ether moiety in the side chain. The latter can be used for direct analytical measurement, e.g., for the determination of phenoxymethyl penicillin in capsules at 268 nm in the method of the U.S. Pharmacopoeia.[247g]

Algebraic methods (e.g., orthogonal polynomials) can be useful in exploiting these weak absorptions to quantitative analytical purposes. Wahbi et al.[262] determined benzyl penicillin and ampicillin at pH 7 in the ranges 247–269 nm and 249–271 nm, respectively. The possibilities of pH-induced difference spectrophotometry are very limited. The method of Davidson and Stenlake[61] for the determination of ampicillin and cloxacillin is described in Section 8.B and in Table 8.1.

More important are the derivative spectrophotometric methods which greatly enhance the possibilities of spectrophotometry for the derivatives with weak bands containing fine structure (see p. 114). For example, second derivative spectrophotometry was used by Murillo et al.[162] for the simultaneous determination of amoxycillin and cefalexin and by Morelli for ampicillin and cloxacillin[152] as well as ampicillin and dicloxacillin[153] in pharmaceutical formulations. In addition to the assay of combined pharmaceutical preparations, derivative spectrophotometry enables even more special problems to be solved. For example, Di Giulio et al.[62] described the determination of several lyposome-entrapped penicillin and cephalosporin derivatives without preliminary separation in opalescent solutions. The amplitudes used in this study were as follows: ampicillin $^3D_{232,238}$; carbenicillin $^3D_{232,237}$; ceftriaxone $^2D_{255,280}$; cefaloridine $^2D_{273,260}$ and cefalexin $^2D_{277,265}$. Kovács-Hadady et al.[117] determined the impurity p-hydroxyphenoxymethyl penicillin in phenoxymethyl penicillin at pH 2 by means of fourth derivative spectra. (The determination of this impurity was also carried out on the nitroso derivative after reaction with nitrous acid.[173])

A large variety of color reactions are available for the quantitative determination of penicillins of which only a short overview is presented here.

A great proportion of these methods are based on complex formation with various metal ions. When penicillins are treated with hydroxylamine in alkaline solution, the β-lactam ring is opened and hydroxamic acid derivatives are formed. These react with iron(III) salts with the formation of red complexes, the measurement of which is the basis of the most widespread colorimetric method for penicillins.[78] The modernized versions of this classical method are used as automated procedures for the determination of penicillins.[123,126,161] The USP XXII prescribes the adaptation of the method to AutoAnalyzer;[247h] here the reagent is iron(III) nitrate and the analytical wavelength 480 nm. Iron(III)-perchlorate is also used for this purpose.[33]

The complex formation of β-lactam antibiotics with copper (II), ion was investigated in detail by Cressman et al.[51,52,166] Smith et al.[237] hydrolyzed ampicillin in acidic media in the presence of traces of copper(II). The complex formation of the hydrolysis products with copper(II) ions is the basis of the methods for the determination of carbenicillin,[113] benzyl penicillin,[267] mecillinam,[137] pivampicillin,[131,132,137] amoxicillin[65,100,191] and ampicillin.[100,137,140] The complex formation reaction of ampicillin with copper(II) in alkaline media is described by Patel et al.[179] Further methods based on the formation of copper complexes are described for penicillin sulphoxides and their esters[204] and for benzyl penicillin.[208]

The reagent of the method of Baker[18] for the determination of the impurity (degradation product) penicilloic acid in penicillins is copper(II) sulphate. The basis of the reaction is that the free penicilloic acid reduces the reagent more rapidly than the intact penicillins. The copper(I) ions formed during the reaction can be measured as the neocuproin complex at 450 nm. This method is suitable for the stability testing of penicillins.

The imidazole-mercury(II) chloride method introduced by Bundgaard and Ilver[37] is one of the most widely used methods for the spectrophotometric determination of natural and semisynthetic penicillins. An advantageous feature of this method is that it can be used even in the presence of decomposition and polymerization products.[38,39] This method was successfully adopted for the determination of pivampicillin and bacampicillin[40] as well as mecillinam and pivmecillinam.[122] The method is official in the British Pharmacopoeia, too.[36e] While using this method for the determination of azlocillin in the presence of its degradation products, it was found that it is advantageous to separate the unreacted drug by TLC prior to the spectrophotometric determination.[192] In another version of the method, imidazole was replaced by 1,2,4-triazole.[88]

Another group of spectrophotometric methods is based on the acidic hydrolysis of penicillins. In the rather complicated decomposition mechanism the reaction carried out in mildly acidic medium leads predominantly to penicillenic acids which can be measured around 325 nm, and this was used for the determination of the parent penicillins.[114,116,181,244,245] Epicillin was determined in capsules and in plasma after acidic hydrolysis.[167] The hydrolysis products can react with 4-dimethylaminobenzaldehyde[84,180] and 4-dimethylaminocinnamaldehyde[195] to form colored condensation products suitable for quantitative analytical purposes. For the assay of amoxicillin capsules, a formaldehyde-hydrochloric acid reagent was used.[239]

In strongly acidic solution at elevated temperatures, penicillins can be oxidized by ammonium vanadate and this was used by Ibrahim et al.[95] and Morelli[154] for the indirect determination of penicillins. If ammonium molybdate is used as the reagent, less drastic conditions are necessary to form the molybdenum blue.[155] Issopoulos used paramolybdate reagent based on the same reaction for the determination of ampicillin, amoxycillin and cephalosporins.[99]

The use of classical reagents is illustrated by making reference to papers describing the application of chromotropic acid,[210,235] 1-fluoro-2,4-dinitrobenzene[230,232] for the determination of penicillins, diazotized sulphanilamide or sulphanilic acid[59,60] for amoxycillin. The automated determination of β-lactam antibiotics using ninhydrin as the reagent is also described.[44]

Charge-transfer complexation using chloranil[125] leading to a stable red complex with benzyl penicillin, chloramine-T hematoxiline reagent which reacts with hydrolyzed penicillins and cephalosporins,[223] and other charge-transfer complex forming reagents[13] were also used for the spectrophotometric analysis of penicillins.

Ion-pair extraction methods can also be used. For example Gowda[86] extracted with chloroform from mildly acidic media the ion pair of penicillin G with the dye Azure A ($\varepsilon_{660\ nm}$ = 6,740). The use of other dyes such as Azure C[164] and crystal violet[29] has also been described.

In contrast to penicillins, *cephalosporins* have an intense absorption band around 260 nm ($\varepsilon \sim$ 8,000) which is in some cases used for their characterization or quantitation. For example the USP XXII prescribes the measurement of the absorbance of cephadroxyl at 264 nm.[247i] Derivative spectrophotometry has found wide application in the field of cephalosporins. El Yazbi and Barary[72] determined cefalexin, cefalotin sodium and cefadrin in the presence of their degradation products while Morelli[147–151] used the first and second derivative spectra with "zero-crossing" technique (see p. 121) for the simultaneous determination of various pairs of cephalosporins in their formulations (cefacetrile-ceftezole, cefoperazone-cefamandole, cefapirin-cefuroxime, cefaloridine-cefalotin). Parra et al.[176] determined cefradine and L-arginine in an injection formulation by measuring $^2D_{258.5}$ and $^2D_{278.5}$ for cefradine and $^2D_{197.5}$ for L-arginine.

Of the numerous methods based on chemical reactions some use the spectral changes during the hydrolytic or other opening reactions of the ring system of the cephalosporins. Papazova et al.[175] determined cefazolin after acid-catalyzed hydrolysis at the maximum of the hydrolysis product, 2-mercapto-5-methylthiadiazole at 320 nm. In the difference spectrophotometric method of Wahbi and Unterhalt[264] the basis of the measurement was the change of the absorbance at 261 nm during the base-catalyzed hydrolysis.

As was described for penicillins, here, too, the β-lactam ring undergoes hydroxylaminolysis upon treatment with hydroxylamine in alkaline medium. This reaction is shown in Equation 10.22 taking cefoxitin as the example.

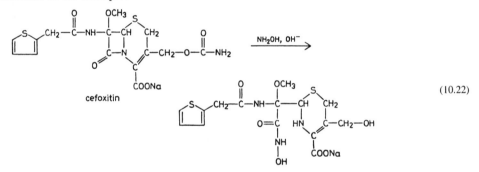

(10.22)

As a consequence of this reaction, the above mentioned band at 266 nm disappears from the spectrum; this is the basis of the difference spectrophotometric method of Pospisilová and Kubes.[186] However, the main application of this reaction is the measurement of the iron(III) complex of the resulting hydroxamic acid derivative (see Section 7.G). It is worth mentioning that nickel II) ions catalyze the reaction.[141] Many applications have been described,[144] among others for the determination of cefadroxil,[135,211] ceftriaxone,[227] aztreonam[209] and other cephalosporins.[34,249] The automated version of the method is official in USP XXII[247h] for the determination of cephalosporins.

Other methods based on complex formation are the indirect method of Mori et al.[160] for the determination of cefalexin and ampicillin where these drugs (due to their sulphur atom) are strongly bound to palladium (II) and hence decrease the absorbance of the palladium(II) o-hydroxyhydroqui-nonephthalein complex at 630 nm and several methods in which the mercury(II) imidazole reagent is used.[135,225,250]

Many other reagents have been used for the determination of cephalosporins. Some of these are ninhydrin for the determination of cefalexin,[174] cefadroxil[134] and other cephalosporins,[130,133] imidazole,[28] ammoniacal silver sulphate for the determination of cefalexin,[139] sodium nitroprus-side,[76] sodium nitrite,[252,253] the Ellman reagent,[226] 1,4-dinitrobenzene,[231] and 1-fluoro-2,4-dinitroani-line.[73] The determination of semi-synthetic cephalosporins as the dihydrolutidine derivatives is also described.[54]

Several indirect methods are also available in which the reducing properties of cephalosporins are exploited for spectrophotometric purposes. In these methods usually the reduced form of the reagent is measured. The reagents include ammonium molybdate,[3,156] paramolybdate,[99,103] molybdophosphoric acid,[101,102] mercury(II)chloride–molybdoarsenic acid[127] and ammonium vanadate (the latter in highly concentrated sulphuric acid medium).[1]

Cycloserine exhibits an absorption maximum at 219 nm ($\varepsilon = 3500$). This short-wavelength band is suitable for the solution of less difficult analytical problems only, e.g., monitoring the dissolution testing of its capsule formulation.[247j] The assay of the capsule formulation in the British Pharmacopoeia is performed after color reaction with sodium nitroprusside and absorbance measurement at 625 nm.[36f] Other reagents used for the determination of cycloserine are chloranil,[263] p-benzoquinone[70] and 9-methoxyacridine.[241] The condensation product of the latter reaction of Stewart and Yoo absorbs at 438 nm.

Gramicidins which belong to the group of peptide antibiotics do not possess characteristic UV absorption. A possibility for the colorimetric determination of gramicidin A is hydrazinolysis and measurement as the 2,4-dinitrophenyl-hydrazone.[193]

Ivashkiv[104] determined cyclic polypeptide antibiotics in the form of ion pairs with bromothymol blue extractable with chloroform at pH 7.5. This method is suitable for the determination of polymyxin B, gramicidin and tyrocidine.

Viomycin was determined in fermentation broths using the reaction with copper(II) chloride in alkaline medium.[93] Diacetylmonoxime reagent in 50% sulphuric acid was used for the determination of capreomycin.[256] Actinomycin C was determined in various liquid dosage forms on the basis of its natural absorption at 436 nm.[109] The reagent for the determination of enniatin A and valinomycin was picric acid in dichloromethane.[16]

5. Aromatic Antibiotics

Benzene Derivatives. Due to its aromatic nitro group, *chloramphenicol* which is the most important drug material in this group of antibiotics exhibits absorption maximum at 274 nm in ethanol (the spectrum can be seen in Figure 3.12). This band is exploited for the determination of chloramphenicol in pharmaceutical formulations in various pharmacopoeias[36g–j,247k] and for the determination of free chromaphenicol in its palmitate after solvent extraction.[36g,247l] The measurement of the absorbance in the 248–314 nm spectral range and the use of orthogonal polynomials enabled the intact drug to be determined in the presence of its degradation product, thus making stability assay possible.[2]

Chemical reactions are also available to transform chloramphenicol to colored derivatives for the spectrophotometric assay. The nitro group can be reduced with zinc/hydrochloric acid to the corresponding amine and after diazotization coupled with N-(1-naphthyl)-ethylenediamine.[85] The reduction can be carried out also with sodium dithionite in alkaline medium.[143] As the reagent, 1-naphthol can also be used.[138] Further possibilities to determine chloramphenicol in the reduced form are the use of guajacol[49] and sodium nitroprusside[58] reagents.

Chloramphenicol forms 2:1 complex with copper(II) in alkaline media which can be used for its determination.[119] As a carboxamide derivative it can be determined by the iron(III)-hydroxamate method, too.[190] Its hydrolyzed form reduces ammonium molybdate to molybdenum blue, and this was exploited by Morelli[146] for the assay of various formulations. The piperidine-8-hydroxyquinoline-vanadate reagent was introduced by Plourde and Braun[185] for the determination of chloramphenicol.

Condensed Aromatic Derivatives. The most important representative of this group, griseofulvin has absorption maxima at 236, 291–292 and 324 nm. The band at 291–292 nm is used for the assay of the bulk drug,[36k] the capsule and the tablet formulation.[247m] For this purpose the more selective difference spectrophotometric determination (acidic vs. neutral) is also used.[36l]

Griseofulvin reacts in alkaline media with ketone reagents such as isonicotinoyl hydrazide[4,251] and thiosemicarbazide[26] to form derivatives suitable for the assay of formulations. The sensitive method based on treatment with sulphuric acid is also worth mentioning.[4]

REFERENCES

1. Abdel-Gawad, F. M.; El-Guindi, N. M.; Ibrahim, M. M. *Egypt. J. Pharm. Sci.* 1988, *29*, 63.
2. Abdel-Hamid, M. E.; Abuirjeie, M. A. *Analyst* 1987, *112*, 895.

3. Abdel-Khalek, M. M.; Mahrous, M. S. *Talanta* 1984, *31*, 635.
4. Abdel-Moety, E. M.; Moustafa, A. A. *Zbl. Pharm. Pharmakother. Laboratoriumsdiagn.* 1988, *127*, 61.
5. Agrawal, J. K.; Halmalkar, S. G.; Vijayavargiya, R. *Microchem. J.* 1976, *21*, 202.
6. Agrawal, Y. K.; Patel, D. R. *Indian J. Pharm. Sci.* 1986, *48*, 92.
7. Alwarthan, A. A.; Al-Tamrah, S. A.; Sultan, S. M. *Analyst* 1991, *116*, 183.
8. Alykov, N. M. *Zh. Anal. Khim.* 1981, *36*, 1706.
9. Alykov, N. M. *Zh. Anal. Khim.* 1984, *39*, 1425.
10. Amer, M. M.; Habeeb, A. A. *Talanta* 1975, *22*, 605.
11. Amer, M. M.; Wahbi, A. M.; Habeeb, A. A. *J. Pharm. Pharmacol.* 1975, *27*, 377.
12. Andrásiné Bárány, Zs.; Gergelyné Zobin, Á.; Szász, Gy. *Acta Pharm. Hung.* 1979, *49*, 265.
13. Askal, H. F.; Saleh, G. A.; Omar, N. M. *Analyst* 1991, *116*, 387.
14. Astrakhanova, M. M.; Kovalenko, L. I. *Farmatsiya* 1982, *31*, 33.
15. Aszalós, A. (Ed.): *Modern Analysis of Antibiotics.* Marcel Dekker: New York, Basel, 1986.
16. Audhya, T. K.; Russel, D. W. *Anal. Lett.* 1973, *6*, 265.
17. Ayad, M.; El-Sadek, M.; Mostaffa, S. *Anal. Lett.* 1986, *19*, 2169.
18. Baker, W. L. *Analyst* 1989, *114*, 1137.
19. Bashkovich, A. P.; Prokopovich, A. V.; Alekseeva, L. E. *Antibiotiki* 1968, *13*, 695.
20. Bashkovich, A. P.; Zimnukhova, E. S. *Antibiotiki* 1972, *17*, 814.
21. Baudet, M. *J. Pharm. Belg.* 1976, *31*, 247.
22. Beltagy, Y. A.; Issa, A.; Rida, S. M. *Pharmazie* 1976, *31*, 258.
23. Bérdy, J. *ICIA Information Bulletin, No. 10.* Etud' imprim: Liege, 1972.
24. Bergisadi, N. *Eczacilik Bul.* 1979, *21*, 40.
25. Bershtein, I. Ya. *Antibiotiki* 1980, *25*, 444.
26. Bhatkar, R. G.; Madkaiker, D. C. *Indian J. Pharm. Sci.* 1980, *42*, 139.
27. Binder, K.; Muenchow, P.; Thonke, M. *Zentralbl. Pharm. Pharmakother. Laboratoriumsdiagn.* 1971, *110*, 833.
28. Bodnar, J. E.; Evans, V. G.; Mays, D. L. *J. Pharm. Sci.* 1977, *66*, 1108.
29. Bonchev, P. R.; Papazova, P. *Microchim. Acta II,* 1975, 503.
30. Bonchev, P. R.; Papazova, P.; Confino, M.; Dimova, D. *Microchim. Acta III,* 1984, 459.
31. Borowiecka, B. *J. Pharm. Pharmacol.* 1976, *28*, 353.
32. Borowiecka, B.; Chojnowski, W.; Pajchel, G. *Farm. Pol.* 1984, *40*, 147.
33. Borowiecka, B.; Pajchel, G.; Chojnowski, W. *Acta Pol. Pharm.* 1986, *43*, 47.
34. Borowiecka, B.; Pajchel, G.; Chojnowski, W. *Acta Pol. Pharm.* 1989, *46*, 463.
35. Bosanquet, A. C. *Cancer Chemother. Pharmacol.* 1986, *17*, 1.
36. *British Pharmacopoeia 1988.* Her Majesty's Stationery Office: London, 1988. a. 539; b. 647; c. 750; d. 39; e. 625; f. 632; g. 118; h. 628; i. 669; j. 927; k. 278; l. 951.
37. Bundgaard, H.; Ilver, K. *J. Pharm. Pharmacol.* 1972, *24*, 790.
38. Bundgaard, H. *J. Pharm. Pharmacol.* 1974, *26*, 385.
39. Bundgaard, H. *Arch. Pharm. Chemi. Sci. Ed.* 1977, *5*, 141.
40. Bundgaard, H. *Arch. Pharm. Chemi. Sci. Ed.* 1979, *6*, 607.
41. Burns, D. T.; Lloyd, G. H.; Watson-Walker, J. *Lab. Pract.* 1968, *17*, 448.
42. Casaccia, P.; Ladogana, A.; Xi, Y. G.; Ingrosso, L.; Pocchiari, M.; Silvestrini, M. C. *Antimicrob. Agents Chemother.* 1991, *35*, 1486.
43. Cavatorta, L.; Rangone, R.; Calanni, C.; Casa, M. *Farmaco. Ed. Prat.* 1970, *25*, 303.
44. Celletti, P.; Moretti, G. P.; Petrangeli, B. *Farmaco. Ed. Prat.* 1972, *27*, 688.
45. Chatterjee, N. R.; Badve, V. M. *Indian J. Pharm. Sci.* 1986, *48*, 18.
46. Chiccarelli, F. S. *J. Pharm. Sci.* 1968, *57*, 1046.
47. Chowdhury, M. K. R.; Goswami, R.; Chakrabarti, P. *Indian J. Exp. Biol.* 1982, *20*, 909.
48. Chowdhury, M. K. R.; Goswami, R.; Chakrabarti, P. *Indian J. Exp. Biol.* 1984, *22*, 447.
49. Cieszynski, T.; Adamski, R.; Gill, M. *Farm. Pol.* 1976, *32*, 1023.
50. Coutts, A. G. *Anal. Proc.* 1985, *22*, 111.
51. Cressman, W. A.; Sugita, E. T.; Doluisio, J. T.; Niebergall, P. J. *J. Pharm. Pharmacol.* 1966, *18*, 801.
52. Cressman, W. A.; Sugita, E. T.; Doluisio, J. T.; Niebergall, P. J. *J. Pharm. Sci.* 1969, *58*, 1471.
53. Csiba, A. *Magy. Kém. Foly.* 1979, *85*, 166.
54. Csiba, A.; Czéhné, I. *Acta Pharm. Hung.* 1979, *49*, 68.
55. Dabrowska, D.; Regosz, A.; Tamkun, L.; Kaminska, E. *Sci. Pharm.* 1984, *52*, 220.

56. Das Gupta, V. *J. Clin. Pharm. Ther.* 1988, *13*, 195.

57. Das Gupta, V.; Stewart, K. R.; Gunter, J. M. *J. Pharm. Sci.* 1983, *72*, 1470.

58. Devani, M.; Shishoo, C. J.; Doshi, K. J.; Shah, A. K. *J. Assoc. Off. Anal. Chem.* 1981, *64*, 557.

59. Dave, J. B.; Banerjee, S. K. *Indian J. Pharm. Sci.* 1986, *48*, 73.

60. Dave, J. B.; Banerjee, S. K. *Indian J. Pharm. Sci.* 1986, *48*, 112.

61. Davidson, A. G.; Stenlake, J. B. *Analyst* 1974, *99*, 476.

62. Di Giulio, A.; Maurizi, G.; Saletti, A.; Amicosante, G.; Mazzeo, P.; Oratore, A. *J. Pharm. Biomed. Anal.* 1989, *7*, 1159.

63. Dijkhuis, I. C. *Pharm. Weekbl.* 1967, *102*, 1308.

64. Divakar, T. E.; Tummura, M. K.; Sastry, C. S. P. *Indian Drugs* 1984, *22*, 28.

65. Doadrio, J. C.; Doadrio, A. *An. R. Acad. Farm.* 1983, *49*, 61.

66. Doulakas, J. *Pharm. Acta Helv.* 1976, *51*, 391.

67. Duda, E. *Anal. Biochem.* 1973, *51*, 651.

68. Egutkin, N. L.; Maidanov, V. V.; Nikitin, Yu. E. *Khim. Pharm. Zh:* 1984, *18*, 241.

69. Eisenmann, W.; Bricker, C. E. *Anal. Chem.* 1949, *21*, 1507.

70. El-Sayed, L.; Mohamed, Z. H.; Wahbi, A. A. *Analyst* 1986, *111*, 915.

71. Elsayed, M. A.; Barary, M.; Mahgoub, H. *Anal. Lett.* 1985, *18*, 1357.

72. El-Yazbi, F. A.; Barary, M. H. *Anal. Lett.* 1985, *18*, 629.

73. Emmanuel, J.; Dias, E. S. *Indian Drugs* 1987, *24*, 357.

74. Emmanuel, J.; Naik, P. N. *Indian Drugs* 1983, *20*, 389.

75. Emmanuel, J.; Shetty, A. R. *Indian Drugs* 1984, *21*, 167.

76. Fogg, A. G.; Abdalla, M. A. *J. Pharm. Biomed. Anal.* 1985, *3*, 315.

77. Fooks, J. R.; McGliveray, I. J.; Strickland, R. D. *J. Pharm. Sci.* 1968, *57*, 314.

78. Ford, J. H. *Anal. Chem.* 1947, *19*, 1004.

79. Ford, J. H.; Prescott, G. C.; Ninman, J. W.; Caban, E. L. *Anal. Chem.* 1953, *25*, 1195.

80. Fujita, Y.; Mori, I.; Kitano, S. *Chem. Pharm. Bull.* 1983, *31*, 1289.

81. Fujita, Y.; Mori, I.; Kitano, S. *Chem. Pharm. Bull.* 1983, *31*, 4016.

82. Fujita, Y.; Mori, I.; Kitano, S. *Chem. Pharm. Bull.* 1984, *32*, 1214.

83. Ganescu, I.; Pleniceanu, M.; Preda, M.; Várhelyi, Cs. *Chem. Anal. (Warsaw)* 1978, *23*, 891.

84. Gasper, T.; Kolsek, J.; Perpar, M. *Z. Anal. Chem.* 1957, *154*, 98.

85. Glombitza, K. -W. *Pharm. Zentralhalle Dtl.* 1967, *106*, 441.

86. Gowda, A. T. *Z. Anal. Chem.* 1989, *334*, 710.

87. Gupta, R. P.; Tiwari, O. P.; Yadav, B. N.; Srivastava, A. A. *Chem. Era* 1980, *16*, 161.

88. Haginaka, J.; Wakai, J.; Yasuda, H.; Uno, T. *Anal. Sci.* 1985, *1*, 73.

89. Hassan, S. S. M.; Amer, M. M.; Ahmed, S. A. *Microchim. Acta III* 1984, 165.

90. Heyes, W. F. *Analytical Profiles of Drug Substances*, 1979, *8*, 399.

91. Hon, P.-K.; Fung, W.-K. *Analyst* 1991, *116*, 751.

92. Hoodless, R. A. *Analyst* 1966, *91*, 333.

93. Horodecka, M.; Oskierko, H.; Roslik, D.; Kowszyk-Gindifer, Z. *Chem. Anal. (Warsaw)* 1967, *12*, 827.

94. Hrdy, O.; Beyrodtova, H. *Cesk. Farm.* 1974, *23*, 122.

95. Ibrahim, E. A.; Beltagy, Y. A.; Abd El-Khalek, M. M. *Talanta* 1977, *24*, 328.

96. Ibrahim, E. A.; Beltagy, Y. A.; Issa, A. S. *Pharmazie* 1972, *27*, 651.

97. Ishidate, M.; Sakaguchi, T. *Pharm. Bull. Japan* 1955, *3*, 147.

98. Issar, M. M.; Singh, H. *Indian J. Technol.* 1982, *20*, 198.

99. Issopoulos, P. B. *J. Pharm. Biomed. Anal.* 1988, *6*, 97.

100. Issopoulos, P. B. *J. Pharm. Biomed. Anal.* 1988, *6*, 321.

101. Issopoulos, P. B. *Analyst* 1988, *113*, 1083.

102. Issopoulos, P. B. *Analyst* 1989, *114*, 237.

103. Issopoulos, P. B. *J. Pharm. Biomed. Anal.* 1989, *7*, 619.

104. Ivashkiv, E. *J. Pharm. Sci.* 1975, *64*, 1401.

105. Janik, B.; Holiat, D. *Acta Pol. Pharm.* 1972, *29*, 169.

106. Jelikic-Stankov, M.; Veselinovic, D. *Pharmazie* 1988, *43*, 49.

107. Jelikic-Stankov, M.; Veselinovic, D. *Analyst* 1989, *114*, 719.

108. Jelikic-Stankov, M.; Veselinovic, D.; Malesev, D.; Radovic, Z. *J. Pharm. Biomed. Anal.* 1989, *7*, 1565.

109. Kachani, Z. F. C. *Z. Analyt. Chem.* 1965, *213*, 427.

110. Kakemi, K.; Uno, T.; Samejima, M. *J. Pharm. Soc. Japan* 1955, *75*, 194, 1955, *75*, 970.

111. Kazenko, O.; Sorenson, O. J.; Wolf, L. M.; Dill, W. A.; Galbraith, M.; Glazko, A. J. *Antibiot. Chemother.* 1957, *7*, 410.

112. Knapczyk, J.; Siedlanowska-Krowczynska, H. *Acta Pol. Pharm.* 1986, *43*, 481.

113. Kolusheva, A.; Papazova, P. *Antibiotiki* 1971, *16*, 1064.

114. Koprivc, L.; Polla, E.; Hranilovic, J. *Acta Pharm. Suec.* 1976, *13*, 421.

115. Korchagin, V. B.; Satarova, D. E.; Kochetkova, E. V.; Vakulenko, N. A. *Antibiotiki* 1975, *20*, 116.

116. Korobkin, V. M.; Korchagin, V. B. *Antibiotiki* 1978, *23*, 178.

117. Kovács-Hadady, K.; Kiss, I. T.; Kiss, M.; Barna-Katona, K. *Analyst* 1988, *113*, 569.

118. Krasnov, E. A.; Fominykh, V. P. *Farmatsiya* 1982, *31*, 75.

119. Krzek, J.; Lechniak, A. *Chem. Anal. (Warsaw)* 1977, *22*, 755.

120. Labza, D. *Acta Pol. Pharm.* 1978, *35*, 335.

121. Labza, D.; Janik, B. *Acta Pol. Pharm.* 1977, *34*, 63.

122. Larsen, C.; Bundgaard, H. *Arch. Pharm. Chemi. Sci. Ed.* 1977, *5*, 1.

123. LeMoigne, B.; Barthes, D.; Fourtillan, J. B.; Hazane, C. *Ann. Farm. Fr.* 1978, *36*, 381.

124. Levine, J.; Garlock, E. D.; Fishbach, H. *J. Amer. Pharmac. Assoc. Sci. Ed.* 1949, *38*, 473.

125. Li, Z.; Wu, L. *Shenyang Yaoxueyuan Xuebao* 1988, *5*, 94.

126. Lin, S. -L.; Sutton, V. J.; Quraishi, M. *J. Assoc. Off. Anal. Chem.* 1979, *62*, 989.

127. Luebbe, C.; Shem, Y.; Demain, A. L. *Appl. Biochem. Biotechnol.* 1986, *12*, 31.

128. Lupashevskaya, D. P.; Bershtein, I. Ya.; Raigorodskaya, V. Ya.; Birman, G. Sh. *Antibiotiki* 1977, *22*, 33.

129. Mahgoub, A. E.; Khairy, E. M.; Kasem, A. *J. Pharm. Sci.* 1974, *63*, 1451.

130. Mahrous, M. S.; Abdel-Khalek, M. M. *Analyst* 1984, *109*, 611.

131. Marchi, E.; Mascellani, G.; Boccali, D. *Farmaco Ed. Prat.* 1973, *28*, 523.

132. Marchi, E.; Mascellani, G.; Boccali, D. *J. Pharm. Sci.* 1974, *63*, 1299.

133. Marelli, L. P. *J. Pharm. Sci.* 1968, *57*, 2172.

134. Marini, D.; Di Cocco, M. E. *Rass. Chim.* 1980, *32*, 231.

135. Marini, D.; Pascucci, E. *Boll. Chim. Farm.* 1980, *119*, 52.

136. Marques, M. R. C.; Hackmann, E. R. N.; Saito, T. *Anal. Lett.* 1989, *22*, 621.

137. Martinez, P. J.; Gutiérrez, P.; Martinez, M. I.; Thomas, J. *Cienc. Ind. Farm.* 1985, *4*, 325.

138. Masterson, D. S. *J. Pharm. Sci.* 1968, *57*, 305.

139. Matousova, O.; Peterkova, M. *Cesk. Farm.* 1979, *28*, 382.

140. Matousova, O.; Peterkova, M.; Kakac, B. *Cesk. Farm.* 1983, *32*, 153.

141. Mays, D. L.; Bangert, F. K.; Cantrell, W. C.; Evans, W. G. *Anal. Chem.* 1975, *47*, 2229.

142. McCormick, J. R. D.; Fox, S. M.; Smith, L. L.; Bitler, B. A.; Reichental, J.; Origoni, V. E.; Muller, W. H.; Winterbottom, R.; Doerchuk, J. *J. Am. Chem. Soc.* 1957, *79*, 2849.

143. Miram, R. *Pharm. Prax. Berl.* 1972, 108.

144. Mohamed, M. E.; Abounassif, M. A.; Al-Khamees, H. A.; Kandil, H. A.; Aboul-Enein, H. Y. *J. Pharm. Belg.* 1988, *43*, 429.

145. Monastero, F.; Means, J. A.; Grenfell, T. C.; Hedger, F. M. *J. Amer. Pharm. Assoc. Sci. Ed.* 1951, *40*, 241.

146. Morelli, B. *J. Pharm. Biomed. Anal.* 1987, *5*, 577.

147. Morelli, B. *Anal. Lett.* 1988, *21*, 43.

148. Morelli, B. *Anal. Lett.* 1988, *21*, 759.

149. Morelli, B. *Analyst* 1988, *113*, 1077.

150. Morelli, B. *J. Pharm. Biomed. Anal.* 1988, *6*, 199.

151. Morelli, B. *J. Pharm. Sci.* 1988, *77*, 615.

152. Morelli, B. *Anal. Chim. Acta* 1988, *209*, 175.

153. Morelli, B. *J. Pharm. Sci.* 1988, *77*, 1042.

154. Morelli, B. *Anal. Lett.* 1987, *20*, 141.

155. Morelli, B.; Mariani, M.; Gesmundo, M. *Anal. Lett.* 1987, *20*, 1429.

156. Morelli, B.; Peluso, P. *Anal. Lett.* 1985, *18*, 1113.

157. Morelli, B.; Peluso, P. *Anal. Lett.* 1985, *18*, 1865.

158. Mori, I.; Fujita, Y.; Fujita, K.; Tanaka, T.; Kawabe, H.; Koshiyama, Y. *Bull. Chem. Soc. Jpn.* 1986, *59*, 2585.

159. Mori, I.; Fujita, Y.; Kawabe, H.; Fujita, K.; Tanaka, T.; Kishimoto, A. *Analyst* 1986, *111*, 1409.

160. Mori, I.; Fujita, Y.; Sakaguchi, K. *Chem. Pharm. Bull.* 1982, *30*, 2599.

161. Munson, J. W.; Papadimitriou, D.; DeLuca, P. P. *J. Pharm. Sci.* 1979, *68*, 1333.

162. Murillo, J. A.; Rodriguez, J.; Lemus, J. M.; Alanon, A. *Analyst* 1990, *115*, 1117.

163. Natarjan, R.; Tayal, J. N. *J. Pharm. Pharmacol.* 1957, *9*, 326.

164. Nayak, A. N.; Ramappa, P. G.; Yathirajan, H. S.; Manjappa, S. *Anal. Chim. Acta* 1982, *134*, 411.

165. Nayler, P.; Whiting, M. C. *J. Chem. Soc.* 1955, 3038.

166. Niebergall, P. J.; Husar, D. A.; Cressman, W. A.; Sugita, E. T.; Doluisio, J. T. *J. Pharm. Pharmacol.* 1966, *18*, 729.

167. Nunez-Vergara, L. J.; Roa, A.; Squella, J. A.; Gonzales-Barbagelata, R. V. *J. Assoc. Off. Anal. Chem.* 1986, *69*, 188.

168. Ochab, S. *Dissnes Pharm. Warsz.* 1965, *17*, 323.

169. Ochab, S.; Borowiecka, B. *Dissnes Pharm. Warsz.* 1967, *19*, 423.

170. Ottis, M.; Malat, M. *Cesk. Pharm.* 1986, *35*, 119.

171. Oxford, A. E. *Nature* 1953, *172*, 395.

172. Palmer, A. *Anal. Proc.* 1985, *22*, 139.

173. Palotás, B.; Szászhegyessy, V.; Horváth, M. *Anal. Biochem.* 1989, *179*, 288.

174. Papazova, P.; Bonchev, P. R.; Kacharova, M. *Microchim. Acta II* 1976, 185.

175. Papazova, P.; Bonchev, P. R.; Kacharova, M. *Talanta* 1983, *30*, 51.

176. Parra, A.; Dolores Gómez, M.; Ródenas, V.; Garcia-Villanova, J.; López, M. L. *J. Pharm. Biomed. Anal.* 1992, *10*, 525.

177. Pasqualucci, C. R.; Vigevani, A.; Radaelli, P.; Gallo, G. G. *J. Pharm. Sci.* 1970, *59*, 685.

178. Pasqualucci, C. R.; Vigevani, A.; Radaelli, P.; Maggi, N. *Farmaco Ed. Prat.* 1969, *24*, 46.

179. Patel, A. A.; Gandhi, T. P.; Patel, P. R.; Patel, M. R.; Patel, V. C. *Indian J. Pharm. Sci.* 1978, *40*, 64.

180. Patel, A. A.; Gandhi, T. P.; Patel, P. R.; Patel, M. R.; Patel, V. C. *Indian J. Pharm. Sci.* 1979, *41*, 40, 229.

181. Pawelczyk, E.; Hermann, T.; Smilowski, B.; Borowitz, P. *Acta Pol. Pharm.* 1979, *36*, 315.

182. Perlman, D. *Science* 1959, *118*, 628.

183. Pernarowski, M.; Searl, R. O.; Naylor, J. *J. Pharm. Sci.* 1969, *58*, 470.

184. Plumier-Vandenbussche, D.; Dryon, L. *Farm. Tijdschr. Belg.* 1976, *53*, 133.

185. Plourde, J. R.; Braun, J. *J. Pharm. Belg.* 1971, *26*, 591.

186. Pospisilová, B.; Kubes, J. *Cesk. Farm.* 1989, *38*, 136.

187. Pradeau, D.; Dauphin, C.; Hamon, M. *Ann. Pharm. Fr.* 1979, *37*, 369.

188. Prescott, G. C. *J. Pharm. Sci.* 1966, *55*, 423.

189. Prodhan, A.; Majumdar, M. K. *Indian Drugs* 1982, *19*, 358.

190. Przyborowski, L. *Acta Pol. Pharm.* 1976, *33*, 223.

191. Przyborowski, L.; Hopkala, H.; Misztal, G. *Farm. Pol.* 1983, *39*, 527.

192. Pujol, M.; Girona, V.; de Bolós, J.; Castillo, M. *Pharm. Acta Helv.* 1987, *62*, 269.

193. Ramachandran, L. K. *Indian J. Biochem.* 1967, *4*, 137.

194. Rao, R. G.; Avadhanulu, A. B.; Giridhar, R. *Indian Drugs* 1989, *26*, 298.

195. Rao, G. R.; Kanjilal, G.; Mohan, R. *Indian Drugs* 1982, *19*, 326.

196. Rao, G. R.; Kanjilal, G.; Mohan, R. *Indian J. Pharm. Sci.* 1980, *42*, 129.

197. Rao, G. R.; Murty, S. S. N.; Rao, V. *Indian Drugs* 1985, *22*, 484.

198. Rao, L. N.; Rao, M. N. *Indian Drugs* 1985, *22*, 548.

199. Ravin, L. J.; James, A. E. *J. Amer. Pharm. Assoc. Sci. Ed.* 1955, *44*, 215.

200. Regosz, A.; Dabrowska, D.; Babilec, H.; Nestoruk, H. *Sci. Pharm.* 1983, *50*, 17.

201. Regosz, A.; Zuk, G. *Pharmazie* 1980, *35*, 24.

202. Rizk, M.; Younis, F. *Anal. Lett.* 1984, *17*, 1803.

203. Rizk, M.; Walash, M. I.; Abou-Ouf, A. A.; Belal, F. *Anal. Lett.* 1981, *14*, 1407.

204. Rogic, D.; Herak, J. J. *Acta Pharm. Jugosl.* 1987, *37*, 135.

205. Roudebush, H. E. *Ann. N. Y. Acad. Sci.* 1965, *120*, 582.

206. Ruczaj, D.; Szawlowska, H.; Lendzion, B. *Chem. Anal. (Warsaw)* 1976, *21*, 1371.

207. Ryan, J. A. *J. Pharm. Sci.* 1984, *73*, 1301.

208. Saha, U. *Analyst* 1986, *111*, 1179.

209. Saha, U. *J. Assoc. Off. Anal. Chem.* 1987, *70*, 686; 1989, *72*, 242.

210. Saha, U.; Roay, D. K. *J. Inst. Chem.* 1982, *54*, 92.

211. Said, S. A.; El Bayoumy, T. *J. Drug. Res.* 1987, *17*, 147.

212. Sakaguchi, T.; Taguchi, K. *Pharm. Bull. Japan* 1955, *3*, 166, 303.

213. Sakaguchi, T.; Taguchi, K. *Japan Analyst* 1957, *6*, 787.

214. Sakaguchi, T.; Taguchi, K.; Suzuki, A. *Japan Analyst* 1957, *6*, 782.

215. Salinas, F.; Nevado, J. J. B.; Espinosa, A. *Analyst* 1989, *114*, 1141.

216. Sampath, S. S.; Robinson, D. H. *J. Pharm. Sci.* 1990, *79*, 428.

217. Sane, R. T.; Kamat, R. P.; Gijare, A. S.; Deshpande, D. S. *Indian Drugs* 1983, *20*, 331.

218. Sane, R. T.; Vaidya, U. M.; Nayak, V. G.; Dhamankar, P. Z.; Joshi, S. K.; Doshi, V. J.; Sawant, S. V.; Malkar, V. B.; Pandit, U. R.; Sathe, A. Z.; Jukar, S.; Nadkarni, A. D. *Indian Drugs* 1982, *19*, 398.

219. Sanghavi, N. M.; Chandramohan, H. S. *Can. J. Pharm. Sci.* 1965, *10*, 59.

220. Sanghavi, N. M.; Katdare, A. V. *Indian J. Pharm. Sci.* 1973, *35*, 87.

221. Sastry, C. S. P.; Divakar, T. E.; Prasad, U. V. *Indian Drugs* 1985, *22*, 604.

222. Sastry, C. S. P.; Divakar, T. E.; Prasad, U. V. *Talanta* 1986, *33*, 164.

223. Sastry, C. S. P.; Satyanarayana, P.; Rao, A. R.; Singh, N. R. P. *Microchim. Acta I* 1989, 17.

224. Sekheta, M. A. F.; Milovanovic, G. A.; Janjic, T. J. *Glas. Hem. Drus. Beograd* 1979, *44*, 447.

225. Sengun, F. I.; Fedai, I. *Chim. Acta Turc.* 1985, *13*, 205.

226. Sengun, F. I.; Fedai, I. *Talanta* 1986, *33*, 366.

227. Sengun, F. I.; Ulas, K. *Talanta* 1986, *33*, 363.

228. Sergeev, G. M.; Dorofeeva, I. V.; Stepanova, L. G. *Zh. Anal. Khim.* 1986, *41*, 1694.

229. Shihabi, Z. K.; Wasilauskad, B. L.; Peacock, J. E. *Therap. Drug Monit.* 1988, *10*, 486.

230. Shingbal, D. M.; Agni, R. M. *Indian Drugs* 1985, *22*, 383.

231. Shingbal, D. M.; Bhangle, S. R. *Indian Drugs* 1987, *24*, 367.

232. Shingbal, D. M.; Naik, R. R. *Indian Drugs* 1985, *22*, 271.

233. Shirokova, L. M.; Charykov, A. K. *Zh. Anal. Khim.* 1981, *36*, 547.

234. Siedlanowska-Krowczynska, H.; Knapczyk, J. *Acta Pol. Pharm.* 1986, *43*, 140.

235. Singh, M. P.; Basu, N.; Roy, D. K.; Mandal, S. K. *Indian J. Exp. Biol.* 1984, *22*, 39.

236. Slavin, A. A.; Etingov, E. D.; Prokopovich, A. V. Brunshtein, I. Z.; Bashkovich, A. P.; Pyshnyi, S. P.; Berhstein, E. M. *Antibiotiki* 1981, *26*, 253.

237. Smith, J. W. G.; De Grey, G. E.; Patel, V. J. *Analyst* 1967, *92*, 247.

238. Smith, R. V.; Harris, R. G.; Sanchez, E.; Maness, D. D.; Martin, A. *Microchem. J.* 1977, *22*, 168.

239. Squella, J. A.; Nunez-Vergara, L. J.; Aros, M. *J. Assoc. Off. Anal. Chem.* 1980, *63*, 1049.

240. Stephens, C. R.; Conover, L. H.; Pasternack, R.; Hochstein, F. A.; Moreland, W. T.; Regna, P. P.; Pilgrim, F. J.; Brunigs, K. J.; Woodward, R. B. *J. Am. Chem. Soc.* 1954, *76*, 3568.

241. Stewart, J. T.; Yoo. G. S. *J. Pharm. Sci.* 1988, *77*, 452.

242. Sulkowska, J.; Budych, A.; Staroscik, R. *Chem. Anal. (Warsaw)* 1985, *30*, 387.

243. Sultan, S. M. *Analyst* 1986, *111*, 97.

244. Szász, Gy.; Budváriné Bárány, Zs.; Gergely, A.; Dávid, A.; Tóth, A. *Acta Pharm. Hung.* 1976, *46*, 248.

245. Szawlowska, H.; Borowitz, P. *Chem. Anal. (Warsaw)* 1978, *23*, 821.

246. Tarbutton, P. *Am. J. Hosp. Pharm.* 1987, *44*, 115.

247. *The United States Pharmacopoeia XXII.* USP Convention Inc.: Rockville, 1990. a. 521; b. 1227; c. 972; d. 86; e. 998, f. 480; g. 646; h. 1474; i. 240; j. 371; k. 271; l. 277; m. 616.

248. Török, I.; Paál, T. *J. Chromatogr.* 1982, *246*, 356.

249. Ulas, K.; Sengun, F. I. *Chim. Acta Turc.* 1985, *13*, 491.

250. Ulas, K.; Sengun, F. I. *Arch. Pharm. Chemi. Sci. Ed.* 1987, *15*, 91.

251. Unterman, W. H. *Rev. Chim. (Bucharest)* 1965, *16*, 286.

252. Uri, J. V.; Jain, T. C. *J. Antibiot.* 1986, *39*, 669.

253. Uri, J. V. *Acta Chim. Hung.* 1991, *128*, 89.

254. Veselinovic, D.; Jelikic-Stankov, M. *Pharmazie* 1987, *42*, 199.

255. Vetuschi, C.; Ragno, G. *Farmaco* 1990, *45*, 757.

256. Voigt, R.; Maa Bared, A. G. *Pharmazie* 1970, *25*, 471.

257. Wahbi, A. A. M.; Al-Khamees, H. A.; Youssef, A. M. A. *J. Assoc. Off. Anal. Chem.* 1988, *71*, 768.

258. Wahbi, A. A. M.; Al-Khamees, H. A.; Youssef, A. M. A. *Analyst* 1988, *113*, 563.

259. Wahbi, A. A. M.; Barary, M.; Mahgoub, H.; Elsayed, M. A. *J. Assoc. Off. Anal. Chem.* 1985, *68*, 1045.

260. Wahbi, A. M.; Abdine, H.; Sadek, A. R. H. *Pharmazie* 1977, *32*, 690.

261. Wahbi, A. M.; Abounassif, M. A.; Al-Kahtani, H. M. G. *Analyst* 1986, *111*, 777.

262. Wahbi, A. M.; Belal, S.; Bedair, M.; Abdine, H. *Pharmazie* 1974, *37*, 641.

263. Wahbi, A. M.; Mohamed, M. E.; Abounassif, M.; Gad-Kariem. E. *Anal. Lett.* 1985, *18*, 261.

264. Wahbi, A. M.; Unterhalt, B. *Fresenius' Z. Anal. Chem.* 1977, *284*, 128.

265. Weygand, F.; Csendes, E. *Ber. Dtsch. Chem. Ges.* 1952, *85*, 45.

266. Yassa, D. A.; Gad ElRub, L. N. *Pharmazie* 1976, *31*, 257.

267. Yusim, R. S.; Nys, P. S.; Kolygina, T. S. *Antibiotiki* 1971, *16*, 39.

268. Zalkovski, S. H.; Sinkula, A. A.; Valvani, S. C. *Physical-Chemical Properties of Drugs,* Marcel Dekker: New York, 1980.

269. Zechmeister, L. (Ed.): *Progress in the Chemistry of Organic Natural Products,* Springer Verlag: Wien, 1963.

O. VITAMINS[37,108]

1. Vitamin A

The spectrum of vitamin A_1 representing the *all-trans* pentaenes is shown as Figure 3.5. The intense absorption band at 325 nm is eminently suitable for the spectrophotometric determination of vitamin A. In the course of its determination in samples of natural origin, mainly in cod-liver oil which is its main natural source, the absorption band is accompanied by background absorption. This must be corrected in order to eliminate the positive error it causes. It is interesting to note that the classical background correction method (base-line method; see Section 4.C.2) which is still in use was developed by Morton and Stubbs just for the determination of vitamin A in cod-liver oil.[69,70] The method based on saponification, extraction, evaporation and dissolution in 2-propanol to measure the absorbance at three wavelengths is adopted in several pharmacopoeias including the latest revision of the United States Pharmacopoeia.[112a] The corrected absorbance which is the basis of the calculation of the concentration of vitamin A is given by Equation 10.23.

$$A_{325,corr} = 6.815 \cdot A_{325} - 2.555 \cdot A_{310} - 4.260 \cdot A_{334} \tag{10.23}$$

Other methods for the elimination of the spectral background are the use of derivative spectrophotometry e.g., for the assay of capsules by Wahbi et al.[123] by measuring $^1D_{306,348}$, or for the determination of vitamin A palmitate in the presence of its degradation products by Abdel Hamid[1] as well as preliminary chromatographic separation which was used for the assay of ointments containing vitamin A.[90]

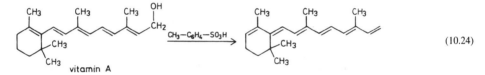

$$\tag{10.24}$$

Of the methods based on chemical reactions, first, a stoichiometric reaction is mentioned. As seen in Equation 10.24 on the catalytic effect of strong acids in benzene medium, vitamin A is dehydrated to anhydrovitamin A which can be measured after the neutralization of the catalyst at 399 nm.[20]

Of much greater importance is the classical Carr-Price method where vitamin A is reacted with antimony trichloride in chloroform medium to form a blue complex ($\lambda_{max} = 620$ nm).[23] The method is very sensitive; the most suitable concentration of vitamin A in the test solution is 0.5–1 μg/ml. A disadvantage of the original version of the method is, however, the poor stability of the color: the absorbance has to be measured within 5–10 seconds after the addition of the reagent. Of the numerous attempts to stabilize the color, the method of Blake and Moran[16] is mentioned: using the mixture of 1,2-dichloroethane and 2,3-dichloropropanol as the solvent, the absorption maximum of the stable product is 555 nm. Only the free primary hydroxyl and the corresponding aldehyde derivatives give positive reaction. Carboxylic acid ester derivatives which do not react in the Carr-Price procedure can be determined with an extremely sensitive method after treatment with 85% phosphoric acid at 453 nm.[103]

The sensitivity of the indirect methods based on the reducing properties of vitamin A using phosphomolybdic acid, phosphotungstic acid and silicotungstic acid reagents is much lower.[98]

2. B Vitamins

Vitamin B_1 (Thiamine). In the ultraviolet spectrum of thiamine in neutral solution, two bands can be found which are suitable for analytical purposes: $\varepsilon_{232\ nm} = 11,640$; $\varepsilon_{266\ nm} = 8,600$. As a consequence of the protonation of the aminopyrimidine moiety in strongly acidic solution, only one broad band appears ($\varepsilon_{246\ nm}$ = 14,300). Measurement at the above wavelengths is widely used for the assay of pharmaceutical formulations containing vitamin B_1. In the case of the analysis of vitamin B complexes, the effect of vitamins B_2 and B_6 can be eliminated by using suitable wavelengths and correction factors,[72] the ratio method of Pernarowsky (see Section 4.F.1), simultaneous equations[10,51] and preliminary chromatographic separation.[52] Using these methods data can, of course, be obtained for the determination of the other components, too.

Of the more modern methods, that of the orthogonal polynomials (Section 4.C.3) was applied to vitamin B_1 by Wahbi et al.[122] by measuring in the spectrum range of 226–292 nm at 6 nm intervals in 0.1 M hydrochloric acid, the background absorbtion caused by the degradation products of vitamin B_1 can be eliminated. Elsayed et al.[30] described a pH-induced difference spectrophotometric method for the determination of vitamin B_1 in the presence of tolbutamide and pyridoxine. The ΔA of a solution in 0.1 M hydrochloric acid was measured against a solution in 0.01 M sodium hydroxide. Derivative spectrophotometry also enables selective determinations. Abdel-Hamid et al.[2] used the negative amplitude $^2D_{254}$ for the selective determination in the presence of degradation products.

Several methods are based on chemical reactions in the analytical chemistry of vitamin B_1. Of these, the fluorimetric procedure based on its oxidation to thiochrome is widely used and probably as a result, the importance of the spectrophotometric methods is less pronounced. Ion-pair extraction methods merit mention in the first place. Das Gupta and Cadwallader[26] used bromothymol blue as the ion-pairing reagent at pH 6.6. After extraction with chloroform the absorbance of the ion pair was determined at 420 nm. The great importance of this method is because as the strongest base of all vitamins, thiamine can be selectively extracted thus creating the basis for its selective determination in multivitamin preparations. The automated version of this method is also described.[82] In another ion-pairing method Mitsui and Fujimura[68] transformed the primary amino group of thiamine to the 2,4-dinitrophenyl derivative prior to the extraction in order to increase the selectivity of the method. The dye was tropeolin 00 and the extraction was carried out with chloroform at pH 3.29. The absorbance was measured at 543 nm after the acidification of the chloroform phase with 10% methanolic hydrochloric acid.

Of the other methods coupling with diazotized aromatic amines such as 6-aminothymol,[41] phenylhydrazine-4-sulphonic acid,[120] formation of charge-transfer complex with chloranil,[6] coupling with 4-aminophenol and measurement at 429 nm[97] can be mentioned. Finally, as a curiosity the kinetic method of Ciesielski[25] merits attention. The basis of the method is the catalytic activity of thiamine on the iodine-azide reaction. The consumption of iodine is measured spectrophotometrically; this is proportional to the concentration of thiamine.

Of the pharmaceutically important derivatives of thiamine, its disulphide and propyl disulphide were determined by Krishnan et al.[57] using 2,6-dichloro-p-benzoquinone-4-chloroimine as the reagent and measuring the absorbances of the derivatives (see p. 179) at 460 and 445 nm, respectively.

Vitamin B_2 (Riboflavin).

The ultraviolet-visible spectrum of the yellow riboflavin affords excellent possibilities for its spectrophotometric determination even in combined vitamin preparations. The positions and intensities of the three bands in aqueous medium suitable for quantitative analytical measurement are as follows: $\varepsilon_{267\ nm}$ = 32,900, ε_{372} = 10,400 and ε_{445} = 12,200. If the most important factor in planning the assay method is sensitivity, then the first, most intense band can be used;[52] if, however, the selectivity is of greater importance, the last band causing the yellow color of the material can be recommended. Measurement at this wavelength guarantees selectivity in the presence of any kind of colorless compounds. For example this wavelength was used by Burger[21] for the determination of vitamin B_2 in the presence of vitamin B_1, B_6, nicotinamide and vitamin C. The method of Ahmad and Rapson[5] for the determination of riboflavin in the presence of its photodegradation products was based on the second and third band with algebraic background correction.

Since the method based on natural absorption can solve a large variety of problems in the analytical chemistry of riboflavin (and further wide possibilities are available based on native fluorescence), spectrophotometric methods based on chemical reactions are of secondary importance only. Only one method is mentioned: periodic acid cleaves the side chain and the formaldehyde thus formed can be determined either by the classical chromotropic acid method or using 3-methyl-benzthiazolin-2-one hydrazone reagent.[99]

Vitamin B$_6$ (Pyridoxine).   The 3-hydroxypyridine-type spectrum of pyridoxine is strongly pH-dependent. Below pH 2, the protonated form is present ($\varepsilon_{290\ nm} = 7,470$) while between pH 6 and 8 the zwitterionic form ($\varepsilon_{254\ nm} = 3,090$ and $\varepsilon_{324} = 5,830$). The analytical application of the spectra taken in strongly alkaline media where pyridoxine is present in the anionic form is rather problematic because of its instability under these circumstances.

Of the above possibilities, the assay of pyridoxine tablet is prescribed in the British Pharmacopoeia in 0.1 M hydrochloric acid.[18a] Jan et al.[51] and Sethi et al.[104] also measure in this solution in the course of the determination of pyridoxine in combinations using the method of simultaneous equations. This solvent is also used for the absorbance measurement after preliminary TLC separation of pyridoxine.[52] Measurement in neutral medium at 324 nm is also used.[21,72] In order to increase the accuracy of the measurement, Burger[21] used the standard addition method. The interference from vitamins B$_2$ and K was eliminated by a correction formula.

The pH-dependence of the spectrum of pyridoxine is the basis of selective spectrophotometric methods. The most frequently used variant is the measurement of the neutral solution against that in 0.1 M hydrochloric acid.[37] In the method of Elsayed et al.[30] for the determination of pyridoxine in the presence of thiamine, the test solution was in 0.01 M sodium hydroxide and the reference in 0.1 N sulphuric acid. In order to eliminate the interference of thiamine, the basis for the calculation was ($A_{310} - A_{300}$). The use of derivative spectrophotometry ($^1D_{296}$) measured in 0.1 N sulphuric acid enables pyridoxine to be determined in the presence of meclozine.[3]

The majority of color reactions adopted for the quantitative determination of pyridoxine belong to the well known family of phenol reactions. Nirmalchandar and Balasubramanian[77] used diazotized 4-nitroaniline as the reagent and measured the resulting azo-dye as the mercury(II) complex. The complex formation causes bathochromic and hyperchromic shift, and in addition it stabilizes the color ($\varepsilon_{530\ nm} = 46,000$). The interference of ascorbic acid can be eliminated by preliminary precipitation as the lead(II) salt. Other vitamins do not interfere and hence this method is suitable for the selective determination of pyridoxine in multi-vitamin preparations. Another method by the same workers[78] is based on oxidative coupling with 4-amino-antipyrine (see Equation 10.25). This results in very stable yellow color but lower sensitivity ($\varepsilon_{420\ nm} = 5,410$). Other vitamins interfere and for this reason pyridoxine has to be separated by ion-exchange chromatography prior to the reaction.

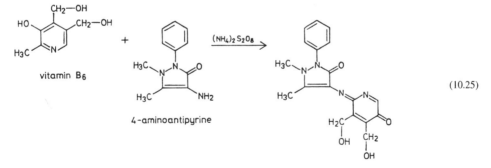

(10.25)

Two further reagents for the determination of pyridoxine are diazotized 4-hydrazino-benzenesulphonic acid[89] leading to an azo-dye and 3,5-dichloro-p-benzoquinonechloroimine[71] which transforms pyridoxine to a blue indophenol. All of the reagents discussed in this section attack the molecule of pyridoxine at the free position 6.

Vitamin B$_{12}$ and Related Cobalamins.   The intense and characteristic UV-visible spectrum of the red-colored vitamin B$_{12}$ (cyanocobalamin) can be seen on p. 27. This spectrum affords excellent possibilities for the determination of cyanocobalamin in pharmaceutical preparations. Of the bands appearing at 278, 361 and 548 nm ($\varepsilon = 15,600$, $28,100$ and $8,540$) the most intense one at 361 nm is generally used in aqueous medium for the assay of bulk vitamin B$_{12}$ and its injection formulation.[18b,112b]

If the other components of multi-vitamin preparations interfere with the assay at 361 nm, it is possible to measure at the long-wavelength band at 548 nm which is a selective wavelength for the determination of cyanocobalamin. If the sensitivity or the selectivity (e.g., in liver preparations) is not sufficient, a variety of chromatographic and extraction techniques[37,108] are available for the sample enrichment and for the elimination of the interfering substances.

The analytical chemistry of the hydroxyl analog of cyanocobalamin (hydroxocobalamin) raises much more interesting problems. Its spectrum is similar to that of cyanocobalamin; an important difference is, however, that as a consequence of the equilibrium (Equation 10.26), in contrast to the situation with cyanocobalamin—the spectrum is pH-dependent.

$$\text{hydroxocobalamin} + H^+ \rightleftharpoons \text{aquocobalamin}^+ \tag{10.26}$$

In mildly acidic media where the spectrophotometric determinations are usually carried out, the material is present in the form of aquocobalamin with an absorption maximum at 351 nm. According to the method of Beyer[14] the composition of the mixtures of cyano- and hydroxocobalamins can be determined using the ratio of the absorbances at 351 and 361 nm (see Section 4.F.1). (The simultaneous determination of the two cobalamins is an important task since hydroxocobalamin is the photodegradation product of cyanocobalamin and, at the same time, cyanocobalamin is the starting material of the synthesis of hydroxocobalamin.) The difference spectrophotometric method of Horváth and Szepesi[46] for the selective determination of hydroxocobalamin is based on the same principle. ΔA_{362} of a solution with pH 8 (glycyne buffer) is measured against a solution with pH 2, and the measurement is based on the selective shift of the spectrum of hydroxocobalamin on the effect of raising the pH. The basis of the method of Celletti et al.[24] for the determination of cyano-, adenosyl- and methylcobalamins is the shift of the long-wavelength band in very strongly acidic media (pH 7 vs. pH 0).

Color reactions are not of great importance in the spectrophotometric analysis of vitamin B_{12}. These are applied if the measurement based on natural absorption is problematic, either because of the very low dose or the interference of accompanying substances. In these cases the assay can be traced back to the wealth of very sensitive and selective methods which are available in the analytical chemistry of cobalt. Prior to the application of these procedures, the organic part of the molecule and the accompanying organic matrix has to be decomposed by digesting with various mixtures of hydrogen peroxide, sulphuric acid, nitric acid and perchloric acid. The liberated cobalt can be determined after color reaction with 1-nitroso-2-naphthol-3,6-disulphonic acid,[37] potassium thiocyanate,[17,19] 4-(5-bromo-2-pyridylazo)-1,3-diaminobenzene,[13] di-(2-pyridyl)-ketone-2-pyridyl hydrazone.[117] A kinetic method is also available based on the catalytic effect of cobalt on the reaction between pyrogallol red and hydrogen peroxide.[66]

Nicotinic Acid and Nicotinamide.   Nicotinic acid and nicotinamide have absorption bands of medium intensity around 262 nm in aqueous or ethanolic media (ε = 2,910 and 2,900). This enables simple analytical problems to be solved e.g., their determination in single component pharmaceutical preparations.[18d,112c] As is characteristic of pyridine derivatives in strongly acidic media, the location of the band remains unchanged but the intensities are markedly increased: ε = 6,900 and 5,540, respectively (see Figure 3.32). On the basis of this hyperchromic shift, the selectivity and sensitivity of the measurement can be increased. Multi-wavelength and especially derivative spectrophotometry are suitable methods for the determination of nicotinic acid and nicotinamide in the presence of other B vitamins.[61,86] Another possibility is their chromatographic separation prior to the spectrophotometric assay.[37,108]

Several methods based on chemical reactions can also be found in the literature for the determination of nicotinic acid and nicotinamide. Applications of the classical König reaction involving the opening of the pyridine ring with cyanogene bromide are the most characteristic examples. As seen in Equation 10.27, this reaction leads to glutaconaldehyde derivatives which can be reacted with aromatic amines to form colored condensation products. As the aromatic amine reagents, aniline, procaine, 4-aminoacetophenone, p-phenylenediamine etc., were used but the aldehyde groups could be condensed with the active methylene group of barbituric acid, too.[37,108] The use of sulphanilic acid (which is shown in Equation 10.27) seems to be the most general; λ_{max} = 450 nm.[80,112d]

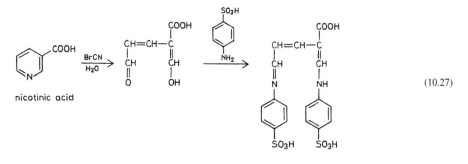

(10.27)

It is worth noting that it is not necessary to use the hazardous cyanogen bromide (or chloride); reaction 10.27 can be performed with the aid of *in situ* generated reagent by using ammonium thiocyanate and chloramine B reagents. Using this variant of the reaction and barbituric acid as the reaction partner, the absorbance maximum was found at 500 nm.[75]

The selective methods for the determination of nicotinamide partly exploit its basic properties (ion-pair extraction method using bromothymol blue as the dye reagent[37]) or use the Hoffmann-degradation of the carboxamide moiety with hypobromite leading to 3-amino-pyridine. In the latter case, the determination can then be based on the reactions of the primary aromatic amino group: diazotization and coupling with N-phenylnaphthylamine[37] or formation of Schiff's base with 4-dimethylaminobenzaldehyde.[79] The method of Nudelman[79] is also specific to nicotinamide. As seen in Equation 10.28 the 1-chloro-2,4-dinitrobenzene reagent reacts with the analyte at the aromatic nitrogen. The resulting quaternary pyridinium derivative rearranges in alkaline medium to a highly conjugated derivative which can be measured at 504 nm.

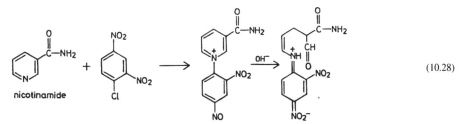

(10.28)

Vitamin B Complexes. In the course of the discussion of the individual members of the vitamin B family, the problem of their determination in combined formulations was mentioned several times. In these formulations, B vitamins (often all of them) are administered together, moreover often in combination with other vitamins. Although in modern drug analysis problems of this type are generally solved by means of high-performance liquid chromatography, computer-assisted methods for multi-component systems can also be used without any separation step.

To illustrate this possibility the method of Petiot et al.[86] is presented for the assay of a combined preparation containing vitamins B_1, B_2 and B_6, nicotinamide and calcium panthothenate. Figure 10.13 shows the spectra of these vitamins in 0.1 N sulphuric acid medium. The determination of the five components was based on absorbance values taken from the entire spectral range between 190 and 400 nm. Better results were obtained if the computer-assisted calculation was based on the first derivative rather than the zero-order spectra (see Figure 10.14).

3. Vitamin C (Ascorbic Acid)

A good review of the analytical chemistry of ascorbic acid is to be found in the paper of Pachla et al.[81] Its ultraviolet spectra are shown in Figure 3.30. Both the undissociated form (0.01 M hydrochloric or sulphuric acid; $\varepsilon_{244\ nm} = 9,850$) and the monoanionic form (pH 7.8; $\varepsilon_{264\ nm} = 15,000$) are suitable for straightforward analytical problems such as the assay of its simple formulations,[64] etc.

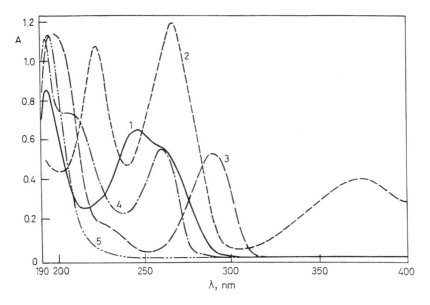

FIGURE 10.13. Spectra of the active ingredients of a multivitamin preparation. Solvent: 0.05 M sulphuric acid. 1: Vitamin B_1, 1.6 mg/100 ml; 2: Vitamin B_2, 2 mg/100 ml; 3: Vitamin B_6, 1.25 mg/100 ml; 4: Nicôtinamide, 1.3 mg/100 ml; 5: Calcium pantothenate, 5 mg/100 ml. (From Petiot, J.; Prognon, P.; Postaire, E.; Larue, M.; Laurencon-Courteille, F.; Pradeau, D. *J. Pharm. Biomed. Anal.* 1990, *8*, 93.[86])

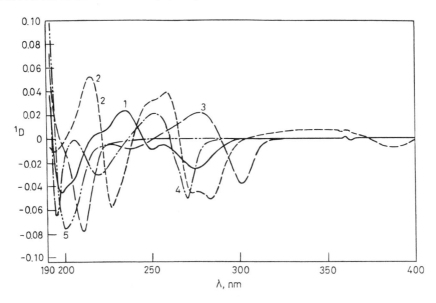

FIGURE 10.14. First derivative spectra of the vitamins in Figure 10.13. See Figure 10.13 for details.

A source of error in these determinations is that in the dissolved state, ascorbic acid is readily oxidized to the spectrophotometrically inactive dehydroascorbic acid. At the same time, however, the ready oxidation of ascorbic acid can be exploited to increase the selectivity of the measurement. This means that in the difference spectrophotometric variant of the method, the reference solution contains ascorbic acid in completely oxidized form while in the sample solution it is protected from the oxidation. In this way the absorbances of well-defined and ill-defined components, which do not change their absorption in the course of the reaction conditions of the oxidation step, are cancelled out. In aqueous solution where the concentration of ascorbic acid is 0.001% at 90°C it is completely oxidized within 2–12 min the oxidizing agent being the oxygen dissolved in the water.

This variant of the difference spectrophotometric method was utilized by Baczyk and Swidzin-ska[11] for the determination of ascorbic acid in pharmaceutical products. In a simpler variant of this method developed by Lau et al.[58] the oxidation by atmospheric oxygen is catalyzed by traces of copper (II) ions. As a result, the reaction is completed at pH 6 at room temperature and after masking the copper ions with EDTA, the difference absorbance can be measured at 267 nm. Other components of multi-vitamin preparations do not interfere. The method has been adopted for the determination of ascorbic acid in beverages, too.[59] Ascorbic acid rapidly decomposes in alkaline solution (causing difficulties in recording the spectrum of the dianionic species) and this was used by Verma et al.[118] for the determination of ascorbic acid in pharmaceutical preparations by the FIA technique.

Another possibility of increasing the selectivity of the measurement is the use of derivative spectrophotometry.[2]

The majority of the numerous methods on the spectrophotometric analysis of ascorbic acid based on chemical reactions are indirect procedures depending on the reducing properties of the analyte. In the first group of these methods, the colored oxidizing agent is decolorized as a result of its reaction with ascorbic acid, and the determination is based on the measurement of the decrease of the absorbance. The classical reagent of this type for the determination of ascorbic acid is the Tillman reagent (2,6-dichloroindophenol) (see Equation 10.29). This reagent is predominantly used as a titrant in the photometric titration variant of the method where the endpoint is detected on the basis of the color of the unreacted titrating agent.[37,108] The spectrophotometric variant of the method is also widely used in clinical and food analysis. The automated[32,116] and kinetic variants[43,44] of the spectrophotometric method have also been described.

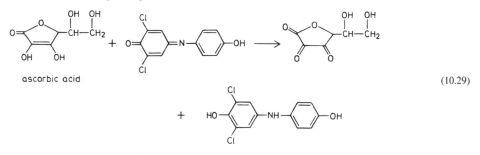

(10.29)

Methods in which the determination of ascorbic acid is based on the decolorization of iodine also belong to this group of methods.

In one of the variants of this method (applied to the FIA technique) Lázaro et al.[60] used a stream of chloramine T into which solutions of potassium iodide, starch and sulphuric acid (blank) and potassium iodide, starch, sulphuric acid and the ascorbic acid to be determined (sample) are injected. In the presence of sulphuric acid, chloramine oxidizes iodide to iodine which forms a blue complex with starch detectable at 650 nm. For this reason blank injections result in large peaks. When the sample solution is injected, the resulting iodine is partly consumed by the ascorbic acid present in the sample leading to smaller peaks. The difference between blank and sample peaks is the measure of the concentration of ascorbic acid in the sample solution. In another FIA method introduced by Hernandez-Mendez et al.,[42] the iodine is generated by means of the iodate-iodide reaction. Ascorbic acid solution injected into this stream results in negative peaks. The manual method of Muralikrishna and Murty[74] was developed by Thorburn Burns et al.[115] to an FIA method. The basis of this procedure is the reduction of vanadotungstophosphoric acid by ascorbic acid detectable at 360 nm.

In another group of spectrophotometric methods based on the reducing properties of ascorbic acid, it reduces the reagents to colored products which can be measured subsequently. One of the most generally used reagents of this type is ammonium molybdate and the measured derivative is molybdenum blue.[29,31,113] Another even more widely used oxidizing agent is iron(III) ion and in the different variants of the method the measurement is based on the formation and spectrophotometry of colored complexes of iron(II) which is generated in a quantity equivalent to ascorbic acid. Selective chelating agents of iron(II) used in these studies are 2-oximinocyclohexanone thiosemicarbazone (ε_{516nm} = 14,900),[95] 1,10-phenanthroline (ε_{510nm} = 22,000),[15] and ferrozin (3-(2-pyridyl-5, 6-bisphenylsulphonic acid-1,2,4-triazine). Using this reagent, an even higher molar absorbtivity is obtainable (ε = 27,900). This method has found wide application.[7,22,53,65] The application of a similar reagent (tripyridyl-s-triazine) in the enzymatic determination of ascorbic acid is described in Section 7.I.[96] Further reagents of this type are copper (II)-2,2′-biquinoline[105] and dimethoxydiquinone.[28] Al-Tamrah[8] used potassium ferricyanide as the oxidizing agent. Its excess is allowed to react with phthalophenone (reduced form of phenolphthalein) and the forming phenolphthalein is measured in alkaline medium at 553 nm.

In a third group of methods ascorbic acid is oxidized to dehydroascorbic acid and the latter is transformed to a spectrophotometrically measurable derivative. The classical method for the determination of ascorbic acid using 2,4-dinitrophenylhydrazine to react with dehydroascorbic acid belongs to this group.[94] Of the numerous versions of this method automated procedures are also available for the determination of ascorbic acid in pharmaceutical products.[85] In the method of Wahba et al.[121] the reagent is phenylhydrazine in 0.1 M hydrochloric acid. The yellow color (λ_{max} = 395 nm) develops without preliminary oxidation, moreover the method is selective for ascorbic acid in the presence of its degradation products.

After being oxidized to dehydroascorbic acid, vitamin C can react with o-phenylenediamine or its derivatives to form quinoxaline derivatives which are fluorescent if suitable functional groups are present, but the spectrophotometric version of the method is also described e.g., by Szepesi[110] who used mercury(II) chloride as the oxidizing agent and 4,5-dimethyl-o-phenylenediamine as the reagent ($\varepsilon_{345\,nm}$ = 10,100). If the oxidation step is omitted the method is suitable for the determination of dehydroascorbic acid impurity in ascorbic acid.

The enzymatic procedures for the determination of ascorbic acid are also based on its redox reactions. In addition to the earlier mentioned method based on ascorbic acid oxidase,[96] the method of Thompson[114] can be mentioned which is based on the inhibitory effect of ascorbic acid on the reaction between hydrogen peroxide and 3,5-dichloro-2-hydroxyphenylsulphonic acid where the progress of the oxidation reaction is monitored via the coupling reaction with 4-aminoantipyrine.

Finally another classical method is mentioned. Condensation of ascorbic acid with diazotized 2-nitro-4-methoxyaniline leads to a dye with absorption maximum at 570 nm. On the basis of this reaction, manual[101,102] and automated[124] methods were developed for the determination of ascorbic acid.

4. D Vitamins[12,36]

In this section the spectrophotometric methods for the determination of D vitamins are summarized. Predominantly vitamin D_2 (ergocalciferol) and D_3 (cholecalciferol) are dealt with. What is described here relates to both vitamins: their natural absorption and also the spectra obtained with chemical reactions are based on the same ring system. The differences in the side chain do not markedly influence the spectra and hence the differentiation between the two vitamins is possible only by chromatographic methods. Of the related derivatives dihydrotachysterol is discussed briefly.

With their trans-cis triene system, D vitamins have a strong absorption band at 265 (for D_2 ε = 18,500). The absorption maximum of the cis diol-type dihydrotachysterol is at 251 nm (ε = 10,100). Even these relatively intense bands allow the solution of simple analytical problems only, since the accompanying related compounds (pre- and pro-vitamins, lumisterols, etc.) and the other fat-soluble vitamins in the formulations interfere with the assay based on natural absorption.

For example dihydrotachysterol can be determined in such a way only in simple tablet formulation.[111a] For its determination in oily injections, purification by solvent extraction and/or TLC separation are necessary together with algebraic background correction.[45,47] The same applies to the determination of vitamin D in oily solutions.[39]

Of the color reactions suitable for the quantitative analysis of D vitamins, the importance of the procedure where they are treated with antimony(III) chloride in chloroform or ethylene dichloride is of great importance. A great advantage of this method leading to an orange complex is its high sensitivity which is about four-fold as compared to the measurement based on natural absorption. From among the many variants of the method, the one which is still in use[112e] can be characterized by adding acetyl chloride to the reaction mixture to stabilize the color.[76] In order to decrease the effect of parasitic absorption the basis of calculation is ($A_{500} - A_{550}$).[109] To increase the selectivity of the method, maleic anhydride is often introduced to the reaction mixture which undergoes Diels-Alder reaction with trans-vitamin D and tachysterol, thus eliminating their interference. A disadvantage of this method is that the other fat-soluble vitamins also give positive reactions; several extraction and column chromatographic separation tests have to be carried out prior to the spectrophotometric determination if the method is intended to be used for the analysis of their oily solutions.[73]

Some other reagents leading to colored products are also available: anisaldehyde–sulphuric acid,[9] trifluoroacetic acid–hydrogen peroxide,[33,34] hydrochloric acid–tetrachloroethane,[38] etc.

5. Vitamin E (Tocopherol)

The absorption band of the phenolic α-tocopherol is not suitable for its direct determination, partly because of its low intensity and partly because of the uncertainties originating from the oxidation of the analyte in solution state. As a consequence of the hypsochromic and hypochromic shifts as compared to the parent free phenol, the spectrum of its acetate is even less suitable for quantitative analytical purposes. It is, however, noteworthy that derivative spectrophotometry was successfully used ($^1D_{291}$) for the determination of tocopherol in soybean oil.[62]

From these facts, methods based on chemical reactions are almost exclusively used in the spectrophotometric analysis of vitamin E. All the important methods are based on the ready oxidation of the quinol-type aromatic ring: its acetate has to be hydrolyzed prior to the application of these methods. (Other typical phenol reagents cannot be applied since all carbon atoms of the phenolic ring are substituted.)

The main reaction scheme of the oxidation of tocopherol is shown in Equation 10.30.

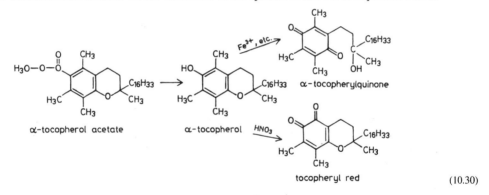

(10.30)

As in the case of the analysis of ascorbic acid, iron(III) salts as oxidizing agents play a predominant role in the presence of complexing agents, which form intensively colored, stable chelates with the resulting iron(II) ions. The classical reagent for this purpose is 2,2'-dipyri-dyl[11b,37,108] but the use of 1,10-phenantroline,[37] 5,6-diphenyl-3-(2-pyridyl)-1,4-triazine[55] and the disulphonic acid derivative of the latter (ferrozine)[4] has also been reported. When the last reagent is used, the absorption maximum appears at 562 nm; $\varepsilon = 28{,}000$. As a typical phenol reagent based on its oxidation, phosphomolybdenic acid can also be used.[37] The above listed reagents react following the upper arrow in Equation 10.30. In these cases reducing agents and vitamin A interfere with the methods and thus, these methods are often combined with preliminary chromatographic separation steps.

If nitric acid is used as the oxidizing agent, the reaction product is tocopheryl red (see the lower arrow in Equation 10.30). This product itself can be measured spectrophotometrically[37,108] but in one of the other variants of the method its ketone groups are condensed with 4-dimethylaminoaniline to form a product which can be measured at 620 nm ($\varepsilon = 4{,}950$).[100]

6. K Vitamins

In this section the following K vitamins, all having the 1,4-naphthoquinone structure are treated: vitamin K_1 (2-methyl-3-phytyl derivative; phytomenadione), vitamin K_3 (2-methyl derivative, menadione, menaphtone) vitamin K_4 (menadiol) which is the reduced form of the latter as well as the sodium bisulphite adduct of menadione.

Due to their 1,4-benzoquinone structure, vitamins K have intense and structured spectra. The maximum values of phytomenadione are as follows: $\varepsilon_{245\ nm} = 17,940$; $\varepsilon_{249} = 18,930$; $\varepsilon_{263} = 17,350$; $\varepsilon_{271\ nm} = 17,850$; $\varepsilon_{327\ nm} = 3,110$ (solvent: i-octane). For quantitative analytical measurements usually the intense band around 249 nm is selected[18e,f,37] but the use of the least intense, at the same time, however, most selective long-wavelength band has also been reported.[49] The spectrum of menadione sodium bisulphite is shifted towards the shorter wavelengths as compared to that of menadione. If this form is to be measured, the pH should be kept below 6 and the test solution should contain excess sodium bisulphite. In alkaline media, the adduct decomposes to form menadione and hence its determination may be traced back to the latter.[37,49,84] Derivative spectrophotometry has also been used successfully. Milch and Szabó[67] measured $^2D_{254.5,248.5}$ for the determination of vitamin K_1 in tablets and injections.

Some color reactions are also used for the determination of vitamins K. Their keto groups react with the classical ketone reagents such as 2,4-dinitrophenylhydrazine[37,112f] and thiosemicarbazide.[106] When the latter method is applied to menadione sodium bisulphite it has to be transformed to menadione by alkaline treatment.

Another type of condensation reaction is used in the methods where the reagents contain active methylene groups such as ethyl cyanoacetate, acetylacetone, diethyl malonate in the presence of ammonia[40] and 3-methyl-l-phenylpyrazolin-5-one.[83] Various colors are obtained; the sensitivities are fairly good.

Some other methods depend on the reversible redox reaction between vitamins K and their reduced derivatives. The menadione-menadiol transformation is shown in Equation 10.31.

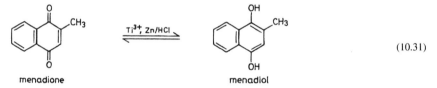

menadione   menadiol

(10.31)

In the method of Vire and Patriarche[119] for the determination of menadione and other vitamins K, the reducing agent is titanium(III)-sulphate and the resulting menadiol is measured as the Ti(IV) complex in the spectral range 510–655 nm. In the method of Reddy et al.[92] menadione is reduced with zinc/hydrochloric acid. It is then re-oxidized to menadione by iodine and the excess of the latter is determined spectrophotometrically.

Finally, it is noteworthy that menadione produces a moderately intense but stable blue color when treated with a zinc(II) chloride concentrated hydrochloric acid reagent. Natural K vitamins do not give positive reaction. Menadiol has to be oxidized to menadione with hydrogen peroxide before the reaction.[37]

7. Other Vitamins

Panthenol and Pantothenic Acid.   Panthenol and pantothenic acid are spectrophotometrically practically inactive substances; chemical reactions, however, afford various possibilities for their spectrophotometric determination. As shown in Equation 10.32, acidic treatment results in the cleavage of the carboxamide bond in the case of both compounds to form pantoyl lactone as the common reaction product.

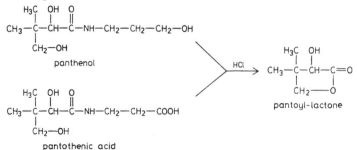

(10.32)

Pantoyl lactone can be determined by the iron(III)-hydroxamate method (see p. 184).[48] Since panthenol and pantothenic acid can be transformed to pantoyl lactone, this method is suitable for their indirect determination, too.[37]

In the course of the acidic (see Equation 10.32) and alkaline hydrolysis of panthenol and pantothenic acid, the other reaction products in addition to the lactone are β-aminopropanol and β-aminopropionic acid, respectively. Their amino groups may also serve as the basis for spectrophotometric procedures using e.g., 1,2-naphthoquinone-4-sulphonic acid[37,108] or ninhydrin[56] as the reagents.

Folic Acid.   The molecule of folic acid contains the spectrophotometrically inactive glutaminic acid and the highly active 4-aminobenzoyl and substituted pteridine units, the latter two being isolated from each other. These are the structural basis of the very intense and structured spectrum of folic acid which is most suitable for spectrophotometric assays ($\varepsilon_{256\,nm} = 26,000$; $\varepsilon_{283} = 25,400$; $\varepsilon_{365} = 9,100$; solvent: 0.1 M sodium hydroxide).

The color reactions of folic acid are usually two-step procedures in which the first step is reductive or oxidative cleavage of the molecule to form amino groups for the color reaction in the second step. As seen in Equation 10.33, two possibilities for the cleavage are oxidation with permanganate in mildly alkaline medium and reduction with zinc/hydrochloric acid.

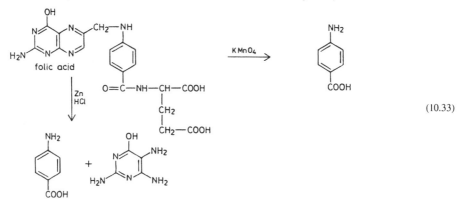

$$(10.33)$$

4-Aminobenzoic acid formed in the course of the permanganate oxidation is diazotized and coupled with N-(1-naphthyl)-ethylenediamine and the resulting azo-dye is measured at 550 nm,[37,108] This classical method is the basis of the determination of folic acid in USP XXI.[111c]

The procedures starting with zinc/hydrochloric acid reduction can produce the color in several ways. One of these is diazotization and coupling to get azo-dyes[18g,93] and others are condensation of the primary amino groups with 4-dimethyl-aminocinnamaldehyde,[93] reaction with ninhydrin[88] or 1,2-naphthoquinone-4-sulphonic acid.[54]

Vitamin P (Rutin).   This flavonoid glycoside possesses an intense and characteristic spectrum ($\varepsilon_{257\,nm} = 19,800$; $\varepsilon_{365\,nm} = 16,800$; solvent: ethanol). Of the two bands, the longer wavelength causing the yellow color of rutin is eminently suitable for the solution of various problems in pharmaceutical analysis. The selectivity of the measurement can be increased further if the measurement is carried out in alkaline medium when this phenolic material is present as the phenolate: in 0.005 M sodium hydroxide the maximum is shifted to 415 nm. Under these conditions other vitamins such as ascorbic acid, pyridoxine, thiamine and nicotinic acid do not interfere with the assay.[107]

Similar bathochromic shift is obtainable if diazotized sulphanilic acid[50] or 4-nitro-2-chloroaniline[91] are used as the reagents. Another color reaction frequently used for its identification and quantitation is the reduction of rutin with nascent hydrogen (zinc/hydrochloric acid or magnesium/ hydrochloric acid). The resulting aromatization of the hetero ring and the formation of oxonium salt take place accompanied by the shift of the maximum to 540 nm.[63] Some other methods for the photometric determination of rutin depend on the formation of metal complexes, e.g., aluminum[35] and nickel[87] chelates.

Because of the great similarity of the spectra of rutin and its aglycone (quercetin) all the reliable methods for the determination of the latter as an impurity in rutin include a chromatographic separation step.

Biotin.   Regarding the possibilities for the spectrophotometric determination of biotin, the publication mentioned in the introduction of this section is reference.[108]

REFERENCES

1. Abdel-Hamid, M. E.; Barary, M. H.; Hassan, E. M.; Elsayed, M. A. *Anal. Letters.* 1986, *19,* 1053.

2. Abdel-Hamid, M. E.; Barary, M. H.; Hassan, E. M.; Elsayed, M. A. *Analyst* 1985, *110,* 831.

3. Abdel-Hamid, M. E.; Barary, M. H.; Korany, M. A.; Hassan, E. M. *Sci. Pharm.* 1985, *53,* 105.

4. Adeniyi, W.; Jaselskis, B. *Talanta* 1980, *27,* 993.

5. Ahmad, I.; Rapson, H. D. C. *J. Pharm. Biomed. Anal.* 1990, *8,* 217.

6. Al-Ghabsha, T. S.; Rahim, S. A.; Al-Sabha, T. N. *Microchem. J.* 1986, *34,* 89.

7. Aly, M. M. *Anal. Chim. Acta* 1979, *106,* 379.

8. Al-Tamrah, S. A. *Anal. Chim. Acta* 1988, *209,* 309.

9. Amer, M. M.; Wahbi, A. M.; Hassan, S. M. *Analyst* 1975, *100,* 238.

10. Ashraf, M.; Rapson, H. D. C. *J. Pharm. (Lahore)* 1979, *1,* 1.

11. Baczyk, S.; Swidzinska, K. *Farm. Pol.* 1975, *31,* 399.

12. Bartos, J.; Pesez, M. *Colorimetric and Fluorimetric Analysis of Steroids,* Academic Press: London, 1976, 78–118.

13. Baudino, O. M.; De Pedernera, M. B. M. *Boll. Chim. Farm.* 1987, *126,* 385.

14. Bayer, J. *Pharmazie* 1964, *19,* 602.

15. Besada, A. *Talanta* 1987, *34,* 731.

16. Blake, J. A.; Moran, J. J. *Can. J. Chem.* 1976, *54,* 1757.

17. Blazis, I.; Malat, M. *Chem. Listy* 1981, *75,* 312.

18. *British Pharmacopoeia 1988,* Her Majesty's Stationery Office: London, 1988. a. 999; b. 163; c. 777; d. 973; e. 443; f. 986; g. 257.

19. Bruno, P. *Anal. Letters* 1981, *14,* 1493.

20. Budowski, P.; Bondi, A. *Analyst* 1957, *82,* 751.

21. Burger, K. *Talanta* 1963, *10,* 573.

22. Butts, W. C.; Mulvihill, H. J. *Clin. Chem.* 1975, *21,* 1493.

23. Carr, F. A.; Price, E. A. *Biochem. J.* 1926, *20,* 497.

24. Celletti, P.; Moretti, G. P.; Petrangeli, B. *Farmaco, Ed. Prat.* 1976, *31,* 413.

25. Ciesielski, W. *Chem. Anal.* 1986, *31,* 469.

26. Das Gupta, V.; Cadwallader, D. E. *J. Pharm. Sci.* 1968, *57,* 112.

27. De Vries, E. J.; Mulder, F. J.; Borsje, B. *J. Assoc. Off. Anal. Chem.* 1977, *60,* 989.

28. Eldawy, M. A.; Tawfik, A. S.; Elshabouri, S. R. *Anal. Chem.* 1975, *47,* 461.

29. El Sayed, E.; Raafat, S. *Talanta* 1979, *26,* 1664.

30. Elsayed, M. A. -H.; Belal, S. F.; Elwalily, A. -F. M.; Abdine, H. *J. Pharm. Sci.* 1979, *68,* 739.

31. Emelyanenko, K. V. *Farm. Zh.* 1986, 73.

32. Garry, P. J.; Owen, G. M.; Lashley, D. W.; Ford, P. C. *Clin. Biochem.* 1974, *7,* 131.

33. Gharbo, S. A.; Gosser, L. A. *Analyst* 1974, *99,* 222.

34. Gharbo, S. A.; Gosser, L. A. *J. Pharm. Sci.* 1975, *64,* 1196.

35. Glasl, H. *Z. Anal. Chem.* 1985, *321,* 325.

36. Görög, S. *Quantitative Analysis of Steroids,* Akadémiai Kiadó Budapest; Elsevier: Amsterdam, 1983, 292–296.

37. Hashmi, M. *Assay of Vitamins in Pharmaceutical Preparations,* Wiley: New York, 1973.

38. Hassan, S. M. *Z. Anal. Chem.* 1978, *293,* 416.

39. Hassan, S. M. *J. Assoc. Off. Anal. Chem.* 1979, *62,* 545.

40. Hassan, S. S. M.; Abd El Fattah, M. M.; Zaki, M. T. M. *Z. Anal. Chem.* 1975, *275,* 115.

41. Hayden, K. J. *Analyst* 1957, *82,* 61.

42. Hernandez-Mendez, J.; Alonso Mateos, A.; Almendral-Parra, M. J.; Garcia De Maria, C. *Anal. Chim. Acta* 1986, *184,* 243.

43. Hiromi, K.; Fujimori, H.; Yamaguchi-Ito, J.; Nakatani, H.; Ohnishi, M.; Tonomura, B. *Chem. Letters* 1977, 1333.

44. Hiromi, K.; Kuwamoto, C.; Ohnishi, M. *Anal. Biochem.* 1980, *100*, 421.
45. Hom, F. S.; Veresh, S. A.; Ebert, W. R. *J. Assoc. Off. Anal. Chem.* 1977, *60*, 48.
46. Horváth, P.; Szepesi, G. *Acta Pharm. Hung.* 1978, *48*, 199.
47. Hrdy, O. *Zbl. Pharm.* 1974, *113*, 1255.
48. Illner, E. *Pharmazie* 1980, *35*, 186.
49. Ismaiel, S. A.; Yassa, D. A. *Pharmazie* 1975, *30*, 407.
50. Jablonowski, W. *Acta Pol. Pharm.* 1975, *32*, 251.
51. Jan, A.; Ali, H.; Ashraf, M. *J. Pharm. (Lahore)*, 1979, *1*, 12.
52. Jarzebinski, J.; Ciszewska, M.; Suchoki, P.; Szrajber, Z. *Acta Pol. Pharm.* 1981, *38*, 461.
53. Jaselskis, B.; Nelapaty, J. *Anal. Chem.* 1972, *44*, 379.
54. Kanjilal, G.; Mahajan, S. N.; Rao, G. R. *Analyst* 1975, *100*, 19.
55. Kanno, C.; Yamauchi, K. *Agric. Biol. Chem.* 1977, *41*, 593.
56. Karawya, M. S.; Ghourab, M. G.; Ibrahim, E. S. *J. Ass. Off. Anal. Chem.* 1974, *57*, 1357.
57. Krishnan, M. V. S.; Mahajan, S. N.; Rao, G. R. *Analyst* 1976, *101*, 601.
58. Lau, O. W.; Luk, S. F.; Wong, K. S. *Analyst* 1987, *112*, 1023.
59. Lau, O. W.; Luk, S. F.; Wong, K. S. *Analyst* 1986, *111*, 665.
60. Lázaro, F.; Rios, A.; Luque de Castro, M. D.; Valcárcel, M. *Analyst* 1986, *111*, 163, 167.
61. Lou, Y.; Wang, Y.; Wang, M.; Chen, D.; Lin, J.; Xu, S.; Hong, D. *Yaowu Fenxi Zazhi* 1983, *3*, 137.
62. Lu, X. *Yaowu Fenxi Zazhi* 1986, *6*, 292.
63. Lukyanchikova, G. I.; Tiraspolskaya, S. G.; Lukyanchikov, M. S. *Farmatsiya* 1984, *33*, 70.
64. Lukyanchikova, G. I.; Tiraspolskaya, S. G.; Lukyanchikov, M. S. *Farmatsiya* 1985, *34*, 60.
65. McGown, E. L.; Rusnak, M. G.; Lewis, C. M.; Tillotson, J. A. *Anal. Biochem.* 1982, *119*, 51.
66. Medina-Escriche, J.; Hernandez-Llorens, M. L.; Llobat-Estelles, M.; Sevillano-Cabeza, A. A. *Analyst* 1987, *112*, 309.
67. Milch, Gy.; Szabó, E. *Analusis* 1988, *16*, 59.
68. Mitsui, T.; Fujimura, Y. *Bunseki Kagaku* 1983, *32*, 264.
69. Morton, R. A.; Stubbs, A. L. *Analyst* 1946, *71*, 348.
70. Morton, R. A.; Stubbs, A. L. *Biochem. J.* 1947, *41*, 525.
71. Moussam, A.-F. A. *Microchim. Acta* 1982, 169.
72. Moussa, A. E.-F. A.; Hassan, S. M. *Pharmazie* 1977, *32*, 50.
73. Mulder, F. J.; De Vries, E. J.; Borsje, B. *J. Assoc. Off. Anal. Chem.* 1977, *60*, 151.
74. Muralikrishna, U.; Murty, J. A. *Analyst* 1989, *114*, 407.
75. Namigohar, F.; Ghanbarpour, A.; Beltran, D. *Ann. Pharm. Fr.* 1981, *39*, 445.
76. Nield, C. H.; Russel, W. C.; Zimmerli, A. *J. Biol. Chem.* 1940, *136*, 73.
77. Nirmalchandar, V.; Balasubramanian, N. *Analyst* 1988, *113*, 1097.
78. Nirmalchandar, V.; Viswanathan, R.; Balasubarmanian, N. *Analyst* 1987, *112*, 653.
79. Nudelman, N. S.; Nudelman, O. *J. Pharm. Sci.* 1976, *65*, 65.
80. *Official Methods of Analysis*, Association of Official Agricultural Chemists, Washington, 1960, 660.
81. Pachla, L. A.; Reynolds, D. L.; Kissinger, P. T. *J. Assoc. Off. Anal. Chem.* 1985, *68*, 1.
82. Park, J. Y. *Anal. Chem.* 1975, *47*, 452.
83. Patel, J. C.; Mehta, R. C.; Shastri, M. R. *Indian J. Pharm. Sci.* 1975, *37*, 141.
84. Pawelczyk, E.; Marciniec, B. *Acta Pol. Pharm.* 1982, *39*, 387.
85. Pelletier, O.; Brassard, R. *J. Assoc. Off. Anal. Chem.* 1975, *58*, 104.
86. Petiot, J.; Prognon, P.; Postaire, E.; Laure, M.; Laurencon-Courteille, F.; Pradeau, D. *J. Pharm. Biomed. Anal.* 1990, *8*, 93.
87. Radovic, Z.; Malesev, D. *Acta Polon. Pharm.* 1987, *44*, 433.
88. Rao, G. R.; Mahajan, S. N.; Kanjilal, G.; Mohan, K. R. *J. Ass Off. Anal. Chem.* 1977, *60*, 531.
89. Rao, G. R.; Sivaramakrishnan, M. V.; Kanjilal, G.; Srivastava, C. M. R. *Indian J. Pharm. Sci.* 1979, *41*, 203.
90. Rapaport, L. I.; Fleish, N. L.; Proshunina, D. V. *Farm. Zh.* 1987, 48.
91. Reddy, M. N.; Sastry, C. S. P.; Sankar, D. G.; Singh, N. R. P. *Indian J. Pharm. Sci.* 1987, *49*, 231.
92. Reddy, M. N.; Sastry, C. S. P.; Viswanadham, N. *J. Inst. Chem.* 1985, *57*, 57.
93. Reddy, M. N.; Viswanadham, N.; Sastry, C. S. P. *Indian Drugs* 1984, *21*, 460.
94. Roe, J. H.; Kuether, C. A. *J. Biol. Chem.* 1943, *147*, 399.
95. Salinas, F.; Diaz, T. G. *Analyst* 1988, *113*, 1657.
96. Samyn, W. *Clin. Chim. Acta.* 1983, *133*, 111.
97. Sane, R. T.; Vaidya, U. M.; Deodhar, K. D. *Indian J. Pharm. Sci.* 1978, *40*, 72.
98. Sastry, C. S. P.; Rajendraprasad, S. N.; Reddy, M. N. *Indian Drugs* 1986, *23*, 242.

  99. Sastry, C. S. P.; Singh, N. R.; Reddy, M. N. *Analysis* 1986, *14*, 355.
 100. Sastry, C. S. P.; Singh, N. R. P.; Reddy, R. V. R. *Indian Drugs* 1986, *23*, 633.
 101. Schmall, M.; Pifer, C. W.; Wolish, E. G. *Anal. Chem.* 1953, *25*, 1486.
 102. Schmall, M.; Pifer, C. W.; Wolish, E. G.; Buschinsky, R.; Gainer, H. *Anal. Chem.* 1954, *26*, 1520.
 103. Schmidiger, O. *Z. Anal. Chem.* 1975, *274*, 382
 104. Sethi, P. D.; Chatterjee, P. K.; Kumar, Y. *Indian J. Pharm. Sci.* 1985, *47*, 118.
 105. Shieh, H. H.; Sweet, T. *Anal. Biochem.* 1979, *96*, 1.
 106. Sidhom, M. B.; El-Kommos, M. E. *J. Assoc. Off. Anal. Chem.* 1982, *65*, 141.
 107. Sivitskaya, O. K.; Rubinskaya, V. G.; Kuleshova, M. I.; Guseva, L. N. *Farmatsiya* 1977, *26*, 81.
 108. Strohecker, R.; Henning, H. M. *Vitamin-Bestimmungen*, Verlag Chemie: Weinheim, 1963.
 109. Stross, P. S.; Brealey, L. *J. Pharm. Pharmacol.* 1955, *7*, 739.
 110. Szepesi, G. *Z. Anal. Chem.* 1973, *265*, 334.
 111. *The United States Pharmacopoeia XXI*. USP Convention Inc.: Rockville, 1985. a. 331; b. 1213: c. 1202.
 112. *The United States Pharmacopoeia XXII*, USP Convention Inc.: Rockville, 1990. a. 1550, b. 363–364, c. 943, d. 1539, e. 1553, f. 819.
 113. Thind, P. S.; Kumar, S. *Indian Drugs* 1984, *21*, 153.
 114. Thompson, R. Q. *Anal. Chem.* 1987, *59*, 1119.
 115. Thorburn Burns, D.; Chimpalee, N.; Chimpalee, D.; Rattanariderom, S. *Anal. Chim. Acta* 1991, *243*, 187.
 116. Vander-Jagt, D. J.; Garry, P. J.; Hunt, W. C. *Clin. Chem.* 1986, *32*, 1004.
 117. Vasilikiotis, G. S.; Kouimtzis, T. A.; Voulgaropoulos, A. *Microchem. J.* 1977, *22*, 479.
 118. Verma, K. K.; Jain, A.; Verma, A.; Chaurasia, A. *Analyst* 1991, *116*, 641.
 119. Vire, J. C.; Patriarche, G. J. *J. Pharm. Belg.* 1976, *31*, 139.
 120. Viswanath, K. K.; Sivaprasad, R. A.; Sivaramakrishnan, M. V. *Indian J. Pharm. Sci.* 1986, *48*, 80.
 121. Wahba, N.; Yassa, D. A.; Labib, R. S. *Analyst* 1974, *99*, 397.
 122. Wahbi, A. M.; Belal, S.; Abdine, H.; Bedair, M. *Analyst* 1981, *106*, 960.
 123. Wahbi, A.-A. M.; Abounassif, M. A.; Alkhatani, H. M. G. *J. Pharm. Biomed. Anal.* 1989, *7*, 39.
 124. Wilson, S.; Guillan, R. A. *Clin. Chem.* 1969, *15*, 282.

P. METALS AND OTHER ELEMENTS OF PHARMACEUTICAL INTEREST

1. Introduction

The spectrophotometric determination of metals in bulk drugs and pharmaceutical preparations is not among the most important tasks in contemporary pharmaceutical analysis. One of the reasons for this is due to a consequence of the prevalence of organic drugs in therapy, the importance of metal-containing drugs has considerably declined in recent decades. Another reason is that in the analytical chemistry, of the currently used metal-containing drugs, complexometric titrations and among the optical methods, atomic absorption spectrophotometry have greatly superseded solution-phase spectrophotometric methods.

In spite of these facts, the importance of spectrophotometric methods based on the formation of colored metal complexes with a large variety of complexing reagents is not negligible even in modern pharmaceutical analysis: several methods of colorimetric metal analysis have been adopted for drug analytical purposes. Of the several publications dealing with the application of complex forming reagents in the spectrophotometric analysis of metals—although they are not necessarily devoted to pharmaceutical aspects—the books of Sandell,[41,42] Snell,[43] Burger[6] and Marczenko[29] are the most important. The best source for literature data from the classical period of the pharmaceutical aspects is the chapter by Medwick[31] in the publication of Higuchi and Brochmann-Hanssen.

This chapter concludes with brief summaries on the spectrophotometric analysis of some other elements of pharmaceutical interest.

2. Iron

The classical reagent for the spectrophotometric analysis of pharmaceutical preparations containing bivalent iron is 2,2′-dipyridyl. The reaction takes place rapidly in slightly acidic medium and the measurement of the red complex is suitable for the determination of small quantities of iron(II) (ε_{522} = 8,700).[29] Using acetate buffer of pH 4.6 this method is widely used in pharmacopoeias, e.g., for the assay of iron(II) gluconate capsule,[46c] iron(II) sorbitate injection,[46d] etc. The method is specific to iron(II); iron(III) does not interfere: moreover iron(II) can be selectively determined in the presence of iron(III). If the analytical task is the determination of the total, iron content in their mixture, iron(III) has to be reduced with ascorbic acid prior to the assay.[44] In such a way the iron(III) content can be calculated as the difference between the two values.

Replacing 2,2'-dipyridyl by structurally related other chelating agents such as 1,10-phenantroline or the 4,7-diphenyl derivative of the latter, the sensitivity increases: ε = 11,000 and 22,400 respectively.[29] Another important family of iron(II) reagents are the triazine derivatives. From among these ferrozine (3-(2-pyridyl)-5,6-bis-(4-sulphophenyl)-1,2,4-triazine disodium) has found application in drug analysis (ε_{562} = 28,000).[25] Of the several other reagents used for the determination of iron(II) in pharmaceutical analysis two are mentioned: isoniazid p-dimethylaminosalicylalde-hyde hydrazone recently introduced by Issopoulos and Economou[22] ($\varepsilon_{471\ nm}$ = 23,000 in micellar medium containing Triton X-100) and 8-hydroxy-7-nitroso-quinoline-5-sulphonic acid which excels with the very long wavelength of its absorption maximum (710 nm).[10]

The most generally applied reagent for the spectrophotometric determination of iron(III) is the thiocyanate ion. In the presence of large excess of ammonium thiocyanate and at the optimum acid concentration of 0.05–0.2 M mineral acid, the molar absorptivity of the red complex is 8,500; this can be increased to 18,000 by adding 50% v/v acetone to the reaction mixture.[29] Both the sensitivity and the selectivity of the method can be increased by extracting the colored complex using various organic solvents.

Trifluoroacetylacetone was also successfully used for the determination of iron(III) in various pharmaceutical formulations (iron sodium gluconate injection, iron saccharate injection, iron-condroitin sulphate capsule, and iron-protein tablet). The enolic form of this reagent forms a stable chelate with iron(III) which can be extracted with hexane and measured at 435 nm.[36]

It is worth mentioning that iron(III) ions form a ternary complex of 1:1:2 ratio with salicylate ions and purpurine (1,2,4-trihydroxyanthraquinone). This can be extracted with 4-methyl-pentan-2-one and measured at 590 nm. The high molar absorptivity of the complex (ε = 47,000) enables 10^{-5}% of iron impurity to be determined in salicylates and acetylsalicylates.[7]

3. Mercury

Undoubtedly, the most important and most widely used reagent for the determination of mercury is dithizone (diphenylthiocarbazone). The outstanding importance of this reagent in metal analysis can be characterized by the fact that two books were written on just this single reagent.[21,23] Among many other metals such as iron, copper, zinc, bismuth, silver, lead, etc., mercury(II) ions form very stable 2:1 complexes with this reagent, with an intense orange color which can be extracted with chloroform or carbon tetrachloride. In the course of the complex formation, either the thioketone or the thioenol form of the reagent reacts, depending on the metal ion and the reaction conditions (see Equation 10.34).

The absorption maximum of dithizone in carbon tetrachloride is at 620 nm (ε = 32,000). The complex formation results in hypsochromic and hyperchromic shifts: in the case of mercury (II) the absorption maximum is at 485 nm (ε = 71,000). The situation is similar with the other metals listed above, also. The selectivity of the measurement can be improved by optimization of the pH. Mercury (II) can be extracted from very strongly acidic solution (more than 1 M mineral acid concentration). Under these circumstances, only silver(I) and copper(II) are extracted from among the metal ions listed above. The selectivity can be further improved by using masking agents. Details of this can be found in the above mentioned publications.[6,21,23,29]

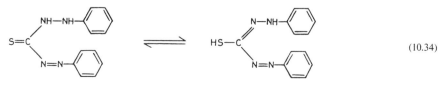

(10.34)

As regards the measuring technique, it has to be noted that the spectra of the excess reagent and the complex to some extent overlap, and for this reason the excess of the reagent has to be removed prior to the measurement by extraction with aqueous alkali. One of the other possibilities is to measure without separation in the presence of the excess reagent at the wavelength of the absorption maximum of the complex and calculate the results on the basis of the calibration graph constructed under identical experimental conditions. A further possibility is to extract the complex, then decompose it by shaking the extract with aqueous potassium iodide, since iodide ions form an even more stable complex with mercury(II). Dithizone liberated during the decomposition of the complex can be measured at 620 nm. An indirect measuring technique is also available where the basis of the measurement is the decrease of the absorbance at 605 nm of dithizone dissolved in dichloromethane due to the slightly acidic solution of the drug, the mercury content of which is to be determined.[11]

In the practice of drug analysis the semi-quantitative variant of the method based on visual comparison of colors is often used since the color of the complex is orange, while the reagent is green. For example in the course of the estimation of mercury as a contaminant in iron(II) fumarate, its quantity is limited on the basis of comparison of the colors of the test solution and a reference solution containing a known quantity of mercury.[46b]

When the dithizone method is applied to organo-mercurial drugs in the case of simple compounds e.g., methylmercury or phenylmercury salts, this can be done directly;[11,24] more complicated organic molecules, however, must undergo oxidative degradation prior to the complexation reaction.[39,40]

In addition to dithizone several other reagents forming colored complexes with mercury have also been introduced. Of these, e.g., 1-salicylidene-5-(2-pyridylmethylidene)-isothiocarbonohydrazide forms a stable 2:1 complex with mercury(II) ions. The absorbance maximum of the complex is at 400 nm in a mixture of dimethylformamide and water with the molar absorptivity of 64,000. As seen, the sensitivity of this method is comparable with that of the dithizone method; an advantage is, however, that no extraction is necessary, and the absorbance of the reagent is negligible at the wavelength of the absorption maximum.[40]

4. Bismuth

The simplest complex-forming reagent for the spectrophotometric determination of bismuth is the iodide ion. Bismuth forms a tetraiodobismuthate complex anion in strongly acidic medium containing potassium iodide in great excess. This anion forms an ion-pair with tetrabutylammonium bromide. This can be extracted with chloroform and measured at 490 nm ($\varepsilon = 11,300$). In order to avoid the oxidation of iodide to iodine, hypo-phosphorous acid is used as the antioxidant.[17] In another variant of the method, the antioxidant is ascorbic acid and the ion-pair is extracted with isobutyl alcohol.[27]

Of the organic reagents used in pharmaceutical analysis for the determination of bismuth, two have already been mentioned in the preceding section dealing with mercury. The absorption maximum of its dithizone complex is at 490 nm ($\varepsilon = 79,200$). The 1-salicylidene-5-(2-pyridylmethylidene)-isothiocarbonohydrazide complex of bismuth absorbs at 421 nm.[32] Much lower is the sensitivity of the methods based on complexes which absorb in the ultraviolet region, the reagents being N-carbamoylmethyliminodiacetic acid[14] ($\varepsilon_{265} = 9,100$) and mucic acid[13] ($\varepsilon_{245} = 6,500$). The use of thiourea as the reagent has also been reported ($\varepsilon_{470 \text{ nm}} = 9,000$).[8]

5. Other Metals

Dithizone is the most important reagent for the pharmaceutically less important metals: silver, copper, lead and zinc. These complexes can be extracted with carbon tetrachloride. Some important characteristics of these complexes are summarized in Table 10.10.

As regards the details, Section 10.P.3 and the references cited there are of interest. It is noteworthy that the color reaction with dithizone is official in the United States Pharmacopoeia for the determination of lead as an impurity in pharmaceuticals.[46f]

Of the above listed metals, copper is generally determined as the yellow complex of copper(II) formed with sodium diethyldithiocarbamate. Its molar absorptivity at the absorption maximum at

TABLE 10.10. Data of Complex Formation of Some Metal Ions with Dithizone

Metal ion	λ_{max},nm	ε	Optimal pH
Cu(II)	550	45,200	1–4
Ag(I)	462	30,500	0
Pb(II)	520	68,600	7–10
Zn(II)	538	92,600	4–11
Hg(II)	485	71,200	0

436 nm after extraction with carbon tetrachloride is 14,000.[6,29] Of the pharmaceutically important potential interfering ions, iron(III) can be masked with ethylenediamine tetraacetic acid, while bismuth can be extracted from the solution in carbon tetrachloride using strong acids. A semi-quantitative form of the diethyldithiocarbamate method based on visual comparison is prescribed in the British Pharmacopoeia as a limit test (10 ppm) for traces of copper in iron[5b,c] and bismuth[5a] preparations. For the assay of pharmaceutical zinc preparations among others, the use of biacetyl-mono-(2-pyridyl)-hydrazone reagent has been described.[1]

Of the metals not included in Table 10.10, *antimony* can also be determined using the complex formation with iodide ions (see the preceding section dealing with bismuth). The methods based on the solvent extraction of ion-pairs formed between $SbCl_6^-$ anion and various cationic dyes of the xanthene and diphenylmethane type, excel with their high sensitivities.[6,29] Another possibility for the determination of antimony(III) salts is the application of the molybdenum blue method making use of their reducing properties.[2]

Arsenic is determined in the United States Pharmacopoeia in such a way that it is first converted to arsine with the aid of zinc/sulphuric acid. This is passed through a solution of silver diethyldithio-carbamate to form a red complex absorbing at 535–540 nm.[46e]

For the determination of *manganese* in pharmaceutical products, the use of 4-(pyridylazo)-resorcinol has been reported.[45]

From among the great many complex-forming agents which have been described as reagents for the spectrophotometric determination of *aluminum* (8-hydroxyquinoline, chrome azurol S, pyrocatechol violet, alizarin S, aluminone, eriochrome cyanine, etc.) the last two were reported for pharmaceutical products. An interesting method for the same purpose is the formation of the hexafluoroacetylacetone complex, its extraction with hexane and measurement by derivative spectrophotometry ($^2D_{311}$).[33]

The rare earth *samarium* is used in therapy as the sulphosalicylate complex. Arsenazo-III was used for its determination in ointment and solution.[15] The importance of the careful optimization of the experimental conditions (pH, solvent) is illustrated in Figure 10.15.

As it is seen, measuring in a 4:1 mixture of methanol and water shows unequivocal advantages compared with the generally used aqueous medium. Not only does the sensitivity increase to about 3.5-fold ($\varepsilon = 164,000$) but at the same time the pH-optimum is also shifted from the value of about 2.5 (which is well known from the literature[6,29]) to about 1. This is very advantageous because at this low pH, the interference from aluminum and EDTA present in the formulation can be avoided.

As a result of the achievements in the development of coordination chemistry in the last two decades, reagents are available for the spectrophotometric determination of even such metals as *potassium,* the determination of which were previously among the most difficult problems in the photometric analysis of metals. Dibenzo-16-crown-6, which belongs to the family of crown ethers, forms with good selectivity a complex cation with potassium ions.

Using suitable colored anions as counter-ions, ion-pairs can be formed which are extractable with organic solvents and can thus serve as the basis for the very selective and sensitive determination of potassium. In the analysis of pharmaceutical products, picrate is usually used as the ion-pair forming reagent, and the extraction is carried out with dichloromethane or 4:1 mixture of dichloromethane and toluene. The absorbance of the extracts is determined at 358 nm.[35,37,38] The use of tropeolin 000 has also been reported (extraction with chloroform, measurment at 482 nm).[26]

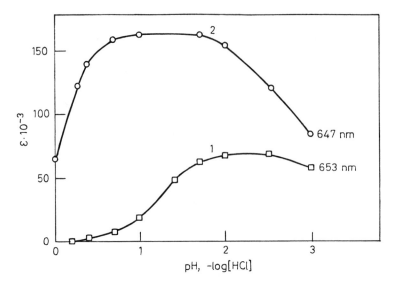

FIGURE 10.15. The absorbance of the samarium(III) arsenazo-III complex at the absorption maximum as a function of pH and hydrochloric acid concentration, respectively. 1. Solvent: water. 2. Solvent: 4:1 v/v mixture of methanol and water.

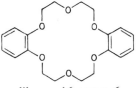

dibenzo-16-crown-6

6. Some Other Elements

As regards the spectrophotometric determination of nonmetals, of the books listed in Section 10.P.1, that of Marczenko[29] and in addition the monograph of Boltz and Howell[4] devoted to this topic are important sources in the literature.

Of the free halogens, *iodine* possesses UV-visible spectra suitable for analytical measurements. Bromine is also colored; its direct spectrophotometric determination, however, is difficult because of its reactivity.

The absorption maximum of iodine in apolar solvents (and in the gaseous phase) is around 515 nm. The molar absorptivity is 925 nm in carbon tetrachloride. In electron donor-type polar or aromatic solvents this violet color changes to various shades of yellow, brown or red as a consequence of the formation of charge-transfer complexes. For example, the maximum in ethanol is at 445 nm ($\varepsilon = 940$). The selectivity and sensitivity of the determination can be greatly increased if iodine is measured as the blue adsorption complex with starch: $\varepsilon_{590\ nm} \sim 40,000$. An indirect approach for the spectrophotometric determination of oxidizing agents is to allow them to react with iodides and measure the absorbance of the resulting stoichiometric quantity of iodine.

Of the halide anions, iodide ion possesses an absorption band in the readily accessible ultraviolet range, with maximum at 226 nm in aqueous medium. The charge-transfer complexes of chloride and bromide ions with iodine absorb in the longer wavelength UV region, thus creating the basis for their determination (with or without solvent extraction). For details the book of Marczenko[29] is cited. Some aspects of this question are briefly discussed in Section 7.I.

Free chlorine and bromine (or chloride and bromide ion after their transformation to chlorine and bromine) can be determined based on their interaction with various dyes. This interaction can be the decolorization of the dye (e.g., methyl red, methyl orange): in these cases, the decrease of the absorbance is proportional to the concentration of the halogen. In other cases the halogenation of the dye results in characteristic spectral changes, and this can also be the basis for quantitative measurements; for example, the yellow phenolsulphonphthalein (phenol red) is changed to violet tetrabromophenolsulphonphthalein (bromophenol blue) during the bromination reaction.[29]

Several ligand exchange reactions are also available for the indirect determination of halide ions. Of this type are the methods of Ichikawa et al. for the determination of chloride,[19] bromide and iodide[20] ions. These methods are based on the reaction of the halide ions with silver(I) 1,5-bis-(6-methylpyrimidyl)-carbazone complex. In the course of this reaction silver halides are precipitated and the violet-colored complexes decompose. The decrease of the absorbance at 530 nm is proportional to the concentration of the halides ($\varepsilon = 12,000$). This reaction can be applied to the determination of covalently bound halogens in pharmaceutical products after combustion in a Schöniger flask.

Combustion in the Schöniger flask is likewise the first step in the determination of the iodine content of thyroid preparations. In the method of Graham[16] the liberated iodine is trapped and determined spectrophotometrically as the triiodide ion. A much more sensitive, automated procedure developed by McGary[30] is based on the catalytic effect of iodine on the cerium(IV) arsenite reaction. The decolorization of cerium(IV) is monitored at 405 nm. Negative peaks are obtained when aliquots of the solution in the Schöniger flask, which contain the iodide, are injected into the stream of cerium(IV) arsenite in an automatic analyzer. The peak height is proportional to the concentration of iodine. The sensitivity of the method enables the quantitative determination of an iodine concentration as low as 2.5 ng/ml.

Covalently-bound *fluorine* is determined after Schöniger combustion making use of the tendency of the resulting fluoride ions to form stable complexes.[4,29] There are two general methods for the quantitation:

1. Decolorization of colored complexes on the effect of ligand exchange reaction and measurement of the decrease of the absorbance. Complexes having been used for this purpose are zirconium(IV) complexes of eriochrome cyanine-R or alizarine S or xylenol orange, thorium(IV) complexes of alizarine S or arsenazo I, etc.
2. Direct measurement at 610 nm which is the maximum of the ternary complex cerium(III)-alizarine complexone-fluoride.

One of the most widely used, classical methods for the non-selective determination of *nitrogen* containing materials is their digestion with concentrated sulphuric acid containing additives in a Kjeldahl flask, followed by basification distillation and titration of the resulting ammonia. Spectrophotometric micromethods are also based on the same general principle: ammonia is determined in the distillate using color reactions, e.g., reaction with potassium iodomercurate (Nessler reagent) or phenol + hypochlorite (indophenol reaction). The recent method of Devani et al.[9] merits attention. The Hantzsch reaction between ammonia, formaldehyde and acetylacetone takes place in acidic solution, and hence no basification and distillation steps are necessary to form the dihydrolutidine derivative which can be measured around 410 nm.

$$NH_3 \;+\; 2\,CH_3{-}\overset{\displaystyle O}{\overset{\|}{C}}{-}CH_2{-}\overset{\displaystyle O}{\overset{\|}{C}}{-}CH_3 \;+\; CH_2O \;\xrightarrow{H^+}\;$$

(10.35)

Nitrate ion can be determined on the basis of nitration of toluene reagent in 56% sulphuric acid. The resulting nitrotoluene is measured at 284 nm.[3]

When pharmaceutical compounds containing *phosphorus* are analyzed, the easier task is the determination of phosphate ions. The most generally used method for their spectrophotometric determination is based on their tendency to form heteropolyanions. For example in acidic media molybdophosphate is formed with the excess of molybdate ions and this is transformed by reducing agents, e.g., hydrazine to phosphor-molybdenum blue which can be measured around 800 nm. Its molar absorptivity strongly depends on the experimental conditions, but it is usually on the order of magnitude of 10,000, thus enabling sensitive determinations to be carried out. This method is applied e.g., by USP XXII for the determination of the phosphate content of corticotropine phosphate injection.[46a] It is important to note that this general reaction is suitable for the indirect determination of various drug substances with reducing properties (mainly phenol derivatives); several examples can be found in Chapter 10.

In order to increase the selectivity, the colored species can be extracted with organic solvents (mainly with butanol). In another version the molybdophosphate is precipitated with an excess of quinoline and the assay is based on the determination of the excess of the reagent in the supernatant liquid at 313 nm.[18] Neither reduction nor precipitation steps are necessary when the molybdovanadatophosphate method is used; the reaction product forming by ammonium metavanadate and ammonium molybdate reagents can be directly measured at 400 nm.[29]

The methods described above are not directly applicable to phosphate ester derivatives; these must be hydrolyzed prior to the assay. This is often carried out by using suitable enzymes. For example corticosteroid 21-phosphates were determined after hydrolysis with phosphatase enzyme using the phosphomolybdate color reaction.[12]

Other types of organophosphorus compounds can be determined by the methods outlined above if their phosphorus content has previously been converted to phosphate. This can be done by applying the Schöniger combustion method,[28] but various wet oxidative digestion methods are also available.

What has been described about the possibilities for the determination of phosphorus applies to a great extent also to *arsenic*, which was already discussed in the previous section. Its most important quantitative photometric determination is the arsenomolybdenum blue method which is analogous to the phosphomolybdenum blue procedure.[4,28,29]

REFERENCES

1. Asuero, A. C.; Rosales, D.; Rodriguez, M. M. *Michrochem. J.* 1984, *30*, 33.

2. Beseda, A.; Gawagious, Y. A.; Tadros, N. B.; Ibrahim, L. F. *Pharmazie* 1987, *42*, 482.

3. Bhatty, M. K.; Townshend, A. *Anal. Chim. Acta.* 1971, *56*, 55.

4. Boltz, D. F.; Howell, J. (Ed.), *Colorimetric Determination on Nonmetals*, 2nd. Ed., Wiley: New York, 1978.

5. *British Pharmacopoeia 1988*, Her Majesty's Stationery Office: London, 1988. a. 77; b. 245; c. 810.

6. Burger, K. *Organic Reagents in Metal Analysis*, Akadémiai Kiadó: Budapest; Pergamon Press: Oxford, 1973.

7. Capitán, F.; Ramirez, A. A.; Linares, C. J. *Analyst* 1986, *111*, 739.

8. Desai, G. S.; Shinde, V. M. *Bull. Chem. Soc. Japan* 1991, *64*, 1951.

9. Devani, M. B.; Shishoo, C. J.; Shah, S. A.; Suhagia, B. N. *J. Ass. Off. Anal. Chem.* 1989, *72*, 953.

10. Eldawy, M. A.; Eishabouri, S. R. *Anal. Chem.* 1975, *47*, 1844.

11. Fleitman, J. S.; Partridge, I. W.; Neu, D. A. *Drug. Dev. Ind. Pharm.* 1991, *17*, 519.

12. Genius, O. B. Dt. *Apoth. Ztg.* 1973, *113*, 297.

13. Gónzalez-Portal, A.; Baluja-Santos, C.; Bermejo-Mártinez, F. *Analyst* 1986, *111*, 547.

14. Gónzalez-Portal, A.; Bermejo-Mártinez, F.; Baluja-Santos, C.; Diez-Rodriguez, M. C. *Microchem. J.* 1985, *31*, 368.

15. Görög, S.; Sütö, J. *Acta Pharm. Hung.* 1981, *51*, 217.

16. Graham, J. H. *J. Pharm. Sci.* 1975, *64*, 1393.

17. Hasebe, K.; Taga, M. *Talanta* 1982, *29*, 1135.

18. Hassan, S. S. M.; Eldesouki, M. H. *Microchim. Acta II.* 1981, 261.

19. Ichiba, H.; Morishita, M.; Katayanagi, M.; Yajima, T. *Chem. Pharm. Bull.* 1986, *34*, 739.

20. Ichiba, H.; Morishita, M.; Yajima, T. *Chem. Pharm. Bull.* 1988, *36*, 5009.

21. Irving, H. M.; *Dithizone*, Chemical Society: London, 1977.

22. Issopoulos, P. B.; Economou, P. T. *J. Anal. Chem.* 1992, *342*, 439.

23. Iwantscheff, G. *Das Dithizon und seine Anwendung in der Mikro und Spureanalyse*, 2 Ausgabe, Verlag Chemie: Weinheim, 1972.

24. Jones, P.; Nickless, G. *Analyst* 1978, *103*, 1121.

25. Juneau, R. *J. Pharm. Sci.* 1977, *66*, 140.

26. Kovalchuk, T. V.; Pyatnitskii, I. V.; Nazarenko, O. Yu.; Koget, T. A.; Tsisar, S. S.; Vetyutneva, N. O. *Farm. Zh.* 1986, 43.

27. Krzek, J.; Al-Mutari, E. *Farm. Pol.* 1980, *36*, 31.

28. Macdonald, A. M. G. *Analyst* 1961, *86*, 3.

29. Marczenko, Z. *Separation and Spectrophotometric Determination of Elements,* Ellis Horwood: Chichester, 1986.

30. McGary, E. D. *J. Pharm. Sci.* 1980, *69*, 948.

31. Medwick, T. Metal-Containing Organic Compounds, in *Pharmaceutical Analysis,* Higuchi, T,; Brochmann-Hanssen, (Ed.), Interscience: New York, 1961.

32. Morales, M. T.; Montana, M. T.; Galan, G.; Gómez-Ariza, J. L. *Farmaco* 1990, *45*, 673.

33. Nobile, L.; Raggi, M. A. *Pharmazie* 1991, *46*, 138.

34. Raggi, M. A.; Cavrini, V.; Di Pietra, A. M. *Pharm. Acta Helv.* 1984, *59*, 225.

35. Raggi, M. A.; Cavrini, V.; Di Pietra, A. M.; Nobile, L. *Acta Pharm. Suec.* 1988, *25*, 133.

36. Raggi, M. A.; Nobile, L.; Perboni, I.; Ramini, M. *Int. J. Pharm.* 1990, *61*, 199.

37. Raggi, M. A.; Nobile, L.; Ramini, M.; Perboni, I. *J. Pharm. Biomed. Anal.* 1989, *7*, 1545.

38. Raggi, M. A.; Nobile, L.; Ramini, M.; Perboni, I. *Farmaco* 1990, *45*, 745.

39. Rolfe, A. C.; Russel, F. R.; Wilkinson, N. T. *Analyst* 1955, *80*, 523.

40. Rosales, D.; Gómez Ariza, J. L. *Anal. Chem.* 1985, *57*, 1411.

41. Sandell, E. B. *Colorimetric Determination of Traces of Metals,* 3rd Ed., Interscience: New York, 1959.

42. Sandell, E. B.; Onishi, H. *Photometric Determination of Traces of Metals,* Wiley: New York, 1978.

43. Snell, F. D. *Photometric and Fluorimetric Methods of Analysis, Metals, Non-metals,* Wiley: New York, 1981.

44. Sullivan, D. J. *J. Ass. Off. Anal. Chem.* 1977, *60*, 1350.

45. Sundaramurthi, N. M.; Shinde, V. M. *Analyst* 1991, *116*, 541.

46. *The United States Pharmacopoeia XXII,* USP Convention Inc.; Rockville, 1990. a. 356; b. 564; c. 566; d. 722; e. 1520; f. 1525.

Q. MISCELLANEOUS

1. Cationic Detergent-Type Quaternary Ammonium Derivatives

As described in detail in Section 6.H, anionic and cationic drugs (and those which can easily be transformed by protonation or deprotonation to the cationic or anionic form) are most generally determined by forming ion-pairs with suitable reagents followed (in the majority of cases) by solvent extraction and the spectrophotometric measurement of the ion-pair. Quaternary ammonium compounds possess especially favorable characteristics for the formation of ion-pairs: in the ionized state, these are independent of the pH and for this reason their extractibility is not pH-dependent, provided that a suitable ion-pairing reagent is used. In addition, the conditions of extractibility are outstanding for cationic surfactants containing bulky apolar groups. On this basis, it is quite natural that the most widespread methods in the spectrophotometric analysis of drugs discussed in this section are based on ion-pair formation.

As the ion-pairing reagents, the anionic dyes listed in Section 7.H can be used. Of the newer applications a few are mentioned here: the determination of cetylpyridinium chloride using methyl orange, Congo Red and picric acid as the reagents,[57] determination of the same and dequalinium bromide with bromocresol purple,[32] determination of septonex and ajatin with Chrome Azurol S[17,26] and measurement of trimethylanilinium benzenesulphonate after extraction with methyl orange.[36] In these examples the extraction was carried out with chloroform from neutral or slightly acidic aqueous solutions. Further examples are presented in Section 7.H.

In the field of ion-pair extraction spectrophotometric analysis of quaternary ammonium compounds, the investigations of Sakai (greatly increasing both the selectivity and the sensitivity of the measurement) merit special discussion. When dibasic acids are used as the ion-pairing reagents (e.g., bromophenol blue, bromocresol green), their dissociation takes place in two steps. If, e.g., the extraction of benzethonium chloride is carried out with bromophenol blue at pH 3.8 where only the sulphonic acid group of the reagent dissociates, the yellow ion pair is extracted with chloroform (λ_{max} = 418 nm). If the pH is raised to 6.7, the other acidic group of phenol type also dissociates. Adding quinine to the reaction mixture at this pH, the dianionic form of the reagent forms a monoanionic ion-pair. This is an excellent reagent for quaternary ammonium compounds: the blue colored ternary ion associate is extractable with chloroform at pH 6.7 ($\varepsilon_{610\,nm}$ = 31,400).[41,42]

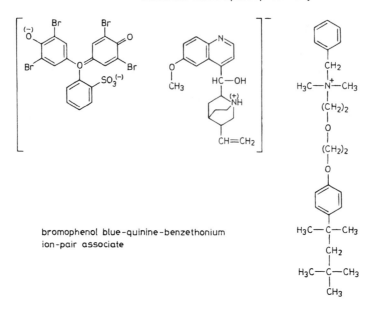

bromophenol blue-quinine-benzethonium
ion-pair associate

The very selective and sensitive determination of benzalkonium salts was also described by Sakai using tetrabromophenolphthalein as the reagent for the extraction with 1,2-dichloroethane at pH 7.5 ($\varepsilon_{610 \text{ nm}}$ = 73,000). An interesting feature of this method is that the spectrophotometric measurement is carried out at 60°C. This is because at this temperature, the interference of the co-extracted tertiary amines are eliminated: namely, the latter show selective thermochromism, while the spectra of the ion-pair associates of quaternary ammonium compounds are independent of the temperature.[43] (See Section 2.K.2 for more details about thermochromism.)

No extraction step is necessary in the procedure of Stevens and Eckardt[48] for the determination of polyquaternium-1 using tryptane blue dye as the reagent. Its absorption maximum at 658 nm in aqueous solution is shifted to 700 nm as a result of ion-pair formation. This enables the selective determination of the water-soluble ion-pair at the maximum of the difference spectrum at 680 nm. Based on the same principle, no extraction was necessary either when benzalkonium chloride was determined with bromothymol blue as the ion-pairing reagent.[31]

The above-mentioned polyquaternium-1 can be determined in various pharmaceutical products using the dye Ponceau S as the reagent in the presence of magnesium ions.[18] A precipitate is formed which can be separated by centrifugation. The unreacted dye can be measured in the supernatant liquid at 510 nm. The concentration of polyquaternium-1 is calculated from the decrease of the absorbance.

The latter method leads to the use of metal complex formation for the determination of quaternary ammonium compounds. In this field the paper of Fujita et al.[15] merits special attention. In this method, ternary complexes are formed with manganese(II) ions and o-hydroxyhydroquinonephthalein. The analyte (various quaternary ammonium compounds with long hydrocarbon chains) and the dye form 2:2:1 complexes with the metal ion. The molar absorptivity of hexadecylpyridinium chloride at pH 9.3 at 575 nm is 38,000. It is worth mentioning that using the same principle, non-quaternary polyamines (spermine, spermidine) could also be determined. In this case the metal component was copper(II) and the pH was set at 4.6. Extremely high sensitivity was obtained: the molar absorptivities were 700,000 and 300,000, respectively.[16]

In an indirect application of metal complex formation for the determination of quaternary ammonium derivatives, thiocyanate ions were used to extract them from aqueous to chloroform phase. The thiocyanate content of the extract can be measured at 500 nm as the iron(III) thiocyanate complex produced by reaction with acetonic iron(III) nitrate reagent, and this absorbance is proportional to the concentration of the quaternary ammonium compound.[53]

The application of charge-transfer complex formation for the determination of quaternary ammonium halides is demonstrated in Section 7.I.

2. Glycerol and its Derivatives

In most of the methods for the determination of glycerol, its reducing properties are exploited for analytical purposes.[27] Using dichromate in sulphuric acid medium glycerol is oxidized to carbon dioxide. The spectrophotometric determination can be based on the measurement of either the unreacted dichromate or the resulting chromium(III). More selective are the methods where the oxidizing agent is sodium periodate. In this reaction, one mole of formic acid and two moles of formaldehyde are produced. The latter can be determined by a variety of color reactions. Of these the chromotropic acid reaction is usually used in the analytical chemistry of glycerol (see p. 179). The optimization of this procedure is described in the papers of Karawya and Ghourab[28] and Sturgeon et al.[49]

Enzymatic methods play a predominant role in the spectrophotometric determination of glycerol in biological samples. In the method of Schneider,[44] glycerol is phosphorylated and subsequently dehydrogenated using the enzymes glycerol kinase and glycerol phosphate dehydrogenase, ATP and NAD (see Section 7.K). The resulting NADH is measured at 505 nm using the iodonitrotetrazolium reagent.

The spectrophotometric determination of the nitric acid ester derivative of glycerol (glycerol trinitrate, nitroglycerin) is usually based on the measurement of the nitric and nitrous acids forming during the hydrolysis of the ester. The acid-catalyzed hydrolysis results in nitric acid; this can be determined by means of nitration of aromatic reagents.

Alkaline hydrolysis is more important. As seen in Equation 10.36, the hydrolysis is accompanied by intramolecular oxido-reduction leading to 2 moles of nitrite ions.

$$
\begin{array}{l}
CH_2-O-NO_2 \\
| \\
CH-O-NO_2 \quad + \quad 5\,OH^- \quad = \quad 2\,NO_2^- \quad + \quad NO_3^- \quad + \quad CH_3COO^- \quad + \\
| \\
CH_2-O-NO_2 \qquad\qquad\qquad\qquad\qquad + \quad HCOO^- \quad + \quad H_2O
\end{array}
\qquad (10.36)
$$

Using reaction 10.36 the assay can be based on the spectrophotometric determination of the resulting nitrite ions. Several variants of the diazotization azo-coupling reaction are suitable for this purpose (see Section 7.D). The content uniformity test for nitroglycerin tablets in USP XXII prescribes an automated version of this technique. The hydrolysis is carried out by strontium hydroxide; the nitrous acid liberated upon acidification reacts with procaine to form the diazonium salt, which is then coupled with N-(1-naphthyl)-ethylenediamine. The resulting azo-dye is measured at 480 nm.[54a]

Other methods are also available using the same general principle. If acridine derivatives (ethacridine, diaminoacridine) are used as the reagents their primary amino group is diazotized in the first step of the reaction, and the excess of the reagent acts as the coupling agent in the second step. This method was used for the determination of nitroglycerin[22,51] and also nitrosorbide.[23] The determination of pentaerithritol tetranitrate was based on reduction with zinc dust followed by nitrosylation of 1-naphthylamine.[46]

3. Sugars, Sugar Alcohols and Polysaccharides

In the classical analytical chemistry of carbohydrates and related derivatives, color reactions played a predominant role. The majority of these methods were based on the reducing properties of the analytes or the condensation reactions of their oxo-groups. Most of these methods can now be regarded as being more or less obsolete and, hence, not even an outlined discussion is attempted here. The books listed in Section 10.A contain detailed discussions and cite the vast quantity of literature to date regarding this field.

Not even the modern analytical chemistry of carbohydrates can omit spectrophotometric methods based on color reactions. The majority of the field of application of these methods is in the domain of clinical analysis and is therefore beyond the scope of this book. For this reason only the basic principles will be dealt with here with some examples; regarding the details further publications are cited.[55,56]

As a consequence of the lack of the oxo-group in their molecules, the color reactions of sugar alcohols are restricted to procedures based on their reducing properties. Sorbitol, mannitol[40] and xylitol[1] can be determined in pharmaceutical products by their oxidation with cerium(IV)-nitrate in strongly acidic medium at elevated temperature and measuring the excess of the oxidizing agent at 420 nm. The above listed compounds as well as calcium gluconate can be determined using periodate reagent: as is described in the preceding section in connection with glycerol, it cleaves these polyhydroxy compounds and the assays can be based on the measurement of the resulting formaldehyde.[1,12]

Because of the great importance of the determination of sugars in clinical analysis, their spectrophotometric analysis (based on the modern versions of classical photometric procedures and on enzymatic reactions) are fully automated. A typical example from the first group is the determination of glucose after its oxidation with ferricyanide in alkaline medium and measurement of the unreacted reagent at 420 nm. The reagent of another important method is o-toluidine[1] which reacts with glucose in a reaction with a rather complicated mechanism (condensation followed by dehydration) to form a green product, measurable at 645 nm.

The most important reagent for the determination of fructose is resorcinol–hydrochloric acid. As seen in Equation 10.37, 5-hydroxymethylfurfural, primarily formed from fructose and hydrochloric acid, reacts with resorcinol. Glucose does not interfere with this assay which was utilized for the determination of fructose in pharmaceutical formulations.[1] The red reaction product can be measured either at 475 or at 296 nm (at the latter wavelength the sensitivity is higher).

Belal et al.[4] developed a method for the determination of glucose, fructose, etc., based on the formation of osazones with phenylhydrazine and subsequent reaction with tetracyanoethylene to form charge-transfer complexes.

The intermediate of the reaction 10.37, (5-hydroxymethylfurfural) has key importance in the spectrophotometric analysis of pharmaceutical products containing sugars and sugar alcohols, since this compound is their characteristic impurity (degradation product). Its natural absorption allows the determination of this impurity in bulk materials and in simple pharmaceutical products[7c,10] ($\varepsilon_{283\ nm} = 16,300$). The alternative of this method used in the British Pharmacopoeia is the above mentioned method using resorcinol as the reagent: this is used as a limit test in USP XXII.[54b] In the case of pharmaceutical products where the other components interfere with the assay, the two difference spectrophotometric methods of Davidson and Dawodu can be used. As seen in Equation 10.37, the aldehyde group of the impurity is reduced with sodium borohydride, thus forming a spectrophotometrically inactive derivative and, hence, the maximum of the difference spectrum is also at 283 nm (see Section 8.D).[9,10] The reagent of the other method is isonicotinoyl hydrazide ($\lambda_{max} = 340$ nm).[11]

Coming back to the spectrophotometric analysis of carbohydrates themselves, it must be stated that enzymatic methods are more developed than the classical methods and their newer versions discussed so far both from the point of view of selectivity and their capability of being automated. The most important enzymatic method for the determination of glucose depends on its oxidation with atmospheric oxygen to gluconic acid in the presence of the enzyme glucose oxidase with simultaneous formation of an equivalent quantity of hydrogen peroxide. In the different variants of the second step of the procedure, the latter reacts with the aid of the enzyme peroxidase with various reagents such as o-dianisidine, 4-aminoantipyrine + phenol to form colored products. (For the equation of the latter reaction see p. 179). If methanol is used as the reagent it is oxidized to formaldehyde which undergoes a ring closure reaction with acetylacetone and ammonia to form a dihydrolutidine derivative (see p. 365). In another enzymatic method glucose is oxidized with NAD with the aid of glucose dehydrogenase. The quantity of the resulting NADH is equivalent to that of glucose and it is measurable directly at 340 nm (see Section 7.K).

For the determination of fructose an enzymatic procedure is available which is analogous to the enzymatic procedure described for glycerol in the preceding section: enzymatic phosphorylation and dehydrogenation and measurement of the forming NADH. Several further possibilities are described in the publications cited above.[5,56,57]

One possibility for the spectrophotometrically important *mucopolysaccharides* (heparin, condroitin sulphate) is the acid-catalyzed or enzymatic degradation of the biopolymer and measurement of the monomers. For example in the method of Fabergas and Casassas[14] for the determination of condroitin sulphate, fluoroglucinol-hydrochloric acid reagent is used. The role of hydrochloric acid is to degrade the polymer and the transformation of the monomer to a furfural derivative which is capable of condensation with fluoroglucinol ($\lambda_{max} = 448$ nm); (for a similar reaction mechanism

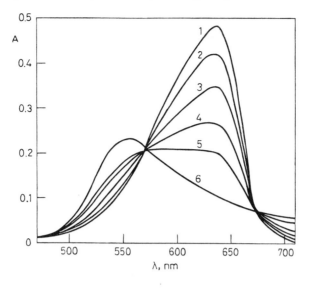

FIGURE 10.16. The effect of heparin of the spectrum of Nile Blue. Solvent: water; cell length: 20 mm; t = 20°C. Concentration of Nile Blue: $4 \cdot 10^{-6}$M. Concentration of heparin (113 IU/mg): 1: 0; 2: $1.6 \cdot 10^{-4}$ g/l; 3: $3.2 \cdot 10^{-4}$ g/l; 4: $4.8 \cdot 10^{-4}$ g/l; 5: $6.4 \cdot 10^{-4}$; 6: $16 \cdot 10^{-4}$ g/l. (From Stuzka, V.; Havlová, J.; Dvorák, J.; Vávra, V. *Cesk Farm.* 1984, *33*, 412.[50])

see Equation 10.37). Spectrophotometric methods depending on acid-catalyzed degradation are available in the analytical chemistry of heparin, too.[45]

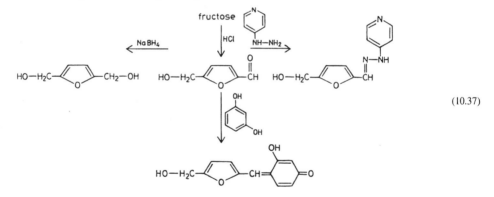

(10.37)

More important are the methods which do not require preliminary hydrolysis and are based on the interaction of the polyanionic heparin with quaternary ammonium cations. Among others cetyltrimethylammonium bromide and lauryl-pyridinium bromide are used as the reagents. The ion pairs formed with heparin are insoluble in water and are soluble in chloroform and as such, enable (after extraction) the determination of heparin in the ultraviolet range. Another possibility is the measurement of the excess reagent in the filtrate after filtering off the precipitate.[24]

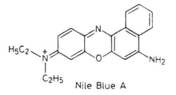

Nile Blue A

In analytical practice methods depending on the interaction of heparin with cationic or neutral dyes are more generally used. In these methods this interaction results in the hypsochromic shift of the spectrum of the dye. As seen in Figure 10.16, the dye Nile Blue A changes its spectrum in the presence of various concentrations of heparin. The absorbance at the absorption maximum of the dye is inversely proportional to the concentration of heparin and is therefore a suitable means for its determination. The results of this method and several analogous methods which are based on the association reaction taking place in homogeneous aqueous solution without any separation steps are in good correlation with those of the biological assay of heparin.[50] Of the numerous dyes showing the above described phenomenon of metachromicity in the presence of heparin, the decrease of absorbance at 620 nm of Azure A is most frequently used for the indirect determination of heparin[30] (see also Section 7.H).

4. Cardiac Glycosides

The spectrophotometric activity of the cardiac glycosides of *Digitalis lanata* is restricted to the α, β-unsaturated lactone group; the band appearing around 215 nm is at a wavelength which is too short to be used in direct spectrophotometric analysis. It is, however, important in monitoring high-performance liquid chromatograms of cardiac glycosides by using a UV detector.

Spectrophotometric analysis of cardiac glycosides is principally based on color reactions. In addition to the general monographs listed in Section 10.A, the books of Bartos and Pesez[3] and Görög[19] contain comprehensive chapters dealing with these methods.

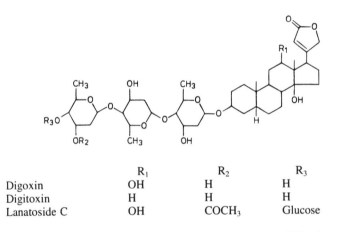

	R_1	R_2	R_3
Digoxin	OH	H	H
Digitoxin	H	H	H
Lanatoside C	OH	$COCH_3$	Glucose

Some characteristics of the numerous methods which seem to still be in use are summarized in Table 10.11. As seen in Table 10.11, some of the methods are based on the steroid skeleton, another group of the methods on the unsaturated lactone ring, while the third (and most important) group is based on the carbohydrate moiety.

Aromatic nitro derivatives attack the molecule of cardiac glycosides at the active methylene group of the lactone ring[8,29,39] while the iron(III) hydroxamate method[47] depends on the hydroxyl-aminolysis of the same. The first step of the reaction of the periodic acid thiobarbiturate methods,[25,33–35] excelling in their extremely high sensitivities is the cleavage of the 2-deoxyhexose type digitoxose units by periodic acid to form malondialdehyde. This reacts with two molecules of thiobarbituric acid with the formation of the violet-colored reaction product. Xanthydrol is also the reagent of the deoxy sugars.[2,52] In connection with the use of dixanthyl urea reagent,[6,37,38] it must be noted that in fact, this is also an indirect application of the xanthydrol method since under the acidic conditions of the reaction, this reagent hydrolyzes to xanthydrol. In this manner the difficulties originating from the instability of the xanthydrol reagents can be avoided. It is worth mentioning that the classical method involving the use of the iron(III) chloride-acetic acid-sulphuric acid reagent is official in the latest revision of the British Pharmacopoeia.[7b]

Finally Section 5.D of this book is referred to in which the method of Horváth[20] is described for the determination of the digoxin content of the drug *Digitalis lanata* based on TLC separation, spot elution and spectrophotometry using the above mentioned xanthydrol reaction. The role of this and numerous other color reactions in the TLC densitometry of digitalis glycosides was also reviewed by Horváth.[21]

TABLE 10.11. Some Spectrophotometric Methods for the Determination of Cardiac Glycosides

Author/Year	Group involved in the reaction	Reagent	λ_{max}, nm	ε	Sample	Chromatographic separation
Evans, 1974	Steroid skeleton	SbCl$_3$	540, 592	4,000	Plant extracts	+
Khafagy, 1974 BP, 1988	Butenolide ring	Picric acid + NaOH	494–495	— pharm. products	Plant extracts,	±|
Rabitzsch, 1969	Butenolide ring	2,2',4,4'-Tetranitrobiphenyl + NaOH	630	26,800	Digitoxin	+
Burns, 1976	Butenolide ring	1,3,5-Trinitrobenzene + NaOH	565	11,400 (Digitoxin)	Standards	—
Solich, 1987, 1988	Butenolide ring + acetyl sugar	NH$_2$OH·HCl + NaOH + Fe(NO$_3$)$_3$	515	1,120 (Lanatoside C)	Pharm. products	—
Mesnard, 1961, 1962; Myrick, 1969; Juhl, 1975	Sugar moiety	HIO$_4$ + arsenous acid + thiobarbituric acid	530–532	158,000*	Pharm. products	—
Bartos, 1963	Sugar moiety	Xanthydrol + 4-toluenesulphonic acid	570	—	Standards	—
Szilágyi, 1977	Sugar moiety	Xanthydrol + HCl	530	46,000	Plant extracts	+
Bican-Fister, 1969	Sugar moiety	Dixanthylurea	535	—	Standards	+
Pötter, 1963, 1972	Sugar moiety	Dixanthylurea	530	64,000 (Digoxin)	Plant extracts	+

*Private communication from P. Horváth.

REFERENCES

1. Baer, A.; Kraemer, *Krankenhauspharmazie* 1983, *4*, 115.
2. Bartos, J. *Ann. Pharm. France* 1963, *21*, 603.
3. Bartos, J.; Pesez, M. *Colorimetric and Fluorimetric Analysis of Steroids*, Academic Press: London, 1976, 219–244.
4. Belal, S.; El Kheir, A. A.; Ayad, M.; El Adl, S. *Microchem. J.* 1988, *37*, 25.
5. Bergmayer, H. U. *Grundlagen der Enzymatischen Analyse*, Verlag Chemie: Weinheim, 1977.
6. Bican-Fister, T.; Merkas, J. *J. Chromat.* 1969, *41*, 91.
7. *British Pharmacopoeia 1988*, Her Majesty's Stationery Office: London 1988.
8. Burns, L. The Determination of the Path of Reaction of Oubain and Digitoxin with Polynitroaromatic Compounds in an Alkaline Medium and the Development of Alternative Methods of Assay for Oubain and Digitoxin, Ph.D. Thesis, Temple University, Philadelphia, 1976.
9. Davidson, A. G. *J. Clin. Pharm. Therap.* 1987, *12*, 11.
10. Davidson, A. G.; Dawodu, T. O. *J. Pharm. Biomed. Anal.* 1987, *5*, 213.
11. Davidson, A. G.; Dawodu, T. O. *J. Pharm. Biomed. Anal.* 1988, *6*, 61.
12. Devani, M. B.; Shishoo, C. J.; Suhagia, B. N.; Shah, S. A. *Indian J. Pharm. Sci.* 1990, *52*, 231.
13. Evans, F. J. *J. Chromat.* 1974, *88*, 411.
14. Fabergas, J. L.; Casassas, E. *Pharm. Acta Helv.* 1981, *56*, 265.
15. Fujita, Y.; Mori, I.; Kitano, S.; Fujita, K. *Anal. Sci.* 1985, *1*, 175.
16. Fujita, Y.; Mori, I.; Kitano, S.; Kamada, Y. *Bunseki Kagaku* 1984, *33*, E103.
17. Gasparic, J.; Filipová, M. *Cesk. Farm.* 1990, *39*, 400.
18. Good, R. M.; Liao, J. C.; Hook, M. J.; Punko, C. L. *J. Ass. Off. Anal. Chem.* 1987, *70*, 979.
19. Görög, S. *Quantitative Analysis of Steroids*, Akadémiai Kiadó, Budapest; Elsevier: Amsterdam, 1983, 375–378.
20. Horváth, P. *Acta Pharm. Hung.* 1982, *52*, 133.
21. Horváth, P. Analysis of Cardiac Glycosides in Plant Extracts, in *Steroid Analysis in the Pharmaceutical Industry*, Görög, S. (Ed.), Ellis Horwood: Chichester, 1989.
22. Ivakhnenko, P. N. *Khim. -Farm. Zh.* 1977, *11*, 93.
23. Ivakhnenko, P. N.; Chigarenko, L. S.; Kilyakova, G. M.; Ventsel, E. S.; Vasilchenko, L. Ya. *Farmatsiya* 1976, *25*, 67.
24. Jarzebinski, J.; Tonska, S. *Acta Pol. Pharm.* 1985, *42*, 101, 374, 571; 1986, *43*, 454.
25. Juhl, W. E. *J. Ass. Off. Anal. Chem.* 1975, *58*, 70.
26. Jurkeviciute, J.; Malat, M. *Cesk. Farm.* 1979, *28*, 379.
27. Kakac, B.; Vejdelek, Z. J. *Handbuch der Kolorimetrie, Band II*, Gustav Fischer Verlag: Jena, 1963, 53–88.
28. Karawya, M. S.; Ghourab, M. G. *J. Ass. Off. Anal. Chem.* 1972, *55*, 1180.
29. Khafagy, S. M.; Girgis, A. N. *Planta Med.* 1974, *25*, 350.
30. Klein, M. D.; Drongowski, R. A.; Linhardt, R. J.; Langer, R. S. *Anal. Biochem.* 1982, *124*, 59.
31. Lowry, J. B. *J. Pharm. Sci.* 1979, *68*, 110.
32. Lukyanchikova, G. I.; Dukkardt, L. N. *Farmatsiya* 1989, *38*, 60.
33. Mesnard, P.; Devaux, G. *Compt. Rend. Acad. Sci.* 1961, *253*, 479.
34. Mesnard, P.; Devaux, G. *Chim. Anal.* 1962, *44*, 287.
35. Myrick, J. W. *J. Pharm. Sci.* 1969, *58*, 1018.
36. Palaniappan, A.; Begg, M. M.; Kaushal, V.; Balasubramanian, M. *Analyst* 1985, *110*, 47.
37. Pötter, H. *Pharmazie* 1963, *18*, 554; 1965, *20*, 737.
38. Pötter, H.; Barisch, H. *Pharmazie* 1972, *27*, 315.
39. Rabitzsch, G.; Jüngling, S. *J. Chromat.* 1969, *41*, 96.
40. Ruckriegel, O.; Pfueller, A. *Krankenhaus-Apoth.* 1979, *29*, 62.
41. Sakai, T. *Anal. Chim. Acta* 1983, *147*, 331.
42. Sakai, T. *Analyst* 1983, *108*, 608.
43. Sakai, T.; Ishida, N. *Bunseki Kagaku* 1978, *27*, 410.
44. Schneider, P. B. *J. Lipid Res.* 1977, *18*, 396.
45. Scott, J. E. *Biochem. J.* 1979, *183*, 91.
46. Shingbal, D. M.; Agni, R. M. *J. Ass. Off. Anal. Chem.* 1984, *67*, 1123.
47. Solich, P.; Karlicek, R.; Jokl, V. *Cesk. Farm.* 1987, *36*, 327; 1988, *37*, 193.
48. Stevens, L. E.; Eckardt, J. I. *Analyst* 1987, *112*, 1619.
49. Sturgeon, R. J.; Deamer, R. L.; Harbison, H. A. *J. Pharm. Sci.* 1979, *68*, 1064.
50. Stuzka, V.; Havlova, J.; Dvorák, J.; Vávra, V. *Cesk. Farm.* 1984, *33*, 412.

51. Svach, M.; Zyka, J. *Z. Anal. Chem.* 1955, *148*, 1.
52. Szilágyi, I.; Zámbó, I.; Török, J. *Planta Med.* 1977, *32*, 60.
53. Taha, A. M.; Abdelkader, M. A.; Abdelfattah, S. *Pharmazie* 1980, *35*, 93.
54. *The United States Pharmacopoeia XXII*, USP Convention Inc.: Rockville, 1990. a. 1474; b. 594; c. 633.
55. Tietz, N. *Fundamentals of Clinical Chemistry*, Saunders: Philadelphia, 1976.
56. Wootton, I. D. P. *Microanalysis in Medical Biochemistry*, Churchill Livingstone: London, 1974.
57. Zarapkar, S. S.; Rele, R. V.; Shah, S. V. M. *Indian Drugs* 1987, *24*, 481.

SUBJECT INDEX

The first part of the Subject Index contains the names of spectrophotometrically measured compounds together with other subjects. Bold and italic numbers, respectively, indicate the pages where the spectra and formulae of the compounds appear. In the case of salts of pharmaceutically active bases the designation of the anions is usually omitted.

In the second part, under the title *Reagents and Derivatization Reactions* the respective subjects are listed separately.

A

Absorptivity 12
Acenocoumarol *80*
Acetamide 47
Acetaminophen, see paracetamol
Acetanilide 57, 61, *62*, *81*, 82, 174
Acetic acid *48*
Acetic anhydride 47
Acetone *48*
Acetonitrile 47
Acetophenone 57, 66
Acetyl chloride 47
Acetylene 47
Acetylsalicylic acid 64, 65, *66*, 113, 127, 184, 203, 205, 206, 225, 229, 255, 256, 261, 272
Acid-base dissociation constants 83, 88–90, 131, 187, 281
Aconitine 270, 275
Acridone 146, **147**
Acriflavinium chloride (tripaflavine) 79, *80*
Acrisorcin 227, 307
Acryl aldehyde 52, 53
ACTH and its sulphoxide 180, 321
Actinomycin 341
Adenine 277
Adrenaline (epinephrine) 60, 101, 192, 214, 246, 249, 251, *252*, 264
Adriamycin 338
Ajatin 367
Ajmalicine 148
Alkaloids 1, 34, 61, 74–77, 139, 178, 185, 188, 234, 269–280
Allantoin 282
Allen correction 99, 307
Allopurino 178
Alloxane 180, 305
Aloin 267
Alprenolol 184, *265*
Aluminum 363
Amidopyrine, see aminopyrine
Amikacin 335
Amiloride 309
Aminacrine 239, 306, *307,* **308**
Amino acids 1, 95, 126, 176, 190, 316–322
1-Aminoadamantane 46
4-Aminoantipyrine *193*, 238, 286

2-Aminobenzoic acid 67, 357
4-Aminobenzoic acid and derivatives 67, 261, 263, 264
Aminoglycoside antibiotics 334, 335
2-Amino-5-nitrobenzophenone *182*, 183
Aminophenazone, see aminopyrine
2-Aminopyridine 72, 127
Aminopyrine 69, 105, **106**, 255, 272, *284*, 285, *286*
4-Aminosalicylic acid 65, 67, 255, 256
Amiodarone 311
Amitriptyline *63*, 82, 129, 239, 305, *306*
Amobarbital 281
Amodiaquine 306, *307*
Amoxcillin 129, 338, 339
Amphetamine 58, **115**, *116*, 222, 246, 247, 249
Amphotericin A,B 231, 336
Ampicillin 184, 204, 205, 338–340
Anhydrotetracycline *337*
Anileridine 227
Aniline 57, 60, 83, *86*, 87, 174
Anisol 57
Antazoline *137*, 300–302
Antibiotics 184, 334–341
Antimony 363
Antipyrine 69, *70*, **113**, 130, 255, 262, *284–286*
Antracyclines 336, 338
Apoatropine 271, 272
Apomorphine 270
Apramycin 335
Arginine 317, 319, 340
Arsenic 363, 366
Artemisia judaica 235
Ascorbic acid, see Vitamin C
Ascorbyl palmitate 226
Aspartame 321
Aspirin, see acetylsalicylic acid
Atenolol 121, *123*
Atracurium 231
Atropic acid 272
Atropine 58, 102, 118, 119, 129, 234, 270–272, 274, 275, 277
Autoanalyzer 192, 235, 236, 276, 339, 340, 369
Auxochromes 46–49, 58–61, 64, 69, 71, 75, 78, 258, 261, 300
Azintamide 239
Azlocillin 339
Azobenzene (cis, trans) 57

REAGENTS AND DERIVATIZATION REACTIONS

U

V

X

Z